Numerical Linear Algebra: Theory and Applications

Larisa Beilina · Evgenii Karchevskii
Mikhail Karchevskii

Numerical Linear Algebra: Theory and Applications

 Springer

Larisa Beilina
Department of Mathematical Sciences
Chalmers University of Technology and
   University of Gothenburg
Gothenburg
Sweden

Mikhail Karchevskii
Department of Computational Mathematics
Kazan Federal University
Kazan, Tatarstan Republic
Russia

Evgenii Karchevskii
Department of Applied Mathematics
Kazan Federal University
Kazan, Tatarstan Republic
Russia

ISBN 978-3-319-86127-2   ISBN 978-3-319-57304-5   (eBook)
DOI 10.1007/978-3-319-57304-5

Printed on acid-free paper

This Springer imprint is published by Springer Nature
The registered company is Springer International Publishing AG
The registered company address is: Gewerbestrasse 11, 6330 Cham, Switzerland

Preface

Problems of numerical linear algebra arise in all fields of modern science. Important examples are computational fluid dynamics, solid mechanics, electrical networks, signal analysis, and optimization. In our book we present an extended basic theory of linear algebra including such topics as matrix algebra, theory of linear systems of equations, spectral theory, vector and matrix norms, combined with the most important direct and iterative numerical methods for solution of linear systems of equations, least squares problems, and eigenproblems. In this book we wanted to combine a solid theoretical background in linear algebra with practical algorithms for numerical solution of linear algebra problems. Most of the presented numerical algorithms are illustrated by computer programs written in MATLAB®, which are given in the electronic supplementary material that can be found on SpringerLink. These programs allow the reader to obtain experience in implementation and evaluation of numerical algorithms for the problems described in the book and give possibility to apply them to the solution of the computer exercises presented in this book. They can also give the reader a better understanding of professional numerical software for the solution of real-life problems in numerical linear algebra.

This book is suitable for use as course material in a one- or two-semester course on numerical linear algebra, matrix computations, or large sparse matrices at the advanced undergraduate or graduate level. We recommend using the material of Chapters 1–7 for courses in the theoretical aspects of linear algebra, or as the first part for a course in numerical linear algebra. In addition to traditional content for courses in linear algebra for students specializing in the physical and mathematical sciences, we include in these chapters some sections that can be useful as course material for special courses on various applications of linear algebra. We hope that this material will also be of interest to scientists. We recommend Chapters 8–12 for courses related to numerical linear algebra, or as the second part of a course in numerical linear algebra. The material of Chapters 8–12 follows the book of Demmel [23]. Compared with [23], we present the numerical material of Chapters 8–12 in a more concise form, which is appropriate to a one-semester course in numerical linear algebra at the undergraduate level. We also enrich our Chapters 8–12 with numerical examples, which can be tested by the MATLAB® and PETSc programs are available in the electronic supplementary material that can be found on SpringerLink.

In the first four chapters we introduce readers to the topic of linear algebra and give the main definitions of complex numbers and polynomials, systems of linear equations, matrices, determinants, vector and inner product spaces, subspaces, linear operators, and eigenvalues and eigenvectors of a linear operator. In Chapter 5 we present canonical forms and factorizations: the singular value decomposition, the Jordan canonical form, matrix pencils and Weierstrass canonical form, the Kronecker canonical form, and their applications in the theory of ordinary differential equations. Chapter 6 discusses vector and matrix norms and Chapter 7 presents the main elements of perturbation theory for basic problems of linear algebra. Chapters 8–11 deal with numerical solution of systems of linear equations, linear least squares problems, and the solution of eigenvalue problems. In Chapter 12 we give a brief introduction to the main iterative methods for the solution of linear systems: Jacobi, Gauss–Seidel, and Successive overrelaxation. We also discuss Krylov subspace methods, the conjugate gradient algorithm, and the preconditioned conjugate gradient algorithm. Compared with other books on the same subject, this book presents a combination of extended material on the rigorous theory of linear algebra together with numerical aspects and implementation of algorithms of linear algebra in MATLAB®. The material of this book was developed from a number of courses which the authors taught repeatedly for a long period at the master's program in engineering mathematics and computational science at Chalmers University of Technology and University of Gothenburg, Sweden, and at Institute of Computational Mathematics and Information Technologies of Kazan Federal University, Russia. Chapters 1–7 were written by Mikhail and Evgenii Karchevskii. Larisa Beilina wrote Chapters 8–12 and the electronic supplementary material that can be found on SpringerLink.

The authors want to thank the following colleagues and students for corrections, proofreading, and contributions to the material of this book: Yu.A. Al'pin, V.B. Andreev, A. Bergqvist, I.D. Beilin, E.V. Chizhonkov, R.Z. Dautov, H. Eklund, N. Ericsson, M. Hoppe, J. Jagers, J. Jansson, B. Galimullin, A.V. Goolin, R.N. Gumerov, A.S. Ilinskii, A. Mutygullin, A. Repina, R.R. Shagidullin, Yu.G. Smirnov, E.L. Stolov, S.I. Soloviov, M.R. Timerbaev, A. Vasilyeva, O. Wickius, E. Åblad.

Gothenburg, Sweden	Larisa Beilina
Kazan, Russia	Evgenii Karchevskii
Kazan, Russia	Mikhail Karchevskii
February 2017	

The original version of this book was revised. An Erratum to this book can be found at https://doi.org/10.1007/978-3-319-57304-5_13

Contents

Chapter 1
Preliminaries

In this chapter we provide the necessary background from the theory of complex numbers. Then we describe the basic properties of polynomials and their roots. We introduce the concept of determinants, establish their properties, and present the basic theory of systems of linear algebraic equations with nonsingular matrices. The main types of rectangular matrices are described.

1.1 Complex Numbers and Polynomials

1.1.1 Complex Numbers: Basic Operations and Formulas

It is well known that not every quadratic equation has a real solution. For example, a simple equation like

$$x^2 + 1 = 0 \qquad (1.1)$$

has no real solution, since the square of a real number is never negative. The situation is changed if we introduce a new number (more precisely, a new symbol). This number is called the *imaginary unit* and is denoted by i. By definition, we have

$$i^2 = -1.$$

Then $\alpha_1 = i$ is a root of equation (1.1). It is natural that

$$(-i)^2 = (-1)^2 i^2 = -1.$$

Electronic supplementary material The online version of this chapter (doi:10.1007/978-3-319-57304-5_1) contains supplementary material, which is available to authorized users.

© Springer International Publishing AG 2017
L. Beilina et al., *Numerical Linear Algebra: Theory and Applications*,
DOI 10.1007/978-3-319-57304-5_1

Then $\alpha_2 = -i$ is the second root of equation (1.1), i.e., this equation has two solutions, similarly to the equation

$$x^2 - 1 = 0.$$

Consider the quadratic equation

$$x^2 + q = 0,$$

where $q > 0$. It is natural to decide that this equation has two roots:

$$\alpha_1 = i\sqrt{q} \quad \text{and} \quad \alpha_2 = -i\sqrt{q}.$$

The numbers of the form ib, where b is real, are called *imaginary numbers*.

Let us now consider a general quadratic equation. For convenience, we write it in the reduced form

$$x^2 - 2px + q = 0. \tag{1.2}$$

Using elementary calculations, we get

$$(x - p)^2 + q - p^2 = 0.$$

Suppose that $q - p^2 > 0$, i.e., the discriminant of equation (1.2) is negative. It is natural to decide that Eq. (1.2) has two roots too:

$$\alpha_1 = p + i\sqrt{q - p^2}, \quad \alpha_2 = p - i\sqrt{q - p^2}. \tag{1.3}$$

The numbers α_1, α_2 in (1.3) have the new form $a + ib$, where a and b are real. They are called *complex numbers*. In the particular case that $b = 0$, the complex number $a + ib$ will be the same as the real number a. If $a = 0$, then the complex number will be the same as the imaginary number ib.

Usually we denote a complex number by the letter z:

$$z = x + iy.$$

The real number x is called the *real part* of z and is denoted by $\operatorname{Re} z$. The real number y is called the *imaginary part* of z and is denoted by $\operatorname{Im} z$. Therefore, we can write

$$z = \operatorname{Re} z + i \operatorname{Im} z.$$

By definition, two complex numbers $z_1 = x_1 + iy_1$ and $z_2 = x_2 + iy_2$ are equal if and only if $x_1 = x_2$ and $y_1 = y_2$.

Let us verify that the numbers α_1, α_2 defined by (1.3) actually satisfy the quadratic equation (1.2). To do this we have to introduce the *algebraic operations* with complex numbers.

The *sum* of the complex numbers $z_1 = x_1 + iy_1$ and $z_2 = x_2 + iy_2$ is the complex number $z = x + iy$, where $x = x_1 + x_2$, $y = y_1 + y_2$, i.e.,

$$\operatorname{Re}(z_1 + z_2) = \operatorname{Re} z_1 + \operatorname{Re} z_2,$$

$$\operatorname{Im}(z_1 + z_2) = \operatorname{Im} z_1 + \operatorname{Im} z_2.$$

The *difference* of the complex numbers z_1 and z_2 is the complex number

$$z = (x_1 - x_2) + i(y_1 - y_2).$$

Clearly, if z is the difference of complex numbers z_1 and z_2, then $z_2 + z = z_1$.

For example, the sum of the complex numbers $z_1 = 1 + i2$ and $z_2 = 3 + i4$ is equal to $z = 4 + i6$, and their difference is $z = -2 - i2$.

The complex number $0 + i0$ is called *zero* and is denoted by the symbol 0.

Multiplication of complex numbers is carried out in the same way as the multiplication of usual binomials using the relation $i^2 = -1$. Therefore, we have

$$z_1 z_2 = (x_1 + iy_1)(x_2 + iy_2) = x_1 x_2 - y_1 y_2 + i(x_1 y_2 + x_2 y_1),$$

i.e., by definition,

$$\operatorname{Re}(z_1 z_2) = \operatorname{Re} z_1 \operatorname{Re} z_2 - \operatorname{Im} z_1 \operatorname{Im} z_2, \tag{1.4}$$

$$\operatorname{Im}(z_1 z_2) = \operatorname{Re} z_1 \operatorname{Im} z_2 + \operatorname{Re} z_2 \operatorname{Im} z_1. \tag{1.5}$$

For example, the product of the complex numbers $z_1 = 1 + i2$, $z_2 = 3 + i4$ is calculated as follows:

$$z_1 z_2 = (1 + i2)(3 + i4) = (1 \times 3 - 2 \times 4) + i(1 \times 4 + 3 \times 2) = -5 + i10.$$

For every complex number z we have $z \times 0 = 0 \times z = 0$.

The reader can easily prove that the operations of addition and multiplication of complex numbers defined above have the same properties as the corresponding operations with real numbers.

1. *Commutativity*: $z_1 + z_2 = z_2 + z_1$, $z_1 z_2 = z_2 z_1$.
2. *Associativity*: $(z_1 + z_2) + z_3 = z_1 + (z_2 + z_3)$, $(z_1 z_2)z_3 = z_1(z_2 z_3)$.
3. *Distributivity*: $(z_1 + z_2)z_3 = z_1 z_3 + z_2 z_3$.

Now using direct substitution, the reader can check that the numbers α_1 and α_2 defined by (1.3) satisfy the quadratic equation (1.2).

Division of complex numbers z_1 and z_2 is defined by the following relationship:

$$z z_2 = z_1. \tag{1.6}$$

Here the complex number z is the *quotient* of z_1 divided by z_2.

If $z_2 \neq 0$, then there exists a unique solution z of Eq. (1.6). Indeed, using (1.4), (1.5), we can write (1.6) in more detail:

$$xx_2 - yy_2 + i(xy_2 + x_2y) = x_1 + iy_1. \tag{1.7}$$

Equating the real and imaginary parts, we get

$$xx_2 - yy_2 = x_1, \tag{1.8}$$
$$xy_2 + yx_2 = y_1. \tag{1.9}$$

The system of equations (1.8), (1.9) has the unique solution

$$x = \frac{x_1x_2 + y_1y_2}{x_2^2 + y_2^2}, \tag{1.10}$$

$$y = \frac{x_2y_1 - x_1y_2}{x_2^2 + y_2^2}. \tag{1.11}$$

Formulas (1.10) and (1.11) define the *rule of division* of complex numbers.

For example, let us divide the complex number $z_1 = 1 + i2$ by $z_2 = 3 + i4$:

$$\frac{z_1}{z_2} = \frac{1 + i2}{3 + i4} = \frac{1 \times 3 + 2 \times 4}{3^2 + 4^2} + i\frac{3 \times 2 - 1 \times 4}{3^2 + 4^2} = \frac{11}{25} + i\frac{2}{25}.$$

For every natural number n, by definition, put $z^n = zz \cdots z$, where the factor is repeated n times, $z^0 = 1$, $z^{-n} = (1/z)^n$.

It is important to note that if the imaginary parts of complex operands are equal to zero, then the operations with complex numbers defined above are identical to the operations with real numbers (check this!). Therefore, we can accept that the field of all complex numbers is an extension of the field of all real numbers.

For a complex number $z = x + iy$, the number $\bar{z} = x - iy$ is called the *complex conjugate* of z. Evidently,

$$\bar{\bar{z}} = z, \quad \overline{z_1 + z_2} = \bar{z}_1 + \bar{z}_2, \quad \overline{z_1z_2} = \bar{z}_1\bar{z}_2. \tag{1.12}$$

Note also that

$$z + \bar{z} = 2x, \quad z - \bar{z} = i2y, \quad z\bar{z} = x^2 + y^2.$$

The real nonnegative number $|z| = \sqrt{z\bar{z}} = \sqrt{x^2 + y^2}$ is called the *absolute value* (or *modulus*) of $z = x + iy$. Obviously,

$$\text{if } |z| = 0, \text{ then } x = 0, \ y = 0, \text{ i.e., } z = 0. \tag{1.13}$$

For all complex numbers z_1, z_2, by elementary calculations we get

$$|z_1 z_2| = |z_1||z_2|. \tag{1.14}$$

For all real numbers x, y, the following inequality is well known:

$$2|xy| \le (x^2 + y^2).$$

Using it, the reader can easily prove that for all complex numbers z_1, z_2, the next inequality holds:

$$|z_1 + z_2| \le |z_1| + |z_2|. \tag{1.15}$$

Relationships (1.13), (1.14), (1.15) show that we can use the absolute values of complex numbers in the same way as the absolute values of real numbers.

Note that $|z_1| = |z_1 - z_2 + z_2| \le |z_1 - z_2| + |z_2|$; therefore, $|z_1| - |z_2| \le |z_1 - z_2|$. Similarly, $|z_2| - |z_1| \le |z_1 - z_2|$. Thus,

$$||z_2| - |z_1|| \le |z_1 - z_2|. \tag{1.16}$$

Recall that every real number x is assumed to correspond to a point on the number line. Analogously, the set of complex numbers may be thought of as a Cartesian plane with "real axis" x and "imaginary axis" y. Thus, $z = x + iy$ may be identified with the point (x, y) in the complex plane. Then complex conjugation is reflection across the real axis, and $|z|$ is the Euclidean distance of z from the origin in the complex plane. (Check this by drawing!)

Recall that the sum of two vectors (x_1, y_1) and (x_2, y_2) is the vector with coordinates $x_1 + x_2$ and $y_1 + y_2$. Therefore, the sum of two complex numbers $z_1 = x_1 + iy_1$ and $z_2 = x_2 + iy_2$ corresponds to the sum of two vectors (make a drawing!). Hence inequalities (1.15), (1.16) can be interpreted as the well-known triangle inequalities for vectors.

In the last two paragraphs we have described the complex plane in terms of rectangular coordinates. The complex plane may also be represented usefully in terms of polar coordinates, in which the position of z in the plane is described in terms of the modulus of z and the angle φ, measured in the counterclockwise direction from the positive real axis. The angle φ belongs to the interval $[0, 2\pi)$ and is called the *argument* of z. The following notation is often used:

$$\varphi = \arg z, \quad \rho = |z|. \tag{1.17}$$

Let us obtain an explicit representation of z using $|z|$ and $\arg z$. We have

$$z = |z| \left(\frac{x}{|z|} + i \frac{y}{|z|} \right).$$

Evidently (make a drawing!),

$$\frac{x}{|z|} = \cos\varphi, \quad \frac{y}{|z|} = \sin\varphi, \tag{1.18}$$

and thus

$$z = \rho(\cos\varphi + i\sin\varphi). \tag{1.19}$$

Relationships (1.17)–(1.19) give the so-called *trigonometric form* of a complex number. This form enables us to take a fresh look at algebraic operations with complex numbers and to obtain several useful formulas.

If $z_1 = \rho_1(\cos\varphi_1 + i\sin\varphi_1)$, $z_2 = \rho_2(\cos\varphi_2 + i\sin\varphi_2)$, then using well-known trigonometric formulas, we have

$$z_1 z_2 = \rho_1\rho_2\left(\cos(\varphi_1 + \varphi_2) + i\sin(\varphi_1 + \varphi_2)\right), \tag{1.20}$$

i.e., to multiply two complex numbers, we multiply their absolute values and add their arguments.

For example, the product of the complex numbers $z_1 = 3\left(\cos(\pi/2) + i\sin(\pi/2)\right)$ and $z_2 = 2\left(\cos(\pi/4) + i\sin(\pi/4)\right)$ is $z_1 z_2 = 6\left(\cos(3\pi/4) + i\sin(3\pi/4)\right)$.

Note that using formula (1.20), we can obtain $\varphi_1 + \varphi_2 \geq 2\pi$. Naturally, because of the periodicity of trigonometric functions, the argument of a complex number is determined modulo 2π. Hence equalities like (1.20) involving arguments will always be interpreted as congruences modulo 2π; i.e., we shall neglect differences of integer multiples of 2π.

Using the trigonometric form of complex numbers and formula (1.20), we can write Eq. (1.6) as

$$\rho\rho_2(\cos(\varphi + \varphi_2) + i\sin(\varphi + \varphi_2)) = \rho_1(\cos\varphi_1 + i\sin\varphi_1), \tag{1.21}$$

whence

$$z = \frac{z_1}{z_2} = \frac{\rho_1}{\rho_2}(\cos(\varphi_1 - \varphi_2) + i\sin(\varphi_1 - \varphi_2)), \tag{1.22}$$

i.e., to divide two complex numbers, we divide their absolute values and subtract their arguments.

For example, using formula (1.22), for the numbers $z_1 = 3(\cos(\pi/2) + i\sin(\pi/2))$ and $z_2 = 2\left(\cos(\pi/4) + i\sin(\pi/4)\right)$, we get $z_1/z_2 = (3/2)(\cos(\pi/4) + i\sin(\pi/4))$.

Let us obtain a formula expressing the powers of a complex number. If we put $z_1 = z_2 = z$ in (1.20), then we get

$$z^2 = zz = \rho^2(\cos 2\varphi + i\sin 2\varphi),$$

and generally,

$$z^n = \rho^n(\cos n\varphi + i\sin n\varphi) \tag{1.23}$$

for every integer n (including zero and negative integers). Formula (1.23) is called *de Moivre's formula*.[1]

Let us now turn to the problem of calculation of an nth root of a complex number $z = \rho(\cos\varphi + i\sin\varphi)$. Here $n \geq 1$ is an integer. This is the problem of calculation of a number $\tilde{z} = \tilde{\rho}(\cos\tilde{\varphi} + i\sin\tilde{\varphi})$ such that

$$\tilde{z}^n = \tilde{\rho}^n(\cos n\tilde{\varphi} + i\sin n\tilde{\varphi}) = \rho(\cos\varphi + i\sin\varphi). \tag{1.24}$$

Evidently, Eq. (1.24) has the following solutions:

$$\tilde{\rho} = \sqrt[n]{\rho}, \quad \tilde{\varphi} = \frac{\varphi}{n} + \frac{2\pi k}{n}, \quad k = 0, 1, \ldots.$$

Here $\sqrt[n]{\rho}$ is the unique real nth root of the real nonnegative number ρ. Hence we see that the n complex numbers

$$z_k = \sqrt[n]{\rho}\,(\cos\varphi_k + i\sin\varphi_k), \quad \varphi_k = \frac{\varphi}{n} + \frac{2\pi k}{n}, \quad k = 0, 1, \ldots, n-1, \tag{1.25}$$

are the nth *roots* of the complex number z. For $k > n-1$ the numbers z_k repeat periodically because of the periodicity of the trigonometric functions.

For example, the four fourth roots of the complex number

$$z = 3\left(\cos\frac{\pi}{2} + i\sin\frac{\pi}{2}\right)$$

are calculated by the following formulas:

$$z_k = \sqrt[4]{3}(\cos\varphi_k + i\sin\varphi_k), \quad \varphi_k = \frac{\pi}{8} + k\frac{\pi}{2}, \quad k = 0, 1, 2, 3.$$

Thus each complex number (except zero) has n distinct nth roots. All these roots lie on the circle of radius $\sqrt[n]{\rho}$ centered at the origin. They divide it into n equal parts.

The following question naturally arises: is it possible to find other roots of a complex number z? The answer is negative. To verify this, the reader can use the results of Section 1.1.3, p. 10, interpreting Eq. (1.24) as a problem of calculation of roots of a polynomial of degree n.

Formula (1.25) is often written in a slightly different form. Let

$$q_k = \cos\frac{2\pi k}{n} + i\sin\frac{2\pi k}{n}, \quad k = 0, 1, 2, \ldots, n-1.$$

Obviously, $q_k^n = 1$ for $k = 0, 1, 2, \ldots, n-1$, i.e., the q_k are nth roots of unity. It is easy to see that

$$z_k = z_0 q_k, \quad k = 0, 1, 2, \ldots, n-1.$$

---

[1] Abraham de Moivre (1667–1754) was a French mathematician.

Therefore, if we calculate the first root

$$z_0 = \sqrt[n]{\rho}\,(\cos\varphi/n + \mathrm{i}\sin\varphi/n)\,,$$

then we obtain all other roots by successive shifts through the angle $2\pi/n$ on the unit circle.

1.1.2 Algebraic Operations with Polynomials

A *polynomial* is a function of the form

$$P_n(z) = a_0 + a_1 z + a_2 z^2 + \cdots + a_n z^n. \tag{1.26}$$

Here $a_0, a_1, \ldots, a_n$ are fixed complex numbers. They are called the *coefficients* of the polynomial. If $a_n \neq 0$, then the integer $n \geq 0$ is called the *degree* of the polynomial, and a_n is called the *leading coefficient* of the polynomial. The variable z can take any complex value.

Two polynomials $P_n(z)$ and $Q_n(z)$ are equal if they have exactly the same coefficients.

If all coefficients of a polynomial are equal to zero, then the polynomial is equal to zero for all z. This polynomial is called the *zero polynomial* and is denoted by 0. The degree of the zero polynomial is undefined.

The sum of two polynomials $P_n(z) + Q_m(z)$ is a polynomial, and its degree is less than or equal to the maximum of n and m, or it is the zero polynomial.

The product of two polynomials $P_n(z)Q_m(z)$ is a polynomial of degree $n + m$.

The addition of the zero polynomial to any polynomial does not change that polynomial. The product of two polynomials is equal to the zero polynomial if and only if one of the factors is the zero polynomial (prove it!).

Let us introduce and investigate the operation of division of polynomials.

Theorem 1.1 *For polynomials $P(z)$ and $Q(z)$ there exist polynomials $q(z)$ and $r(z)$ such that*

$$P(z) = Q(z)q(z) + r(z), \tag{1.27}$$

and either $r = 0$ or the degree of $r(z)$ is less than the degree of $Q(z)$. Moreover, the polynomials $q(z)$ and $r(z)$ are uniquely determined by the polynomials $P(z)$ and $Q(z)$.

Proof First we suppose that either $P = 0$ or the degree of $P(z)$ is less than the degree of $Q(z)$. Then the unique solution of equation (1.27) is $q = 0$ and $r(z) = P(z)$.

Assume now that $P(z)$ is a polynomial of degree n, $Q(z)$ has degree m, and $n \geq m$. To simplify the notation we suppose that the leading coefficient of the polynomial $Q(z)$ is equal to one. If we suppose that the leading coefficient is an arbitrary nonzero number, then the formulas written below should be modified in an obvious way. Thus we take

$$P(z) = a_n z^n + a_{n-1} z^{n-1} + \cdots + a_0,$$
$$Q(z) = z^m + b_{m-1} z^{m-1} + \cdots + b_0,$$
$$q(z) = c_{n-m} z^{n-m} + c_{n-m-1} z^{n-m-1} + \cdots + c_0,$$
$$r(z) = d_{m-1} z^{m-1} + d_{m-2} z^{m-2} + \cdots + d_0.$$

The coefficients of polynomials $P(z)$ and $Q(z)$ are known. Let us calculate the coefficients of $q(z)$ and $r(z)$. Collecting all coefficients with the same power of z on the right-hand side of (1.27) and equating them with the corresponding coefficients of the polynomial $P(z)$, we get

$$
\begin{aligned}
a_n &= c_{n-m}, \\
a_{n-1} &= c_{n-m-1} + c_{n-m} b_{m-1}, \\
a_{n-2} &= c_{n-m-2} + c_{n-m-1} b_{m-1} + c_{n-m} b_{m-2}, \\
&\cdots\cdots\cdots \\
a_m &= c_0 + c_1 b_{m-1} + c_2 b_{m-2} + \cdots + c_m b_0, \\
a_{m-1} &= d_{m-1} + c_0 b_{m-1} + c_1 b_{m-2} + \cdots + c_{m-1} b_0, \\
&\cdots\cdots\cdots \\
a_0 &= d_0 + c_0 b_0.
\end{aligned}
\tag{1.28}
$$

The obtained relationships form a system of linear equations for the coefficients of the polynomials $q(z)$ and $r(z)$. This system is easily solved and uniquely defines the coefficients of these polynomials. First the coefficients c_j are calculated in the order of descending indices:

$$
\begin{aligned}
c_{n-m} &= a_n, \\
c_{n-m-1} &= a_{n-1} - c_{n-m} b_{m-1}, \\
c_{n-m-2} &= a_{n-2} - c_{n-m-1} b_{m-1} - c_{n-m} b_{m-2}, \\
&\cdots\cdots\cdots \\
c_0 &= a_m - c_1 b_{m-1} - c_2 b_{m-2} - \cdots - c_m b_0.
\end{aligned}
\tag{1.29}
$$

Then, using the calculated values of c_j, the coefficients d_j are calculated by the following formulas:

$$
\begin{aligned}
d_{m-1} &= a_{m-1} - c_0 b_{m-1} - c_1 b_{m-2} - \cdots - c_{m-1} b_0, \\
d_{m-2} &= a_{m-2} - c_0 b_{m-2} - c_1 b_{m-3} - \cdots - c_{m-2} b_0, \\
&\cdots\cdots\cdots \\
d_0 &= a_0 - c_0 b_0.
\end{aligned}
\tag{1.30}
$$

Note that $c_{n-m} \neq 0$, since $a_n \neq 0$, but the coefficients of $r(z)$ generally speaking can be equal to zero.

We suppose here that $2m \leq n$. If $2m > n$, then formulas (1.28)–(1.30) should be modified in an obvious way.

The method of calculation of the coefficients of polynomials $q(z)$ and $r(z)$, which is described in the proof, is called *Horner's rule*.[3] It is commonly used for algebraic calculations.

Formula (1.27) defines the operation of *division* of a polynomial $P(z)$ by a polynomial $Q(z)$. The polynomials $q(z)$ and $r(z)$ are called respectively the *quotient* and *remainder* of the division. If $r = 0$, then we say that $Q(z)$ *divides* $P(z)$, or $Q(z)$ is a *factor* of $P(z)$.

Remark 1.1 It follows from (1.29) and (1.30) that if $P(z)$ and $Q(z)$ are polynomials with real coefficients, then the coefficients of the polynomials $q(z)$ and $r(z)$ are real numbers.

As an example of application of Horner's rule, let us divide

$$P_4(z) = 2z^4 - 3z^3 + 4z^2 - 5z + 6 \quad \text{by} \quad Q_2(z) = z^2 - 3z + 1,$$

i.e., let us calculate polynomials

$$q_2(z) = c_2 z^2 + c_1 z + c_0 \quad \text{and} \quad r(z) = d_1 z + d_0$$

such that

$$P_4(z) = Q_2(z) q_2(z) + r(z).$$

In this example we have $n = 4$ and $m = 2$. First, using (1.29), we calculate the coefficients c_2, c_1, and c_0:

$$
\begin{aligned}
c_2 &= a_4 = 2, \\
c_1 &= a_3 - c_2 b_1 = -3 - 2(-3) = 3, \\
c_0 &= a_2 - c_1 b_1 - c_2 b_0 = 4 - 3(-3) - 2 \times 1 = 11.
\end{aligned}
$$

Then, using (1.30), we calculate the coefficients d_1 and d_0:

$$
\begin{aligned}
d_1 &= a_1 - c_0 b_1 - c_1 b_0 = -5 - 11(-3) - 3 \times 1 = 25, \\
d_0 &= a_0 - c_0 b_0 = 6 - 11 \times 1 = -5.
\end{aligned}
$$

Thus,

$$q_2(z) = 2z^2 + 3z + 11, \quad r(z) = 25z - 5.$$

The question naturally arises whether the corresponding coefficients of the polynomials $P_n(z)$ and $Q_n(z)$ are equal if the values of those polynomials are equal for all

---

[2] Here and below, the symbol □ indicates the end of a proof.

[3] William George Horner (1786–1837) was a British mathematician.

z. In other words, are all coefficients of a polynomial equal to zero if that polynomial is equal identically to zero? The answer is yes, but we will prove it somewhat later. Oddly enough, a simple proof is based on some results of the theory of systems of linear algebraic equations (see Section 1.2.3, p. 27).

1.1.3 Roots of Polynomials and Their Properties

A *root* of a polynomial $P_n(z)$ is a complex number α such that $P_n(\alpha) = 0$.

Theorem 1.2 (Bézout's[4] theorem). *Let α be a complex number, $n \geq 1$. Then $z - \alpha$ divides the polynomial $P_n(z) - P_n(\alpha)$.*

Proof Using Theorem 1.1, p. 10, we obtain $P_n(z) - P_n(\alpha) = q_{n-1}(z)(z - \alpha) + r$, where r is either a number (a polynomial of degree zero) or the zero polynomial. If in the last equality we take $z = \alpha$, then $r = 0$, i.e., $P_n(z) - P_n(\alpha) = q_{n-1}(z)(z-\alpha)$. □

The next corollary immediately follows from Bézout's theorem.

Corollary 1.1 *A complex number α is a root of a polynomial $P_n(z)$ if and only if $z - \alpha$ divides $P_n(z)$.*

A complex number α is called a root of *multiplicity* $k \geq 1$ of a polynomial $P_n(z)$ if $(z - \alpha)^k$ divides $P_n(z)$:

$$P_n(z) = (z - \alpha)^k q_{n-k}(z),$$

but $z - \alpha$ does not divide $q_{n-k}(z)$, i.e., the number α is not a root of the polynomial $q_{n-k}(z)$.

A root of multiplicity one is called *simple*.

A polynomial is called *normalized* if the original polynomial has been divided by the leading coefficient. Evidently, each root of the original polynomial is a root of the normalized polynomial, and conversely, each root of the normalized polynomial is a root of the original polynomial. To simplify the notation, properties of roots of polynomials are usually investigated for polynomials in normalized form.

Theorem 1.3 (The fundamental theorem of algebra). *Each polynomial*

$$P_n(z) = z^n + a_{n-1}z^{n-1} + \cdots + a_0, \quad n \geq 1,$$

has at least one root.

Proof As usual, we denote by x_1, x_2 the Cartesian coordinates of points in the plane $\mathbb{R}^2$. Let $x = (x_1, x_2) \in \mathbb{R}^2$ and let $z = x_1 + ix_2$ be the corresponding complex

---

[4]Étienne Bézout (1730–1783) was a French mathematician.

number. The equality $f(x) = |P_n(z)|$ determines the function f of two real variables. This function is nonnegative for all $x \in \mathbb{R}^2$. If there exists a point $y = (y_1, y_2)$ such that $f(y) = 0$, then the complex number $\alpha = y_1 + iy_2$ is a root of the polynomial P_n. To prove the existence of such a point y, first of all, we check that the function f is continuous on $\mathbb{R}^2$.

From (1.16), p. 5, it follows that

$$|f(\tilde{x}) - f(x)| = ||P_n(\tilde{z})| - |P_n(z)|| \leq |P_n(\tilde{z}) - P_n(z)|$$

for any two points x and $\tilde{x}$. Here $\tilde{z} = \tilde{x}_1 + i\tilde{x}_2$. Put $h = \tilde{z} - z$. Then

$$P_n(\tilde{z}) = P_n(z + h) = (z + h)^n + a_{n-1}(z + h)^{n-1} + \cdots + a_1(z + h) + a_0. \quad (1.31)$$

Using the binomial formula, we see that

$$(z + h)^k = z^k + C_k^1 z^{k-1} h + \cdots + C_k^{k-1} z h^{k-1} + h^k$$

for each integer $k \geq 1$. Further, we group like terms on the right-hand side of (1.31) and get

$$P_n(z + h) = P_n(z) + c_1 h + c_2 h^2 + \cdots + c_{n-1} h^{n-1} + h^n, \quad (1.32)$$

where the coefficients $c_1, \ldots, c_{n-1}$ depend only on z and on the coefficients of the polynomial P_n. Using (1.14), (1.15), pp. 4, 5, it is easy to verify that

$$|f(\tilde{x}) - f(x)| \leq |P_n(z + h) - P_n(z)| \leq L(|h| + |h|^2 + \cdots + |h|^n), \quad (1.33)$$

where L depends only on $|z|$ and on the modulus of the coefficients of the polynomial P_n. The right-hand side of inequality (1.33) is less than any preassigned positive number if the distance $|h|$ between points $\tilde{x}$ and x is small enough. This means that the function f is continuous.

Suppose that $f(0) = |a_0| > 0$. Otherwise, zero is a root of the polynomial. Let B_R be the open disk of radius R centered at the origin. Denote by S_R the circle that constitutes the boundary of B_R. Take $x \in S_R$. Write $f(x)$ in the form

$$f(x) = |z^n - (-a_{n-1}z^{n-1} - \cdots - a_0)|.$$

Using (1.16), p. 5, we get

$$f(x) \geq |z|^n - |a_{n-1}||z|^{n-1} - \cdots - |a_0| = R^n - |a_{n-1}|R^{n-1} - \cdots - |a_0|$$
$$= R^n(1 - |a_{n-1}|R^{-1} - \cdots - |a_0|R^{-n}).$$

The right-hand side of the last inequality tends to infinity as $R \to \infty$. Therefore, if R is big enough, then

$$f(x) \geq 2f(0) \quad \text{for all} \quad x \in S_R. \quad (1.34)$$

As we have seen, the function f is continuous on the whole plane. Hence, by the extreme value theorem, in the closure of B_R there exists a point y where the function f attains its minimum. Evidently, $f(y) \le f(0)$, and using (1.34), we see that $y \notin S_R$, i.e., y is an interior point of B_R. We assume that $f(y) > 0$. Otherwise, $\alpha = y_1 + iy_2$ is a root of the polynomial P_n.

Let $h = h_1 + ih_2$. If $|h|$ is small enough, then $\tilde{y} = (y_1 + h_1, y_2 + h_2) \in B_R$. By definition, $f(\tilde{y}) = |P_n(\alpha + h)|$. Using (1.32), we get

$$P_n(\alpha + h) = P_n(\alpha) + c_1 h + c_2 h^2 + \cdots + h^n,$$

where the coefficients $c_1, \ldots, c_{n-1}$ depend only on α and on the coefficients of the polynomial P_n. Since by assumption $P_n(\alpha) \ne 0$, we can write

$$\frac{P_n(\alpha + h)}{P_n(\alpha)} = 1 + d_1 h + \cdots + d_n h^n.$$

Among the numbers $d_1, \ldots, d_n$, at least the number d_n is not equal to zero. Suppose that $d_k \ne 0$ and $d_j = 0$, $j = 1, \ldots, k - 1$. Then for every $c \ne 0$ we have

$$\frac{P_n(\alpha + h)}{P_n(\alpha)} = 1 + \frac{d_k}{c^k}(ch)^k + \frac{d_{k+1}}{c^{k+1}}(ch)^{k+1} + \cdots + \frac{d_n}{c^n}(ch)^n. \tag{1.35}$$

Choose the number c such that $c^k = -d_k$ (see p. 6) and put $v = ch$. Then

$$\frac{f(\tilde{y})}{f(y)} = \frac{|P_n(\alpha + h)|}{|P_n(\alpha)|} = |1 - v^k + v^k b(v)|,$$

where

$$b(v) = \frac{d_{k+1}}{c^{k+1}}v + \cdots + \frac{d_n}{c^n}v^{n-k}.$$

Choose now the number h such that $0 < v < 1$ and $|b(v)| \le 1/2$. For this v, evidently

$$\frac{f(\tilde{y})}{f(y)} \le 1 - \frac{v^k}{2} < 1,$$

but that is impossible, since in the closure of B_R the function f attains its minimum value at the point y. Thus we have a contradiction. Therefore, $f(y) = 0$, i.e., the number $\alpha = y_1 + iy_2$ is a root of the polynomial P_n. $\qquad\square$

Theorem 1.4 *Each polynomial of degree $n \ge 1$ has n roots (counted with their multiplicities).*

Proof Let $P_n(z) = z^n + a_{n-1}z^{n-1} + \cdots + a_0$, $n \geq 1$. By the fundamental theorem of algebra, the polynomial P_n has a root. Denote this root by α_1, and suppose that its multiplicity is equal to $k_1 \geq 1$. Then

$$P_n(z) = (z - \alpha_1)^{k_1} q_{n-k_1}(z).$$

If $k_1 = n$, then evidently, $q_{n-k_1} = 1$. Otherwise, the polynomial $q_{n-k_1}(z)$ has a root. Denote it by α_2. Clearly, the number α_2 is a root of the polynomial P_n, and by construction, $\alpha_2 \neq \alpha_1$. Suppose that the multiplicity of α_2 as a root of the polynomial q_{n-k_1} is equal to k_2. Then

$$q_{n-k_1}(z) = (z - \alpha_2)^{k_2} q_{n-k_1-k_2}(z),$$

and hence
$$P_n(z) = (z - \alpha_1)^{k_1} (z - \alpha_2)^{k_2} q_{n-k_1-k_2}(z).$$

Obviously, the number k_2 is the multiplicity of α_2 as the root of the polynomial P_n. Continuing this process, we get

$$P_n(z) = (z - \alpha_1)^{k_1} (z - \alpha_2)^{k_2} \cdots (z - \alpha_m)^{k_m}, \tag{1.36}$$

where the integers $k_1, k_2, \ldots, k_m$ are greater than or equal to one, and

$$k_1 + k_2 + \cdots + k_m = n.$$

$\square$

Theorem 1.5 *No polynomial P_n of degree $n \geq 1$ can have more than n roots.*

Proof Indeed, let $P_n(\alpha) = 0$, and suppose that the root α is not equal to any roots $\alpha_1, \alpha_2, \ldots, \alpha_m$, which were defined in the proof of the previous theorem. By the corollary of Bézout's theorem, we have $P_n(z) = (z - \alpha)q_{n-1}(z)$. Therefore, using (1.36), we get
$$(z - \alpha_1)^{k_1} (z - \alpha_2)^{k_2} \cdots (z - \alpha_m)^{k_m} = (z - \alpha)q_{n-1}(z).$$

For $z = \alpha$, the right-hand side of the last equality is equal to zero, but the left-hand side is not equal to zero. This contradiction means that the polynomial P_n has only the roots $\alpha_1, \alpha_2, \ldots, \alpha_m$. $\square$

1.1.4 Viète's Formulas

Let $P_n(z) = a_0 + a_1 z + a_2 z^2 + \cdots + a_n z^n$ be a polynomial of degree $n \geq 1$. Suppose that the polynomial P_n has roots $\alpha_1, \alpha_2, \ldots, \alpha_m$ of multiplicities $k_1, k_2, \ldots, k_m$, respectively, and $k_1 + k_2 + \cdots + k_m = n$. Using the results of Section 1.1.3, we can write the polynomial P_n in the form

$$P_n(z) = A(z - \alpha_1)^{k_1}(z - \alpha_2)^{k_2} \cdots (z - \alpha_m)^{k_m},$$

where A is a constant.

Let now P_n be a normalized polynomial of degree $n \geq 1$. Let us numerate the roots of P_n by integers 1, 2, ..., n, repeating each root according to its multiplicity, and write (1.36) in the form

$$P_n(z) = (z - \alpha_1)(z - \alpha_2) \cdots (z - \alpha_n).$$

Removing parentheses, collecting all coefficients with the same power of z on the right-hand side of the last equality, and equating them with the corresponding coefficients of the polynomial P_n, we get the following formulas, which relate the coefficients of P_n to sums and products of its roots:

$$a_{n-1} = -(\alpha_1 + \alpha_2 + \cdots + \alpha_n),$$

$$a_{n-2} = \alpha_1\alpha_2 + \alpha_1\alpha_3 + \cdots + \alpha_{n-1}\alpha_n,$$

$$\cdots\cdots\cdots\cdots\cdots\cdots\cdots$$

$$a_0 = (-1)^n \alpha_1\alpha_2 \cdots \alpha_n.$$

The rule of construction of these formulas is obvious: the number of factors increases by one per line, and then all the possible products of different factors are added up in each line. They are called *Viète's formulas*.[5]

1.1.5 Polynomials with Real Coefficients

Suppose that all coefficients of a polynomial $P_n(z) = z^n + a_{n-1}z^{n-1} + \cdots + a_0$ of degree $n \geq 1$ are real, and α is a root of P_n. Then the complex conjugate $\overline{\alpha}$ is also a root of P_n. Indeed, if $P_n(\alpha) = 0$, then $\overline{P_n(\alpha)} = 0$. Further, using (1.12), p. 4, we see that $\overline{P_n(\alpha)} = \overline{\alpha}^n + a_{n-1}\overline{\alpha}^{n-1} + \cdots + a_0 = P_n(\overline{\alpha})$.

Assume now that $\alpha_1, \alpha_2, \ldots, \alpha_s$ are all the real roots of the polynomial P_n. Denote by $k_1, k_2, \ldots, k_s$ their multiplicities and put

$$r = k_1 + k_2 + \cdots + k_s, \quad Q_r(z) = (z - \alpha_1)^{k_1}(z - \alpha_2)^{k_2} \cdots (z - \alpha_s)^{k_s}.$$

---

[5]François Viète (1540–1603) was a French mathematician.

Then

$$P_n(z) = Q_r(z)R_{n-r}(z). \tag{1.37}$$

Evidently, all coefficients of the polynomial Q_r are real, and therefore all coefficients of the polynomial R_{n-r} are also real (see the remark on p. 10). By construction, the polynomial R_{n-r} can have only complex roots. Note that for all z and α we have

$$(z - \alpha)(z - \overline{\alpha}) = z^2 + pz + q,$$

where $p = -\alpha - \overline{\alpha} = -2\,\mathrm{Re}\,\alpha$, $q = \alpha\overline{\alpha} = |\alpha|^2$ are real numbers. Hence if α is a complex root of the polynomial P_n, and therefore is a root of R_{n-r}, then using (1.37), we obtain

$$P_n(z) = Q_r(z)(z^2 + pz + q)R_{n-r-2}(z).$$

Since the numbers p and q are real, the polynomial R_{n-r-2} has only real coefficients. Continuing this process, we see that

$$P_n(z) = (z-\alpha_1)^{k_1}(z-\alpha_2)^{k_2}\cdots(z-\alpha_s)^{k_s}(z^2+p_1z+q_1)\cdots(z^2+p_tz+q_t). \tag{1.38}$$

Here s is the number of distinct real roots of the polynomial P_n, and t is the number of all pairs of complex conjugate roots of P_n.

From (1.38) it follows immediately that every polynomial with real coefficients of odd degree has at least one real root.

Assuming that z in equality (1.38) is real, we see that every polynomial with real coefficients can be written in the form of a product of real linear and quadratic factors.

For example, it is easy to see that the number $\alpha = -3$ is a root of the polynomial

$$P_3(z) = a_3z^3 + a_2z^2 + a_1z + a_0 = z^3 - 6z + 9.$$

Let us divide $P_3(z)$ by $Q_1(z) = z + b_0 = z + 3$, i.e., let us calculate a polynomial

$$q_2(z) = c_2z^2 + c_1z + c_0$$

such that $P_3(z) = Q_1(z)q_2(z)$. We perform calculations by Horner's rule and see that $q_2(z) = z^2 - 3z + 3$. The remainder is equal to zero, since $z + 3$ divides the polynomial $P_3(z)$:

$$P_3(z) = (z + 3)\left(z^2 - 3z + 3\right).$$

Clearly, the number $\alpha = -3$ is not a root of the polynomial $q_2(z)$. Hence α is a simple root of the polynomial $P_3(z)$. To find the other two roots we have to solve the quadratic equation $z^2 - 3z + 3 = 0$. The discriminant of this equation is equal to -3, and therefore, it does not have real roots. Thus we have written the polynomial $P_3(z)$ of degree three with real coefficients in the form of a product of its real linear and quadratic factors.

1.2 Systems of Linear Equations, Matrices, Determinants

1.2.1 Permutations

Let us consider the set of n integers $M_n = \{1, 2, 3, \ldots, n\}$. We can arrange these integers in different orders. Each arrangement of M_n in some definite order is called a *permutation* of M_n. For example, the following permutations exist:

$$1, 2, 3, \ldots, n \tag{1.39}$$

and

$$2, 1, 3, \ldots, n. \tag{1.40}$$

In general, a permutation of M_n can be written in the form

$$n_1, n_2, \ldots, n_n, \tag{1.41}$$

where $n_1, n_2, \ldots, n_n$ are the integers $1, 2, 3, \ldots, n$ in some order.

The number of distinct permutations of the set M_n is usually denoted by P_n. Let us prove by induction that P_n is equal to the product of the first n integers: $1 \times 2 \times 3 \cdots n$, which is written $n!$ and referred to as "n factorial." We obviously have $P_1 = 1!$ and $P_2 = 2!$. Suppose that $P_{n-1} = (n-1)!$. Take a permutation of M_{n-1} and unite it with the element n. We can put n in the first place, in the second place, and so on. The last possible place for the element n is the nth. Hence, using each permutation of M_{n-1}, we can construct n permutations of M_n. Since by the induction hypothesis we have $P_{n-1} = (n-1)!$, we see that

$$P_n = n!. \tag{1.42}$$

We say that two elements n_i and n_j in the permutation (1.41) form an *inversion* if $i < j$ but $n_i > n_j$. For example, the permutation (1.39) has no inversions, while the permutation (1.40) involves a single inversion, formed by the elements n_1 and n_2.

The number of inversions in a given permutation is called the *signature* of the permutation and is denoted by $\sigma(n_1, n_2, \ldots, n_n)$.

A permutation is called *even* or *odd* according to whether the signature of the permutation is even or odd. For example, the permutation (1.39) is even, and the permutation (1.40) is odd.

If any two elements in a permutation are interchanged, we say that a *transposition* has been made in the permutation. Every transposition is uniquely determined by the serial numbers of the two interchanged elements. For example, the permutation (1.40) is transformed from the permutation (1.39) by the transposition $(1, 2)$.

Theorem 1.6 *A transposition changes the parity of a permutation.*

Proof It is enough to check that the numbers of inversions in the permutations

$$n_1, n_2, \ldots, n_{i-1}, n_i, n_{i+1}, \ldots, n_{j-1}, n_j, n_{j+1}, \ldots, n_n, \tag{1.43}$$

$$n_1, n_2, \ldots, n_{i-1}, n_j, n_{i+1}, \ldots, n_{j-1}, n_i, n_{j+1}, \ldots, n_n \tag{1.44}$$

differ by an odd number. Let us introduce the sets

$$B_1 = \{n_1, n_2, \ldots, n_{i-1}\}, \quad B_2 = \{n_{i+1}, \ldots, n_{j-1}\}, \quad B_3 = \{n_{j+1}, \ldots, n_n\},$$

and denote by B_{ks}^+ (B_{ks}^-) the number of elements in the set B_k that are greater (less) than n_s, $s = i, j$. Clearly, $B_{ks}^+ + B_{ks}^- = \text{card}(B_k)$ for all $k = 1, 2, 3$ and $s = i, j$. Here card(B_k) is the number of elements in the set B_k. The transposition (i, j) described in (1.43), (1.44) changes only the pairs that include n_i or n_j. Hence, it is enough to calculate the numbers of inversions in the permutations (1.43), (1.44) corresponding to the pairs that include n_i or n_j. Evidently, the number of such inversions in the permutation (1.43) is equal to

$$B_I = B_{1i}^+ + B_{2i}^- + B_{3i}^- + B_{1j}^+ + B_{2j}^+ + B_{3j}^- + I(n_i, n_j),$$

where $I(n_i, n_j)$ is the number of inversions in the pair n_i, n_j, and for the permutation (1.44), this number is equal to

$$B_{II} = B_{1j}^+ + B_{2j}^- + B_{3j}^- + B_{1i}^+ + B_{2i}^+ + B_{3i}^- + I(n_j, n_i).$$

Obviously,

$$\begin{aligned}
B_I - B_{II} &= B_{2i}^- - B_{2i}^+ + B_{2j}^+ - B_{2j}^- + I(n_i, n_j) - I(n_j, n_i) \\
&= B_{2i}^- - B_{2i}^+ + B_{2j}^+ - B_{2j}^- \pm 1 = B_{2i}^- - B_{2i}^+ + B_{2j}^+ - B_{2j}^- \pm 2(B_{2i}^+ + B_{2j}^-) \pm 1 \\
&= B_{2i}^- + B_{2i}^+ + B_{2j}^+ + B_{2j}^- - 2(B_{2i}^+ + B_{2j}^-) \pm 1 \\
&= 2\text{card}(B_2) - 2(B_{2i}^+ + B_{2j}^-) \pm 1.
\end{aligned}$$

Thus the number $B_I - B_{II}$ is odd.   □

Theorem 1.7 *For every natural number n, the number of all even permutations of the set $M_n = \{1, 2, 3, \ldots, n\}$ is equal to the number of all odd permutations of M_n.*

Proof It follows from Theorem 1.6 that by the operation of transposition, each even permutation is transformed to an odd permutation. The converse is also true. Therefore, there is a one-to-one correspondence between the set of all even permutations and the set of all odd permutations of M_n. These two sets are finite, and thus the numbers of even and odd permutations are equal.   □

1.2.2 Determinants and Their Basic Properties

A *square matrix* of order n is a square array consisting of n rows and n columns:

$$A = \begin{pmatrix} a_{11} & a_{12} & \ldots & a_{1n} \\ a_{21} & a_{22} & \ldots & a_{2n} \\ & \cdots\cdots & & \\ a_{n1} & a_{n2} & \ldots & a_{nn} \end{pmatrix}. \tag{1.45}$$

Here $a_{ij}, i, j = 1, 2, \ldots, n$, are (generally speaking) complex numbers.

The *determinant* of the matrix A is the number

$$|A| = \sum_{n_1 n_2 \ldots n_n} (-1)^{\sigma(n_1, n_2, \ldots, n_n)} a_{1n_1} a_{2n_2} \cdots a_{nn_n}. \tag{1.46}$$

We also use the following notation:

$$|A| = \det(A) = \Delta = \begin{vmatrix} a_{11} & a_{12} & \ldots & a_{1n} \\ a_{21} & a_{22} & \ldots & a_{2n} \\ & \cdots\cdots & & \\ a_{n1} & a_{n2} & \ldots & a_{nn} \end{vmatrix}. \tag{1.47}$$

Observe that the determinant of a matrix A of order n is the sum of $n!$ summands. The summands are all possible products of n elements of the matrix A such that each term in (1.46) contains one element from each row and one element from each column of the matrix A. Each product enters into the determinant with a + sign if the permutation $n_1, n_2, \ldots, n_n$ is even and with a − sign if it is odd.

Using Theorem 1.7 we see that the number of terms in (1.46) with a + sign is equal to the number of terms with a − sign.

We say that the elements $a_{1n_1}, a_{2n_2}, \ldots, a_{nn_n}$ form the *diagonal* of the matrix A. The diagonal is called *even* if the permutation $n_1, n_2, \ldots, n_n$ is even and *odd* if this permutation is odd.

It is useful to note that

$$\begin{vmatrix} a_{11} & a_{12} \\ a_{21} & a_{22} \end{vmatrix} = a_{11}a_{22} - a_{12}a_{21}, \tag{1.48}$$

$$\begin{vmatrix} a_{11} & a_{12} & a_{13} \\ a_{21} & a_{22} & a_{23} \\ a_{31} & a_{32} & a_{33} \end{vmatrix} = a_{11}a_{22}a_{33} + a_{12}a_{23}a_{31} + a_{13}a_{21}a_{32}$$

$$- a_{13}a_{22}a_{31} - a_{12}a_{21}a_{33} - a_{11}a_{23}a_{32}.$$

The reader can prove the following equality using only the definition of the determinant:

$$
\begin{vmatrix}
1 & 0 & \dots & 0 \\
a_{21} & a_{22} & \dots & a_{2n} \\
& \dotfill & & \\
a_{n1} & a_{n2} & \dots & a_{nn}
\end{vmatrix}
=
\begin{vmatrix}
a_{22} & a_{23} & \dots & a_{2n} \\
a_{32} & a_{33} & \dots & a_{3n} \\
& \dotfill & & \\
a_{n2} & a_{n3} & \dots & a_{nn}
\end{vmatrix}.
\tag{1.49}
$$

Note that the left-hand side of (1.49) contains a determinant of a matrix of order n, and the right-hand side contains a determinant of a matrix of order $n - 1$.

Denote by a_k the kth row of the matrix A, i.e., $a_k = (a_{k1}, a_{k2}, \ldots, a_{kn})$. The sum of two rows $f = (f_1, f_2, \ldots, f_n)$ and $g = (g_1, g_2, \ldots, g_n)$, by definition, is the row

$$
f + g = (f_1 + g_1, f_2 + g_2, \ldots, f_n + g_n).
$$

The product of a number α and a row f is the row

$$
\alpha f = (\alpha f_1, \alpha f_2, \ldots, \alpha f_n).
$$

If all elements of a row are equal to zero, then we say that this row is *zero* and write $0 = (0, 0, \ldots, 0)$.

Often it is useful to consider the determinant to be a function of its rows, i.e., we may write $\Delta = \Delta(a_1, a_2, \ldots, a_n)$. Similarly, we can consider the determinant to be a function of its columns.

Let us formulate and prove the basic properties of determinants.

1. If all elements of a row (or a column) of a determinant $|A|$ are equal to zero, then the determinant is equal to zero. The proof immediately follows from the fact that in this case, each diagonal of the matrix A contains a zero element.

2. The determinant is a linear function of each row (or column) separately with the others fixed, namely,

$$
\Delta(a_1, a_2, \ldots, a_k + b_k, \ldots, a_n)
$$

$$
= \Delta(a_1, a_2, \ldots, a_k, \ldots, a_n) + \Delta(a_1, a_2, \ldots, b_k, \ldots, a_n),
$$

$$
\Delta(a_1, a_2, \ldots, \alpha a_k, \ldots, a_n) = \alpha \Delta(a_1, a_2, \ldots, a_k, \ldots, a_n),
$$

where α is a number. This statement immediately follows from formula (1.46).

3. If two rows (or two columns) of a determinant $|A|$ are identical, then the determinant vanishes. Suppose that row a_i is equal to row a_j, $i < j$. We can represent the set of all diagonals of the matrix A as the union of all pairs of the form

$$
a_{1n_1}, a_{2n_2}, \ldots, a_{in_i}, \ldots, a_{jn_j}, \ldots, a_{nn_n},
\tag{1.50}
$$

$$
a_{1n_1}, a_{2n_2}, \ldots, a_{in_j}, \ldots, a_{jn_i}, \ldots, a_{nn_n}.
\tag{1.51}
$$

The diagonals (1.50), (1.51) have opposite parity, since the corresponding permutations have the form (1.43), (1.44), and each of them is transformed into the other by the transposition (i, j). The product of all elements in the diagonal (1.50) is equal to the product for the diagonal (1.51), since by assumption, $a_{in_i} = a_{jn_i}$, $a_{in_j} = a_{jn_j}$. Therefore, in (1.46), the sum of each pair of terms corresponding to diagonals (1.50) and (1.51) is equal to zero. Thus $|A| = 0$. In the same way we can prove that if two columns of a determinant are identical, then the determinant is equal to zero.

4. If two rows (or columns) of a determinant Δ are interchanged, then the resulting determinant has the value $-\Delta$. To simplify the notation, we prove the statement for the rows a_1 and a_2. The proof for any other rows is analogous. By Property 3, we have $\Delta(a_1 + a_2, a_1 + a_2, a_3, \ldots, a_n) = 0$. On the other hand, using Properties 2 and 3, we see that

$$
\begin{aligned}
0 &= \Delta(a_1 + a_2, a_1 + a_2, a_3, \ldots, a_n) \\
&= \Delta(a_1, a_1 + a_2, a_3, \ldots, a_n) + \Delta(a_2, a_1 + a_2, a_3, \ldots, a_n) \\
&= \Delta(a_1, a_1, a_3, \ldots, a_n) + \Delta(a_1, a_2, a_3, \ldots, a_n) \\
&\quad + \Delta(a_2, a_1, a_3, \ldots, a_n) + \Delta(a_2, a_2, a_3, \ldots, a_n) \\
&= \Delta(a_1, a_2, a_3, \ldots, a_n) + \Delta(a_2, a_1, a_3, \ldots, a_n),
\end{aligned}
$$

i.e., $\Delta(a_1, a_2, a_3, \ldots, a_n) = -\Delta(a_2, a_1, a_3, \ldots, a_n)$.

5. The value of a determinant remains unchanged if to any row is added the product of any number and another row. The same is true for columns. Let us prove as above the statement for the first two rows. Then by Properties 2 and 3, we have

$$
\begin{aligned}
\Delta(a_1 + \alpha a_2, a_2, a_3, \ldots, a_n) \\
= \Delta(a_1, a_2, a_3, \ldots, a_n) + \alpha \Delta(a_2, a_2, a_3, \ldots, a_n) \\
= \Delta(a_1, a_2, a_3, \ldots, a_n).
\end{aligned}
$$

6. Now we prove the formulas of expansion of determinants in terms of rows and columns, which is often used in calculations. Let us introduce the unit row:

$$
i_k = (\underbrace{0, \ldots, 0}_{k-1}, 1, \underbrace{0, \ldots, 0}_{n-k}).
$$

Using Property 2, we easily get

$$
\Delta(a_1, a_2, a_3, \ldots, a_n) = \sum_{k=1}^{n} a_{1k} \Delta(i_k, a_2, a_3, \ldots, a_n). \tag{1.52}
$$

The determinant $\Delta(i_k, a_2, a_3, \ldots, a_n)$ is called the *cofactor* of the element a_{1k} in the determinant $\Delta(a_1, a_2, a_3, \ldots, a_n)$ and is denoted by A_{1k}. Using this notation, we write (1.52) in the form

$$\Delta(a_1, a_2, a_3, \ldots, a_n) = \sum_{k=1}^{n} a_{1k} A_{1k}.$$

For the same reason,

$$\Delta(a_1, a_2, a_3, \ldots, a_n) = \sum_{k=1}^{n} a_{ik} A_{ik}, \quad i = 1, 2, \ldots, n. \tag{1.53}$$

Here A_{ik} is the cofactor of a_{ik}, i.e., the determinant obtained when the row a_i in the determinant $\Delta(a_1, a_2, a_3, \ldots, a_n)$ is replaced by the unit row i_k. Formula (1.53) is known as the expansion of the determinant $\Delta(a_1, a_2, a_3, \ldots, a_n)$ in terms of the row a_i.

Note that $\sum_{k=1}^{n} a_{ik} A_{lk} = 0$ for $l \neq i$. Indeed, the left-hand side of the last equality is the expansion of the determinant with $a_i = a_l$, but we know that such a determinant is equal to zero. Combining this equality with (1.53), we get

$$\sum_{k=1}^{n} a_{ik} A_{lk} = |A|\delta_{il}, \quad i, l = 1, 2, \ldots, n, \tag{1.54}$$

where

$$\delta_{il} = \begin{cases} 0, & i \neq l, \\ 1, & i = l, \end{cases} \tag{1.55}$$

is the *Kronecker delta*.[6]

Often it is more convenient to write the determinant A_{lk} in another form. The reader can easily prove that $A_{lk} = (-1)^{k+l} M_{lk}$, where A_{lk} is the cofactor and M_{lk} is the determinant obtained from the determinant $|A|$ by deleting the lth row and kth column (hint: rearranging rows and columns, write A_{lk} in the form of the determinant on the left-hand side of Eq. (1.49)). The determinant M_{lk} is called the *minor* of the element a_{lk}.

Note the following formula for the expansion of the determinant in terms of its columns:

$$\sum_{k=1}^{n} a_{ki} A_{kl} = |A|\delta_{il}, \quad i, l = 1, 2, \ldots, n. \tag{1.56}$$

---

[6]Leopold Kronecker (1823–1891) was a German mathematician.

As an example, let us calculate the fifth-order determinant

$$\Delta = \begin{vmatrix} -2 & 5 & 0 & -1 & 3 \\ 1 & 0 & 3 & 7 & -2 \\ 3 & -1 & 0 & 5 & -5 \\ 2 & 6 & -4 & 1 & 2 \\ 0 & -3 & -1 & 2 & 3 \end{vmatrix}.$$

First we zero out the third column except for the last entry. To do this, we multiply the last row by three and add the result to the second row. After that, we multiply the last row by four and subtract the result from the fourth row. As a result, we get

$$\Delta = \begin{vmatrix} -2 & 5 & 0 & -1 & 3 \\ 1 & -9 & 0 & 13 & 7 \\ 3 & -1 & 0 & 5 & -5 \\ 2 & 18 & 0 & -7 & -10 \\ 0 & -3 & -1 & 2 & 3 \end{vmatrix}.$$

Expanding this determinant in terms of the third column, we obtain

$$\Delta = (-1)^{3+5}(-1) \begin{vmatrix} -2 & 5 & -1 & 3 \\ 1 & -9 & 13 & 7 \\ 3 & -1 & 5 & -5 \\ 2 & 18 & -7 & -10 \end{vmatrix}.$$

Now we zero out the first column except for the second element. To do this, we multiply the second row by two and add the result to the first row. After that, we multiply the second row by three and subtract the result from the third row. And finally, we multiply the second row by two and subtract the result from the last row. As a result, we get

$$\Delta = - \begin{vmatrix} 0 & -13 & 25 & 17 \\ 1 & -9 & 13 & 7 \\ 0 & 26 & -34 & -26 \\ 0 & 36 & -33 & -24 \end{vmatrix}.$$

Expanding this determinant in terms of the first column, we obtain

$$\Delta = -(-1)^{2+1} \begin{vmatrix} -13 & 25 & 17 \\ 26 & -34 & -26 \\ 36 & -33 & -24 \end{vmatrix} = \begin{vmatrix} -13 & 25 & 17 \\ 26 & -34 & -26 \\ 36 & -33 & -24 \end{vmatrix}.$$

Let us calculate this third-order determinant, expanding it in terms of the third row:

$$\Delta = 36 \begin{vmatrix} 25 & 17 \\ -34 & -26 \end{vmatrix} - (-33) \begin{vmatrix} -13 & 17 \\ 26 & -26 \end{vmatrix} + (-24) \begin{vmatrix} -13 & 25 \\ 26 & -34 \end{vmatrix}$$

$$= 36(-72) - (-33)(-104) + (-24)(-208) = -1032.$$

7. The matrix

$$A^T = \begin{pmatrix} a_{11} & a_{21} & \ldots & a_{n1} \\ a_{12} & a_{22} & \ldots & a_{n2} \\ & \ldots\ldots & \\ a_{1n} & a_{2n} & \ldots & a_{nn} \end{pmatrix} \tag{1.57}$$

is called the *transpose* of the matrix A. Observe that in (1.57) we write the columns of A as the rows of A^T.

The determinants of matrices A and A^T are equal.

Let us prove this statement by induction on the order of the determinant. For a second-order determinant the statement is evidently true. We shall suppose that this equality holds for every determinant of order $n - 1$ and prove it for an arbitrary determinant $|A|$ of order n. Expanding $|A|$ in terms of the first row, we get

$$|A| = a_{11} M_{11} - a_{12} M_{12} + \cdots + (-1)^{n+1} a_{1n} M_{1n}. \tag{1.58}$$

Expanding the determinant $|A^T|$ along the first column, we obtain

$$|A^T| = a_{11} M_{11}^T - a_{12} M_{21}^T + \cdots + (-1)^{n+1} a_{1n} M_{n1}^T. \tag{1.59}$$

Here M_{ij}^T is the minor of the element in the position i, j in the determinant $|A^T|$. By the inductive assumption, $M_{ij}^T = M_{ji}$. Thus $|A^T| = |A|$.

8. We say that the rows of a matrix A are *linearly dependent* if there exist scalars $\alpha_1, \alpha_2, \ldots, \alpha_n$, not all zero, such that

$$\alpha_1 a_1 + \alpha_2 a_2 + \cdots + \alpha_n a_n = 0. \tag{1.60}$$

In the contrary case, i.e., when (1.60) implies $\alpha_1 = \cdots = \alpha_n = 0$, the rows are called *linearly independent*. Evidently, if the rows of a matrix are linearly independent, then none of them are equal to zero. The concept of linear dependence of the columns of a matrix is introduced similarly.

The determinant of a matrix A is equal to zero if and only if the rows of the matrix A are linearly dependent. Let us prove this statement.

Suppose that the rows of A are linearly dependent, and for definiteness suppose that equality (1.60) holds for $\alpha_1 \neq 0$. Using the property of linearity of the determinant in the first row, and then using Properties 5 and 1, we get

$$\alpha_1 \Delta(a_1, a_2, \ldots, a_n) = \Delta(\alpha_1 a_1, a_2, \ldots, a_n)$$
$$= \Delta(\alpha_1 a_1 + \alpha_2 a_2 + \cdots + \alpha_n a_n, a_2, \ldots, a_n) = 0.$$

Thus $\Delta(a_1, a_2, \ldots, a_n) = 0$, since $\alpha_1 \neq 0$.

Let us prove the converse statement, namely, that if the rows of a matrix are linearly independent, then the determinant of the matrix is not equal to zero. Suppose that $|A| = 0$. Let us consider all determinants of order $n - 1$ obtained from the determinant $|A|$ by deleting a row and a column. If each of them is equal to zero, then we consider the determinants of order $n - 2$, and so on. Finally, either all elements of the matrix A are equal to zero (then the assertion is obviously true) or there is a nonzero determinant of order $k \geq 1$ obtained from the determinant $|A|$ by deleting some $n - k$ rows and some $n - k$ columns, and all determinants of order greater than k are equal to zero. Denote this determinant by d_k. Since after interchanging rows and columns of a determinant only its sign is changed, we can assume without loss of generality that d_k consists of elements of the first k rows and the first k columns of the matrix A.

Consider the determinant d_{k+1}, which consists of the elements of the first $k + 1$ rows and the first $k+1$ columns of the matrix A. By assumption, $d_{k+1} = 0$. Expanding d_{k+1} in terms of the last column, we get

$$\alpha_1 a_{1k+1} + \alpha_2 a_{2k+1} + \cdots + \alpha_k a_{kk+1} + d_k a_{k+1k+1} = 0. \tag{1.61}$$

We emphasize that $d_k \neq 0$, and the numbers $\alpha_1, \alpha_2, \ldots, \alpha_k$ are cofactors of corresponding elements of the last column of the determinant d_{k+1}.

By interchanging columns in the determinant $|A|$ we can construct the last column of the determinant d_{k+1} using the following elements:

$$a_{1j}, \ a_{2j}, \ \ldots, \ a_{kj}, \ a_{k+1j}, \quad j = k + 2, \ k + 3, \ \ldots, \ n.$$

By assumption, each of the constructed determinants d_{k+1} is equal to zero. Expanding each of them in terms of its last column, we get

$$\alpha_1 a_{1j} + \alpha_2 a_{2j} + \cdots + \alpha_k a_{kj} + d_k a_{k+1j} = 0, \quad j = k + 2, \ k + 3, \ \ldots, \ n.$$

Finally, if we replace the $(k + 1)$th column of the determinant $|A|$ with its column number $j \leq k$, then we obtain the zero determinant (since the two columns of this determinant are equal). For the same reason we have $d_{k+1} = 0$. Expanding each d_{k+1} thus constructed in terms of its last column, we get

$$\alpha_1 a_{1j} + \alpha_2 a_{2j} + \cdots + \alpha_k a_{kj} + d_k a_{k+1j} = 0, \quad j = 1, \ 2, \ \ldots, \ k.$$

Thus

$$\alpha_1 a_{1j} + \alpha_2 a_{2j} + \cdots + \alpha_k a_{kj} + d_k a_{k+1j} + 0 \times a_{k+2j} + \cdots + 0 \times a_{nj} = 0,$$

where $j = 1, 2, \ldots, n$; $d_k \neq 0$, i.e., the rows of the matrix A are linearly dependent.

Remark 1.2 Since $|A^T| = |A|$, it is clear that the determinant of a matrix A is equal to zero if and only if the columns of the matrix A are linearly dependent.

Below are two examples of calculation of determinants that are often used in applications.

1. The determinant of a triangular matrix. A matrix A is called *upper triangular* if $a_{ij} = 0$ for $i > j$. A matrix A is called *lower triangular* if $a_{ij} = 0$ for $i < j$.

If a matrix A is triangular, then

$$|A| = a_{11}a_{22} \cdots a_{nn}. \tag{1.62}$$

Let us prove this statement for an upper triangular matrix. Then (1.62) holds also for each lower triangular matrix A, since $|A| = |A^T|$, and A^T is an upper triangular matrix.

For determinants of order one and two, equality (1.62) is evidently true. Let us check (1.62) by induction on the order of the determinant. Suppose that (1.62) holds for each determinant of order $n - 1$ and consider the following determinant:

$$|A| = \begin{vmatrix} a_{11} & a_{12} & a_{13} & \cdots & a_{1n} \\ 0 & a_{22} & a_{23} & \cdots & a_{2n} \\ 0 & 0 & a_{33} & \cdots & a_{3n} \\ \cdots & \cdots & \cdots & \cdots & \cdots \\ 0 & 0 & 0 & \cdots & a_{nn} \end{vmatrix}.$$

Expanding $|A|$ in terms of the first column, we get

$$|A| = a_{11} \begin{vmatrix} a_{22} & a_{23} & \cdots & a_{2n} \\ 0 & a_{33} & \cdots & a_{3n} \\ \cdots & \cdots & \cdots & \cdots \\ 0 & 0 & \cdots & a_{nn} \end{vmatrix}.$$

The determinant on the right-hand side in the last equality has order $n - 1$, and by the inductive assumption it is equal to $a_{22}a_{33} \cdots a_{nn}$. Thus (1.62) holds.

2. The *Vandermonde*[7] *determinant* of order n is the determinant defined as follows:

$$d = \begin{vmatrix} 1 & 1 & 1 & \cdots & 1 \\ a_1 & a_2 & a_3 & \cdots & a_n \\ a_1^2 & a_2^2 & a_3^2 & \cdots & a_n^2 \\ \cdots & \cdots & \cdots & \cdots & \cdots \\ a_1^{n-1} & a_2^{n-1} & a_3^{n-1} & \cdots & a_n^{n-1} \end{vmatrix}.$$

Let us prove that for every $n \geq 2$, the Vandermonde determinant is equal to the product of all possible differences $a_i - a_j$ with $1 \leq j < i \leq n$, namely,

---

[7]Alexandre-Théophile Vandermonde (1735–1796) was a French musician and mathematician.

$$d = \prod_{1 \le j < i \le n} (a_i - a_j).$$

The assertion is obviously true for $n = 2$. Let us use the method of mathematical induction. Suppose that the assertion is true for the Vandermonde determinant of order $n - 1$, i.e.,

$$\begin{vmatrix} 1 & 1 & \ldots & 1 \\ a_2 & a_3 & \ldots & a_n \\ \ldots & \ldots & \ldots & \ldots \\ a_2^{n-2} & a_3^{n-2} & \ldots & a_n^{n-2} \end{vmatrix} = \prod_{2 \le j < i \le n} (a_i - a_j).$$

Consider the determinant d. Multiply the penultimate row by a_1 and subtract from the last row; then multiply the $(n - 2)$th row by a_1 and subtract from the $(n - 1)$th row, and so on. As a result, we get

$$\begin{pmatrix} 1 & 1 & 1 & \ldots & 1 \\ 0 & a_2 - a_1 & a_3 - a_1 & \ldots & a_n - a_1 \\ 0 & a_2^2 - a_1 a_2 & a_3^2 - a_1 a_3 & \ldots & a_n^2 - a_1 a_n \\ \ldots & \ldots & \ldots & \ldots & \ldots \\ 0 & a_2^{n-1} - a_1 a_2^{n-2} & a_3^{n-1} - a_1 a_3^{n-2} & \ldots & a_n^{n-1} - a_1 a_n^{n-2} \end{pmatrix}.$$

Expanding d in terms of the first column, we obtain the following determinant of order $n - 1$:

$$d = \begin{vmatrix} a_2 - a_1 & a_3 - a_1 & \ldots & a_n - a_1 \\ a_2^2 - a_1 a_2 & a_3^2 - a_1 a_3 & \ldots & a_n^2 - a_1 a_n \\ \ldots & \ldots & \ldots & \ldots \\ a_2^{n-1} - a_1 a_2^{n-2} & a_3^{n-1} - a_1 a_3^{n-2} & \ldots & a_n^{n-1} - a_1 a_n^{n-2} \end{vmatrix}.$$

Note that $a_2 - a_1$ is a common multiple of all elements of the first column, $a_3 - a_1$ is a common multiple of all elements of the second column, and so on. Therefore,

$$d = (a_2 - a_1)(a_3 - a_1)\ldots(a_n - a_1) \begin{vmatrix} 1 & 1 & \ldots & 1 \\ a_2 & a_3 & \ldots & a_n \\ \ldots & \ldots & \ldots & \ldots \\ a_2^{n-2} & a_3^{n-2} & \ldots & a_n^{n-2} \end{vmatrix},$$

where the last multiplier is the Vandermonde determinant of order $n - 1$. Thus,

$$d = (a_2 - a_1)(a_3 - a_1)\ldots(a_n - a_1) \prod_{2 \le j < i \le n} (a_i - a_j) = \prod_{1 \le j < i \le n} (a_i - a_j).$$

1.2.3   Cramer's Rule

In this section we consider *systems of linear algebraic equations* in which the number of unknowns is equal to the number of equations:

$$
\begin{aligned}
a_{11}x_1 + a_{12}x_2 + \cdots + a_{1n}x_n &= b_1, \\
a_{21}x_1 + a_{22}x_2 + \cdots + a_{2n}x_n &= b_2, \\
&\cdots\cdots\cdots \\
a_{n1}x_1 + a_{n2}x_2 + \cdots + a_{nn}x_n &= b_n.
\end{aligned}
\tag{1.63}
$$

The square matrix

$$
A = \begin{pmatrix}
a_{11} & a_{12} & \dots & a_{1n} \\
a_{21} & a_{22} & \dots & a_{2n} \\
\dots & \dots & \dots & \dots \\
a_{n1} & a_{n2} & \dots & a_{nn}
\end{pmatrix},
\tag{1.64}
$$

which consists of the coefficients of these equations, is called the *matrix of system* (1.63). In this section we assume that $|A| \neq 0$. In this case, the matrix A is called *nonsingular* (if $|A| = 0$, then the matrix A is called *singular*). The array of numbers $b_1, b_2, \ldots, b_n$ is called the column of the right-hand side of system (1.63). If the right-hand side of the system is zero, i.e., $b_i = 0$ for all $i = 1, 2, \ldots, n$, then system (1.63) is called *homogeneous*. A homogeneous system of equations always has a solution. For example, we can take $x_1, x_2, \ldots, x_n = 0$. This solution is called *trivial*.

Theorem 1.8 *Every homogeneous system of linear algebraic equations with a square nonsingular matrix has only the trivial solution.*

Proof Assume the contrary. Then there exist scalars $x_1, x_2, \ldots, x_n$, not all zero, such that

$$
\begin{aligned}
a_{11}x_1 + a_{12}x_2 + \cdots + a_{1n}x_n &= 0, \\
a_{21}x_1 + a_{22}x_2 + \cdots + a_{2n}x_n &= 0, \\
&\cdots\cdots\cdots \\
a_{n1}x_1 + a_{n2}x_2 + \cdots + a_{nn}x_n &= 0.
\end{aligned}
\tag{1.65}
$$

This means that the columns of the matrix A are linearly dependent; hence we get $|A| = 0$, which is impossible for the nonsingular matrix A. Therefore, our assumption of the existence of a nontrivial solution of system (1.65) with a nonsingular matrix is incorrect. $\square$

Remark 1.3 System (1.65) has a nontrivial solution if and only if $|A| = 0$. This statement immediately follows from Remark 1.2.

Theorem 1.9 *For a given right-hand side, system (1.63) with a nonsingular matrix A cannot have two different solutions.*

Proof Assume the contrary, and let $x_1, x_2, \ldots, x_n$ and $y_1, y_2, \ldots, y_n$ be two different solutions of system (1.63), i.e.,

$$
\begin{aligned}
a_{11}x_1 + a_{12}x_2 + \cdots + a_{1n}x_n &= b_1, \\
a_{21}x_1 + a_{22}x_2 + \cdots + a_{2n}x_n &= b_2, \\
&\cdots\cdots\cdots \\
a_{n1}x_1 + a_{n2}x_2 + \cdots + a_{nn}x_n &= b_n,
\end{aligned}
\tag{1.66}
$$

and

$$
\begin{aligned}
a_{11}y_1 + a_{12}y_2 + \cdots + a_{1n}y_n &= b_1, \\
a_{21}y_1 + a_{22}y_2 + \cdots + a_{2n}y_n &= b_2, \\
&\cdots\cdots\cdots \\
a_{n1}y_1 + a_{n2}y_2 + \cdots + a_{nn}y_n &= b_n.
\end{aligned}
\tag{1.67}
$$

Put $z_1 = x_1 - y_1$, $z_2 = x_2 - y_2$, ..., $z_n = x_n - y_n$. Subtracting term by term the corresponding equations in (1.66) and (1.67), we see that $z_1, z_2, \ldots, z_n$ is the solution of the homogeneous system (1.65). Then by Theorem 1.8, it follows that $z_1 = z_2 = \cdots = z_n = 0$, i.e., the assumption of the existence of two different solutions of system (1.63) is incorrect.   □

Theorem 1.10 *For every right-hand side, system* (1.63) *with a nonsingular matrix A has a solution.*

Proof Let us construct the solution of system (1.63) in the form

$$
x_i = c_{i1}b_1 + c_{i2}b_2 + \cdots + c_{in}b_n, \quad i = 1, 2, \ldots, n.
\tag{1.68}
$$

Here c_{ik}, $i, k = 1, 2, \ldots, n$, are unknown coefficients. Substituting (1.68) into the equations of system (1.63) and collecting on the left-hand side of these equations all coefficients of the same b_i, we get

$$
\begin{aligned}
b_1(a_{i1}c_{11} + a_{i2}c_{21} + \cdots + a_{in}c_{n1}) \\
+ b_2(a_{i1}c_{12} + a_{i2}c_{22} + \cdots + a_{in}c_{n2}) \\
+ \cdots + b_i(a_{i1}c_{1i} + a_{i2}c_{2i} + \cdots + a_{in}c_{ni}) \\
+ \cdots + b_n(a_{i1}c_{1n} + a_{i2}c_{2n} + \cdots + a_{in}c_{nn}) = b_i,
\end{aligned}
\tag{1.69}
$$

where $i = 1, 2, \ldots, n$. Clearly, if we can find coefficients c_{ik} such that the following conditions hold, then formulas (1.68) will give us a solution of system (1.63):

$$
a_{i1}c_{1k} + a_{i2}c_{2k} + \cdots + a_{in}c_{nk} = \delta_{ik}, \quad i, k = 1, 2, \ldots, n,
\tag{1.70}
$$

where δ_{ik} is the Kronecker delta. Comparing (1.70) and (1.54), p. 21, we see that Eq. (1.70) holds if we put

$$
c_{ik} = \frac{A_{ki}}{|A|}, \quad i, k = 1, 2, \ldots, n.
\tag{1.71}
$$

Substituting (1.71) into (1.68), we get the following formulas for the solution of system (1.63):

$$x_i = (A_{1i}b_1 + A_{2i}b_2 + \cdots + A_{ni}b_n)/|A|, \quad i = 1, 2, \ldots, n. \tag{1.72}$$

Using the expansion of a determinant in terms of a column, we can write (1.72) in more compact form:

$$x_i = \frac{\Delta_i}{\Delta}, \quad i = 1, 2, \ldots, n. \tag{1.73}$$

Here $\Delta = |A|$, Δ_i is the determinant obtained when the ith column of $|A|$ is replaced by the right-hand side of system (1.63). □

Formulas (1.73) are called *Cramer's*[8] *formulas* (or *Cramer's rule*). As an example, we now solve the following system of equations using Cramer's rule:

$$x_1 + x_2 + x_3 = -2,$$
$$x_1 + 3x_2 - 2x_4 = -4,$$
$$2x_1 + x_3 - x_4 = -1,$$
$$2x_2 - x_3 - 3x_4 = -3.$$

Let us calculate the corresponding determinants:

$$\Delta = \begin{vmatrix} 1 & 1 & 1 & 0 \\ 1 & 3 & 0 & -2 \\ 2 & 0 & 1 & -1 \\ 0 & 2 & -1 & -3 \end{vmatrix} = 4, \quad \Delta_1 = \begin{vmatrix} -2 & 1 & 1 & 0 \\ -4 & 3 & 0 & -2 \\ -1 & 0 & 1 & -1 \\ -3 & 2 & -1 & -3 \end{vmatrix} = 4, \quad \Delta_2 = \begin{vmatrix} 1 & -2 & 1 & 0 \\ 1 & -4 & 0 & -2 \\ 2 & -1 & 1 & -1 \\ 0 & -3 & -1 & -3 \end{vmatrix} = -4,$$

$$\Delta_3 = \begin{vmatrix} 1 & 1 & -2 & 0 \\ 1 & 3 & -4 & -2 \\ 2 & 0 & -1 & -1 \\ 0 & 2 & -3 & -3 \end{vmatrix} = -8, \quad \Delta_4 = \begin{vmatrix} 1 & 1 & 1 & -2 \\ 1 & 3 & 0 & -4 \\ 2 & 0 & 1 & -1 \\ 0 & 2 & -1 & -3 \end{vmatrix} = 4.$$

By Cramer's rule, we get $x_1 = \Delta_1/\Delta = 1$, $x_2 = \Delta_2/\Delta = -1$, $x_3 = \Delta_3/\Delta = -2$, and $x_4 = \Delta_4/\Delta = 1$.

For numerical computations, Cramer's formulas are used very rarely. Systems of linear equations are usually solved numerically by different variants of Gaussian elimination or by iterative methods (see p. 41 in this chapter and Chapters 8, 12).

As an example of application of Cramer's rule, let us construct the so-called *Lagrange*[9] *interpolation formula*.

Theorem 1.11 *Let $z_0, z_1, \ldots, z_n$ be distinct numbers, and let $h_0, h_1, \ldots, h_n$ be arbitrary numbers. Then there exists one and only one polynomial*

---

[8]Gabriel Cramer (1704–1752) was a Swiss mathematician.
[9]Joseph-Louis Lagrange (1736–1813) was a French mathematician.

$$P_n(z) = a_0 + a_1 z + a_2 z^2 + \cdots + a_n z^n$$

such that

$$P_n(z_j) = h_j, \quad j = 0, 1, \ldots, n. \tag{1.74}$$

Proof The set of conditions (1.74) is the system of linear equations for the coefficients of the polynomial P_n. The determinant of this system is the Vandermonde determinant (see p. 25). Obviously, it is not equal to zero; hence system (1.74) has a unique solution for every right-hand side. $\square$

It is clear now that if a polynomial of order n is equal to zero at least $n + 1$ points, then all its coefficients are equal to zero.

It is not hard to construct the polynomial satisfying conditions (1.74) in an explicit form. Namely, the Lagrange interpolation formula gives the solution of this problem:

$$P_n(z) = h_0 \Phi_0(z) + h_1 \Phi_1(z) + \cdots + h_n \Phi_n(z), \tag{1.75}$$

where Φ_j is a polynomial of order n satisfying the following conditions:

$$\Phi_j(z_k) = 0, \quad k = 0, 1, \ldots, j - 1, j + 1, \ldots, n, \tag{1.76}$$

$$\Phi_j(z_j) = 1, \tag{1.77}$$

for $j = 0, 1, 2 \ldots, n$.

As we have seen in Section 1.1.3, p. 10, every polynomial is uniquely determined up to a constant factor by all its roots. Therefore,

$$\Phi_j(z) = A_j(z - z_0)(z - z_1) \cdots (z - z_{j-1})(z - z_{j+1}) \cdots (z - z_n).$$

Using (1.77), we see that

$$A_j = \frac{1}{(z_j - z_0)(z_j - z_1) \cdots (z_j - z_{j-1})(z_j - z_{j+1}) \cdots (z_j - z_n)},$$

i.e.,

$$\Phi_j(z) = \frac{(z - z_0)(z - z_1) \cdots (z - z_{j-1})(z - z_{j+1}) \cdots (z - z_n)}{(z_j - z_0)(z_j - z_1) \cdots (z_j - z_{j-1})(z_j - z_{j+1}) \cdots (z_j - z_n)},$$

where $j = 0, 1, 2, \ldots, n$.

1.2.4   Matrices: Basic Operations and Transformations

We have introduced the concept of a square matrix (see p. 18). *A rectangular $m \times n$ matrix* is a rectangular array consisting of m rows and n columns:

$$
A = \begin{pmatrix} a_{11} & a_{12} & \ldots & a_{1n} \\ a_{21} & a_{22} & \ldots & a_{2n} \\ & \ldots \ldots \\ a_{m1} & a_{m2} & \ldots & a_{mn} \end{pmatrix}. \tag{1.78}
$$

Here a_{ij}, $i = 1, 2, \ldots, m$, $j = 1, 2, \ldots, n$, are (generally speaking) complex numbers. Sometimes, the dimensions of the matrix A are indicated explicitly, and we use the following notation: $A(m, n)$ or $A^{m \times n}$.

In the particular case $m = n$, we have a square matrix of order n. We denote the set of all rectangular $m \times n$ matrices by $M_{m,n}$. We denote the set of all square matrices of order n by M_n.

There are two more special cases. If $m = 1$ and n is arbitrary, we have the *row matrix*

$$
x = (x_1, x_2, \ldots, x_n). \tag{1.79}
$$

A row matrix is often called a *row*, and we say that this row has *length n*. If $n = 1$ and m is arbitrary, we have the *column matrix*

$$
x = \begin{pmatrix} x_1 \\ x_2 \\ \vdots \\ x_m \end{pmatrix}. \tag{1.80}
$$

This matrix is also called a *column* of *length m*. Note that the second subscript in the notation for elements of rows and columns is usually not used. Rows and columns are often called *vectors*.

Let us describe some special types of square matrices.

The elements $a_{11}, a_{22}, \ldots, a_{nn}$ of a square matrix A of order n constitute its *main diagonal*. A *diagonal matrix* is a square matrix all of whose elements outside the main diagonal are equal to zero:

$$
D = \begin{pmatrix} d_{11} & 0 & \ldots & 0 \\ 0 & d_{22} & \ldots & 0 \\ & \ldots \ldots \\ 0 & 0 & \ldots & d_{nn} \end{pmatrix}. \tag{1.81}
$$

We also denote this matrix by

$$
D = \text{diag}(d_{11}, d_{22}, \ldots, d_{nn}).
$$

A diagonal matrix in which $d_{ii} = 1$ for $i = 1, 2, \ldots, n$ is called the *identity matrix*, and is denoted by I:

$$I = \begin{pmatrix} 1 & 0 & \ldots & 0 \\ 0 & 1 & \ldots & 0 \\ & \ldots \ldots & \\ 0 & 0 & \ldots & 1 \end{pmatrix}. \tag{1.82}$$

A *permutation matrix* is a matrix obtained from the identity matrix by interchanging the ith and the kth columns, it is denoted by P_{ik}. For example, the following three matrices are the permutation matrices of order 3:

$$P_{12} = \begin{pmatrix} 0 & 1 & 0 \\ 1 & 0 & 0 \\ 0 & 0 & 1 \end{pmatrix}, \quad P_{13} = \begin{pmatrix} 0 & 0 & 1 \\ 0 & 1 & 0 \\ 1 & 0 & 0 \end{pmatrix}, \quad P_{23} = \begin{pmatrix} 1 & 0 & 0 \\ 0 & 0 & 1 \\ 0 & 1 & 0 \end{pmatrix}.$$

Let us recall that a *lower triangular* matrix L is a square matrix all of whose elements above the main diagonal are equal to zero,

$$L = \begin{pmatrix} l_{11} & 0 & \ldots & 0 \\ l_{21} & l_{22} & \ldots & 0 \\ & \ldots \ldots & \\ l_{n1} & l_{n2} & \ldots & l_{nn} \end{pmatrix}, \tag{1.83}$$

and an *upper triangular* matrix U is a square matrix all of whose elements below the main diagonal are equal to zero,

$$U = \begin{pmatrix} u_{11} & u_{12} & \ldots & u_{1n} \\ 0 & u_{22} & \ldots & u_{2n} \\ & \ldots \ldots & \\ 0 & 0 & \ldots & u_{nn} \end{pmatrix}. \tag{1.84}$$

A triangular matrix of the form

$$L_k = \begin{pmatrix} 1 & \cdots & 0 & 0 & \cdots & 0 \\ & \ldots \ldots & & & \\ 0 & \cdots & l_{k,k} & 0 & \cdots & 0 \\ 0 & \cdots & l_{k+1,k} & 1 & \cdots & 0 \\ & \ldots \ldots & & & \\ 0 & \cdots & l_{n,k} & 0 & \cdots & 1 \end{pmatrix} \tag{1.85}$$

is called an *elementary lower triangular matrix*. Note that this matrix differs from the identity matrix only in elements of the kth column.

Let us introduce the operations of matrix addition and multiplication of matrices by scalars.

The *product of an $m \times n$ matrix A by a scalar α* is the matrix

$$
\alpha A = \begin{pmatrix}
\alpha a_{11} & \alpha a_{12} & \ldots & \alpha a_{1n} \\
\alpha a_{21} & \alpha a_{22} & \ldots & \alpha a_{2n} \\
\multicolumn{4}{c}{\ldots\ldots\ldots} \\
\alpha a_{m1} & \alpha a_{m2} & \ldots & \alpha a_{mn}
\end{pmatrix}.
$$

Each element of the matrix αA is the product of the corresponding element of the matrix A by the number α.

The *sum* of two $m \times n$ matrices A and B is the $m \times n$ matrix C with the elements $c_{ij} = a_{ij} + b_{ij}$. In matrix notation we write this as $C = A + B$.

The *zero matrix* is the matrix all of whose elements are equal to zero. It is denoted by 0.

The reader can easily verify the following properties of the two operations that we have just introduced:

1. $A + 0 = A$,
2. $(A + B) + C = A + (B + C)$,
3. $A + B = B + A$,
4. $(\alpha + \beta)A = \alpha A + \beta A$.

Note that the sum of two upper (lower) triangular matrices is an upper (lower) triangular matrix.

By definition, the *product* of a row x and a column y of the same length n is the number

$$
(x_1, x_2, \ldots, x_n) \begin{pmatrix} y_1 \\ y_2 \\ \vdots \\ y_n \end{pmatrix} = \sum_{k=1}^{n} x_k y_k. \tag{1.86}
$$

In other words, this product is obtained by multiplying together corresponding elements in the row x and the column y and then adding the products.

For example,

$$
(5 \; -1 \; 3 \; 1) \begin{pmatrix} -1 \\ -2 \\ 3 \\ 4 \end{pmatrix} = 5 \times (-1) + (-1) \times (-2) + 3 \times 3 + 1 \times 4 = 10.
$$

The *product Ax of an $m \times n$ matrix A by a column x* of length n is the column y of length m with the elements

$$
y_i = \sum_{j=1}^{n} a_{ij} x_j, \quad i = 1, 2, \ldots, m.
$$

In matrix notation, it is written as follows:

$$y = Ax.$$

We can write the same in detail:

$$\begin{pmatrix} y_1 \\ y_2 \\ \vdots \\ y_m \end{pmatrix} = \begin{pmatrix} a_{11} & a_{12} & \dots & a_{1n} \\ a_{21} & a_{22} & \dots & a_{2n} \\ & \dots\dots\dots & \\ a_{m1} & a_{m2} & \dots & a_{mn} \end{pmatrix} \begin{pmatrix} x_1 \\ x_2 \\ \vdots \\ x_n \end{pmatrix}.$$

Observe that the ith element of the column y is the product of the ith row of the matrix A and the column x.

For example,

$$\begin{pmatrix} 0 & -3 & 1 \\ 2 & 1 & 5 \\ -4 & 0 & -2 \end{pmatrix} \begin{pmatrix} 3 \\ -2 \\ 2 \end{pmatrix} = \begin{pmatrix} 8 \\ 14 \\ -16 \end{pmatrix}.$$

It immediately follows from the definition that for all scalars α, β and vectors x, y (of suitable length), we have

$$A(\alpha x + \beta y) = \alpha Ax + \beta Ay. \tag{1.87}$$

Therefore, we say that the operation of matrix-vector multiplication is linear.

The *product of a row x of length m by an $m \times n$ matrix A* is the row y of length n with the elements

$$y_j = \sum_{i=1}^{m} a_{ij} x_i, \quad j = 1, 2, \dots, n.$$

In matrix notation, it is written as follows:

$$y = xA.$$

We can write the same in detail:

$$(y_1, y_2, \dots, y_n) = (x_1, x_2, \dots, x_m) \begin{pmatrix} a_{11} & a_{12} & \dots & a_{1n} \\ a_{21} & a_{22} & \dots & a_{2n} \\ & \dots\dots\dots & \\ a_{m1} & a_{m2} & \dots & a_{mn} \end{pmatrix}.$$

The jth element of the row y is the product of the row x and the jth column of the matrix A.

For example,

$$
(5\ 1\ 0\ -3)\begin{pmatrix} 2 & 0 \\ 1 & -4 \\ 3 & 1 \\ 0 & -1 \end{pmatrix} = (11\ -1).
$$

It immediately follows from the definition that for all scalars α, β and rows x, y (of suitable length), we have

$$(\alpha x + \beta y)A = \alpha x A + \beta y A. \tag{1.88}$$

This means that the operation of multiplication of a row by a matrix is linear.

Using the operations that we have just introduced, we can write the system (1.63) of n linear equations with n unknowns either in the form

$$Ax = b, \tag{1.89}$$

where A is a given square matrix, b is a given vector, and x is an unknown vector, or in the form

$$xA^T = b, \tag{1.90}$$

where b is a given row and x is an unknown row. The form (1.89) is used more often.

Suppose that A is an $m \times n$ matrix and B is an $n \times p$ matrix. Then the $m \times p$ matrix C with the elements

$$c_{ij} = \sum_{q=1}^{n} a_{iq} b_{qj}, \quad i = 1, 2, \ldots, m, \ j = 1, 2, \ldots, p,$$

is called the *product* of the matrices A and B, which is written as $C = AB$, or in detail as follows:

$$
\begin{pmatrix} c_{11} & c_{12} & \ldots & c_{1p} \\ c_{21} & c_{22} & \ldots & c_{2p} \\ & \ldots\ldots & \\ c_{m1} & c_{m2} & \ldots & c_{mp} \end{pmatrix} = \begin{pmatrix} a_{11} & a_{12} & \ldots & a_{1n} \\ a_{21} & a_{22} & \ldots & a_{2n} \\ & \ldots\ldots & \\ a_{m1} & a_{m2} & \ldots & a_{mn} \end{pmatrix} \begin{pmatrix} b_{11} & b_{12} & \ldots & b_{1p} \\ b_{21} & b_{22} & \ldots & b_{2p} \\ & \ldots\ldots & \\ b_{n1} & b_{n2} & \ldots & b_{np} \end{pmatrix}.
$$

It is useful to observe that each column in the matrix C is calculated as the product of the matrix A by the corresponding column of the matrix B. Similarly, each row in the matrix C is calculated as the product of the corresponding row of the matrix A by the matrix B. Note also that each element c_{ij} is the product of the ith row of the matrix A and the jth column of the matrix B.

For instance,

$$\begin{pmatrix} 5 & -1 & 3 & 1 \\ 2 & 0 & -1 & 4 \end{pmatrix} \begin{pmatrix} -1 & 3 & 0 \\ -2 & 1 & 1 \\ 3 & 0 & -2 \\ 4 & 1 & 2 \end{pmatrix} = \begin{pmatrix} 10 & 15 & -5 \\ 11 & 10 & 10 \end{pmatrix}.$$

The matrix product depends on the order of the factors. For example,

$$\begin{pmatrix} 1 & 2 \\ 3 & 2 \end{pmatrix} \begin{pmatrix} 1 & 2 \\ 1 & 1 \end{pmatrix} = \begin{pmatrix} 3 & 4 \\ 5 & 8 \end{pmatrix},$$

$$\begin{pmatrix} 1 & 2 \\ 1 & 1 \end{pmatrix} \begin{pmatrix} 1 & 2 \\ 3 & 2 \end{pmatrix} = \begin{pmatrix} 7 & 6 \\ 4 & 4 \end{pmatrix}.$$

Two matrices A, B *commute* (with each other), or are *commuting matrices*, if

$$AB = BA.$$

Commuting matrices exist. For example,

$$\begin{pmatrix} 7 & -12 \\ -4 & 7 \end{pmatrix} \begin{pmatrix} 26 & 45 \\ 15 & 26 \end{pmatrix} = \begin{pmatrix} 26 & 45 \\ 15 & 26 \end{pmatrix} \begin{pmatrix} 7 & -12 \\ -4 & 7 \end{pmatrix} = \begin{pmatrix} 2 & 3 \\ 1 & 2 \end{pmatrix}.$$

For every square matrix A, we have

$$AI = IA = A.$$

Let us verify the following properties of the operation of matrix multiplication:

1. $(A + B)C = AC + BC$,
2. $C(A + B) = CA + CB$,
3. $A(BC) = (AB)C$.

Clearly, the dimensions of these matrices are matched such that all operations here make sense.

It is easy to see that Properties 1 and 2 follow from (1.88), (1.87), respectively. To prove Property 3, note that each element of the matrix $D = A(BC)$ is a number of the form $d_{ij} = a_i(Bc_j)$, where a_i is the ith row of the matrix A, and c_j is the jth column of the matrix C. The elements of the matrix $F = (AB)C$ are the numbers $f_{ij} = (a_i B)c_j$. Therefore, it is enough to prove that $x(By) = (xB)y$ for each row x and column y (obviously, their lengths have to correspond to the dimensions of the matrix B). Suppose that the matrix B has m rows and n columns. By elementary calculations, we get

$$x(By) = \sum_{i=1}^{m} x_i \sum_{j=1}^{n} b_{ij} y_j = \sum_{i=1}^{m} \sum_{j=1}^{n} b_{ij} x_i y_j, \tag{1.91}$$

and similarly,

$$(xB)y = \sum_{j=1}^{n} y_j \sum_{i=1}^{m} b_{ij} x_i = \sum_{j=1}^{n} \sum_{i=1}^{m} b_{ij} x_i y_j. \tag{1.92}$$

The sums (1.91), (1.92) differ only in the order of their summands, and hence are equal.

The proof of the following statements is left to the reader.

1. Let P_{ik} be a permutation matrix. Then the vector $P_{ik} x$ is obtained from the vector x by interchanging the elements x_i and x_k.
2. The matrix $P_{ik} A$ is obtained from the matrix A by interchanging the ith and kth rows. Hint: this is a consequence of the previous statement.
3. If matrices L, M are lower triangular, then the matrix LM is lower triangular, and the same is true for upper triangular matrices.
4. Every lower triangular matrix L can be represented in the form of the product of the elementary lower triangular matrices L_k, namely,

$$L = L_1 L_2 \cdots L_{n-1} L_n. \tag{1.93}$$

Hint: make calculations according to the following placement of parentheses:

$$L = L_1(L_2 \cdots (L_{n-2}(L_{n-1} L_n) \cdots),$$

i.e., first premultiply L_n by L_{n-1}, then premultiply the product by L_{n-2}, and so on.
5. For every square matrix A, the following equalities hold:

$$\det(P_{ik} A) = \det P_{ik} \det A = -\det A. \tag{1.94}$$

Let us prove that for each square matrix A and elementary lower triangular matrix L_k, the following equality holds:

$$\det(L_k A) = l_{kk} \det A. \tag{1.95}$$

Indeed, let $a = (a_1, a_2, \ldots, a_n)$ be a vector. By elementary calculations we get

$$
L_k a =
\begin{pmatrix}
a_1 \\
a_2 \\
\vdots \\
a_{k-1} \\
l_{k,k} a_k \\
l_{k+1,k} a_k + a_{k+1} \\
l_{k+2,k} a_k + a_{k+2} \\
\vdots \\
l_{n,k} a_k + a_n
\end{pmatrix}.
$$

Each column of the matrix $L_k A$ has this form. Therefore, we can factor out from the kth row of $\det(L_k A)$ the common factor l_{kk}. Then we can see that the determinant thus obtained is equal to $\det A$ if we multiply its kth row by l_{jk} and subtract the result from the jth row for $j = k + 1, k + 2, \ldots, n$. As a result, we get (1.95).

Now using (1.95), (1.93), and (1.62), the reader can easily verify the equality

$$
\det(LA) = \det L \ \det A \tag{1.96}
$$

for every square matrix A and lower triangular matrix L, and also the analogous equality for every upper triangular matrix R:

$$
\det(RA) = \det R \ \det A. \tag{1.97}
$$

Let us discuss the concept of the transpose of a rectangular matrix, whose definition is analogous to the definition of transpose of a square matrix (see p. 23).

Clearly, the transpose of an $m \times n$ matrix is an $n \times m$ matrix. For instance, the transpose of a row matrix is a column matrix, and conversely.

The basic properties of the operation of transposition of rectangular matrices are the following.

1. For every matrix A we have $(A^T)^T = A$.
2. For all numbers α, β and matrices A, B of the same dimensions, we have $(\alpha A + \beta B)^T = \alpha A^T + \beta B^T$. This means that the operation of transposition is linear.
3. If the product AB is defined, then (a) the product $B^T A^T$ is also defined, and (b) the following equality is true:

$$
(AB)^T = B^T A^T.
$$

All properties except 3 (b) follow immediately from the definition (prove them!).

Let us prove Property 3 (b). The (i, j)th element of the matrix $(AB)^T$ is the product of the jth row of the matrix A and the ith column of the matrix B. The (i, j)th element of the matrix $B^T A^T$ is the product of the ith row of the matrix B^T

and the jth column of the matrix A^T. The ith row of the matrix B^T is equal to the ith column of the matrix B, and the jth column of the matrix A^T is equal to the jth row of the matrix A. This completes the proof of Property 3 (b).

We shall next investigate the operation of inversion of square matrices using results of Section 1.2.3, p. 27. Let us recall that if $|A| \neq 0$, then the matrix A is called nonsingular, and if $|A| = 0$, then the matrix A is called singular. We prove the following two statements using Remark 1.3.

If the square matrices A, B are both nonsingular, then the matrix $C = AB$ is nonsingular. To prove this, it is enough to show that the homogeneous system of linear equations

$$ABx = 0 \qquad\qquad (1.98)$$

has only the trivial solution. Indeed, $Bx = 0$, since A is nonsingular, and hence $x = 0$, since B is nonsingular.

If one of the square matrices A, B is singular, then the matrix $C = AB$ is also singular. Indeed, in this case it is enough to show that system (1.98) has a nontrivial solution. Suppose that the matrix B is singular. Then there exists a vector $x \neq 0$ such that $Bx = 0$; therefore, $ABx = 0$.

Now let the matrix A be singular and the matrix B nonsingular. Then there exists a vector $y \neq 0$ such that $Ay = 0$. Since B is nonsingular, the system $Bx = y$ has a unique solution x. The vector x is not equal to zero, since $y \neq 0$. Again we get $ABx = 0$ for $x \neq 0$.

A matrix X is called a *right inverse* of the matrix A if

$$AX = I. \qquad\qquad (1.99)$$

A matrix Y is called a *left inverse* of the matrix A if

$$YA = I. \qquad\qquad (1.100)$$

If the matrix A is singular, then it does not have a right inverse. Indeed, if there existed a right inverse X, then we would have

$$\det(AX) = \det(I) = 1.$$

On the other hand, $\det(AX) = 0$, since A is singular. It is proved similarly that a singular matrix does not have a left inverse.

If $\det(A) \neq 0$, then there exists one and only one right inverse of the matrix A. Indeed, let us denote by x_k the kth column of the matrix X, and by i_k the kth column of the matrix I. Using (1.99), we get the following systems of linear equations:

$$Ax_k = i_k, \quad k = 1, 2, \ldots, n. \qquad\qquad (1.101)$$

Each of these n systems has a unique solution x_k, since the matrix A is nonsingular. The proof of the existence and uniqueness of a left inverse of the nonsingular matrix A is similar.

In fact, the left inverse and the right inverse of a nonsingular matrix A are equal to each other. Indeed, if $YA = I$, then $YAX = X$, but $AX = I$, and hence $Y = X$.

Thus if (1.99) holds, then the matrix X is called the *inverse* of A. The inverse matrix of a matrix A is denoted by A^{-1}, and by definition,

$$AA^{-1} = I.$$

Let us write the inverse matrix in an explicit form. To do this, we introduce the *adjugate* matrix. The adjugate matrix $\tilde{A}$ of A is the transpose of the matrix of cofactors of the elements of the matrix A. Namely,

$$\tilde{A} = \begin{pmatrix} A_{11} & A_{21} & \cdots & A_{n1} \\ A_{12} & A_{22} & \cdots & A_{n2} \\ & & \cdots\cdots\cdots & \\ A_{1n} & A_{2n} & \cdots & A_{nn} \end{pmatrix},$$

where A_{ij} denotes the cofactor of the element a_{ij} in A.

Now we can write formulas (1.54), p. 21, in matrix form:

$$A\tilde{A} = |A|I. \tag{1.102}$$

Therefore, if $|A| \neq 0$, then the matrix

$$A^{-1} = |A|^{-1}\tilde{A} \tag{1.103}$$

is the inverse of the matrix A.

For example, let us calculate the inverse of the matrix

$$A = \begin{pmatrix} 3 & -1 & 0 \\ -2 & 1 & 1 \\ 2 & -1 & 4 \end{pmatrix}.$$

Expanding the determinant of A in terms of the first row, we get $|A| = 5$. The cofactors of the elements of the matrix A are calculated in the following way:

$$A_{11} = \begin{vmatrix} 1 & 1 \\ -1 & 4 \end{vmatrix} = 5, \quad A_{12} = -\begin{vmatrix} -2 & 1 \\ 2 & 4 \end{vmatrix} = 10, \quad A_{13} = \begin{vmatrix} -2 & 1 \\ 2 & -1 \end{vmatrix} = 0,$$

$$A_{21} = -\begin{vmatrix} -1 & 0 \\ -1 & 4 \end{vmatrix} = 4, \quad A_{22} = \begin{vmatrix} 3 & 0 \\ 2 & 4 \end{vmatrix} = 12, \quad A_{23} = -\begin{vmatrix} 3 & -1 \\ 2 & -1 \end{vmatrix} = 1,$$

$$A_{31} = \begin{vmatrix} -1 & 0 \\ 1 & 1 \end{vmatrix} = -1, \quad A_{32} = -\begin{vmatrix} 3 & 0 \\ -2 & 1 \end{vmatrix} = -3, \quad A_{33} = \begin{vmatrix} 3 & -1 \\ -2 & 1 \end{vmatrix} = 1.$$

Using (1.103), we obtain

$$A^{-1} = \frac{1}{|A|} \begin{pmatrix} A_{11} & A_{21} & A_{31} \\ A_{12} & A_{22} & A_{32} \\ A_{13} & A_{23} & A_{33} \end{pmatrix} = \begin{pmatrix} 1 & 4/5 & -1/5 \\ 2 & 12/5 & -3/5 \\ 0 & 1/5 & 1/5 \end{pmatrix}.$$

Below are some properties of the operation of inversion of matrices.

1. The matrix A^{-1} is nonsingular and $(A^{-1})^{-1} = A$. This statement is an evident consequence of the equality $AA^{-1} = I$.
2. If the matrices A, B are nonsingular, then $(AB)^{-1} = B^{-1}A^{-1}$. Indeed, the following equalities hold: $AB(B^{-1}A^{-1}) = A(BB^{-1})A^{-1} = AA^{-1} = I$.
3. If the matrix A is nonsingular, then A^T is nonsingular and $(A^T)^{-1} = (A^{-1})^T$. The matrix A^T is nonsingular, since the equality $|A^T| = |A|$ holds. Using Property 3 (b), p. 38, we see that $(A^T)(A^{-1})^T = (A^{-1}A)^T = I^T = I$, i.e., the matrix $(A^{-1})^T$ is the inverse of A^T.

The proof of the following statements is left to the reader.

1. If matrices $A_1, A_2,..., A_p$ are nonsingular, then

$$(A_1 A_2 \cdots A_p)^{-1} = A_p^{-1} A_{p-1}^{-1} \cdots A_1^{-1}. \tag{1.104}$$

2. If P_{ik} is a permutation matrix, then

$$P_{ik}^{-1} = P_{ik}. \tag{1.105}$$

3. If L_k is an elementary lower triangular matrix such that $l_{kk} \neq 0$, then

$$L_k^{-1} = \begin{pmatrix} 1 \ldots & 0 & 0 \ldots 0 \\ \cdots\cdots\cdots \\ 0 \ldots & 1/l_{k,k} & 0 \ldots 0 \\ 0 \ldots & -l_{k+1,k}/l_{k,k} & 1 \ldots 0 \\ \cdots\cdots\cdots \\ 0 \ldots & -l_{n,k}/l_{k,k} & 0 \ldots 1 \end{pmatrix}. \tag{1.106}$$

4. If L is a lower triangular matrix such that all elements of its main diagonal are nonzero, then the inverse L^{-1} exists and is a lower triangular matrix. The analogous statement is true for upper triangular matrices.

1.2.5 Gaussian Elimination

In this section we consider an algorithm for solving the system of linear equations

$$Ax = b \tag{1.107}$$

with a square nonsingular matrix A. This algorithm is called *Gaussian elimination.*[10]
Like many other methods, it is based on the following statement.

Let B be a given nonsingular matrix. Then the system of linear equations

$$BAx = Bb \tag{1.108}$$

is equivalent to system (1.107), namely, each solution of system (1.108) is a solution
of (1.107), and conversely, each solution of system (1.107) is a solution of (1.108).
Indeed, if x is a solution of system (1.108), then

$$B(Ax - b) = 0,$$

but the matrix B is nonsingular; hence, $Ax - b = 0$. The converse is obvious.

Usually the matrix B is chosen such that the matrix BA is "simpler" than A and the
solution of system (1.108) is easier to calculate than the solution of system (1.107). In
Gaussian elimination, the matrix B is the product of special lower triangular matrices
such that the matrix BA is upper triangular. In this case, problem (1.108) is trivial.

Let us describe the Gaussian elimination algorithm. In the first step we take in
the first column of the matrix A the element with the largest absolute value. Suppose
that this element is a_{i1}. It is not equal to zero. Indeed, if $a_{i1} = 0$, then all elements in
the first column of A are equal to zero, and $|A| = 0$, but we have assumed that the
matrix A is nonsingular.

Then we multiply both sides of system (1.107) by the permutation matrix P_{i1}. We
denote this matrix by P_1 (note that $P_1 = I$ if in the first column of A, the element
with the largest absolute value is a_{11}) and get

$$A_1x = b_1, \tag{1.109}$$

where $A_1 = P_1A$, $b_1 = P_1b$. Observe that the matrix A_1 is obtained from the
matrix A by interchanging the first and the ith rows, and the column b_1 is obtained
from the column b by interchanging the first and the ith elements. We denote by $a_{kl}^{(1)}$
the elements of the matrix A_1, and by $b_k^{(1)}$ the elements of the column b_1. By construction, $a_{11}^{(1)} \neq 0$.

After that, we multiply both sides of system (1.109) by the elementary lower
triangular matrix

---

[10]Johann Carl Friedrich Gauss (1777–1855) was a German mathematician.

$$L_1 = \begin{pmatrix} l_{1,1} & 0 & 0 & \dots & 0 & 0 \\ l_{2,1} & 1 & 0 & \dots & 0 & 0 \\ & & \dots\dots\dots & & \\ l_{n-1,1} & 0 & 0 & \dots & 1 & 0 \\ l_{n,1} & 0 & 0 & \dots & 0 & 1 \end{pmatrix},$$ (1.110)

where $l_{11} = 1/a_{11}^{(1)}$, $l_{21} = -a_{21}^{(1)}/a_{11}^{(1)}$, ..., $l_{n1} = -a_{n1}^{(1)}/a_{11}^{(1)}$, and get

$$A_2 x = b_2,$$ (1.111)

where $b_2 = L_1 b_1$,

$$A_2 = L_1 A_1 = \begin{pmatrix} 1 & a_{12}^{(2)} & a_{13}^{(2)} & \dots & a_{1n}^{(2)} \\ 0 & a_{22}^{(2)} & a_{23}^{(2)} & \dots & a_{2n}^{(2)} \\ & & \dots\dots\dots & & \\ 0 & a_{n2}^{(2)} & a_{n3}^{(2)} & \dots & a_{nn}^{(2)} \end{pmatrix}.$$ (1.112)

Multiplication of the matrices L_1 and A_1 is equivalent to the following transformation of the matrix A_1: all elements of the first row of the matrix A_1 are divided by $a_{11}^{(1)}$; after that, for all $i = 2, \dots, n$, the first row is multiplied by $a_{i1}^{(1)}$ and is subtracted from the ith row of A_1. Similarly, the elements of the column b_2 are calculated by the following formulas: $b_1^{(2)} = b_1^{(1)}/a_{11}^{(1)}$, $b_i^{(2)} = b_i^{(1)} - b_1^{(2)} a_{i1}^{(1)}$, where $i = 2, \dots, n$.

Note that all elements of the first column of the matrix A_2 except the first element are equal to zero. Now we take the element with the largest absolute value among the elements $a_{22}^{(2)}$, $a_{32}^{(2)}$, ..., $a_{n2}^{(2)}$. Suppose that this element is $a_{i2}^{(2)}$. It is not equal to zero. Indeed, if $a_{i2}^{(2)} = 0$, then all the numbers $a_{22}^{(2)}$, $a_{32}^{(2)}$, ..., $a_{n2}^{(2)}$ would be equal to zero, and expanding the determinant of A_2 in terms of the first column, we would get $\det A_2 = 0$. On the other hand, we see that

$$\det A_2 = l_{11} \det(P_1 A) = \det(P_1 A)/a_{11}^{(1)} = \pm \det(A)/a_{11}^{(1)} \neq 0,$$

since L_1 is an elementary lower triangular matrix and P_1 is either the identity matrix or a permutation matrix.

Then we multiply both sides of system (1.111) by the permutation matrix $P_2 = P_{2i}$ (in other words, we interchange the second row and the ith row of the matrix A_2) and get

$$\tilde{A}_2 x = P_2 L_1 P_1 b,$$ (1.113)

where

$$\tilde{A}_2 = P_2 A_2 = \begin{pmatrix} 1 & a_{12}^{(2)} & a_{13}^{(2)} & \dots & a_{1n}^{(2)} \\ 0 & \tilde{a}_{22}^{(2)} & \tilde{a}_{23}^{(2)} & \dots & \tilde{a}_{2n}^{(2)} \\ & & \dots\dots\dots & & \\ 0 & \tilde{a}_{n2}^{(2)} & \tilde{a}_{n3}^{(2)} & \dots & \tilde{a}_{nn}^{(2)} \end{pmatrix}.$$

Multiplying both sides of (1.113) by the elementary lower triangular matrix

$$
L_2 = \begin{pmatrix}
1 & 0 & 0\,0\ldots 0\,0 \\
0 & l_{2,2} & 0\,0\ldots 0\,0 \\
0 & l_{3,2} & 1\,0\ldots 0\,0 \\
& & \cdots\cdots \\
0 & l_{n-1,2} & 0\,0\ldots 1\,0 \\
0 & l_{n,2} & 0\,0\ldots 0\,1
\end{pmatrix},
$$

where $l_{22} = 1/\tilde{a}_{22}^{(2)}$, $l_{32} = -\tilde{a}_{32}^{(2)}/\tilde{a}_{22}^{(2)}$, ..., $l_{n2} = -\tilde{a}_{n2}^{(2)}/\tilde{a}_{22}^{(2)}$, we get

$$
A_3 x = L_2 P_2 L_1 P_1 b,
$$

where $A_3 = L_2 \tilde{A}_2 = L_2 P_2 L_1 P_1 A$. It is easy to see that

$$
A_3 = \begin{pmatrix}
1 & a_{12}^{(2)} & a_{13}^{(2)} & \ldots & a_{1n}^{(2)} \\
0 & 1 & a_{23}^{(3)} & \ldots & a_{2n}^{(3)} \\
0 & 0 & a_{33}^{(3)} & \ldots & a_{3n}^{(3)} \\
& & \cdots\cdots\cdots \\
0 & 0 & a_{n3}^{(3)} & \ldots & a_{nn}^{(3)}
\end{pmatrix}.
$$

It is important to note that all elements of the second column of the matrix A_3 except the first two elements are equal to zero.

Continuing this process, we finally get the system of linear equations

$$
Ux = f \tag{1.114}
$$

(which obviously is equivalent to the original system), where

$$
U = L_n P_n L_{n-1} P_{n-1} \cdots L_1 P_1 A, \tag{1.115}
$$

$$
f = L_n P_n L_{n-1} P_{n-1} \cdots L_1 P_1 b,
$$

and, what is important,

$$
U = \begin{pmatrix}
1 & a_{12}^{(2)} & a_{13}^{(2)} & \ldots & a_{1n-1}^{(2)} & a_{1n}^{(2)} \\
0 & 1 & a_{23}^{(3)} & \ldots & a_{2n-1}^{(3)} & a_{2n}^{(3)} \\
0 & 0 & 1 & \ldots & a_{3n-1}^{(4)} & a_{3n}^{(4)} \\
& & & \cdots\cdots\cdots \\
0 & 0 & 0 & \ldots & 1 & a_{n-1,n}^{(n)} \\
0 & 0 & 0 & \ldots & 0 & 1
\end{pmatrix} \tag{1.116}
$$

is a triangular matrix with ones on the main diagonal.

Problem (1.114) is not difficult. From the last equation of system (1.114) we see that $x_n = f_n$. Using the penultimate equation, we get

$$x_{n-1} = f_{n-1} - a_{n-1,n}^{(n)} x_n, \tag{1.117}$$

and so on. Finally, using the first equation, we obtain

$$x_1 = f_1 - a_{1,2}^{(2)} x_2 - a_{1,3}^{(2)} x_3 - \cdots - a_{1,n}^{(2)} x_n. \tag{1.118}$$

Thus the Gaussian elimination algorithm can be divided into two parts. In the first part (sometimes called *forward elimination*), we reduce a given system to a system with a triangular matrix U. In the second part (sometimes called *backward substitution*), we solve this system with the triangular matrix obtained in the first part.

Remark 1.4 We choose the elements with the largest absolute values for forward elimination to improve numerical stability. These elements are called *pivot elements*, and the corresponding interchanging of rows is called *pivoting* (see Section 8.1, p. 249, for a detailed description). If we do not worry about rounding errors of numerical computations, then we can use as a pivot element any nonzero element of the column in each step of forward elimination.

Gaussian elimination also makes it possible to compute the determinant of a square matrix. Using (1.115), (1.104), and (1.105), we get

$$A = P_1 L_1^{-1} P_2 L_2^{-1} \cdots P_n L_n^{-1} U. \tag{1.119}$$

From (1.119), (1.94), and (1.96) it follows that

$$\det A = \det(P_1 L_1^{-1} P_2 L_2^{-1} \cdots P_n L_n^{-1} U) = \prod_{i=1}^{n} \det P_i \prod_{i=1}^{n} \det L_i^{-1}$$

$$= \pm \prod_{i=1}^{n} \det L_i^{-1}. \tag{1.120}$$

Here we have taken into account that $\det U = 1$. Using (1.106), it is easy to see that

$$\det L_i^{-1} = \tilde{a}_{ii}^{(i)};$$

hence

$$\det A = \pm a_{11}^{(1)} \tilde{a}_{22}^{(2)} \cdots \tilde{a}_{nn}^{(n)}. \tag{1.121}$$

Thus the determinant of the matrix A is equal to the product of all the pivot elements up to sign. The sign is determined by the number of interchanges of rows in forward elimination. If this number is even, then the sign is plus; if this number is odd, then the sign is minus.

Let us estimate the number of arithmetic operations required to solve a system of linear equations by Gaussian elimination. In the first step of forward elimination, the matrix L_1 is constructed. This requires n operations. Then the matrix L_1 is multiplied by the matrix A_1. It is easy to verify that the multiplication of the matrix L_1 by a column requires $2(n-1)+1 = 2n-1$ operations. The total number of columns is n. Therefore, the multiplication of the matrix L_1 by the matrix A_1 requires $2n^2 - n$ operations. After that, the matrix L_1 is multiplied by the column $P_1 b$. Thus the first step of forward elimination requires $2n^2 + n - 1$ operations.

It is easy to see that in the second step of forward elimination, the product $L_2 \tilde{A}_2$ is calculated by the multiplication of matrices of order $n - 1$. Hence the second step requires $2(n-1)^2 + (n-2)$ operations, and all steps of forward elimination require $2(1^2 + 2^2 + \cdots + n^2) + (1 + 2 + \cdots + (n-1))$ operations. It is well known that $1 + 2 + \cdots + n - 1 = n(n-1)/2$, $1^2 + 2^2 + \cdots + n^2 = n(n+1)(2n+1)/6$. Thus forward elimination requires

$$n(n+1)(2n+1)/3 + n(n-1)/2 \approx 2n^3/3$$

arithmetic operations. Note that we neglect the terms of order n^2, on the assumption that n is big enough. It is easy to see that the calculations by formulas (1.117), (1.118) require $2(n-1)+2(n-2)+\cdots+2 = 2(1+\cdots+n-1) = n(n-1) \approx n^2$ operations, and finally, we can conclude that to solve a system of n equations in n unknowns, the Gaussian elimination algorithm requires approximately $2n^3/3$ operations.

Note that Cramer's formulas require, as is easy to calculate, $n^2 n!$ arithmetic operations, which is much bigger. For example, if $n = 20$, then $n^2 n! \approx 9.7 \times 10^{20}$ and $2n^3/3 \approx 5.3 \times 10^3$.

For example, let us solve the following system of linear equations by Gaussian elimination:

$$
\begin{aligned}
3x_1 + 6x_2 + 15x_3 &= 60, \\
3x_1 + 2x_2 + 9x_3 &= 34, \\
9x_1 + 6x_2 - 3x_3 &= 12.
\end{aligned}
$$

First of all, we write down the matrix and the column of the right-hand side of the system:

$$
A = \begin{pmatrix} 3 & 6 & 15 \\ 3 & 2 & 9 \\ 9 & 6 & -3 \end{pmatrix}, \quad b = \begin{pmatrix} 60 \\ 34 \\ 12 \end{pmatrix}.
$$

The element with the largest absolute value in the first column of the matrix A is $a_{31} = 9$. In accordance with the algorithm described above, the matrix A_1 and the column b_1 look as follows:

$$A_1 = \begin{pmatrix} 9 & 6 & -3 \\ 3 & 2 & 9 \\ 3 & 6 & 15 \end{pmatrix}, \quad b_1 = \begin{pmatrix} 12 \\ 34 \\ 60 \end{pmatrix}$$

(we have interchanged the first row of the matrix A with its third row and the first element of the column b with its third element).

Now we divide the first row of the matrix A_1 by 9, multiply the result by 3, and subtract that result from the second and the third rows; also we divide the first element of the column b_1 by 9, multiply the result by 3, and subtract that result from the second and the third elements of b_1. As a result, we get

$$A_2 = \begin{pmatrix} 1 & 2/3 & -1/3 \\ 0 & 0 & 10 \\ 0 & 4 & 16 \end{pmatrix}, \quad b_2 = \begin{pmatrix} 4/3 \\ 30 \\ 56 \end{pmatrix}.$$

The element with the largest absolute value among the elements $a_{22}^{(2)}$, $a_{32}^{(2)}$ is $a_{32}^{(2)}$. Therefore, we interchange the second and third rows of the matrix A_2, and also the second and third elements of the column b_2, and obtain

$$\tilde{A}_2 = \begin{pmatrix} 1 & 2/3 & -1/3 \\ 0 & 4 & 16 \\ 0 & 0 & 10 \end{pmatrix}, \quad \tilde{b}_2 = \begin{pmatrix} 4/3 \\ 56 \\ 30 \end{pmatrix}.$$

We divide the second row of the matrix $\tilde{A}_2$ and the second element of the column $\tilde{b}_2$ by 4, and get

$$\tilde{\tilde{A}}_2 = \begin{pmatrix} 1 & 2/3 & -1/3 \\ 0 & 1 & 4 \\ 0 & 0 & 10 \end{pmatrix}, \quad \tilde{\tilde{b}}_2 = \begin{pmatrix} 4/3 \\ 14 \\ 30 \end{pmatrix}.$$

Finally, we divide the last row of $\tilde{\tilde{A}}_2$ and the last element of $\tilde{\tilde{b}}_2$ by 10, and get

$$A_3 = \begin{pmatrix} 1 & 2/3 & -1/3 \\ 0 & 1 & 4 \\ 0 & 0 & 1 \end{pmatrix}, \quad b_3 = \begin{pmatrix} 4/3 \\ 14 \\ 3 \end{pmatrix}.$$

The forward elimination is done.

Using the back substitution algorithm, we calculate $x_3 = 3$, after that $x_2 = 14 - 3 \times 4 = 2$, and finally, $x_1 = 4/3 - (2/3) \times 2 + (1/3) \times 3 = 1$.

As we have seen above, the determinant of the matrix A is equal to the product of all the pivot elements up to sign. In this example, the pivot elements are 9, 4, and 10. The number of interchanges of rows in forward elimination was two. Therefore, the determinant is equal to the product of the pivot elements: $\det(A) = 360$.

1.2.6 The Determinant of the Product of Matrices

Theorem 1.12 *The determinant of the product of arbitrary square matrices A and B (of the same dimension) is equal to the product of their determinants:*

$$\det(AB) = \det A \det B. \tag{1.122}$$

Proof If the matrix A is singular, then the matrix AB is also singular (see p. 39), and in this case, equality (1.122) obviously holds.

If the matrix A is nonsingular, then, using (1.119), we get

$$AB = P_1 L_1^{-1} P_2 L_2^{-1} \cdots P_n L_n^{-1} U B.$$

In this product, each factor except B is either a permutation matrix or a triangular matrix; hence

$$\det(AB) = \prod_{i=1}^{n} \det P_i \prod_{i=1}^{n} \det L_i^{-1} \det(U) \det B = \prod_{i=1}^{n} \det P_i \prod_{i=1}^{n} \det L_i^{-1} \det B,$$

but we have (see (1.120))

$$\prod_{i=1}^{n} \det P_i \prod_{i=1}^{n} \det L_i^{-1} = \det A.$$

Thus equality (1.122) holds. $\qquad\square$

From (1.122) it follows immediately that if the matrix A is nonsingular, then $\det(A^{-1}) = 1/\det A$.

1.2.7 Basic Matrix Types

In this section we describe some types of matrices that are often used in different problems of linear algebra. Here we consider some basic properties of these matrices. A more detailed study will be done in the next chapters.

Let A be a rectangular matrix. The *Hermitian*[11] *adjoint* A^* of A is defined by $A^* = (\bar{A})^T$, where $\bar{A}$ is the componentwise conjugate. Another name for the Hermitian adjoint of a matrix is the conjugate transpose. It is easy to see that $(A^*)^* = A$, $(\alpha A)^* = \bar{\alpha} A^*$, $(A + B)^* = A^* + B^*$, $(AB)^* = B^* A^*$.

A square matrix A is called *Hermitian* if $A = A^*$. It is called *skew-Hermitian* if $A = -A^*$. The determinant of a Hermitian matrix is a real number. Indeed,

---

[11]Charles Hermite (1822–1901) was a French mathematician.

since $\det(A^*) = \det((\overline{A})^T) = \det(\overline{A}) = \overline{\det(A)}$, we see that $\det(A) = \overline{\det(A)}$ if the matrix A is Hermitian.

Every square matrix A can be represented in the form

$$A = H_1 + iH_2, \tag{1.123}$$

where H_1, H_2 are Hermitian matrices and i is the imaginary unit. The matrices H_1, H_2 are uniquely determined by the matrix A. Indeed, representation (1.123) follows from the obvious identity

$$A = \frac{1}{2}(A + A^*) + i\frac{1}{2i}(A - A^*)$$

and the following easily verifiable equalities:

$$(A + A^*)^* = A + A^*, \quad \left(\frac{1}{i}(A - A^*)\right)^* = \frac{1}{i}(A - A^*).$$

If we assume that in addition to (1.123) there exists an additional representation

$$A = \tilde{H}_1 + i\tilde{H}_2$$

with Hermitian matrices $\tilde{H}_1$, $\tilde{H}_2$, then

$$(H_1 - \tilde{H}_1) + i(H_2 - \tilde{H}_2) = 0. \tag{1.124}$$

The Hermitian adjoint of the left part of equality (1.124) is also equal to zero:

$$(H_1 - \tilde{H}_1) - i(H_2 - \tilde{H}_2) = 0. \tag{1.125}$$

Adding together the corresponding terms in (1.124) and (1.125), we get $H_1 = \tilde{H}_1$; hence $H_2 = \tilde{H}_2$. Thus representation (1.123) is unique.

A *real matrix* is a matrix all of whose elements are real numbers.

A real Hermitian matrix is called *symmetric*. For every symmetric matrix we have $A = A^T$. A real square matrix is called *skew-symmetric* if $A = -A^T$.

For every real square matrix, the following representation holds:

$$A = A_1 + A_2, \tag{1.126}$$

where A_1 is a symmetric matrix and A_2 is skew-symmetric. Arguing as above, we see that this representation is unique and

$$A_1 = \frac{1}{2}(A + A^T), \quad A_2 = \frac{1}{2}(A - A^T).$$

A square matrix A is called *unitary* if $AA^* = I$, in other words, if $A^{-1} = A^*$. It follows from the definition that the absolute value of the determinant of every unitary matrix is equal to one. The product of two unitary matrices is a unitary matrix (prove it!).

An important example of a unitary matrix is a diagonal matrix with diagonal elements $q_1, q_2, ..., q_n$ such that the absolute value of each element is equal to one. The reader can easily prove that this matrix is unitary.

Let A be a unitary matrix of order n. Sometimes we say that the rectangular matrix B that consists of m, $m < n$, columns of the square unitary matrix A is unitary. Clearly, $B^*B = I_m$, where I_m is the identity matrix of order m.

A real unitary matrix is called *orthogonal*. The determinant of every orthogonal matrix is either plus one or minus one. Two examples of orthogonal matrices are the permutation matrix P_{kl} and the second-order matrix

$$Q_2(\varphi) = \begin{pmatrix} \cos\varphi & -\sin\varphi \\ \sin\varphi & \cos\varphi \end{pmatrix},$$

where φ is a real number.

A square matrix A is called *normal* if $AA^* = A^*A$, that is, if A commutes with its Hermitian adjoint. It is easy to see that Hermitian, skew-Hermitian, and unitary matrices are normal.

For example, the matrix $A = \begin{pmatrix} 1 & -1 \\ 1 & 1 \end{pmatrix}$ is normal, but it belongs to none of above-mentioned matrix types.

1.2.8 Block Matrices and Basic Operations with Block Matrices

It is useful in many cases to interpret a matrix as having been broken into sections called blocks or submatrices, that is, to represent it in the form

$$A = \begin{pmatrix} A_{11} & A_{12} & \ldots & A_{1n} \\ A_{21} & A_{22} & \ldots & A_{2n} \\ & \ldots\ldots\ldots & \\ A_{m1} & A_{m2} & \ldots & A_{mn} \end{pmatrix}, \tag{1.127}$$

where the elements A_{ij} are themselves matrices. Note that all blocks in (1.127) belonging to one row have the same number of rows, and all blocks in one column have the same number of columns. Every matrix may be interpreted as a *block matrix* in different ways, with each interpretation defined by how its rows and columns are partitioned. For example,

$$\begin{pmatrix} 1 & 8 & 7 & 6 \\ 3 & 5 & 0 & 2 \\ 1 & 4 & 9 & 3 \end{pmatrix}, \quad \begin{pmatrix} 1 & 8 & 7 & 6 \\ 3 & 5 & 0 & 2 \\ 1 & 4 & 9 & 3 \end{pmatrix}, \quad \begin{pmatrix} 1 & 8 & 7 & 6 \\ 3 & 5 & 0 & 2 \\ 1 & 4 & 9 & 3 \end{pmatrix}.$$

It is easy to see that with block matrices we can operate by the same formal rules as with ordinary matrices. If in addition to matrix (1.127) we introduce the matrix

$$B = \begin{pmatrix} B_{11} & B_{12} & \dots & B_{1n} \\ B_{21} & B_{22} & \dots & B_{2n} \\ & \dots\dots\dots & \\ B_{m1} & B_{m2} & \dots & B_{mn} \end{pmatrix} \tag{1.128}$$

such that for each pair of indices i, j the dimensions of blocks A_{ij}, B_{ij} coincide, then the matrix $C = A + B$ can be represented as the block matrix with blocks $C_{ij} = A_{ij} + B_{ij}, i = 1, \dots, m, j = 1, \dots, n$. Suppose that

$$B = \begin{pmatrix} B_{11} & B_{12} & \dots & B_{1p} \\ B_{21} & B_{22} & \dots & B_{2p} \\ & \dots\dots\dots & \\ B_{n1} & B_{n2} & \dots & B_{np} \end{pmatrix}. \tag{1.129}$$

Then the matrix $C = AB$ can be represented as the block matrix with blocks

$$C_{ij} = \sum_{q=1}^{n} A_{iq} B_{qj}, \quad i = 1, 2, \dots, m, \; j = 1, 2, \dots, p. \tag{1.130}$$

This, of course, requires that each product $A_{iq} B_{qj}$ exist, i.e., that the horizontal dimension of each block A_{iq} coincide with the vertical dimension of the corresponding block B_{qj}.

Let us obtain some useful formulas for calculation of determinants of block matrices. We begin with the simplest case. Suppose that

$$A = \begin{pmatrix} I & A_{12} \\ 0 & A_{22} \end{pmatrix} \tag{1.131}$$

is a 2×2 block matrix, I is the identity matrix, A_{22} is a square matrix, and A_{12} is a (generally speaking) rectangular matrix. Then we get

$$|A| = |A_{22}|. \tag{1.132}$$

Similarly, if

$$A = \begin{pmatrix} A_{11} & A_{12} \\ 0 & I \end{pmatrix}, \tag{1.133}$$

where A_{11} is a square matrix, then

$$|A| = |A_{11}|. \tag{1.134}$$

Theorem 1.13 *Suppose that*

$$A = \begin{pmatrix} A_{11} & A_{12} \\ 0 & A_{22} \end{pmatrix}, \tag{1.135}$$

where A_{11}, A_{22} are square matrices. Then

$$|A| = |A_{11}||A_{22}|. \tag{1.136}$$

Proof First we prove that if the matrix A_{11} is singular, then $|A| = 0$. Denote by n_1 the order of the matrix A_{11}, and by n_2 the order of A_{22}. If $|A_{11}| = 0$, then there exists a vector $x_1 \neq 0$ of length n_1 such that $A_{11}x_1 = 0$. Then for the nonzero vector $x = (x_1, 0, \ldots, 0)$ of length $n_1 + n_2$, we obviously have $Ax = 0$. Therefore, $|A| = 0$. Thus we have proved that if $|A_{11}| = 0$, then equality (1.136) trivially holds. Now let $|A_{11}| \neq 0$. It is easy to see that

$$\begin{pmatrix} A_{11} & A_{12} \\ 0 & A_{22} \end{pmatrix} = \begin{pmatrix} A_{11} & 0 \\ 0 & I \end{pmatrix} \begin{pmatrix} I & A_{11}^{-1}A_{12} \\ 0 & A_{22} \end{pmatrix}. \tag{1.137}$$

Hence

$$|A| = \begin{vmatrix} A_{11} & 0 \\ 0 & I \end{vmatrix} \begin{vmatrix} I & A_{11}^{-1}A_{12} \\ 0 & A_{22} \end{vmatrix}.$$

Combining the last equality, (1.132), and (1.134), we finally obtain (1.136). □

The proof of the following statements is left to the reader.

1. Suppose that

$$A = \begin{pmatrix} A_{11} & A_{12} & A_{13} & \ldots & A_{1n} \\ 0 & A_{22} & A_{23} & \ldots & A_{2n} \\ 0 & 0 & A_{33} & \ldots & A_{3n} \\ \ldots & \ldots & \ldots & \ldots & \ldots \\ 0 & 0 & 0 & \ldots & A_{nn} \end{pmatrix} \tag{1.138}$$

is a block triangular matrix, where A_{ii}, $i = 1, 2, \ldots, n$, are arbitrary square matrices. Then $|A| = |A_{11}||A_{22}| \cdots |A_{nn}|$.

2. Suppose that

$$A = \begin{pmatrix} A_{11} & A_{12} \\ A_{21} & A_{22} \end{pmatrix}$$

is a block matrix, A_{11}, A_{22} are square matrices, and $|A_{11}| \neq 0$. Then

$$|A| = |A_{11}||A_{22} - A_{21}A_{11}^{-1}A_{12}|. \tag{1.139}$$

Hint: calculate the product

$$\begin{pmatrix} A_{11} & A_{12} \\ A_{21} & A_{22} \end{pmatrix} \begin{pmatrix} I & -A_{11}^{-1} A_{12} \\ 0 & I \end{pmatrix}.$$

Relationship (1.139) can be considered a generalization of formula (1.48) for calculating a second-order determinant.

A matrix A of the form (1.138) is called *block upper triangular*. A *block lower triangular* matrix can be defined similarly. A matrix A of the form

$$A = \begin{pmatrix} A_{11} & 0 & 0 & \ldots & 0 \\ 0 & A_{22} & 0 & \ldots & 0 \\ 0 & 0 & A_{33} & \ldots & 0 \\ \ldots & \ldots & \ldots & \ldots & \ldots \\ 0 & 0 & 0 & \ldots & A_{nn} \end{pmatrix},$$

where A_{ii}, $i = 1, 2, \ldots, n$, are arbitrary square matrices, is called *block diagonal*. In this case, we also use the following notation: $A = \mathrm{diag}(A_{11}, A_{22}, \ldots, A_{nn})$.

Chapter 2
Vector Spaces

In courses on analytic geometry, basic operations with vectors in three-dimensional Euclidean space are studied. If some basis in the space is fixed, then a one-to-one correspondence between geometric vectors and ordered triples of real numbers (the coordinates of vectors in the basis) is determined, and algebraic operations with vector coordinates can be substituted for geometric operations with the vectors themselves.

Similar situations arise in many other areas of mathematics and its applications when objects under investigation are described by *tuples* (finite ordered lists) of real (or complex) numbers. Then the concept of a multidimensional coordinate space as the set of all tuples with algebraic operations on those tuples naturally arises.

In this chapter we will systematically construct and investigate such spaces. First of all, we will introduce the space $\mathbb{R}^n$ of all n-tuples of real numbers and the space $\mathbb{C}^n$ of all n-tuples of complex numbers. We will begin with definitions and basic properties of these spaces, since later we will introduce and study more general vector spaces. All results that we will obtain for general spaces hold for the vector spaces $\mathbb{R}^n$ and $\mathbb{C}^n$. We will provide also a variety of useful examples of specific bases in finite-dimensional spaces.

2.1 The Vector Spaces $\mathbb{R}^n$ and $\mathbb{C}^n$

2.1.1 The Vector Space $\mathbb{R}^n$

The vector space $\mathbb{R}^n$ is the set of all n-tuples $x = (x_1, x_2, \ldots, x_n)$ of real numbers, where $n \geq 1$ is a given integer. Elements of the space $\mathbb{R}^n$ are called *vectors* or *points*; the numbers $x_k, k = 1, 2, \ldots, n$, are called the *components* of the vector x.

© Springer International Publishing AG 2017
L. Beilina et al., *Numerical Linear Algebra: Theory and Applications*,
DOI 10.1007/978-3-319-57304-5_2

Two vectors $x, y \in \mathbb{R}^n$ are *equal* if and only if $x_k = y_k$ for all $k = 1, 2, \ldots, n$. The vector all of whose components are zero is called the *zero* vector and is denoted by 0. A vector of the form

$$i_k = (\underbrace{0, \ldots, 0}_{k-1}, 1, \underbrace{0, \ldots, 0}_{n-k}),$$

where the kth component is equal to one and all other components are zero, is called a *standard unit vector*. In the space $\mathbb{R}^n$ there are exactly n standard unit vectors: $i_1, i_2, \ldots, i_n$.

The following linear operations are introduced on the space $\mathbb{R}^n$: *scalar multiplication* (the multiplication of a vector by a scalar) and *vector addition*. Namely, for every real number α and $x, y \in \mathbb{R}^n$, by definition, we put

$$\alpha x = (\alpha x_1, \alpha x_2, \ldots, \alpha x_n),$$

$$x + y = (x_1 + y_1, x_2 + y_2, \ldots, x_n + y_n).$$

These linear operations satisfy the following properties. In the list below, x, y, z are arbitrary vectors in $\mathbb{R}^n$, and α, β are arbitrary real numbers.

1. *Commutativity* of vector addition: $x + y = y + x$.
2. *Associativity* of vector addition: $(x + y) + z = x + (y + z)$.
3. The zero vector is the *identity element* of vector addition: $x + 0 = x$.
4. For every vector x there exists a unique *inverse element* such that $x + (-x) = 0$, where by definition, $-x = (-1)x$.
5. *Distributivity* of scalar multiplication with respect to vector addition:

$$\alpha(x + y) = \alpha x + \alpha y.$$

6. *Distributivity* of scalar multiplication with respect to scalar addition:

$$(\alpha + \beta)x = \alpha x + \beta x.$$

7. Associativity of scalar multiplication: $(\alpha\beta)x = \alpha(\beta x)$.
8. The number 1 is the *identity element* of scalar multiplication: $1x = x$.

Properties 1–8 are called the *vector space axioms*. They follow immediately from the definition of linear operations with elements of the space $\mathbb{R}^n$. It is easy to see that Axioms 1–8 correspond exactly to the properties of linear operations with vectors in three-dimensional Euclidean space.

It is important to note that $\mathbb{R}^1$ is a vector space, but at the same time it is the set of all real numbers. As usual, we denote $\mathbb{R}^1$ by $\mathbb{R}$.

2.1.2 The Vector Space $\mathbb{C}^n$

The vector space $\mathbb{C}^n$ is the set of all n-tuples $x = (x_1, x_2, \ldots, x_n)$ of complex numbers, where $n \geq 1$ is a given integer. The elements of the space $\mathbb{C}^n$ are called *vectors* or *points*; the numbers $x_k, k = 1, 2, \ldots, n$, are called the *components* of the vector x.

Two vectors $x, y \in \mathbb{C}^n$ are *equal* if and only if $x_k = y_k$ for all $k = 1, 2, \ldots, n$. The vector all of whose components are zero is called the *zero* vector and is denoted by 0. The vector i_k whose kth component is equal to one and all other components are equal to zero is called a *standard unit vector*. In the space $\mathbb{C}^n$ there are exactly n standard unit vectors: $i_1, i_2, \ldots, i_n$.

The linear operations of *scalar multiplication* and *vector addition* are introduced on the space $\mathbb{C}^n$ in the usual way: for every complex number α and all $x, y \in \mathbb{C}^n$, by definition, we put

$$\alpha x = (\alpha x_1, \alpha x_2, \ldots, \alpha x_n),$$

$$x + y = (x_1 + y_1, x_2 + y_2, \ldots, x_n + y_n).$$

Note that we have actually already met this linear space. We can interpret the set of all $m \times n$ matrices with operations of matrix addition and multiplication of matrices by scalars (see p. 32) as the space $\mathbb{C}^{mn}$ of all vectors of length mn. The vectors were written in the form of rectangular arrays, but from the point of view of linear operations with vectors, this fact does not matter.

Properties 1–8 above hold also for the linear operations on the space $\mathbb{C}^n$.

Note that $\mathbb{C}^1$ is a vector space, but at the same time it is the set of all complex numbers. As usual, we denote $\mathbb{C}^1$ by $\mathbb{C}$.

2.2 Abstract Vector Spaces

2.2.1 Definitions and Examples

Two concepts more general than the spaces $\mathbb{R}^n$ and $\mathbb{C}^n$ are widely used in many areas of mathematics. These generalizations are abstract vector spaces: real and complex. A *real vector space* **X** is a set that is closed under the operations of vector addition and scalar multiplication that satisfy the axioms listed below. Elements of **X** are called *vectors*. The operation of *vector addition* takes any two elements $x, y \in \mathbf{X}$ and assigns to them a third element $z = x + y \in \mathbf{X}$, which is called the sum of the vectors x and y. The operation of *scalar multiplication* takes any real number α and any element $x \in \mathbf{X}$ and gives an element $\alpha x \in \mathbf{X}$, which is called the product of α and x.

In order for **X** to be a real vector space, the following *vector space axioms*, which are analogous to Properties 1–8 of the space $\mathbb{R}^n$ (see p. 56), must hold for all elements $x, y, z \in \mathbf{X}$ and real numbers α, β.

1. *Commutativity* of vector addition: $x + y = y + x$.
2. *Associativity* of vector addition: $(x + y) + z = x + (y + z)$.
3. There exists a unique element $0 \in \mathbf{X}$, called the *zero element* of the space $\mathbf{X}$, such that $x + 0 = x$ for all $x \in \mathbf{X}$.
4. For every element $x \in \mathbf{X}$ there exists a unique element $x' \in \mathbf{X}$, called the *additive inverse* of x, such that $x + x' = 0$.[1]
5. *Distributivity* of scalar multiplication with respect to vector addition:

$$\alpha(x + y) = \alpha x + \alpha y.$$

6. *Distributivity* of scalar multiplication with respect to scalar addition:

$$(\alpha + \beta)x = \alpha x + \beta x.$$

7. Associativity of scalar multiplication: $(\alpha\beta)x = \alpha(\beta x)$.
8. The number 1 is the *identity element* of scalar multiplication: $1x = x$.

If in the definition of the space $\mathbf{X}$ multiplication by complex numbers is allowed, then $\mathbf{X}$ is called a *complex vector space*. It is assumed that Axioms 1–8, where $\alpha, \beta \in \mathbb{C}$, hold.

The proof of the following statements is left to the reader (here $\mathbf{X}$ is an arbitrary vector space):

1. $-0 = 0$ (here 0 is the zero element of $\mathbf{X}$);
2. $\alpha 0 = 0$ for every scalar α;
3. $0x = 0$ for every vector $x \in \mathbf{X}$;
4. if $\alpha x = 0$, $x \in \mathbf{X}$, then at least one of the factors is zero;
5. $-x = (-1)x$ for every $x \in \mathbf{X}$;
6. $y + (x - y) = x$ for every $x, y \in \mathbf{X}$, where, by definition, $x - y = x + (-y)$.

In the remainder of the book we denote vector spaces by the capital letters $\mathbf{X}$, $\mathbf{Y}$, $\mathbf{Z}$. Unless otherwise stated, vector spaces are complex. The definitions and the results are mostly true for real spaces, too. Cases in which distinctions must be made in the interpretation of results for real spaces are specially considered.

The proof that the following sets are vector spaces is left to the reader.

1. The set $\mathbf{V}_3$ of all geometric vectors of three-dimensional Euclidean space with the usual definitions of operations of multiplication of a vector by a real scalar and vector addition is a real vector space.
2. The set of all real-valued functions of a real variable is a real vector space if the sum of two functions and the product of a function and a real number are defined as usual.
3. The set of all real-valued functions that are defined and continuous on the closed segment $[a, b]$ of the real axis is a real vector space. This space is denoted by

---

[1]The vector x' is usually denoted by $-x$.

$C[a, b]$. Hint: recall that the sum of two continuous functions is a continuous function, and the product of a continuous function and a real number is a continuous function.
4. The set of all functions in the space $C[a, b]$ that are equal to zero at a fixed point $c \in [a, b]$ is a real vector space.
5. The set of all polynomials with complex coefficients with the usual definitions of the sum of two polynomials and the product of a polynomial and a complex number is a complex vector space.
6. The set $\mathbf{Q}_n$ of all polynomials of order no more than n, where $n \geq 0$ is a given integer, joined with the zero polynomial, is a complex vector space. Hint: as we have seen in Section 1.1.2, p. 8, the sum of two polynomials is either a polynomial of degree no more than the maximum degree of the summands or the zero polynomial.

The reader can answer the next two questions.

1. Consider the set of all positive functions defined on the real axis and introduce on this set the operation of vector addition as the multiplication of two functions $f \cdot g$ and the operation of scalar multiplication as the calculation of the power function f^α. Is this set a vector space?
2. Consider the set of all even functions defined on the segment $[-1, 1]$ and introduce on this set the operation of vector addition as the multiplication of two functions and the operation of scalar multiplication as usual multiplication of a function by a scalar. Is this set a vector space?

2.2.2 Linearly Dependent Vectors

Two vectors a and b in a vector space $\mathbf{X}$ are said to be *linearly dependent (proportional)* if there exist numbers α and β, not both zero, such that

$$\alpha a + \beta b = 0.$$

Clearly, in this case we have either $a = \gamma b$ or $b = \delta a$, where γ, δ are some numbers.

For example, if $k \neq l$, then the standard unit vectors $i_k, i_l \in \mathbb{C}^n$, are nonproportional (prove it!).

Vectors $x_1 = (1 + i, 3, 2 - i, 5)$, $x_2 = (2, 3 - 3i, 1 - 3i, 5 - 5i) \in \mathbb{C}^4$ are proportional, since $2/(1 + i) = (3 - 3i)/3 = (1 - 3i)/(2 - i) = (5 - 5i)/5 = 1 - i$.

Let us generalize the concept of linear dependence of two vectors. A set of vectors $\{a_i\}_{i=1}^m = \{a_1, a_2, ...,a_m\}$, $m \geq 1$, in a vector space $\mathbf{X}$ is said to be *linearly dependent* if there exist numbers $x_1, x_2,..., x_m$, not all zero, such that

$$x_1 a_1 + x_2 a_2 + \cdots + x_m a_m = 0. \tag{2.1}$$

For instance, the set of vectors

$$a_1 = \begin{pmatrix} 5 \\ 2 \\ 1 \end{pmatrix}, \quad a_2 = \begin{pmatrix} -1 \\ 3 \\ 3 \end{pmatrix}, \quad a_3 = \begin{pmatrix} 9 \\ 7 \\ 5 \end{pmatrix}, \quad a_4 = \begin{pmatrix} 3 \\ 8 \\ 7 \end{pmatrix}$$

in the space $\mathbb{R}^3$ is linearly dependent, since for $x_1 = 4$, $x_2 = -1$, $x_3 = -3$, $x_4 = 2$ we have

$$x_1 a_1 + x_2 a_2 + x_3 a_3 + x_4 a_4 = 4 \begin{pmatrix} 5 \\ 2 \\ 1 \end{pmatrix} - \begin{pmatrix} -1 \\ 3 \\ 3 \end{pmatrix} - 3 \begin{pmatrix} 9 \\ 7 \\ 5 \end{pmatrix} + 2 \begin{pmatrix} 3 \\ 8 \\ 7 \end{pmatrix} = \begin{pmatrix} 0 \\ 0 \\ 0 \end{pmatrix} = 0.$$

It is useful to note that there are many other sets of coefficients x_1, x_2, x_3, x_4 such that the linear combination $x_1 a_1 + x_2 a_2 + x_3 a_3 + x_4 a_4$ is equal to zero. For example,

$$2a_1 + a_2 - a_3 = 2 \begin{pmatrix} 5 \\ 2 \\ 1 \end{pmatrix} + \begin{pmatrix} -1 \\ 3 \\ 3 \end{pmatrix} - \begin{pmatrix} 9 \\ 7 \\ 5 \end{pmatrix} = 0,$$

$$3a_2 + a_3 - 2a_4 = 3 \begin{pmatrix} -1 \\ 3 \\ 3 \end{pmatrix} + \begin{pmatrix} 9 \\ 7 \\ 5 \end{pmatrix} - 2 \begin{pmatrix} 3 \\ 8 \\ 7 \end{pmatrix} = 0.$$

It is useful to write the definition of linear dependence of vectors in matrix form. We use the following notation. Let $\mathcal{A}_m = \{a_1, a_2, \ldots, a_m\}$ be a finite ordered list of vectors in the space **X**. For $x \in \mathbb{C}^m$, by definition, we put

$$\mathcal{A}_m x = x_1 a_1 + x_2 a_2 + \cdots + x_m a_m.$$

Then we can say that the vectors $a_1, a_2, \ldots, a_m$ are *linearly dependent* if there exists a nonzero vector $x \in \mathbb{C}^m$ such that

$$\mathcal{A}_m x = 0.$$

A vector $a \in \mathbf{X}$ is a *linear combination* of vectors $b_1, b_2, \ldots, b_p$, $p \geq 1$, if there exists a vector $x \in \mathbb{C}^p$ such that

$$a = x_1 b_1 + x_2 b_2 + \cdots + x_p b_p. \tag{2.2}$$

We can write this in matrix form:

$$a = \mathcal{B}_p x.$$

A linear combination of vectors is called *nontrivial* if at least one of the numbers x_1, $x_2, \ldots, x_p$ in (2.2) is not equal to zero.

The proof of the following two theorems is left to the reader.

Theorem 2.1 *A set of vectors is linearly dependent if it contains a linearly dependent subset, in particular if it contains the zero vector.*

Theorem 2.2 *A set of vectors $\{a_i\}_{i=1}^m$ is linearly dependent if and only if it contains a vector a_k that can be represented as a linear combination of other vectors of the set $\{a_i\}_{i=1}^m$.*

Suppose that each vector of the set $\{a_i\}_{i=1}^m$ is a linear combination of the vectors $\{b_i\}_{i=1}^p$, i.e.,

$$a_k = \sum_{j=1}^p x_{jk} b_j, \quad k = 1, 2, \ldots, m. \tag{2.3}$$

We can write (2.3) in matrix form:

$$\mathcal{A}_m = \mathcal{B}_p X(p, m), \tag{2.4}$$

where the kth column of the matrix X consists of the coefficients x_{jk} of the kth linear combination in (2.3).

The following property of transitivity holds. If each vector of the set $\{a_i\}_{i=1}^m$ is a linear combination of the vectors $\{b_i\}_{i=1}^p$, and each vector of $\{b_i\}_{i=1}^p$ is a linear combination of the vectors $\{c_i\}_{i=1}^q$, then each vector of $\{a_i\}_{i=1}^m$ is a linear combination of $\{c_i\}_{i=1}^q$. Indeed, using matrix notation, we can write

$$\mathcal{A}_m = \mathcal{B}_p X(p, m), \quad \mathcal{B}_p = \mathcal{C}_q Y(q, p).$$

Substituting $\mathcal{C}_q Y(q, p)$ for $\mathcal{B}_p$ in the first equality, we get

$$\mathcal{A}_m = \mathcal{C}_q Z(q, m),$$

where

$$Z(q, m) = Y(q, p) X(p, m).$$

We say that the two sets of vectors $\{a_i\}_{i=1}^m$ and $\{b_i\}_{i=1}^p$ are *equivalent* if there exist matrices $X(p, m)$ and $Y(m, p)$ such that

$$\mathcal{A}_m = \mathcal{B}_p X(p, m), \quad \mathcal{B}_p = \mathcal{A}_m Y(m, p), \tag{2.5}$$

i.e., each vector of the set $\mathcal{A}_m$ is a linear combination of the vectors of the set $\mathcal{B}_p$ and conversely.

Using the property of transitivity, the reader can easily prove the next statement. Suppose that the sets $\{a_i\}_{i=1}^m$ and $\{b_i\}_{i=1}^p$ are equivalent and the vector $x \in \mathbf{X}$ is a linear combination of the vectors $\{a_i\}_{i=1}^m$. Then x can be represented as a linear combination of the vectors $\{b_i\}_{i=1}^p$.

2.2.3   Linearly Independent Sets of Vectors

A set of vectors $\mathcal{A}_m = \{a_i\}_{i=1}^m$ in a vector space $\mathbf{X}$ is said to be *linearly independent* if $\mathcal{A}_m x = 0$ implies $x = 0$.

Linearly independent sets exist. Let us give some simple examples.

1. Each vector $a \neq 0$ forms a linearly independent set, which consists of one vector.
2. If $m \leq n$, then the standard unit vectors $i_1, i_2, \ldots, i_m \in \mathbb{C}^n$ are linearly independent. Indeed, for every $x \in \mathbb{C}^m$, the vector

$$x_1 i_1 + x_2 i_2 + \cdots + x_m i_m \in \mathbb{C}^n$$

   has the form $(x_1, x_2, \ldots, x_m, 0, \ldots, 0)$ and is equal to zero if and only if $x = 0$.
3. The set of vectors $\varphi_0(z) \equiv 1$, $\varphi_1(z) = z$, ..., $\varphi_k(z) = z^k$, where z is a complex number and $k \geq 0$ is a given integer, is linearly independent in the vector space of polynomials (see p. 59). This statement immediately follows from the fact that if a polynomial is equal to zero, then all its coefficients are equal to zero (see p. 31).

The next theorem is an evident consequence of Theorem 2.1.

Theorem 2.3 *Every subset of a linearly independent set $\{a_i\}_{i=1}^m$ is linearly independent.*

Theorem 2.4 *Every set $\{a_1, a_2, ..., a_n, b\}$ of $n+1$ vectors in the space $\mathbb{C}^n$ is linearly dependent.*

Proof Suppose that the set of vectors $\{a_i\}_{i=1}^n$ is linearly independent. Denote by A the matrix whose columns are the vectors a_i, $i = 1, 2, \ldots, n$. Clearly, $\det A \neq 0$, and the system of linear equations $Ax = b$ has a solution x. Therefore,

$$x_1 a_1 + \cdots + x_n a_n = b,$$

i.e., the set of vectors $\{a_1, a_2, ..., a_n, b\}$ is linearly dependent. $\square$

It follows immediately from Theorem 2.4 that every set $\{a_i\}_{i=1}^m \in \mathbb{C}^n$, $m > n$, is linearly dependent.

Theorem 2.5 *Suppose that the set of vectors $\mathcal{A}_m = \{a_i\}_{i=1}^m$ in the space $\mathbf{X}$ is linearly independent and each vector of the set $\mathcal{A}_m$ is a linear combination of the vectors $\mathcal{B}_p = \{b_i\}_{i=1}^p$. Then $m \leq p$.*

Proof Assume the contrary, i.e., let $m > p$. By definition, there exists a $p \times m$ matrix X such that $\mathcal{A}_m = \mathcal{B}_p X$. Therefore, for every $y \in \mathbb{C}^m$ we have $\mathcal{A}_m y = \mathcal{B}_p X y$. The columns of the matrix X form a set of vectors in the space $\mathbb{C}^p$. The number of vectors in this set is $m > p$, and hence it is linearly dependent. Thus there exists a vector $y \in \mathbb{C}^m$ that is not equal to zero such that $Xy = 0$, but then $\mathcal{A}_m y = 0$, which means, contrary to the assumption, that the set of vectors $a_1, a_2, \ldots, a_m$ is linearly dependent. □

Corollary 2.1 *Every two linearly independent equivalent sets of vectors have the same number of vectors.*

The reader is invited to prove the next theorem (hint: use the reasoning of the proof of Theorem 2.5).

Theorem 2.6 *Suppose that the set $\{a_k\}_{k=1}^m$ is linearly independent and each vector of the set $\{b_k\}_{k=1}^m$ is a linear combination of the vectors $\{a_k\}_{k=1}^m$, i.e., there exists a square matrix X of order m such that $\mathcal{B}_m = \mathcal{A}_m X$. The set $\{b_k\}_{k=1}^m$ is linearly independent if and only if the matrix X is nonsingular.*

It is important to note that in Theorem 2.6 the matrix X is uniquely determined by the sets $\mathcal{A}_m$ and $\mathcal{B}_m$. Indeed, if we assume that there exists another matrix $\widetilde{X}$ such that $\mathcal{B}_m = \mathcal{A}_m \widetilde{X}$, then $\mathcal{A}_m(\widetilde{X} - X) = 0$, and $\widetilde{X} = X$, since the set $\mathcal{A}_m$ is linearly independent.

2.2.4 The Rank of a Set of Vectors

Let $\{a_i\}_{i=1}^m$ be a given set of vectors in the space **X**. Suppose that not all vectors $\{a_i\}_{i=1}^m$ are equal to zero. Then this set necessarily contains a linearly independent subset of vectors. In particular, the set $\{a_i\}_{i=1}^m$ itself can be linearly independent.

A linearly independent subset $\{a_{i_k}\}_{k=1}^r \subset \{a_i\}_{i=1}^m$ is called *maximal* if including any other vector of the set $\{a_i\}_{i=1}^m$ would make it linearly dependent.

For example, let us consider the following set of vectors:

$$a_1 = \begin{pmatrix} 2 \\ -2 \\ -4 \end{pmatrix}, \quad a_2 = \begin{pmatrix} 1 \\ 9 \\ 3 \end{pmatrix}, \quad a_3 = \begin{pmatrix} -2 \\ -4 \\ 1 \end{pmatrix}, \quad a_4 = \begin{pmatrix} 3 \\ 7 \\ -1 \end{pmatrix} \quad (2.6)$$

in the space $\mathbb{R}^3$. Evidently, the vectors a_1, a_2 are linearly independent and form a maximal linearly independent subset, since the determinants

$$\begin{vmatrix} 2 & 1 & -2 \\ -2 & 9 & -4 \\ -4 & 3 & 1 \end{vmatrix}, \quad \begin{vmatrix} 2 & 1 & 3 \\ -2 & 9 & 7 \\ -4 & 3 & -1 \end{vmatrix},$$

which consist of the components of the vectors a_1, a_2, a_3 and a_1, a_2, a_4, respectively, are equal to zero. Therefore, the sets of vectors a_1, a_2, a_3 and a_1, a_2, a_4 are linearly dependent.

Generally speaking, the set $\{a_i\}_{i=1}^m$ can contain several maximal linearly independent subsets, but the following result is true.

Theorem 2.7 *Every two maximal linearly independent subsets of the set* $\{a_i\}_{i=1}^m$ *contain the same number of vectors.*

Proof It follows from the definition of a maximal linearly independent subset that each vector of the set $\{a_i\}_{i=1}^m$ is a linear combination of vectors of a maximal linearly independent subset $\{a_{i_k}\}_{k=1}^r$. Obviously,

$$a_{i_k} = a_{i_k} + \sum_{i=1, i \neq i_k}^{m} 0 a_i;$$

hence the converse is also true. Therefore, the set $\{a_i\}_{i=1}^m$ and each of its maximal linearly independent subsets are equivalent. Thus, using Corollary 2.1, we claim that every two maximal linearly independent subsets of the set $\{a_i\}_{i=1}^m$ contain the same number of vectors.   □

This result allows us to introduce the following concept. The *rank* of a set of vectors in the space $\mathbf{X}$ is the number of vectors in each of its maximal linearly independent subsets.

For example, the rank of the set of vectors (2.6) is equal to two.

The number of linearly independent vectors in the space $\mathbb{C}^n$ is no more than n. Therefore, the rank of every set of vectors in $\mathbb{C}^n$ is less than or equal to n.

Clearly, a set of vectors $\{a_i\}_{i=1}^m$ in a vector space $\mathbf{X}$ is linearly independent if and only if its rank is equal to m.

2.3 Finite-Dimensional Vector Spaces. Bases

2.3.1 *Bases in the Space* $\mathbb{C}^n$

A linearly independent set $\{e_k\}_{k=1}^n$ (which consists of n vectors) is called a *basis* in the space $\mathbb{C}^n$. The standard unit vectors $\{i_k\}_{k=1}^n$ form the *standard* (or the *natural*) basis in the space $\mathbb{C}^n$.

It follows from Property 8, p. 23, of determinants that a set $\{e_k\}_{k=1}^n \subset \mathbb{C}^n$ is a basis if and only if the matrix $\mathcal{E}_n$, the columns of which are formed by the vectors $e_1, e_2, \ldots, e_n$, is nonsingular.

In the proof of Theorem 2.4, p. 62, we established that if $\{e_k\}_{k=1}^n$ is a basis in the space $\mathbb{C}^n$, then each vector $x \in \mathbb{C}^n$ can be represented as a linear combination

$$x = \xi_1 e_1 + \xi_2 e_2 + \cdots + \xi_n e_n. \tag{2.7}$$

The coefficients in the linear combination (2.7) are uniquely determined by the vector x and satisfy the following system of linear algebraic equations with the nonsingular matrix $\mathcal{E}_n$:

$$\mathcal{E}_n \xi = x. \tag{2.8}$$

Here $\xi = (\xi_1, \xi_2, \ldots, \xi_n)$ is the column of coefficients of the expansion of x with respect to the basis $\{e_k\}_{k=1}^n$.

2.3.2 Finite-Dimensional Spaces. Examples

A vector space $\mathbf{X}$ is called *finite-dimensional* if there exist vectors

$$\mathcal{E}_n = \{e_1, e_2, \ldots, e_n\}$$

that form a linearly independent set in the space $\mathbf{X}$ and such that each vector $x \in \mathbf{X}$ can be represented as a linear combination

$$x = \sum_{k=1}^n \xi_k e_k = \mathcal{E}_n \xi, \quad \xi \in \mathbb{C}^n. \tag{2.9}$$

The set of vectors $\{e_k\}_{k=1}^n$ is called a *basis* of the space $\mathbf{X}$. The number n is called the *dimension* of $\mathbf{X}$, and we denote by $\mathbf{X}_n$ this n-dimensional vector space. The coefficients $\xi_1, \xi_2, \ldots, \xi_n$ in the expansion (2.9) are called the *coordinates* of x with respect to the basis $\{e_k\}_{k=1}^n$.

The coordinates of each vector $x \in \mathbf{X}_n$ are uniquely determined by the basis $\{e_k\}_{k=1}^n$. Indeed, suppose that in addition to (2.9) there exists an expansion $x = \mathcal{E}_n \tilde{\xi}$. Then $\mathcal{E}_n(\xi - \tilde{\xi}) = 0$. Therefore, $\xi = \tilde{\xi}$, since the set of vectors $\{e_k\}_{k=1}^n$ is linearly independent.

Theorem 2.8 *In an n-dimensional vector space* $\mathbf{X}_n$*, every system* $\tilde{\mathcal{E}}_n = \{\tilde{e}_k\}_{k=1}^n$ *consisting of n linearly independent vectors is a basis.*

Proof It is enough to show that each vector $x \in \mathbf{X}_n$ can be represented as a linear combination

$$x = \tilde{\mathcal{E}}_n \tilde{\xi}. \tag{2.10}$$

By the definition of an n-dimensional vector space, a basis $\mathcal{E}_n$ exists in $\mathbf{X}_n$. Therefore, each vector of the set $\tilde{\mathcal{E}}_n$ can be represented as a linear combination of the vectors of $\mathcal{E}_n$. In other words, there exists a square matrix T of order n such that $\tilde{\mathcal{E}}_n = \mathcal{E}_n T$. The matrix T is nonsingular (see p. 63). Since $\mathcal{E}_n$ is a basis, there exists a vector $\xi \in \mathbb{C}^n$ such that $x = \mathcal{E}_n \xi$. Since the matrix T is nonsingular, there exists a vector $\tilde{\xi} \in \mathbb{C}^n$ such that $\xi = T\tilde{\xi}$. Thus we get the relationship $x = \mathcal{E}_n T\tilde{\xi} = \tilde{\mathcal{E}}_n \tilde{\xi}$ of the form (2.10). $\square$

If a vector space is not finite-dimensional, then the space is called *infinite-dimensional*.

Let us give some examples of finite-dimensional and infinite-dimensional vector spaces.

1. Three arbitrary non-coplanar vectors form a basis in the space $\mathbf{V}_3$. The space $\mathbf{V}_3$ is three-dimensional.
2. Evidently, the spaces $\mathbb{C}^n$, $\mathbb{R}^n$ are n-dimensional.
3. The set $\mathbf{Q}_n$ of all polynomials of order no more than n is finite-dimensional. Its dimension is equal to $n + 1$. For example, the set of vectors $\{1, z, \ldots, z^n\}$, where z is a complex variable, is a basis in $\mathbf{Q}_n$.
4. The vector space of all polynomials is infinite-dimensional. Indeed, for an arbitrarily large integer k, the set of vectors $\{1, z, \ldots, z^k\}$ is linearly independent in this space.
5. The space $C[a, b]$ is infinite-dimensional, since it contains polynomials with real coefficients of arbitrary order.

2.3.3   Change of Basis

Let $\mathcal{E}_n = \{e_k\}_{k=1}^n$, $\tilde{\mathcal{E}}_n = \{\tilde{e}_k\}_{k=1}^n$ be bases in a vector space $\mathbf{X}_n$. As we have shown, the sets $\mathcal{E}_n$ and $\tilde{\mathcal{E}}_n$ are equivalent, and there exist square matrices T and $\tilde{T}$ of order n such that

$$\mathcal{E}_n = \tilde{\mathcal{E}}_n \tilde{T}, \quad \tilde{\mathcal{E}}_n = \mathcal{E}_n T. \tag{2.11}$$

The matrix T is called the *change of basis matrix* from $\mathcal{E}_n$ to $\tilde{\mathcal{E}}_n$. The matrices T and $\tilde{T}$ are mutually inverse. Indeed, substituting $\mathcal{E}_n T$ for $\tilde{\mathcal{E}}_n$ in the first equality in (2.11), we obtain $\mathcal{E}_n = \mathcal{E}_n T\tilde{T}$. Thus we get

$$T\tilde{T} = I, \tag{2.12}$$

since the vectors of the basis $\mathcal{E}_n$ are linearly independent (see the remark after Theorem 2.6, p. 63).

Suppose that we know the vector ξ of the coordinates of an element $x \in \mathbf{X}_n$ with respect to the basis $\mathcal{E}_n$, and we also know the change of basis matrix T from $\mathcal{E}_n$ to the basis $\tilde{\mathcal{E}}_n$. Let us construct a formula for calculating the vector $\tilde{\xi}$ of coordinates of the same element x with respect to the basis $\tilde{\mathcal{E}}_n$. Using (2.9), we see that $x = \mathcal{E}_n \xi$, but $\mathcal{E}_n = \tilde{\mathcal{E}}_n \tilde{T} = \tilde{\mathcal{E}}_n T^{-1}$ (see (2.11), (2.12)). Therefore, $x = \tilde{\mathcal{E}}_n T^{-1}\xi$, which means that

$$\tilde{\xi} = T^{-1}\xi. \tag{2.13}$$

For example, suppose that vectors e_1, e_2, e_3 form a basis in a three-dimensional space $\mathbf{X}_3$. Let us consider the vectors

$$
\begin{aligned}
\tilde{e}_1 &= 5e_1 - e_2 - 2e_3, \\
\tilde{e}_2 &= 2e_1 + 3e_2, \\
\tilde{e}_3 &= -2e_1 + e_2 + e_3.
\end{aligned}
$$

Writing these equalities in matrix form, we get $\tilde{\mathcal{E}} = \mathcal{E}T$, where

$$
\tilde{\mathcal{E}} = \{\tilde{e}_1, \tilde{e}_2, \tilde{e}_3\}, \quad \mathcal{E} = \{e_1, e_2, e_3\}, \quad T = \begin{pmatrix} 5 & 2 & -2 \\ -1 & 3 & 1 \\ -2 & 0 & 1 \end{pmatrix}.
$$

It is easy to see that $\det T = 1$; hence the matrix T is nonsingular. Therefore, the vectors $\tilde{e}_1, \tilde{e}_2, \tilde{e}_3$ also form a basis in the space $\mathbf{X}_3$. Let us consider the vector $a = e_1 + 4e_2 - e_3$. The coordinates of the vector a with respect to the basis $\mathcal{E}$ are the numbers $\xi_1 = 1, \xi_2 = 4, \xi_3 = -1$, i.e., $a = \mathcal{E}\xi$, where $\xi = (\xi_1, \xi_2, \xi_3)$. Now we calculate the coordinates of the same vector, but with respect to the basis $\tilde{\mathcal{E}}$. Calculating the matrix T^{-1}, we get

$$
T^{-1} = \begin{pmatrix} 3 & -2 & 8 \\ -1 & 1 & -3 \\ 6 & -4 & 17 \end{pmatrix},
$$

and therefore,

$$
\tilde{\xi} = T^{-1}\xi = \begin{pmatrix} 3 & -2 & 8 \\ -1 & 1 & -3 \\ 6 & -4 & 17 \end{pmatrix} \begin{pmatrix} 1 \\ 4 \\ -1 \end{pmatrix} = \begin{pmatrix} -13 \\ 6 \\ -27 \end{pmatrix},
$$

i.e., $a = -13\tilde{e}_1 + 6\tilde{e}_2 - 27\tilde{e}_3$. Thus we have calculated the coordinate representation of the vector a with respect to the basis $\tilde{\mathcal{E}}$.

Note that infinitely many bases exist in the space $\mathbf{X}_n$. Indeed, if $\mathcal{E}_n$ is a basis, then the set of vectors $\tilde{\mathcal{E}}_n = \mathcal{E}_n T$, where T is an arbitrary nonsingular matrix, also is a basis (see Theorem 2.6, p. 63).

Below are some examples of bases in the space of polynomials of order no more than n with complex coefficients, which are often used in applications.

1. The *natural basis* for this space is the set of vectors $\{1, z, \ldots, z^n\}$, where z is a complex variable.
2. The polynomials

$$
\Phi_j(z) = \frac{(z - z_0)(z - z_1) \cdots (z - z_{j-1})(z - z_{j+1}) \cdots (z - z_n)}{(z_j - z_0)(z_j - z_1) \cdots (z_j - z_{j-1})(z_j - z_{j+1}) \cdots (z_j - z_n)},
$$

$j = 0, 1, 2, \ldots, n$, where $z_0, z_1, \ldots, z_n$ are arbitrary distinct complex numbers, also form a basis in the space of polynomials (see p. 29). Such a basis is called a *Lagrange basis*.

3. Let us prove that the polynomials

$$\varphi_0(z) \equiv 1, \; \varphi_1(z) = (z - z_0), \; \varphi_2(z) = (z - z_0)(z - z_1), \ldots,$$
$$\varphi_n(z) = (z - z_0)(z - z_1) \cdots (z - z_{n-1}),$$
$$\tag{2.14}$$

where $z_0, z_1, \ldots, z_{n-1}$ are arbitrary distinct complex numbers, form a basis. As in the case of a Lagrange basis, it is enough to check that for the numbers $z_0, z_1, \ldots, z_{n-1}$, and a number z_n that does not coincide with any of the numbers $z_0, z_1, \ldots, z_{n-1}$, the system of equations

$$c_0\varphi_0(z_j) + c_1\varphi_1(z_j) + \cdots + c_n\varphi_n(z_j) = h_j, \quad j = 0, 1, 2, \ldots, n, \tag{2.15}$$

has a unique solution for every choice of $h_0, h_1, \ldots h_n$. This fact is evident, since system (2.15) is triangular,

$$c_0 = h_0,$$
$$c_0 + c_1(z_1 - z_0) = h_1,$$
$$c_0 + c_1(z_2 - z_0) + c_2(z_2 - z_0)(z_2 - z_1) = h_2, \tag{2.16}$$
$$\cdots\cdots\cdots\cdots\cdots\cdots\cdots\cdots\cdots\cdots\cdots\cdots$$
$$c_0 + c_1(z_n - z_0) + \cdots + c_n(z_n - z_0)(z_n - z_1) \cdots (z_n - z_{n-1}) = h_n,$$

and all diagonal coefficients are different from zero. The basis defined in (2.14) is called a *Newton basis*.[2]

---

[2] Sir Isaac Newton (1642–1727) was an English physicist and mathematician.

Chapter 3
Inner Product Spaces

As discussed in the previous chapter, vector spaces are analogous to three-dimensional Euclidean space $\mathbf{V}_3$ of geometric vectors (directed line segments). However, such important concepts as the length of a vector and the angle between two vectors were not introduced for abstract spaces. In three-dimensional Euclidean space, using the lengths of two vectors and the angle between them, we can calculate the inner product (the dot product) of these vectors. Many geometric problems in the space $\mathbf{V}_3$ are solved with help of the dot product.

The concept of an inner product on an abstract space will be introduced axiomatically in this chapter. After that, the concepts of the length of a vector and the angle between two vectors will be introduced based on the concept of the inner product. Then we will investigate the concept of orthogonal bases. Some important examples of orthogonal bases in finite-dimensional spaces, particularly in polynomial spaces, will be constructed. The basic properties of subspaces of unitary spaces will be described. We begin our considerations with inner products on the spaces $\mathbb{R}^n$ and $\mathbb{C}^n$.

3.1 Inner Products on $\mathbb{R}^n$ and $\mathbb{C}^n$

An *inner product* on the space $\mathbb{R}^n$ is a function that assigns to each pair of vectors $x, y \in \mathbb{R}^n$ a real number (x, y) and satisfies the following *axioms* (which correspond to the properties of the inner product of vectors in three-dimensional Euclidean space):

1. $(x, x) \geq 0$ for all $x \in \mathbb{R}^n$; $(x, x) = 0$ if and only if $x = 0$;
2. $(x, y) = (y, x)$ for all $x, y \in \mathbb{R}^n$;
3. $(\alpha x + \beta y, z) = \alpha(x, z) + \beta(y, z)$ for all $x, y, z \in \mathbb{R}^n$ and for all $\alpha, \beta \in \mathbb{R}$.

Clearly, the next property follows from Axioms 2 and 3:

4. $(x, \alpha y + \beta z) = \alpha(x, y) + \beta(x, z)$ for all $x, y, z \in \mathbb{R}^n$ and for all $\alpha, \beta \in \mathbb{R}$.

© Springer International Publishing AG 2017
L. Beilina et al., *Numerical Linear Algebra: Theory and Applications*,
DOI 10.1007/978-3-319-57304-5_3

An inner product on the space $\mathbb{R}^n$ can be specified in infinitely many ways. For example, we can put

$$(x, y) = \sum_{k=1}^{n} x_k y_k.$$

This inner product on the space $\mathbb{R}^n$ is called *standard*. We can construct a variety of inner products if we put

$$(x, y) = \sum_{k=1}^{n} \rho_k x_k y_k, \tag{3.1}$$

where $\rho_1, \rho_2, \ldots, \rho_n$ are positive numbers. Varying these numbers, we get different inner products. The verification of Axioms 1–3 is trivial for both examples.

By the *length* of a vector $x \in \mathbb{R}^n$ we mean the nonnegative number $|x| = \sqrt{(x, x)}$. It can be shown that the length of vectors in $\mathbb{R}^n$ satisfies the following properties[1] (which correspond to the properties of the length of vectors in three-dimensional Euclidean space):

1. $|x| \geq 0$ for all $x \in \mathbb{R}^n$; $|x| = 0$ if and only if $x = 0$;
2. $|\alpha x| = |\alpha||x|$ for all $x \in \mathbb{R}^n$ and for all $\alpha \in \mathbb{R}$;
3. $|x + y| \leq |x| + |y|$ for all $x, y \in \mathbb{R}^n$.

Inequality 3 is called the *triangle inequality* (or *Minkowski's[2] inequality*).

The vector space $\mathbb{R}^n$, together with a specified inner product on it, is often called the *Euclidean space* $\mathbb{R}^n$. It is important to note that by specifying the inner product on the space $\mathbb{R}^n$ in different ways, we get different Euclidean spaces. The space $\mathbb{R}^n$ together with the standard inner product is called the *real coordinate space*. This space plays an important role in many areas of mathematics and its applications. For instance, it is systematically used in calculus for the study of functions of several real variables.

An *inner product* on the space $\mathbb{C}^n$ is a function that assigns to each pair of vectors $x, y \in \mathbb{C}^n$ a (generally speaking) complex number (x, y) and satisfies the following *axioms*:

1. $(x, x) \geq 0$ for all $x \in \mathbb{C}^n$; $(x, x) = 0$ if and only if $x = 0$;
2. $(x, y) = \overline{(y, x)}$ for all $x, y \in \mathbb{C}^n$; recall that the overline means the complex conjugate and note that the inner product on the complex vector space $\mathbb{C}^n$ is not commutative, unlike the inner product on the real space $\mathbb{R}^n$
3. $(\alpha x + \beta y, z) = \alpha(x, z) + \beta(y, z)$ for all $x, y, z \in \mathbb{C}^n$ and for all $\alpha, \beta \in \mathbb{C}$.

Clearly, the next property follows from Axioms 2 and 3:

4. $(x, \alpha y + \beta z) = \overline{\alpha}(x, y) + \overline{\beta}(x, z)$ for all $x, y, z \in \mathbb{C}^n$ and for all $\alpha, \beta \in \mathbb{C}$.

---

[1] The verification of inequality 3 will be carried out in Section 3.2.2, p. 72.

[2] Hermann Minkowski (1864–1909) was a German mathematician.

The vector space $\mathbb{C}^n$, together with a specified inner product on it, is often called the *unitary space* $\mathbb{C}^n$.

An inner product on the space $\mathbb{C}^n$ can be specified in infinitely many ways. For example, we can put

$$(x, y) = \sum_{k=1}^{n} x_k \overline{y}_k.$$

This inner product on the space $\mathbb{C}^n$ is called *standard*. The verification of Axioms 1–3 is trivial. Inner products on $\mathbb{C}^n$ can also be specified similarly to (3.1).

The *length* of a vector $x \in \mathbb{C}^n$ is defined by the relationship $|x| = \sqrt{(x, x)}$. Properties of the form 1–3, p. 70, hold.

3.2 Abstract Inner Product Spaces

3.2.1 Definitions and Examples

An *inner product* on an abstract real vector space $\mathbf{X}$ is a function that assigns to each pair of vectors $x, y \in \mathbf{X}$ a real number (x, y) and satisfies axioms of the form 1–3, p. 69 (which are called the *inner product axioms for a real vector space*). An *inner product space* is a vector space together with a specified inner product on that space. A real inner product space is often called a *Euclidean space*.

An *inner product* on the complex vector space $\mathbf{X}$ is a function that assigns to each pair of vectors $x, y \in \mathbf{X}$ a (generally speaking) complex number (x, y) and satisfies axioms of the form 1–3, p. 70 (which are called the *inner product axioms for a complex vector space*). A complex inner product space is often referred to as a *unitary space*.

The reader can verify the inner product axioms for the following examples.

1. The space $\mathbf{V}_3$ together with the usual inner product (the dot product) is a real inner product space.
2. Let p be an integrable and positive function on an interval (a, b) of the real axis. Let us specify an inner product on the space $C[a, b]$ by the formula

$$(f, g) = \int_a^b p(x) f(x) g(x) \mathrm{d}x, \quad f, g \in C[a, b]. \tag{3.2}$$

The space $C[a, b]$ together with the inner product (3.2) is a Euclidean space.
3. Let us specify an inner product on the space $\mathbf{Q}_n$. We assign to each pair of elements

$$P_n(z) = a_0 + a_1 z + \cdots + a_n z^n, \quad Q_n(z) = b_0 + b_1 z + \cdots + b_n z^n$$

of the space $\mathbf{Q}_n$ the complex number

$$(P_n, Q_n) = \sum_{j=0}^{n} \rho_j a_j \bar{b}_j,$$

where $\rho_0, \rho_1, \ldots, \rho_n$ are given positive numbers. Together with this inner product, the space $\mathbf{Q}_n$ is a unitary space.

An inner product can be specified on every finite-dimensional vector space $\mathbf{X}_n$ of dimension $n \geq 1$. Indeed, let $\{e_k\}_{k=1}^{n}$ be a basis in $\mathbf{X}_n$ and let $x = \sum_{k=1}^{n} \xi_k e_k$, $y = \sum_{k=1}^{n} \eta_k e_k$ be elements of the space $\mathbf{X}_n$. We can take as an inner product on $\mathbf{X}_n$ the function

$$(x, y) = \sum_{k=1}^{n} \xi_k \bar{\eta}_k, \quad x, \, y \in \mathbf{X}_n. \tag{3.3}$$

It is easy to see that the function (3.3) satisfies the inner product axioms.

3.2.2 The Cauchy–Schwarz Inequality

Suppose that a and b are vectors in the three-dimensional Euclidean space $\mathbf{V}_3$, and the vectors $a - b$ and b are orthogonal, i.e., $(a - b, b) = 0$.[3] Then by the Pythagorean[4] theorem,

$$|a|^2 = |a - b|^2 + |b|^2. \tag{3.4}$$

Now suppose that a and b are vectors in an abstract inner product space $\mathbf{X}$ such that $(a - b, b) = 0$. If we put $|v| = \sqrt{(v, v)}$ for all vectors $v \in \mathbf{X}$, then the Pythagorean identity of the form (3.4) holds for vectors in $\mathbf{X}$. Indeed, using elementary calculations, we get

$$
\begin{aligned}
|a|^2 = (a, a) &= (a - b + b, a - b + b) \\
&= (a - b, a - b) + (b, b) + (a - b, b) + (b, a - b) \\
&= (a - b, a - b) + (b, b) + (a - b, b) + \overline{(a - b, b)} \\
&= (a - b, a - b) + (b, b) = |a - b|^2 + |b|^2.
\end{aligned}
$$

Theorem 3.1 (Cauchy–Schwarz[5] inequality). *Let* $\mathbf{X}$ *be an inner product space. For all vectors* $x, \, y \in \mathbf{X}$, *the following inequality holds:*

---

[3] We can say that the vector b is the projection of the vector a on the line that is parallel to the vector b.

[4] Pythagoras of Samos (570–495 B.C.) was an Ionian Greek philosopher and mathematician.

[5] Augustin-Louis Cauchy (1789–1857) was a French mathematician, Karl Hermann Amandus Schwarz (1843–1921) was a German mathematician.

$$|(x, y)|^2 \leq (x, x)(y, y). \tag{3.5}$$

The two sides in (3.5) are equal if and only if x and y are proportional.

Proof If $y = 0$, then inequality (3.5) transforms to a trivial equality, and for each vector $x \in \mathbf{X}$, the vectors x and y are proportional, since $0x + y = 0$. For this reason, we suppose that $y \neq 0$, and put $e = |y|^{-1}y$. Clearly, $(e, e) = 1$ and

$$(x - (x, e)e, (x, e)e) = 0.$$

Hence in the identity (3.4) we can take $a = x$, $b = (x, e)e$ and get

$$|x|^2 = |x - (x, e)e|^2 + |(x, e)|^2.$$

Therefore, $|x|^2 \geq |(x, e)|^2$. The last inequality is equivalent to (3.5). Now we suppose that $|x|^2 = |(x, e)|^2$, i.e., the two sides in (3.5) are equal. Then $|x - (x, e)e|^2 = 0$, and therefore, $x = (x, e)e$, i.e., $x = ((x, y)/|y|^2)y$, and thus the vectors x and y are proportional. Conversely, if the vectors x and y are proportional, then it is easy to see that the two sides in (3.5) are equal. □

The number $|x| = \sqrt{(x, x)}$ is called the *length* of the vector $x \in \mathbf{X}$. Inequality (3.5) often is written in the form

$$|(x, y)| \leq |x||y| \quad \text{for all} \quad x, y \in \mathbf{X}. \tag{3.6}$$

The length of vectors in an abstract inner product space satisfies properties that are analogous to the properties of the length of vectors in three-dimensional Euclidean space, namely:

1. $|x| \geq 0$ for all $x \in \mathbf{X}$; $|x| = 0$ if and only if $x = 0$;
2. $|\alpha x| = |\alpha||x|$ for all $x \in \mathbf{X}$ and for all $\alpha \in \mathbb{C}$;
3. $|x + y| \leq |x| + |y|$ for all $x, y \in \mathbf{X}$.

Inequality 3 is called the *triangle inequality* (or *Minkowski's inequality*).

It is evident that Properties 1 and 2 hold. Let us prove that the triangle inequality follows from the Cauchy–Schwarz inequality. Indeed,

$$|x + y|^2 = (x + y, x + y) = |x|^2 + 2\text{Re}(x, y) + |y|^2.$$

Using (3.6), we see that $|\text{Re}(x, y)| \leq |x||y|$, and therefore,

$$|x + y|^2 \leq |x|^2 + 2|x||y| + |y|^2 = (|x| + |y|)^2.$$

The last inequality is equivalent to inequality 3.

By analogy with three-dimensional Euclidean space $\mathbf{V}_3$, we say that two vectors x, $y \in \mathbf{X}$ are orthogonal if $(x, y) = 0$.

For example, if $k \neq l$, then the vectors i_k and $i_l \in \mathbb{C}^n$ are orthogonal with respect to the standard inner product.

It follows from inequality (3.6) that if $\mathbf{X}$ is a real inner product space, then

$$(x, y)/|x||y| \in [-1, 1]$$

for all nonzero vectors $x, y \in \mathbf{X}$. This fact leads us to introduce the concept of the angle between two vectors in $\mathbf{X}$. Namely, we assume that the cosine of the angle between $x, y \in \mathbf{X}$ is equal to $(x, y)/|x||y|$.

3.2.3   The Gram Matrix

Let $\{a_i\}_{i=1}^m$ be a set of vectors in an inner product space $\mathbf{X}$. The *Gram*[6] *matrix* of the set $\{a_i\}_{i=1}^m$ is the square matrix of order m of the form

$$G = \begin{pmatrix} (a_1, a_1) & (a_2, a_1) & \ldots & (a_m, a_1) \\ (a_1, a_2) & (a_2, a_2) & \ldots & (a_m, a_2) \\ \ldots\ldots\ldots\ldots\ldots \\ (a_1, a_m) & (a_2, a_m) & \ldots & (a_m, a_m) \end{pmatrix}. \tag{3.7}$$

Note that since $(a_k, a_l) = \overline{(a_l, a_k)}$, the Gram matrix of any set of vectors is Hermitian (see p. 48).

Theorem 3.2 *A set of vectors $\{a_i\}_{i=1}^m$ is linearly independent if and only if its Gram matrix is nonsingular.*

Proof Suppose that the Gram matrix G of a set of vectors $\{a_i\}_{i=1}^m$ is nonsingular. Then the set $\{a_i\}_{i=1}^m$ is linearly independent. Indeed, if

$$x_1 a_1 + x_2 a_2 + \cdots + x_m a_m = 0,$$

then

$$(x_1 a_1 + x_2 a_2 + \cdots + x_m a_m, a_k) = 0, \quad k = 1, 2, \ldots, m.$$

Hence,

$$x_1 (a_1, a_k) + x_2 (a_2, a_k) + \cdots + x_m (a_m, a_k) = 0, \quad k = 1, 2, \ldots, m. \tag{3.8}$$

System (3.8) is a homogeneous system of linear algebraic equations for the unknowns $x_1, x_2, \ldots, x_m$ with the matrix G. Since the Gram matrix G is nonsingular, system (3.8) has the trivial solution only. Thus, $x_1 = \cdots = x_m = 0$. Conversely, suppose that a set of vectors $\{a_i\}_{i=1}^m$ is linearly independent. Let us construct a linear

---

[6]Jørgen Pedersen Gram (1850–1916) was a Danish mathematician.

combination of the columns of the matrix G with some coefficients $x_1, x_2, \ldots, x_m$. Equating this linear combination to zero, we get

$$x_1(a_1, a_k) + x_2(a_2, a_k) + \cdots + x_m(a_m, a_k) = 0, \quad k = 1, \ldots, m. \tag{3.9}$$

Multiplying both sides of the kth equality in (3.9) by $\overline{x}_k$ and then adding term by term all obtained equalities, we get

$$\left(\sum_{k=1}^{m} x_k a_k, \sum_{k=1}^{m} x_k a_k \right) = 0.$$

Therefore,

$$x_1 a_1 + x_2 a_2 + \cdots + x_m a_m = 0. \tag{3.10}$$

Since the set of vectors $\{a_i\}_{i=1}^{m}$ is linearly independent, it follows from (3.10) that $x_1 = \cdots = x_m = 0$. Thus we see that if a linear combination of the columns of the matrix G is equal to zero, then all the coefficients in this linear combination are equal to zero. This means that the columns of the matrix G are linearly independent, i.e., the matrix G is nonsingular. $\qquad\qquad\square$

Let us examine for linear dependence the vectors

$$x_1 = (1, 3, 3, 1, -2), \quad x_2 = (3, 3, 1, -3, 2), \quad x_3 = (1, 3, -1, 1, 3)$$

in the space $\mathbb{R}^5$. For this purpose we introduce the standard inner product on $\mathbb{R}^5$ and calculate the third-order Gram matrix $G = \{(x_i, x_j)\}_{i,j=1}^{3}$. By elementary calculations we get

$$G = \begin{pmatrix} 24 & 8 & 2 \\ 8 & 32 & 14 \\ 2 & 14 & 21 \end{pmatrix}, \quad \det(G) = 2^4 \, 650,$$

i.e., the vectors x_1, x_2, x_3 are linearly independent.

3.2.4 Orthogonal Sets of Vectors. Gram–Schmidt Orthogonalization Process

A set of vectors $\{a_i\}_{i=1}^{m}$ is called *orthogonal* if all the vectors a_i, $i = 1, 2, \ldots, m$, are nonzero and $(a_i, a_k) = 0$ for $i \neq k$. The Gram matrix of every orthogonal set is diagonal and nonsingular. Evidently, every orthogonal set is linearly independent. A set of vectors $\{a_i\}_{i=1}^{m}$ is called *orthonormal* if $(a_i, a_k) = \delta_{ik}$ for $i, k = 1, 2, \ldots, m$. The Gram matrix of every orthonormal set is the identity matrix. The length of each vector in an orthonormal set is equal to one.

The change of basis matrix from one orthonormal basis $\{e_k\}_{k=1}^n$ to another ortho-normal basis $\{\tilde{e}_k\}_{k=1}^n$ in an inner product space is unitary. Indeed, writing the equality

$$\tilde{\mathcal{E}}_n = \mathcal{E}_n T \tag{3.11}$$

in detail, we get $\tilde{e}_k = \sum_{j=1}^n t_{jk} e_j$, $k = 1, 2, \ldots, n$. Therefore,

$$\left(\sum_{j=1}^n t_{jk} e_j, \sum_{j=1}^n t_{jl} e_j \right) = (\tilde{e}_k, \tilde{e}_l) = \delta_{kl}, \quad k, l = 1, 2, \ldots, n,$$

since the basis $\tilde{\mathcal{E}}_n$ is orthonormal. Now we transform the left-hand side of the last equality using the orthonormality of the set $\mathcal{E}_n$ and obtain

$$\sum_{j=1}^n t_{jk} \bar{t}_{jl} = \delta_{kl}, \quad k, l = 1, 2, \ldots, n.$$

This means that the matrix T is unitary (see p. 50).

It is important to note that the inverse statement is also true, which can be shown by arguments similar to those given above. Namely, if the basis $\mathcal{E}_n$ is orthonormal and the matrix T is unitary, then the basis $\tilde{\mathcal{E}}_n = \mathcal{E}_n T$ is also orthonormal.

Theorem 3.3 (Gram–Schmidt[7] orthogonalization). *Every linearly independent set $\{a_i\}_{i=1}^m$ is equivalent to an orthonormal set $\{b_i\}_{i=1}^m$, and the vector b_1 may be chosen proportional to the vector a_1.*

Proof Put $h_1 = a_1$ and $h_2 = x_{2,1} h_1 + a_2$. The vector h_1 is not equal to zero, since the vector a_1 is not equal to zero, as an element of a linearly independent set. For every coefficient $x_{2,1}$, the vector h_2 is not equal to zero, since h_2 is a linear combination of linearly independent vectors, and one of the coefficients in this linear combination is not equal to zero (it is equal to one). Now we define the number $x_{2,1}$ such that the vector h_2 is orthogonal to the vector h_1. Writing this condition, we get $0 = x_{2,1}(h_1, h_1) + (a_2, h_1)$, and hence $x_{2,1} = -(a_2, h_1)/(h_1, h_1)$. Thus we have constructed the vectors h_1 and h_2 such that $(h_1, h_2) = 0$ and $h_1, h_2 \neq 0$. Suppose that we have constructed the vectors $h_1, h_2, \ldots, h_k$ such that $h_1, h_2, \ldots, h_k \neq 0$ and $(h_i, h_j) = 0$ for $i \neq j$, $i, j = 1, \ldots, k$. We are looking for a vector h_{k+1} of the form

$$h_{k+1} = x_{k+1,1} h_1 + x_{k+1,2} h_2 + \cdots + x_{k+1,k} h_k + a_{k+1}. \tag{3.12}$$

The vector h_{k+1} is not equal to zero for all coefficients $x_{k+1,1}, \ldots, x_{k+1,k}$. Indeed, by construction, each vector $h_1, h_2, \ldots, h_k$ is a linear combination of the vectors $\{a_i\}_{i=1}^m$, and the linear combination h_j consists of the vectors of the set $\{a_i\}_{i=1}^m$ whose indices

---

[7]Erhard Schmidt (1876–1959) was a German mathematician.

i are less than or equal to j. Therefore, the vector h_{k+1} is a linear combination of the linearly independent vectors $a_1, a_2, \ldots, a_{k+1}$, and the vector a_{k+1} is included in this linear combination with a coefficient that is equal to one.

We define the numbers $x_{k+1,1}, x_{k+1,2}, \ldots, x_{k+1,k}$ such that the vector h_{k+1} is orthogonal to the vectors $h_1, h_2, \ldots, h_k$. Consistently fulfilling these conditions, we get

$$x_{k+1,1} = -(a_{k+1}, h_1)/(h_1, h_1),$$

$$x_{k+1,2} = -(a_{k+1}, h_2)/(h_2, h_2), \ldots,$$

$$x_{k+1,k} = -(a_{k+1}, h_k)/(h_k, h_k).$$

Continuing this process, we construct an orthogonal set of nonzero vectors $\{h_i\}_{i=1}^m$. If we take

$$b_i = (|h_i|)^{-1} h_i, \quad i = 1, \ldots, m, \tag{3.13}$$

then we get the orthonormal set of vectors $\{b_i\}_{i=1}^m$.

As we have established, each vector of the set $\{h_i\}_{i=1}^m$ is a linear combination of the vectors $\{a_i\}_{i=1}^m$. Formula (3.12) shows that each vector of the set $\{a_i\}_{i=1}^m$ is a linear combination of the vectors $\{h_i\}_{i=1}^m$. Formula (3.13) shows that the sets $\{b_i\}_{i=1}^m$ and $\{h_i\}_{i=1}^m$ are equivalent. Thus all three considered sets are equivalent.

Finally, we note that the vectors a_1 and b_1 are proportional, since by construction, $b_1 = (|a_1|)^{-1} a_1$. □

Remark 3.1 The proof of Theorem 3.3 is constructive. It includes a description of the algorithm for construction of an orthonormal set of vectors that is equivalent to a given linearly independent set of vectors. This algorithm is called the *Gram–Schmidt orthogonalization process*. Note that for numerical realizations, the Gram–Schmidt orthogonalization process is used very rarely, since it is strongly influenced by rounding errors.

Let us assume, for example, that the polynomials $Q_0(x) \equiv 1$, $Q_1(x) = x$, $Q_2(x) = x^2$ of a real variable x are given. Using the Gram–Schmidt orthogonalization process, we construct polynomials P_0, P_1, P_2 of respective degrees zero, one, and two that are orthonormal with respect to the inner product defined by the formula

$$(f, g) = \int_{-1}^{1} f(x) g(x) \mathrm{d}x.$$

Calculating according to the Gram–Schmidt orthogonalization process, we get

$$\tilde{P}_0 = Q_0 \equiv 1,$$

$$\tilde{P}_1(x) = Q_1(x) - \tilde{P}_0(x) \int\limits_{-1}^{1} Q_1(x)\tilde{P}_0(x)dx \left(\int\limits_{-1}^{1} \tilde{P}_0^2(x)dx \right)^{-1} = x,$$

$$\tilde{P}_2(x) = Q_2(x) - \tilde{P}_0(x) \int\limits_{-1}^{1} Q_2(x)\tilde{P}_0(x)dx \left(\int\limits_{-1}^{1} \tilde{P}_0^2(x)dx \right)^{-1}$$

$$- \tilde{P}_1(x) \int\limits_{-1}^{1} Q_2(x)\tilde{P}_1(x)dx \left(\int\limits_{-1}^{1} \tilde{P}_1^2(x)dx \right)^{-1} = x^2 - 1/3,$$

$$P_0(x) = \tilde{P}_0(x) \left(\int\limits_{-1}^{1} \tilde{P}_0^2(x)dx \right)^{-1/2} = 1/\sqrt{2}, \quad P_1(x) = \tilde{P}_1(x) \left(\int\limits_{-1}^{1} \tilde{P}_1^2(x)dx \right)^{-1/2}$$

$$= x\sqrt{3/2},$$

$$P_2(x) = \tilde{P}_2(x) \left(\int\limits_{-1}^{1} \tilde{P}_2^2(x)dx \right)^{-1/2} = \frac{1}{2}\sqrt{\frac{5}{2}}(3x^2 - 1).$$

In the same way, we can construct the polynomials $P_3(x)$, ..., $P_n(x)$ of degree greater than two, applying the Gram–Schmidt orthogonalization process to the polynomials $1, x, x^2, \ldots, x^n$ for a given positive integer n. The polynomials

$$P_0(x), \ P_1(x), \ldots, P_n(x), \ldots$$

are called the *Legendre*[8] *polynomials*. The following so-called *Rodrigues's formula*[9] is valid:

$$P_k(x) = \sqrt{\frac{2k+1}{2}} \frac{1}{k!2^k} \frac{d^k}{dx^k}(x^2 - 1)^k, \quad k = 0, 1, \ldots. \tag{3.14}$$

Using Rodrigues's formula and the formula for integration by parts, the reader can prove that

$$\int\limits_{-1}^{1} P_k(x)P_l(x)dx = 0 \quad k \neq l, \ k, l = 0, 1, 2, \ldots. \tag{3.15}$$

---

[8]Adrien-Marie Legendre (1752–1833) was a French mathematician.
[9]Benjamin Olinde Rodrigues (1794–1851) was a French mathematician.

Remark 3.2 Let f_1 be a given nonzero vector in an inner product space $\mathbf{X}_n, n > 1$. Clearly, there exists a vector f_2 that is not proportional to f_1. If $n > 2$, we may take a vector f_3 such that the vectors f_1, f_2, f_3 are linearly independent. Continuing this process, if $n > 3$, we get a basis in the space $\mathbf{X}_n$ that includes the vector f_1. Applying after that the Gram–Schmidt orthogonalization process, we can construct an orthonormal basis that includes a vector that is proportional to the vector f_1.

3.2.5 The Expansion of a Vector with Respect to a Basis in an Inner Product Space

Let $\mathbf{X}_n$ be an inner product space and $\{e_k\}_{k=1}^n$ a basis of $\mathbf{X}_n$. The coefficients of the expansion of a vector $x \in \mathbf{X}_n$ with respect to the basis $\{e_k\}_{k=1}^n$ can be computed as the solution of a system of linear equations with a Hermitian nonsingular matrix. Indeed, successively calculating the inner product of both sides of the equality

$$\xi_1 e_1 + \xi_2 e_2 + \cdots + \xi_n e_n = x$$

with the vectors $e_1, e_2, ..., e_n$, we get the following system of linear equations:

$$(e_1, e_1)\xi_1 + (e_2, e_1)\xi_2 + \cdots + (e_n, e_1)\xi_n = (x, e_1),$$

$$(e_1, e_2)\xi_1 + (e_2, e_2)\xi_2 + \cdots + (e_n, e_2)\xi_n = (x, e_2),$$

$$\dots\dots\dots\dots\dots\dots\dots\dots\dots\dots\dots\dots\dots\dots\dots$$

$$(e_1, e_n)\xi_1 + (e_2, e_n)\xi_2 + \cdots + (e_n, e_n)\xi_n = (x, e_n).$$

The matrix of this system is the Gram matrix of the basis $\{e_k\}_{k=1}^n$. If the basis is orthogonal, then the matrix is diagonal, and the solution of the system can be easily calculated:

$$\xi_k = \frac{(x, e_k)}{(e_k, e_k)}, \quad k = 1, 2, \ldots, n. \tag{3.16}$$

The coefficients (3.16) are called the *Fourier*[10] *coefficients* of the vector x with respect to the orthogonal set of vectors $\{e_k\}_{k=1}^n$. Note that if the basis $\{e_k\}_{k=1}^n$ is orthonormal, then for every vector $x \in \mathbf{X}_n$ we have the following expansion:

$$x = \sum_{k=1}^n (x, e_k) e_k. \tag{3.17}$$

---

[10]Jean Baptiste Joseph Fourier (1768–1830) was a French mathematician and physicist.

3.2.6   The Calculation of an Inner Product

Let x and y be vectors in an inner product space $\mathbf{X}_n$. Suppose that we know the vectors $\xi, \eta \in \mathbb{C}^n$ of coefficients of the expansions of x and y with respect to a basis $\mathcal{E}_n$, i.e., $x = \mathcal{E}_n \xi$ and $y = \mathcal{E}_n \eta$. Then

$$(x, y) = \left(\sum_{k=1}^{n} \xi_k e_k, \sum_{k=1}^{n} \eta_k e_k \right) = \sum_{k,l=1}^{n} \xi_k \overline{\eta}_l (e_k, e_l) = (G\xi, \eta), \qquad (3.18)$$

where G is the Gram matrix of the basis $\mathcal{E}_n$, and the parentheses on the right-hand side of equality (3.18) denote the standard inner product on the space $\mathbb{C}^n$. Therefore, for the calculation of the inner product (x, y) it is enough to know the coefficients of the expansions of the vectors x and y with respect to a basis and the Gram matrix of that basis.

If the basis is orthonormal, then

$$(x, y) = \sum_{k=1}^{n} \xi_k \overline{\eta}_k. \qquad (3.19)$$

Thus the inner product of vectors can be computed as the standard inner product of the coefficients of the expansions of these vectors with respect to any orthonormal basis.

3.2.7   Reciprocal Basis Vectors

Let $\mathcal{E}_n = \{e_k\}_{k=1}^n$ be a basis in an inner product space $\mathbf{X}_n$. It is easy to see that the equations

$$(e^i, e_j) = \delta_{ij}, \quad i, j = 1, 2, \ldots, n, \qquad (3.20)$$

uniquely define the linearly independent vectors $e^1, e^2, \ldots, e^n$. The basis $\mathcal{E}^n = \{e^k\}_{k=1}^n$ is *reciprocal* to the original one. Clearly, the original and the reciprocal bases coincide if and only if the basis $\mathcal{E}_n$ is orthonormal. Let G be the Gram matrix of the basis $\mathcal{E}_n$, and let $\widetilde{G}$ be the Gram matrix of the basis $\mathcal{E}^n$. Using elementary calculations, we get $\mathcal{E}_n = \mathcal{E}^n G$, $\mathcal{E}^n = \mathcal{E}_n \widetilde{G}$. Therefore, $\widetilde{G} = G^{-1}$. The coefficients of the expansions $x = \mathcal{E}_n \xi$, $y = \mathcal{E}^n \eta$ are the following: $\xi^k = (x, e^k)$, $\eta_k = (y, e_k)$, $k = 1, 2, \ldots, n$, and

$$(x, y) = \sum_{k=1}^{n} \xi^k \overline{\eta}_k.$$

The numbers $\xi^1, \xi^2, \ldots, \xi^n$ are called the *contravariant components* of the vector x, and the numbers $\eta_1, \eta_2, \ldots, \eta_n$ are called the *covariant components* of the vector y.

3.2.8 Examples of Orthogonal Bases

Let us begin with examples of orthogonal bases in the space $\mathbb{C}^n$.

1. The standard basis $\{i_k\}_{k=1}^n$ is orthonormal with respect to the standard inner product on $\mathbb{C}^n$ (prove it!).

2. The *Fourier basis*. It is convenient now to number the basis vectors and their components from 0 to $n-1$. Recall that the complex numbers

$$q_k = \cos\frac{2\pi k}{n} + i\sin\frac{2\pi k}{n}, \quad k = 0, 1, \ldots, n-1,$$

are the nth roots of unity (see p. 7). As usual, i denotes the imaginary unit. Let us introduce the set of vectors $\{\varphi_k\}_{k=0}^{n-1}$ whose components are calculated by the following formula:

$$(\varphi_k)_j = q_k^j, \quad j, k = 0, 1, \ldots, n-1. \tag{3.21}$$

The set of vectors $\{\varphi_k\}_{k=0}^{n-1}$ is orthogonal with respect to the standard inner product on the space $\mathbb{C}^n$. Indeed, first of all, we note that $q_k = q_1^k$, $\bar{q}_k = q_1^{-k}$. Therefore, calculating the inner product (φ_k, φ_l), we get

$$(\varphi_k, \varphi_l) = \sum_{j=0}^{n-1} q_1^{(k-l)j} = 1 + (q_1^p) + (q_1^p)^2 + \cdots + (q_1^p)^{n-1}, \tag{3.22}$$

where $p = k - l$. For $k = l$, i.e., for $p = 0$, we have $(\varphi_k, \varphi_k) = n$. If $p \neq 0$, then the sum on the right-hand side of (3.22) is a geometric progression with ratio q_1^p, and since $|p| = |k - l| < n$, we see that $q_1^p \neq 1$. Using the formula for the sum of the first n terms of a geometric progression, we obtain

$$\sum_{j=0}^{n-1} (q_1^p)^j = \frac{(q_1^p)^n - 1}{q_1^p - 1}, \tag{3.23}$$

but $(q_1^n)^p = q_1^{pn} = 1$, whence $(\varphi_k, \varphi_l) = 0$ for $k \neq l$.

Using (3.16), we see that the Fourier coefficients ξ of every vector $x \in \mathbb{C}^n$ with respect to the basis (3.21) are calculated by the formulas

$$\xi_k = \frac{(x, \varphi_k)}{(\varphi_k, \varphi_k)} = \frac{1}{n}\sum_{j=0}^{n-1} x_j q_k^{-j}, \quad k = 0, 1, \ldots, n-1. \tag{3.24}$$

The components of the vector x are calculated as follows:

$$x_j = \sum_{k=0}^{n-1} \xi_k q_k^j, \quad j = 0, 1, \ldots, n-1. \tag{3.25}$$

The basis $\{\varphi_k\}_{k=0}^{n-1}$ is usually called the *Fourier basis*. It is widely used in digital (audio and video) signal processing.

In real-life applications, the number n (the length of the processed signal) is very large, and therefore, special algorithms for calculation of sums of the form (3.25) and (3.24) are used. They are called the fast Fourier transform (FFT).

Below are examples of orthogonal bases in the space $\mathbf{P}_n$ of polynomials with real coefficients. Let us consider the set of all polynomials of the form

$$P_n(x) = a_n x^n + a_{n-1} x^{n-1} + \cdots + a_0,$$

where the coefficients $a_0, a_1, \ldots, a_n$ are real numbers, x is a real variable, $n \geq 0$ is a given integer. Evidently, this set joined with the zero polynomial is a real vector space with the usual definitions of the operations of addition of two polynomials and multiplication of a polynomial by a real number.

1. The *Legendre polynomials*. If we specify the inner product on the space $\mathbf{P}_n$ by the formula

$$(f, g) = \int_{-1}^{1} f(x)g(x)\, dx \quad \text{for all} \quad f, g \in \mathbf{P}_n, \tag{3.26}$$

then the Legendre polynomials $P_0, P_1, \ldots, P_n$ (see (3.14), (3.15), p. 78) form an orthogonal basis in the space $\mathbf{P}_n$.

2. The *Chebyshev*[11] *polynomials*. Now we specify the inner product on the space $\mathbf{P}_n$ using the relationship

$$(f, g) = \int_{-1}^{1} f(x)g(x)\frac{1}{\sqrt{1-x^2}}\, dx \quad \text{for all} \quad f, g \in \mathbf{P}_n. \tag{3.27}$$

The Chebyshev polynomials are defined by the recurrence relation

$$T_0(x) \equiv 1, \ T_1(x) = x, \tag{3.28}$$

$$T_{k+1}(x) = 2x T_k(x) - T_{k-1}(x), \ k = 1, 2, \ldots. \tag{3.29}$$

Here k is the degree of the polynomial $T_k(x)$.

---

[11]Pafnuty Lvovich Chebyshev (1821–1894) was a Russian mathematician.

Let us construct an explicit formula for the Chebyshev polynomials. We look for the value of the polynomial $T_k(x)$ in the form $T_k(x) = \lambda^k$. Substituting λ^k for $T_k(x)$ in the recurrence relation (3.29), we get

$$\lambda^{k+1} = 2x\lambda^k - \lambda^{k-1}.$$

Therefore, if $\lambda \neq 0$, then λ satisfies the quadratic equation

$$\lambda^2 - 2x\lambda + 1 = 0.$$

This equation has the following roots: $\lambda_{1,2} = x \pm \sqrt{x^2 - 1}$. Hence the functions

$$T_k^{(1)}(x) = (x + \sqrt{x^2 - 1})^k, \quad T_k^{(2)}(x) = (x - \sqrt{x^2 - 1})^k,$$

and as a consequence, the functions

$$T_k(x) = c_1 T_k^{(1)}(x) + c_2 T_k^{(2)}(x), \quad k = 0, 1, \ldots,$$

satisfy (3.29). Here c_1 and c_2 are arbitrary real numbers. The numbers c_1 and c_2 are defined by conditions (3.28):

$$c_1 + c_2 = 1,$$

$$(c_1 + c_2)x + (c_1 - c_2)\sqrt{x^2 - 1} = x.$$

Therefore, $c_1 = c_2 = 1/2$, i.e., the polynomials

$$T_k(x) = \frac{1}{2}\left(x + \sqrt{x^2 - 1}\right)^k + \frac{1}{2}\left(x - \sqrt{x^2 - 1}\right)^k, \quad k = 0, 1, 2, \ldots,$$

satisfy (3.29) and (3.28). For $|x| \leq 1$, the Chebyshev polynomials can be written in a more compact form. In this case, we can put $x = \cos\varphi$. Then

$$T_k(x) = \frac{1}{2}(\cos\varphi + i\sin\varphi)^k + \frac{1}{2}(\cos\varphi - i\sin\varphi)^k,$$

and using de Moivre's formula (see (1.23), p. 6), we get $T_k(x) = \cos k\varphi$, and hence

$$T_k(x) = \cos(k\arccos x). \tag{3.30}$$

The Chebyshev polynomials are orthogonal with respect to the inner product (3.27). Indeed, using (3.30), we can write

$$(T_k, T_l) = \int_{-1}^{1} \frac{\cos(k\arccos x)\cos(l\arccos x)}{\sqrt{1 - x^2}}\, dx.$$

If we put $x = \cos\varphi$, then using elementary calculations, we get

$$(T_k, T_l) = \int\limits_0^\pi \cos k\varphi \cos l\varphi \, d\varphi = 0, \quad k \neq l.$$

Thus the Chebyshev polynomials $T_0, T_1, \ldots, T_n$ form an orthogonal basis with respect to the inner product (3.27) on the space $\mathbf{P}_n$ of polynomials with real coefficients.

3.3 Subspaces

3.3.1 The Sum and Intersection of Subspaces

A set L of elements in a vector space $\mathbf{X}$ is a *subspace* of $\mathbf{X}$ if $\alpha x + \beta y \in L$ for all $x, y \in L$ and for all complex numbers α, β. *Trivial* examples of subspaces are the following: the space $\mathbf{X}$ itself is a subspace; the set consisting only of the zero vector is a subspace. Every subspace L includes the zero vector, since by definition, for every $x \in L$ the vector $0x$ belongs to L.

The proof of the two following theorems is left to the reader.

Theorem 3.4 *Let $a_1, a_2, \ldots, a_m$, $m \geq 1$, be given vectors in a vector space $\mathbf{X}$. The set of all linear combinations $x_1 a_1 + x_2 a_2 + \cdots + x_m a_m$ is a subspace of $\mathbf{X}$. This subspace is called the* span *of $a_1, a_2, \ldots, a_m$ and is denoted by $\mathrm{span}\{a_1, a_2, \ldots, a_m\}$.*

Theorem 3.5 *Let a_1, a_2 be given vectors in a vector space $\mathbf{X}$, and $a_2 \neq 0$. The set L of all vectors of the form $a_1 + \alpha a_2$, where $\alpha \in \mathbb{C}$, is called the* line passing through *the point a_1 and parallel to the vector a_2. The set L is a subspace if and only if the vectors a_1, a_2 are linearly dependent.*

Let L_1, L_2 be subspaces of a vector space $\mathbf{X}$. The set L of all vectors of the form $a_1 + a_2$, where $a_1 \in L_1, a_2 \in L_2$, is called the *sum of the subspaces L_1 and L_2* and is denoted by $L = L_1 + L_2$. The set L is a subspace. Indeed, let $x, y \in L$. This means that there exist vectors $a_1, b_1 \in L_1, a_2, b_2 \in L_2$ such that $x = a_1 + a_2$ and $y = b_1 + b_2$. Let α, β be arbitrary complex numbers. Then

$$\alpha x + \beta y = \alpha(a_1 + a_2) + \beta(b_1 + b_2) = (\alpha a_1 + \beta b_1) + (\alpha a_2 + \beta b_2).$$

Since L_1 is a subspace, the vector $\alpha a_1 + \beta b_1$ belongs to L_1. Similarly, the vector $\alpha a_2 + \beta b_2$ belongs to L_2. Therefore, the vector $\alpha x + \beta y$ belongs to L.

The *intersection of the subspaces L_1 and L_2*, i.e., the set $L_1 \cap L_2$ of all vectors that are elements of both L_1 and L_2, is also a subspace of $\mathbf{X}$. Indeed, let there be given vectors $x, y \in L_1 \cap L_2$. For every complex number α, the vector αx belongs to both L_1 and L_2, i.e., $\alpha x \in L_1 \cap L_2$. Similarly, for every β, the vector βy belongs to

$L_1 \cap L_2$. Hence evidently, $\alpha x + \beta y \in L_1 \cap L_2$. A set of vectors $\{e_k\}_{k=1}^m \subset L$ is a *basis of a subspace L* if those vectors are linearly independent and every vector $x \in L$ can be represented as a linear combination of the vectors $\{e_k\}_{k=1}^m$. The number m is called the *dimension of the subspace L* and is denoted by dim L.

The subspace consisting only of the zero vector is called the *zero subspace* and is denoted by $\{0\}$. As usual, we assume that $\dim\{0\} = 0$.

The reader is now invited to describe all possible subspaces of the space $\mathbf{V}_3$.

A subspace L of a finite-dimensional space $\mathbf{X}_n$ coincides with $\mathbf{X}_n$ if and only if dim $L = n$. This statement immediately follows from the fact that every set of n linearly independent vectors in the space $\mathbf{X}_n$ forms a basis in this space (see Theorem 2.8, p. 65).

Evidently, a given basis $\{e_k\}_{k=1}^m$ of a subspace $L \subset \mathbf{X}_n$ can be joined with some vectors to complete a basis $\{e_k\}_{k=1}^n$ of the space $\mathbf{X}_n$. Similarly, if L_1 and L_2 are subspaces, and $L_1 \subset L_2$, then dim $L_1 \le$ dim L_2, and every basis of L_1 can be joined with some elements of L_2 to complete the basis in the subspace L_2.

The sum of the subspaces L_1 and L_2 is called *direct* if the components $x_1 \in L_1$ and $x_2 \in L_2$ of each vector $x = x_1 + x_2 \in (L_1 + L_2)$ are uniquely determined. The direct sum of subspaces L_1 and L_2 is denoted by $L_1 \oplus L_2$.

Theorem 3.6 *The sum of two subspaces L_1 and L_2 is direct if and only if it follows from the equality*

$$x_1 + x_2 = 0, \quad x_1 \in L_1, \ x_2 \in L_2, \qquad (3.31)$$

that $x_1 = 0$, $x_2 = 0$.

Proof Suppose that $x_1 = 0$, $x_2 = 0$ follow from (3.31). Let us prove that the components $x_1 \in L_1$ and $x_2 \in L_2$ of each vector $x = x_1 + x_2 \in (L_1 + L_2)$ are uniquely determined. Suppose that there exists one more expansion of the vector x, i.e., we have $x = \tilde{x}_1 + \tilde{x}_2, \tilde{x}_1 \in L_1, \ \tilde{x}_2 \in L_2$. Then evidently, $(x_1 - \tilde{x}_1) + (x_2 - \tilde{x}_2) = 0$. Since $x_1 - \tilde{x}_1 \in L_1$, $x_2 - \tilde{x}_2 \in L_2$, we see that $x_1 - \tilde{x}_1 = 0$, $x_2 - \tilde{x}_2 = 0$, and therefore, $x_1 = \tilde{x}_1$, $x_2 = \tilde{x}_2$. Conversely, suppose that the components $x_1 \in L_1$ and $x_2 \in L_2$ of each vector $x = x_1 + x_2 \in (L_1 + L_2)$ are uniquely determined, and let $x_1 + x_2 = 0$ for some $x_1 \in L_1$, $x_2 \in L_2$. Since $0 + 0 = 0$, we have $x_1 = x_2 = 0$. □

Theorem 3.7 *The sum of two subspaces L_1 and L_2 is direct if and only if*

$$L_1 \cap L_2 = \{0\}.$$

Proof Let $L_1 \cap L_2 = \{0\}$, $x_1 + x_2 = 0$, $x_1 \in L_1$, $x_2 \in L_2$. Since $x_1 = -x_2$, we have $x_1 \in L_2$. Hence $x_1 \in L_1 \cap L_2$. Therefore, $x_1 = 0$, and evidently, $x_2 = 0$. Conversely, let $x \in L_1 \cap L_2$. Then $x \in L_1, x \in L_2$, and moreover, it is obvious that $x + (-x) = 0$. Since the sum of L_1 and L_2 is direct, using Theorem 3.6, we get $x = 0$, and thus $L_1 \cap L_2 = \{0\}$. □

The reader may supply a proof of the next theorem.

Theorem 3.8 *Let L be a subspace of a finite-dimensional vector space* $\mathbf{X}_n$*. Then there exists a subspace* $M \subset \mathbf{X}_n$ *such that* $\mathbf{X}_n = L \oplus M$.

Let L_1 and L_2 be subspaces of an inner product space. If $(x, y) = 0$ for all $x \in L_1$ and $y \in L_2$, then we say that the subspaces L_1 and L_2 are *orthogonal* and write $L_1 \perp L_2$. The sum of orthogonal subspaces is called an *orthogonal sum*.

Every orthogonal sum is direct. Indeed, let $L_1 \perp L_2, x_1 \in L_1, x_2 \in L_2, x_1 + x_2 = 0$. Since the vectors x_1 and x_2 are orthogonal, using the Pythagorean identity, we see that $|x_1 + x_2|^2 = |x_1|^2 + |x_2|^2$. Hence, $|x_1|^2 + |x_2|^2 = 0$, and $x_1 = x_2 = 0$.

The concepts of the direct sum and the orthogonal sum are applied in a natural way to the case of any finite number of subspaces. Namely, the sum of subspaces $L_1, L_2, \ldots, L_k$ in an inner product space is called *orthogonal* if $L_i \perp L_j$ for $i \neq j, i, j = 1, 2, \ldots, k$. Theorem 3.6 is easily generalized to the case of any finite number of subspaces.

The reader can prove that every orthogonal sum of a finite number of subspaces is *direct*, i.e., the components $x_j \in L_j, j = 1, 2, \ldots, k$, of every vector x in the sum are uniquely determined.

The reader can also answer the next question. Is it true that the sum of subspaces $L_1 + L_2 + \cdots + L_k, k > 2$, is direct if their intersection is the zero subspace?

3.3.2   The Dimension of the Sum of Subspaces

Theorem 3.9 *If* $L = L_1 \oplus L_2 \oplus \cdots \oplus L_k$ *is the direct sum of finite-dimensional subspaces* $L_1, L_2, \ldots, L_k$ *of a vector space* $\mathbf{X}$*, then*

$$\dim L = \dim L_1 + \dim L_2 + \cdots + \dim L_k. \tag{3.32}$$

Proof Let us prove the theorem for the case $k = 2$. For an arbitrary k, the proof is analogous. Let

$$f_1, f_2, \ldots, f_p; \quad g_1, g_2, \ldots, g_q \tag{3.33}$$

be bases of the subspaces L_1 and L_2, respectively. Then the union of these two sets is a basis of the subspace $L_1 \oplus L_2$. Indeed, for every $x \in L_1 \oplus L_2$, we have $x = x_1 + x_2$, where

$$x_1 = \alpha_1 f_1 + \alpha_2 f_2 + \cdots + \alpha_p f_p \in L_1, \quad x_2 = \beta_1 g_1 + \beta_2 g_2 + \cdots + \beta_q g_q \in L_2,$$

and if $x = 0$, then $x_1 = x_2 = 0$, since the sum $L_1 \oplus L_2$ is direct. Hence all the numbers $\alpha_1, \alpha_2, \ldots, \alpha_p, \beta_1, \beta_2, \ldots, \beta_q$ are equal to zero, since $\{f_k\}_{k=1}^p, \{g_k\}_{k=1}^q$ are bases. Thus the set of vectors (3.33) is linearly independent. It is clear now that $\dim(L_1 \oplus L_2) = p + q$. □

Theorem 3.10 *If L_1 and L_2 are arbitrary finite-dimensional subspaces of a vector space $\mathbf{X}$, then*

$$\dim(L_1 + L_2) = \dim L_1 + \dim L_2 - \dim(L_1 \cap L_2). \qquad (3.34)$$

Proof Obviously, the space $G = L_1 \cap L_2$ is finite-dimensional. Suppose that a set $\mathcal{G}_l = \{g_i\}_{i=1}^{l}$ is a basis of G, the union of $\mathcal{G}_l$ and vectors $\mathcal{F}_k = \{f_i\}_{i=1}^{k}$ is a basis of the subspace L_1, and the union of $\mathcal{G}_l$ and the vectors $\mathcal{H}_m = \{h_i\}_{i=1}^{m}$ is a basis of the subspace L_2. Let F be the span of $\mathcal{F}_k$ and let H be the span of $\mathcal{H}_m$. We shall prove that

$$L_1 + L_2 = F + G + H. \qquad (3.35)$$

Indeed, if $x \in L_1 + L_2$, then $x = x_1 + x_2$, where $x_1 \in L_1$, $x_2 \in L_2$. Clearly, $x_1 = f + g_-$, $x_2 = h + g_+$, where $f \in F$, $h \in H$, $g_+, g_- \in G$. Therefore, $x = f + g + h$, where $g = g_+ + g_- \in G$. Thus, $x \in F + G + H$. It is easier to prove that if $x \in F + G + H$, then $x \in L_1 + L_2$. The sum on the right-hand side of (3.35) is direct. In fact, suppose that $f + g + h = 0$, where $f \in F$, $g \in G$, $h \in H$. Let us show that $f, g, h = 0$. We have $f + g = -h$. Clearly, $-h \in L_2$, and $f + g \in L_1$, and therefore, $f + g \in G$, $h \in G$. If we put $h + g = \widetilde{g}$, then $f + \widetilde{g} = 0$ and $\widetilde{g} \in G$. Since the set of vectors $\mathcal{F}_k \cup \mathcal{G}_l$ is linearly independent, we obtain $f = 0$, $\widetilde{g} = 0$. Similarly, $h = 0$, $g = 0$. Using Theorem 3.9, we get $\dim(L_1 + L_2) = \dim(F \oplus G \oplus H) = k + l + m$, but $\dim L_1 = k + l$, $\dim L_2 = l + m$, and $\dim(L_1 \cap L_2) = l$. Finally, let us note that $k + l + m = (k + l) + (l + m) - l$. $\qquad \square$

Corollary 3.1 *Suppose that L_1, L_2 are subspaces of an n-dimensional space $\mathbf{X}_n$, and $\dim L_1 + \dim L_2 > n$. Then $L_1 \cap L_2 \neq \{0\}$.*

Proof Since $L_1 + L_2$ is a subspace of $\mathbf{X}_n$, we get $\dim(L_1 + L_2) \leq n$, and using (3.34), we see that $\dim(L_1 \cap L_2) = \dim L_1 + \dim L_2 - \dim(L_1 + L_2) \geq 1$. $\qquad \square$

3.3.3 The Orthogonal Projection of a Vector onto a Subspace

Let L be a subspace of an inner product space $\mathbf{X}$ and let x be a vector in $\mathbf{X}$. A vector $y \in L$ is the *best approximation* of x if

$$|x - y| \leq |x - z| \quad \text{for all} \quad z \in L. \qquad (3.36)$$

Theorem 3.11 *Let L be a finite-dimensional subspace of $\mathbf{X}$. Then for every $x \in \mathbf{X}$, there exists a unique best approximation of x in L.*

Proof If $L = \{0\}$, then the unique best approximation of x is the zero vector. Therefore we assume that $L \neq \{0\}$. Let $y, z \in L$. If we write z in the form $z = y + h$, where $h \in L$, then

$$(x - z, x - z) = (x - y - h, x - y - h)$$
$$= (x - y, x - y) - (x - y, h) - (h, x - y) + (h, h).$$

Hence if $(x - y, h) = 0$ for all $h \in L$, then condition (3.36) holds. Conversely, if (3.36) holds, then

$$-(x - y, h) - (h, x - y) + (h, h) \geq 0 \quad \text{for all} \quad h \in L.$$

Substituting $h_1 = ((x - y, h)/|h|^2)h$ for h, we get $-|(x - y, h)|^2/|h|^2 \geq 0$, therefore, $(x - y, h) = 0$. Thus, $y \in L$ is the best approximation of $x \in \mathbf{X}$ if and only if

$$(x - y, h) = 0 \quad \text{for all} \quad h \in L. \tag{3.37}$$

In other words, the vector $x - y$ is orthogonal to the subspace L. Geometrically, this conclusion is quite obvious (make a drawing!). If a vector y satisfying condition (3.37) exists, then it is uniquely determined by the vector x. Indeed, let there exist an additional vector $\tilde{y} \in L$ such that $(x - \tilde{y}, h) = 0$ for all $h \in L$. Then $(y - \tilde{y}, h) = 0$ for all $h \in L$. If we take $h = y - \tilde{y}$, then we get $y = \tilde{y}$.

We shall prove now that a vector $y \in L$ satisfying condition (3.37) exists. Let $\{e_k\}_{k=1}^m$ be a basis of the subspace L. Condition (3.37) is equivalent to the following:

$$(x - y, e_k) = 0, \quad k = 1, 2, \ldots, m. \tag{3.38}$$

We seek the vector y in the form $y = \sum_{i=1}^{m} \eta_i e_i$. It follows from (3.38) that

$$\left(\sum_{i=1}^{m} \eta_i e_i, e_k \right) = (x, e_k), \quad k = 1, 2, \ldots, m.$$

The last condition gives a system of linear equations with unknowns $\eta_1, \eta_2, \ldots, \eta_m$:

$$\sum_{i=1}^{m} \eta_i (e_i, e_k) = (x, e_k), \quad k = 1, 2, \ldots, m. \tag{3.39}$$

The matrix of this system is the Gram matrix of the basis $\{e_k\}_{k=1}^m$. This matrix is nonsingular (see Theorem 3.2, p. 74). Therefore, system (3.39) has a unique solution for each $x \in \mathbf{X}$, i.e., condition (3.37) uniquely determines the vector y. □

Remark 3.3 If the basis $\{e_k\}_{k=1}^m$ of the subspace L is orthonormal, then the vector y can be easily calculated, namely, in this case we get $y = \sum_{k=1}^{m} (x, e_k) e_k$.

It is natural that the vector y satisfying condition (3.37) is called the *orthogonal projection* of the vector x onto the subspace L and that the vector $z = x - y$ is called the *perpendicular* dropped from the point x to the subspace L.

Note that $(x - y, y) = 0$, since $y \in L$. Therefore, the Pythagorean identity (see Section 3.2.2, p. 72) holds:

$$|x|^2 = |x - y|^2 + |y|^2. \tag{3.40}$$

It follows from (3.40) that $|y|^2 \le |x|^2$. This is the so-called *Bessel's*[12] *inequality*, which shows that the length of the projection of a vector is less than or equal to the length of the vector.

If the set of vectors $\{e_k\}_{k=1}^m$ is orthonormal, then Bessel's inequality has the form

$$\sum_{k=1}^m |(x, e_k)|^2 \le |x|^2 \quad \text{for all} \quad x \in \mathbf{X}. \tag{3.41}$$

The two sides in (3.41) are equal if and only if $x \in L$, i.e., if $x = \sum_{k=1}^m (x, e_k)e_k$.

Note that the Cauchy–Schwarz inequality (3.5), p. 73, can be interpreted as a special case of Bessel's inequality (3.41) in which the orthonormal set of vectors consists of only one vector $e_1 = |y|^{-1}y$, $y \ne 0$.

For example, let L be the subspace of the space $\mathbb{R}^4$ spanned by the vectors $a_1 = (-3, 0, 7, 6)$, $a_2 = (1, 4, 3, 2)$, and $a_3 = (2, 2, -2, -2)$. Let us calculate the orthogonal projection of the vector $x = (14, -3, -6, -7)$ onto the subspace L and the perpendicular dropped from the point x to the subspace L.

The vectors a_1 and a_2 are linearly independent, and the vector a_3 is the linear combination of a_1 and a_2, namely, $a_3 = (-1/2)a_1 + (1/2)a_2$. Hence the vectors a_1, a_2 form a basis of the subspace L. The components η_1, η_2 of the vector y (which is the projection of x onto L) with respect to the basis a_1, a_2 can be computed as the solution of the system of equations

$$\eta_1(a_1, a_1) + \eta_2(a_2, a_1) = (x, a_1), \tag{3.42}$$
$$\eta_1(a_1, a_2) + \eta_2(a_2, a_2) = (x, a_2). \tag{3.43}$$

Computing the inner products, we get $(a_1, a_1) = 9 + 49 + 36 = 94$, $(a_2, a_1) = 30$, $(a_2, a_2) = 30$, $(x, a_1) = -126$, $(x, a_2) = -30$. Solving system (3.42), (3.43), we obtain $\eta_1 = -3/2$, $\eta_2 = 1/2$, i.e., $y = (-3/2)a_1 + (1/2)a_2 = (5, 2, -9, -8)$ is the orthogonal projection of the vector x onto the subspace L, and $z = x - y = (9, -5, 3, 1)$ is the perpendicular dropped from the point x to the subspace L.

A bad choice of basis in the subspace L can cause great computational difficulties in the practical calculation of the element of best approximation. Here is a relevant example. Let us specify the inner product in the space $C[0, 1]$ of continuous functions, using the formula

---

[12]Friedrich Wilhelm Bessel (1784–1846) was a German mathematician and astronomer.

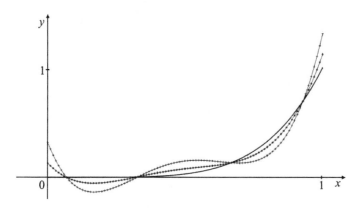

Fig. 3.1 Example of an almost linearly dependent basis. The plot of the function φ is indicated by the solid line; the plots of the approximating polynomial are indicated by the symbols "+" (for $\varepsilon = 5 \cdot 10^{-4}$) and "*" (for $\varepsilon = 2 \cdot 10^{-4}$)

$$(f, g) = \int\limits_0^1 f(x)g(x)\mathrm{d}x, \quad f, \, g \in C[0, 1]. \tag{3.44}$$

We shall consider the five-dimensional subspace of $C[0, 1]$ spanned by the basis that consists of the functions $\varphi_0(x) \equiv 1, \varphi_1(x) = x, \varphi_2(x) = x^2, \varphi_3(x) = x^3, \varphi_4(x) = x^4$ and calculate the best approximation of the function $\varphi(x) = x^5$.

The Gram matrix in this case is easily calculated:

$$\int\limits_0^1 \varphi_k(x)\varphi_l(x)\mathrm{d}x = 1/(k + l + 1), \quad k, \, l = 0, \, 1, \, \ldots, \, 4. \tag{3.45}$$

Evidently, the right-hand-side column of the system of linear equations (3.39) is equal to $(1/6, \, 1/7, \, 1/8, \, 1/9, \, 1/10)^T$. We assume that the last element of the right-hand-side column was calculated with a computational error, and substitute the number $(1/10) + \varepsilon$ for $1/10$.

Figure 3.1 shows a plot of the function $\varphi(x)$ and plots of the approximating polynomial $P_4(x) = \eta_0 + \eta_1 x + \eta_2 x^2 + \eta_3 x^3 + \eta_4 x^4$ for different values of ε. We see that significant errors in the approximation of the function φ correspond to small errors of the computation of the right-hand side (which are inevitable in practice). The reason for this effect is that the selected basis is almost linearly dependent. To verify this, just look at the plots of the functions x^p, $p = 1, 2, \ldots$, on the interval $[0, 1]$. These plots are similar even if the numbers p are not very large. Therefore, the matrix of system (3.39) is almost singular (it is also said to be ill-conditioned).

The matrix with elements (3.45), i.e., the matrix of the form

$$H_n = \left\{ \frac{1}{i+j-1} \right\}_{i,j=1}^{n} \tag{3.46}$$

is called a *Hilbert*[13] *matrix*. It is often applied in various areas of mathematics. Even for $n > 10$, this matrix is so ill conditioned that the corresponding system is not practically solvable by a computer.

Remark 3.4 Usually, orthogonal bases (for example, the Legendre polynomials or the Chebyshev polynomials; see pp. 79, 82) are used for approximations of functions by polynomials. In this case, system (3.39) is diagonal.

3.3.4 The Orthogonal Decomposition of an Inner Product Space

Let L be a subspace of an inner product space $\mathbf{X}$. The set of all vectors in $\mathbf{X}$ that are orthogonal to L is called the orthogonal complement of the subspace L and is denoted by $L^{\perp}$. The reader can easily prove that $L^{\perp}$ is a subspace of the space $\mathbf{X}$.

Theorem 3.12 (orthogonal decomposition). *Let L be a finite-dimensional subspace of an inner product space $\mathbf{X}$ and let $L^{\perp}$ be the orthogonal complement of the subspace L. Then the space $\mathbf{X}$ is the orthogonal sum of the subspaces L and $L^{\perp}$, i.e.,*

$$\mathbf{X} = L \oplus L^{\perp}. \tag{3.47}$$

Proof Using Theorem 3.11, we see that for each $x \in \mathbf{X}$ there exists $y \in L$ such that $(x - y, h) = 0$ for all $h \in L$. Therefore, $z = x - y \in L^{\perp}$ and $x = y + z$, which means (see Section 3.3.1, p. 84) that decomposition (3.47) is valid. $\qquad \square$

Let $e \in \mathbf{X}$, $e \neq 0$. Denote by π_e the set of all vectors in the space $\mathbf{X}$ that are orthogonal to e. It is easy to see that π_e is a subspace of $\mathbf{X}$. This subspace is called the *hyperplane* orthogonal to the vector e.

Theorem 3.13 *Let x be an arbitrary vector and e a nonzero vector in an inner product space $\mathbf{X}_n$. Then there exist a vector $y \in \pi_e$ and a number μ such that*

$$x = \mu e + y. \tag{3.48}$$

---

[13]David Hilbert (1862–1943) was a German mathematician.

The number μ and the vector y are uniquely determined by the vector x. Moreover,

$$|x - y| \leq |x - z| \quad \text{for all} \quad z \in \pi_e, \tag{3.49}$$

i.e., y is the element of best approximation of x in the subspace π_e.

The reader is invited to prove Theorem 3.13 (hint: use the idea of the proof of Theorem 3.12).

Chapter 4
Linear Operators

In this chapter we introduce the concept of a linear operator defined on a linear space. We study basic properties of linear operators acting on finite-dimensional linear spaces. We give a detailed investigation of their spectral properties. Special attention is paid to the study of the structure of the main classes of linear operators in finite-dimensional Euclidean and unitary spaces.

4.1 Linear Operators and Their Basic Properties

4.1.1 Basic Definitions. Operations with Operators

Let **X**,**Y** be linear spaces. We say that φ is a map from **X** to **Y** and write $\varphi : \mathbf{X} \to \mathbf{Y}$ if a unique vector $\varphi(x) \in \mathbf{Y}$ is defined for every $x \in \mathbf{X}$. We also say in this case that the *function* φ with values in the space **Y** is defined on the space **X** and write $x \to \varphi(x)$. We note that at the same time, not every vector in **Y** is necessarily the result of the mapping φ of some vector $x \in \mathbf{X}$.

We say that a map φ is *linear* if for every $x, y \in \mathbf{X}$ and for scalars α, β, we have

$$\varphi(\alpha x + \beta y) = \alpha \varphi(x) + \beta \varphi(y). \tag{4.1}$$

In linear algebra, almost all mappings are linear, and they are called *linear operators*, or in some contexts simply *operators*. Usually operators are denoted by capital letters. For example, the relationship (4.1) for a linear operator $\mathcal{A}$ will be written as

$$\mathcal{A}(\alpha x + \beta y) = \alpha \mathcal{A}x + \beta \mathcal{A}y.$$

Using the definition of a linear mapping, we see that $\mathcal{A}0 = 0$ for every operator $\mathcal{A}$.

© Springer International Publishing AG 2017
L. Beilina et al., *Numerical Linear Algebra: Theory and Applications*,
DOI 10.1007/978-3-319-57304-5_4

If an operator maps a space $\mathbf{X}$ into the same space $\mathbf{X}$, then we say that it acts on the space $\mathbf{X}$ or that the operator is a *transformation* of the space $\mathbf{X}$.

If some basis $\{e_j\}_{j=1}^n$ is fixed in a finite-dimensional space $\mathbf{X}_n$, then to define a linear operator $\mathcal{A}$ on $\mathbf{X}_n$, it is enough to describe the action of the operator on all the basis vectors, since for every vector $x = \sum_{j=1}^n \xi_j e_j$, we have $\mathcal{A}x = \sum_{j=1}^n \xi_j \mathcal{A}e_j$.

Operations with operators.

We define a *linear combination* of two operators $\mathcal{A} : \mathbf{X} \to \mathbf{Y}$ and $\mathcal{B} : \mathbf{X} \to \mathbf{Y}$ as a mapping $\alpha\mathcal{A} + \beta\mathcal{B} : \mathbf{X} \to \mathbf{Y}$ given by

$$(\alpha\mathcal{A} + \beta\mathcal{B})x = \alpha(\mathcal{A}x) + \beta(\mathcal{B}x) \quad \text{for all} \quad x \in \mathbf{X}, \tag{4.2}$$

where α and β are scalars. We define the *product* of two operators $\mathcal{A} : \mathbf{X} \to \mathbf{Y}$ and $\mathcal{B} : \mathbf{Y} \to \mathbf{Z}$ as the mapping $\mathcal{B}\mathcal{A} : \mathbf{X} \to \mathbf{Z}$ given by

$$\mathcal{B}\mathcal{A}x = \mathcal{B}(\mathcal{A}x) \quad \text{for all} \quad x \in \mathbf{X}. \tag{4.3}$$

The reader can easily prove that $\alpha\mathcal{A} + \beta\mathcal{B}$ and $\mathcal{B}\mathcal{A}$ are linear operators.

The product of a finite number of operators is defined in the same way. The reader will have no difficulty in showing that if the product of operators $\mathcal{C}, \mathcal{B}, \mathcal{A}$ is defined, then $\mathcal{C}\mathcal{B}\mathcal{A} = \mathcal{C}(\mathcal{B}\mathcal{A}) = (\mathcal{C}\mathcal{B})\mathcal{A}$.

Examples of linear operators.

1. The *null operator* $0 : \mathbf{X} \to \mathbf{Y}$ is defined by $0x = 0$ for all $x \in \mathbf{X}$. This operator transforms every vector of the space $\mathbf{X}$ into the zero vector of the space $\mathbf{Y}$.

2. The *identity operator* $I : \mathbf{X} \to \mathbf{X}$ is defined by $Ix = x$ for all $x \in \mathbf{X}$. This operator transforms every vector of the space $\mathbf{X}$ into itself.

3. The projection operator. Let the linear space $\mathbf{X}$ be a direct sum of subspaces L and M. Then every vector $x \in \mathbf{X}$ can be written in the form $x = x_1 + x_2$, where $x_1 \in L$, $x_2 \in M$, and the vectors x_1, x_2 are uniquely defined by the vector x. Let us define the operator $\mathcal{P} : \mathbf{X} \to L$ such that $\mathcal{P}x = x_1$. The operator $\mathcal{P}$ is called the *projection operator* onto the subspace L (in parallel with the subspace M). If $\mathbf{X}$ is an inner product space and can be represented as an orthogonal sum of subspaces L and M, then the operator $\mathcal{P}$ is called the *operator of the orthogonal projection*.

Let us prove that the operator $\mathcal{P}$ is linear. Suppose $x = \mathcal{P}x + x_2$ and $y = \mathcal{P}y + y_2$, where $x, y \in \mathbf{X}$, $x_2, y_2 \in M$. Then for all scalars α, β we have

$$\alpha x + \beta y = \alpha\mathcal{P}x + \beta\mathcal{P}y + \alpha x_2 + \beta y_2.$$

Since L and M are subspaces, we have $\alpha\mathcal{P}x + \beta\mathcal{P}y \in L$, $\alpha x_2 + \beta y_2 \in M$, and thus, $\mathcal{P}(\alpha x + \beta y) = \alpha\mathcal{P}x + \beta\mathcal{P}y$.

In the same manner we can introduce the linear operator $\mathcal{Q}$ that projects the space $\mathbf{X}$ onto the subspace M. We can easily obtain the following equalities: $\mathcal{P} + \mathcal{Q} = I$, $\mathcal{PQ} = 0$, $\mathcal{QP} = 0$, $\mathcal{P}^2 = \mathcal{P}$, $\mathcal{Q}^2 = \mathcal{Q}$. Generally, if the space $\mathbf{X}$ is a direct sum of several subspaces,

$$\mathbf{X} = L_1 \oplus L_2 \oplus \cdots \oplus L_k,$$

and $\mathcal{P}_i$ is the projection operator onto the subspace L_i, $i = 1, 2, ..., k$, then

$$\mathcal{P}_1 + \mathcal{P}_2 + \cdots + \mathcal{P}_k = I, \quad \mathcal{P}_i^2 = \mathcal{P}_i, \quad \mathcal{P}_i \mathcal{P}_j = 0 \text{ for } i \neq j, \tag{4.4}$$

where $i, j = 1, 2, ..., k$.

4. *Matrix–vector multiplication.* Let $A(m, n)$ be a rectangular matrix. Define the map $A : \mathbb{C}^n \to \mathbb{C}^m$ by the rule

$$y = Ax. \tag{4.5}$$

Matrix–vector multiplication is a linear operation (see Section 1.2.4, p. 34). Therefore, the operator $A : \mathbb{C}^n \to \mathbb{C}^m$ defined by (4.5) is linear.

4.1.2 The Inverse Operator

A linear operator $\mathcal{A} : \mathbf{X} \to \mathbf{Y}$ is called *invertible* if there exists a map $\mathcal{B} : \mathbf{Y} \to \mathbf{X}$ such that

$$\mathcal{B}\mathcal{A}x = x \quad \text{for all} \quad x \in \mathbf{X}, \tag{4.6}$$

$$\mathcal{A}\mathcal{B}y = y \quad \text{for all} \quad y \in \mathbf{Y}. \tag{4.7}$$

A map $\mathcal{B} : \mathbf{Y} \to \mathbf{X}$ satisfying (4.6) and (4.7) is called an *inverse of the map* $\mathcal{A}$.

Let us check that if an inverse $\mathcal{B}$ of the map $\mathcal{A}$ exists, then $\mathcal{B}$ is a linear operator. Let $y_1, y_2 \in \mathbf{Y}$, $\alpha, \beta \in \mathbb{C}$. Take $x_1 = \mathcal{B}y_1$, $x_2 = \mathcal{B}y_2$. Then $\mathcal{A}x_1 = \mathcal{A}\mathcal{B}y_1 = y_1$ and $\mathcal{A}x_2 = \mathcal{A}\mathcal{B}y_2 = y_2$. Therefore,

$$\mathcal{B}(\alpha y_1 + \beta y_2) = \mathcal{B}(\alpha \mathcal{A}x_1 + \beta \mathcal{A}x_2)$$
$$= \mathcal{B}\mathcal{A}(\alpha x_1 + \beta x_2) = \alpha x_1 + \beta x_2 = \alpha \mathcal{B}y_1 + \beta \mathcal{B}y_2.$$

We claim that if a linear operator $\mathcal{A} : \mathbf{X} \to \mathbf{Y}$ is invertible, then that operator is a bijective map acting from the space $\mathbf{X}$ to the space $\mathbf{Y}$. In fact, first let x_1, x_2 be two vectors in $\mathbf{X}$ such that $x_1 \neq x_2$. Then $\mathcal{A}x_1 \neq \mathcal{A}x_2$. Indeed, if we assume that $\mathcal{A}x_1 = \mathcal{A}x_2$, then $\mathcal{B}\mathcal{A}x_1 = \mathcal{B}\mathcal{A}x_2$, and hence $x_1 = x_2$. Second, if $y \in \mathbf{Y}$, then for $x = \mathcal{B}y$ we have $\mathcal{A}x = \mathcal{A}\mathcal{B}y = y$, i.e., every vector $y \in \mathbf{Y}$ is the result of the mapping $\mathcal{A}$ of some vector $x \in \mathbf{X}$.

It is easy to see that if a linear operator $\mathcal{A}$ is invertible, then it has a unique inverse (check it!). The *inverse operator* to the operator $\mathcal{A}$ is denoted by $\mathcal{A}^{-1}$. By definition, if the operator $\mathcal{A}^{-1}$ exists, then $(\mathcal{A}^{-1})^{-1} = \mathcal{A}$.

Examples.

1. The identity operator is invertible with $I^{-1} = I$.
2. Obviously, the null operator acting from $\mathbf{X}_n$ into $\mathbf{Y}_m$, where $m \geq 1$, is a noninvertible operator.
3. If $L \neq \mathbf{X}$, then the projection operator $\mathcal{P} : \mathbf{X} \to L$ is a noninvertible operator.
4. Every square matrix A of order n defines a linear operator acting on the space $\mathbb{C}^n$. If the matrix A is nonsingular, then the operator is invertible. The inverse operator is defined by the inverse matrix A^{-1} (see p. 40).

Let $\mathcal{A} : \mathbf{X} \to \mathbf{Y}$, $\mathcal{B} : \mathbf{Y} \to \mathbf{Z}$ be invertible operators. The reader can easily show that the operator $\mathcal{B}\mathcal{A}$ is invertible and $(\mathcal{B}\mathcal{A})^{-1} = \mathcal{A}^{-1}\mathcal{B}^{-1}$.

4.1.3   The Coordinate Representation Operator

Let $\mathbf{X}_n$ be an n-dimensional linear space, and let $\mathcal{E}_n = \{e_k\}_{k=1}^{n}$ be a basis for $\mathbf{X}_n$. Define an operator that maps the space $\mathbb{C}^n$ onto the space $\mathbf{X}_n$ by the rule

$$x = \mathcal{E}_n \xi, \quad \xi \in \mathbb{C}^n. \tag{4.8}$$

Evidently, this operator is linear, and we denote it by $\mathcal{E}$.

If $x \in \mathbf{X}_n$ is a given vector, then there exists a unique representation $x = \sum_{k=1}^{n} \xi_k e_k$, because $\mathcal{E}_n$ is a basis. The scalars ξ_k, $k = 1, 2, \ldots, n$, are the coordinates of x with respect to the basis $\mathcal{E}_n$, and the vector $\xi \in \mathbb{C}^n$ is the unique *coordinate representation* of x. Given the basis $\mathcal{E}_n$, the following linear mapping from $\mathbf{X}_n$ to $\mathbb{C}^n$ is well defined:

$$x \to \xi, \quad \text{where} \quad x = \mathcal{E}_n \xi.$$

We call this map the *coordinate representation operator* and denote it by $\mathcal{E}^{-1}$.

Using the definitions of the operators $\mathcal{E}$ and $\mathcal{E}^{-1}$, we get

$$\mathcal{E}^{-1}\mathcal{E}\xi = \xi \quad \text{for all} \quad \xi \in \mathbb{C}^n, \quad \mathcal{E}\mathcal{E}^{-1}x = x \quad \text{for all} \quad x \in \mathbf{X}_n,$$

i.e., the operators $\mathcal{E}$ and $\mathcal{E}^{-1}$ are mutually inverse.

Usually, to calculate a coordinate representation of x, it is necessary to solve a v system of linear algebraic equations with a square nonsingular matrix (see pp. 64, 67, 79). If $\mathbf{X}_n$ is an inner product space and $\mathcal{E}_n$ is an orthonormal basis, then the coordinate representation of x can be calculated more easily (see (3.17), p. 80, and the examples on pp. 81, 82).

4.1.4   Isomorphism of Finite-Dimensional Linear Spaces

We say that two linear spaces $\mathbf{X}$, $\mathbf{Y}$ are *isomorphic* if there exists an invertible linear operator such that $\mathcal{A} : \mathbf{X} \to \mathbf{Y}$. In other words, linear spaces $\mathbf{X}$ and $\mathbf{Y}$ are isomorphic if there exists a linear bijective correspondence between $\mathbf{X}$ and $\mathbf{Y}$. We also say in this case that the mapping $\mathcal{A}$ of $\mathbf{X}$ onto $\mathbf{Y}$ is an *isomorphism* between $\mathbf{X}$ and $\mathbf{Y}$.

Obviously, isomorphisms between linear spaces have the property of *transitivity*, i.e., if $\mathbf{X}$, $\mathbf{Z}$ are isomorphic and $\mathbf{Y}$, $\mathbf{Z}$ are isomorphic, then $\mathbf{X}$, $\mathbf{Y}$ are also isomorphic.

Theorem 4.1 *All finite-dimensional complex linear spaces of the same dimension are isomorphic to each other.*

Proof By transitivity, it is sufficient to prove that every n-dimensional complex linear space $\mathbf{X}_n$ is isomorphic to the space $\mathbb{C}^n$. Let $\mathcal{E}_n$ be a basis of $\mathbf{X}_n$. Then the coordinate representation operator $\mathcal{E}^{-1}$ realizes an isomorphism between $\mathbf{X}_n$ and $\mathbb{C}^n$ (see Section 4.1.3). $\qquad\square$

For the same reason, all n-dimensional real linear spaces are isomorphic to the space $\mathbb{R}^n$.

Theorem 4.2 *If two finite-dimensional linear spaces $\mathbf{X}$, $\mathbf{Y}$ are isomorphic, then they have the same dimension.*

Proof Let $\{e_k\}_{k=1}^n$ be a basis of $\mathbf{X}$. Suppose that a linear operator $\mathcal{A}$ is a bijective map from the space $\mathbf{X}$ to the space $\mathbf{Y}$. Writing $\sum_{k=1}^n \alpha_k \mathcal{A} e_k = 0$, we obtain $\mathcal{A} \sum_{k=1}^n \alpha_k e_k = 0$. Acting on both sides of the last equality by the operator $\mathcal{A}^{-1}$, we get $\sum_{k=1}^n \alpha_k e_k = 0$. Therefore, $\alpha_1, \alpha_2, \ldots, \alpha_n = 0$, i.e., the vectors $\{\mathcal{A} e_k\}_{k=1}^n$ are linearly independent elements of the space $\mathbf{Y}$. Hence the dimension of the space $\mathbf{X}$ is greater than or equal to n. Exchanging the roles of the spaces $\mathbf{X}$ and $\mathbf{Y}$, we get that they have the same dimension. $\qquad\square$

Consequently, we have the following result.

Theorem 4.3 *Two finite-dimensional complex (or real) linear spaces are isomorphic if and only if they have the same dimension.*

If linear spaces $\mathbf{X}$, $\mathbf{Y}$ are isomorphic, then there exists a bijective correspondence between the linear operations with elements in $\mathbf{X}$ and the linear operations with elements in $\mathbf{Y}$. Particularly, if a complex (or real) linear space $\mathbf{X}$ is finite-dimensional, then by introducing a basis for $\mathbf{X}$, the linear operations with elements in $\mathbf{X}$ can be replaced by the linear operations with vectors in the space $\mathbb{C}^n$ (or $\mathbb{R}^n$).

4.1.5 The Matrix of a Linear Operator

Let $\mathcal{A} : \mathbf{X}_n \to \mathbf{Y}_m$ be a linear operator. Suppose that $\mathcal{E}_n = \{e_k\}_{k=1}^n$ is a basis of $\mathbf{X}_n$ and $\mathcal{Q}_m = \{q_k\}_{k=1}^m$ is a basis of $\mathbf{Y}_m$. For each $i = 1, 2, \ldots, n$, the vector $\mathcal{A}e_i$ can be uniquely expanded in terms of the basis $\mathcal{Q}_m$:

$$\mathcal{A}e_i = \sum_{j=1}^m a_{ji}^{(eq)} q_j, \quad i = 1, 2, \ldots, n. \tag{4.9}$$

Consider the matrix

$$A_{eq} = \begin{pmatrix} a_{11}^{(eq)} & a_{12}^{(eq)} & \ldots & a_{1n}^{(eq)} \\ a_{21}^{(eq)} & a_{22}^{(eq)} & \ldots & a_{2n}^{(eq)} \\ \ldots\ldots\ldots\ldots\ldots \\ a_{m1}^{(eq)} & a_{m2}^{(eq)} & \ldots & a_{mn}^{(eq)} \end{pmatrix} \tag{4.10}$$

(the ith column of A_{eq} consists of the coordinates of $\mathcal{A}e_i$ with respect to the basis $\mathcal{Q}_m$). The matrix A_{eq} is called the *matrix of the operator* $\mathcal{A}$. This matrix is uniquely determined by the operator $\mathcal{A}$ and by the bases $\mathcal{E}_n$, $\mathcal{Q}_m$. We denote an operator and the corresponding matrix by the same letter in different typefaces. Subscripts in the notation of the matrix of an operator indicate which bases were used to construct the matrix.

Note that we can write relations (4.9) more concisely:

$$\mathcal{A}\mathcal{E}_n = \mathcal{Q}_m A_{eq}. \tag{4.11}$$

Suppose that $x = \mathcal{E}_n\xi \in \mathbf{X}_n, \xi \in \mathbb{C}^n$. We can expand the vector $\mathcal{A}x$ in terms of the basis $\mathcal{Q}_m$: $\mathcal{A}x = \mathcal{Q}_m\eta, \eta \in \mathbb{C}^m$. Then, using (4.11), we get

$$\mathcal{Q}_m\eta = \mathcal{A}x = \mathcal{A}\mathcal{E}_n\xi = \mathcal{Q}_m A_{eq}\xi,$$

and therefore,

$$\eta = A_{eq}\xi. \tag{4.12}$$

Relationship (4.12) shows the dependence between the coordinates of the vectors x and $\mathcal{A}x$ with respect to the bases of the linear spaces $\mathbf{X}_n$ and $\mathbf{Y}_m$, respectively.

It follows from (4.12) that if the matrix A_{eq} of the operator $\mathcal{A}$ is known, then we can construct the vector $\mathcal{A}x \in \mathbf{Y}_m$ corresponding to the vector $x \in \mathbf{X}_n$ in the following way.

1. Calculate the coordinates $\xi \in \mathbb{C}^n$ of x with respect to the basis $\mathcal{E}_n$. Using the coordinate representation operator $\mathcal{E}^{-1}$, we can write $\xi = \mathcal{E}^{-1}x$ (see Section 4.1.3).

2. Using (4.12), calculate the coordinates $\eta \in \mathbb{C}^m$ of $y = \mathcal{A}x \in \mathbf{Y}_m$ with respect to the basis $\mathcal{Q}_m$.

3. Calculate the vector y by the formula $y = \mathcal{Q}\eta$. Here $\mathcal{Q}$ is the operator defined by the rule analogous to (4.8).

The above implies that using the operators $\mathcal{E}$ and $\mathcal{Q}$ constructed by the bases $\mathcal{E}_n$ and $\mathcal{Q}_m$, we can write (4.11) in the following equivalent forms:

$$A_{eq} = \mathcal{Q}^{-1}A\mathcal{E} \quad \text{or} \quad A = \mathcal{Q}A_{eq}\mathcal{E}^{-1}. \tag{4.13}$$

To be precise, equalities (4.13) mean that

$$A_{eq}\xi = \mathcal{Q}^{-1}A\mathcal{E}\xi \quad \text{for all} \quad \xi \in \mathbb{C}^n, \qquad Ax = \mathcal{Q}A_{eq}\mathcal{E}^{-1}x \quad \text{for all} \quad x \in \mathbf{X}_n. \tag{4.14}$$

Equalities (4.13), (4.14) are illustrated by the following diagrams:

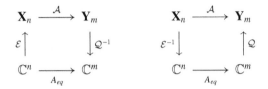

Therefore, if some bases $\mathcal{E}_n$ and $\mathcal{Q}_m$ were chosen for the spaces $\mathbf{X}_n$ and $\mathbf{Y}_m$, then to each linear operator $A : \mathbf{X}_n \to \mathbf{Y}_m$ there uniquely corresponds the linear operator $A_{eq} : \mathbb{C}^n \to \mathbb{C}^m$. This is the matrix–vector multiplication operator defined by the rule (4.12), where A_{eq} is the matrix of the operator A with respect to the bases $\mathcal{E}_n$ and $\mathcal{Q}_m$. Conversely, the linear operator $A : \mathbf{X}_n \to \mathbf{Y}_m$ that is defined by the equality $A = \mathcal{Q}A\mathcal{E}^{-1}$ uniquely corresponds to each $m \times n$ matrix A.

If $A : \mathbf{X}_n \to \mathbf{X}_n$, then

$$A\mathcal{E}_n = \mathcal{E}_n A_e, \tag{4.15}$$

or

$$A_e = \mathcal{E}^{-1}A\mathcal{E}, \tag{4.16}$$

where A_e is the matrix of the operator A with respect to the basis $\mathcal{E}_n$.

We note that there are two obvious cases in which the matrix of the linear operator $A : \mathbf{X}_n \to \mathbf{Y}_m$ does not depend on the choice of the bases for $\mathbf{X}_n$ and $\mathbf{Y}_m$.

1. The matrix of the null operator for any choice of bases for $\mathbf{X}_n$ and $\mathbf{Y}_m$ is the zero matrix.

2. The matrix of the identity operator with respect to any basis of the space $\mathbf{X}_n$ is the identity matrix.

By definition of the matrix of a linear operator we have

$$(\alpha A + \beta B)_{eq} = \alpha A_{eq} + \beta B_{eq} \tag{4.17}$$

for all linear operators $\mathcal{A}, \mathcal{B} : \mathbf{X}_n \to \mathbf{Y}_m$ and $\alpha, \beta \in \mathbb{C}$, i.e., the linear operations with their matrices correspond to the linear operations with their operators.

A similar statement under certain conditions is true for the product of two operators. Let $\mathcal{A} : \mathbf{X}_n \to \mathbf{Y}_m$, $\mathcal{B} : \mathbf{Y}_m \to \mathbf{Z}_p$ be linear operators. Suppose that $\{e_k\}_{k=1}^n$, $\{q_k\}_{k=1}^m$, and $\{r_k\}_{k=1}^p$ are the bases for the spaces $\mathbf{X}_n$, $\mathbf{Y}_m$, and $\mathbf{Z}_p$, respectively. Let A_{eq} be the matrix of the operator $\mathcal{A}$, B_{qr} the matrix of the operator $\mathcal{B}$, $(BA)_{er}$ the matrix of the operator $\mathcal{BA} : \mathbf{X}_n \to \mathbf{Z}_p$. Let us show that

$$(BA)_{er} = B_{qr} A_{eq}, \tag{4.18}$$

i.e., the matrix of the product of two operators is equal to the product of the matrices of those operators. Indeed, using (4.13), we get

$$(BA)_{er} = \mathcal{R}^{-1}\mathcal{BAE} = \mathcal{R}^{-1}\mathcal{R}B_{qr}\mathcal{Q}^{-1}\mathcal{Q}A_{eq}\mathcal{E}^{-1}\mathcal{E} = B_{qr}A_{eq}.$$

It is important to note that the same basis $\{q_k\}_{k=1}^m \subset \mathbf{Y}_m$ was used for the definition of the matrices of the operators $\mathcal{A}$ and $\mathcal{B}$. Usually we assume that the matching condition for these bases is satisfied.

Let us consider two examples.

1. We define the linear operator $\mathcal{A} : \mathbb{C}^4 \to \mathbb{C}^4$ by the rule $\mathcal{A}x = (x_2, x_1, x_3 + x_4, x_4)$, where $x = (x_1, x_2, x_3, x_4) \in \mathbb{C}^4$. Our goal is to calculate the matrix of the operator $\mathcal{A}$ with respect to the natural basis (see p. 64) of the space $\mathbb{C}^4$. It is easy to see that $\mathcal{A}i_1 = (0, 1, 0, 0) = i_2$, $\mathcal{A}i_2 = (1, 0, 0, 0) = i_1$, $\mathcal{A}i_3 = (0, 0, 1, 0) = i_3$, and $\mathcal{A}i_4 = (0, 0, 1, 1) = i_3 + i_4$. Hence the matrix of the operator $\mathcal{A}$ is

$$\begin{pmatrix} 0 & 1 & 0 & 0 \\ 1 & 0 & 0 & 0 \\ 0 & 0 & 1 & 1 \\ 0 & 0 & 0 & 1 \end{pmatrix}.$$

2. Let us denote by $\mathbf{Q}_2$ the linear space of all polynomials with complex coefficients and with degree at most 2. Let us define the linear operator $\mathcal{T} : \mathbf{Q}_2 \to \mathbf{Q}_2$ by the rule $\mathcal{T}q_2(z) = q_2(z + h)$, where $q_2 \in \mathbf{Q}_2$. Here h is a fixed complex number (a shift). Our goal is to calculate the matrix of the operator $\mathcal{T}$ with respect to the basis of the space $\mathbf{Q}_2$ that consists of the polynomials $\varphi_0(z) \equiv 1$, $\varphi_1(z) = z$, and $\varphi_2(z) = z^2$. We see that $\mathcal{T}\varphi_0 = \varphi_0$, $\mathcal{T}\varphi_1 = h\varphi_0 + \varphi_1$, $\mathcal{T}\varphi_2 = h^2\varphi_0 + 2h\varphi_1 + \varphi_2$. Hence the matrix of the operator $\mathcal{T}$ is equal to

$$\begin{pmatrix} 1 & h & h^2 \\ 0 & 1 & 2h \\ 0 & 0 & 1 \end{pmatrix}.$$

Therefore, if $q_2(z) = a_0 + a_1z + a_2z^2$, then $\mathcal{T}q_2(z) = b_0 + b_1z + b_2z^2$, where

$$\begin{pmatrix} b_0 \\ b_1 \\ b_2 \end{pmatrix} = \begin{pmatrix} 1 & h & h^2 \\ 0 & 1 & 2h \\ 0 & 0 & 1 \end{pmatrix} \begin{pmatrix} a_0 \\ a_1 \\ a_2 \end{pmatrix} = \begin{pmatrix} a_0 + ha_1 + h^2 a_2 \\ a_1 + 2ha_2 \\ a_2 \end{pmatrix}.$$

The matrix A_{eq} of the linear operator $\mathcal{A} : \mathbf{X}_n \to \mathbf{Y}_m$ is determined by the bases $\{e_k\}_{k=1}^n$ and $\{q_k\}_{k=1}^m$ of the spaces $\mathbf{X}_n$ and $\mathbf{Y}_m$. Suppose now that we take new bases $\{\tilde{e}_k\}_{k=1}^n$ and $\{\tilde{q}_k\}_{k=1}^m$ in $\mathbf{X}_n$ and $\mathbf{Y}_m$. The linear operator $\mathcal{A}$ will be represented by a new matrix with respect to these bases. Let us check what relations exist between different matrices representing the same operator. Denote by $A_{\tilde{e}\tilde{q}}$ the matrix of the operator $\mathcal{A}$ with respect to the bases $\{\tilde{e}_k\}_{k=1}^n$ and $\{\tilde{q}_k\}_{k=1}^m$. Suppose that we know the matrices of the bases change (see Section 2.3.3, p. 66), i.e.,

$$\tilde{\mathcal{E}}_n = \mathcal{E}_n T, \quad \tilde{\mathcal{Q}}_m = \mathcal{Q}_m R. \tag{4.19}$$

Using (4.13), we obtain $\mathcal{A} = \mathcal{Q} A_{eq} \mathcal{E}^{-1}$, $A_{\tilde{e}\tilde{q}} = \tilde{\mathcal{Q}}^{-1} \mathcal{A} \tilde{\mathcal{E}}$. Therefore,

$$A_{\tilde{e}\tilde{q}} = \tilde{\mathcal{Q}}^{-1} \mathcal{Q} A_{eq} \mathcal{E}^{-1} \tilde{\mathcal{E}}.$$

Taking into account (4.19), we get $\tilde{\mathcal{E}}_n \xi = \mathcal{E}_n T \xi$ for every $\xi \in \mathbb{C}^n$. Hence $\tilde{\mathcal{E}} = \mathcal{E} T$. This implies that $\mathcal{E}^{-1} \tilde{\mathcal{E}} = T$. Likewise, $\tilde{\mathcal{Q}}^{-1} \mathcal{Q} = R^{-1}$. Consequently,

$$A_{\tilde{e}\tilde{q}} = R^{-1} A_{eq} T. \tag{4.20}$$

Two matrices A and B are called *equivalent* if $A = CBD$ for some nonsingular matrices C and D.

Consider an important special case. If the linear operator $\mathcal{A}$ maps the space $\mathbf{X}_n$ into the same space $\mathbf{X}_n$, then

$$A_{\tilde{e}} = T^{-1} A_e T. \tag{4.21}$$

Square matrices B and C are called *similar matrices* if there exists a nonsingular matrix D such that

$$B = D^{-1} C D. \tag{4.22}$$

We also say that the matrix C is transformed into the matrix B by a *similarity transformation*. Relation (4.21) shows that the matrices of the same operator $\mathcal{A} : \mathbf{X}_n \to \mathbf{X}_n$ are similar with respect to the different bases.

4.1.6 The Matrix of the Inverse Operator

For every nonsingular matrix D we have $\det(D^{-1}) = 1/\det(D)$. Hence similar matrices have the same determinant. Because of that, we say that the determinant of the matrix of a linear operator $\mathcal{A} : \mathbf{X}_n \to \mathbf{X}_n$ is the *determinant of this operator* and write $\det(\mathcal{A})$. The determinant is an *invariant* of the linear operator, i.e., it is the same for every basis in $\mathbf{X}_n$.

We say that a linear operator $\mathcal{A} : \mathbf{X}_n \to \mathbf{X}_n$ is nonsingular if $\det(\mathcal{A}) \neq 0$. Every nonsingular operator $\mathcal{A}$ is invertible. Indeed, let $\{e_k\}_{k=1}^n$ be a basis in $\mathbf{X}_n$. Define an operator $\mathcal{B}$ by the relationship

$$\mathcal{B} = \mathcal{E} A_e^{-1} \mathcal{E}^{-1}.$$

Since $\mathcal{A} = \mathcal{E} A_e \mathcal{E}^{-1}$, we have $\mathcal{B}\mathcal{A} = \mathcal{A}\mathcal{B} = \mathcal{E} I \mathcal{E}^{-1} = I$. Therefore, the operator $\mathcal{B}$ is the inverse operator to the operator $\mathcal{A}$.

The above implies that for every basis of the space $\mathbf{X}_n$, the matrix of the inverse operator $\mathcal{A}^{-1}$ is the inverse matrix to the matrix of the operator $\mathcal{A}$.

Theorem 4.4 *If a linear operator* $\mathcal{A} : \mathbf{X}_n \to \mathbf{X}_n$ *is invertible, then it is nonsingular.*

Theorem 4.5 *A linear operator* $\mathcal{A} : \mathbf{X}_n \to \mathbf{X}_n$ *is invertible if and only if the equation* $\mathcal{A}x = 0$ *has only the trivial solution* $x = 0$.

The proof of Theorems 4.4 and 4.5 is left to the reader.

4.1.7   Linear Spaces of Linear Operators

Consider the set of all linear operators from $\mathbf{X}_n$ to $\mathbf{Y}_m$. The definitions of addition of linear operators and their multiplication by scalars were introduced in Section 4.1.1. It is easy to prove that these operations satisfy the linear space axioms. Thus the set of all linear operators from $\mathbf{X}_n$ to $\mathbf{Y}_m$ is a *linear space*.

Using results of Section 4.1.5, we can conclude that this linear space is isomorphic to the linear space of all $m \times n$ matrices. The isomorphism can be defined by relationship (4.11). The dimension of the linear space of all linear operators from $\mathbf{X}_n$ to $\mathbf{Y}_m$ is equal to mn.

We obtain a *real linear space* of operators if the linear spaces $\mathbf{X}_n$ and $\mathbf{Y}_m$ are real and linear operators can be multiplied only by real scalars.

4.1.8   The Image and the Kernel of a Linear Operator

Let $\mathcal{A}$ be a linear operator acting from a linear space $\mathbf{X}$ into a linear space $\mathbf{Y}$. The *image* of $\mathcal{A}$, denoted by $\mathrm{Im}(\mathcal{A})$, is the subset of $\mathbf{Y}$ consisting of all vectors that can be represented in the form $y = \mathcal{A}x$ for some $x \in \mathbf{X}$. The *kernel* of $\mathcal{A}$, denoted by $\mathrm{Ker}(\mathcal{A})$, is the subset of $\mathbf{X}$ consisting of all vectors x such that $\mathcal{A}x = 0$.

Theorem 4.6 *The set* $\mathrm{Im}(\mathcal{A})$ *is a linear subspace of the space* $\mathbf{Y}$.

Proof If $y_1, y_2 \in \mathrm{Im}(\mathcal{A})$, then there exist vectors $x_1, x_2 \in \mathbf{X}$ such that $y_1 = \mathcal{A}x_1$ and $y_2 = \mathcal{A}x_2$. Therefore, for all $\alpha, \beta \in \mathbb{C}$ we have $\alpha y_1 + \beta y_2 = \alpha \mathcal{A}x_1 + \beta \mathcal{A}x_2$.

Since the operator $\mathcal{A}$ is linear, we have $\alpha y_1 + \beta y_2 = \mathcal{A}(\alpha x_1 + \beta x_2)$. This means that $\alpha y_1 + \beta y_2 \in \text{Im}(\mathcal{A})$. $\qquad\square$

The proof of the following theorem is left to the reader.

Theorem 4.7 *The set* $\text{Ker}(\mathcal{A})$ *is a linear subspace of the space* $\mathbf{X}$.

The dimension of the subspace $\text{Im}(\mathcal{A}) \subset \mathbf{Y}_m$ is called the *rank* of the operator $\mathcal{A}$ and is denoted by $\text{rank}(\mathcal{A})$. The dimension of the kernel of $\mathcal{A}$ is called the *defect* of the operator $\mathcal{A}$ and is denoted by $\text{def}(\mathcal{A})$.

Theorem 4.8 *For every linear operator* $\mathcal{A} : \mathbf{X}_n \to \mathbf{Y}_m$, *the following equality holds:*

$$\text{rank}(\mathcal{A}) + \text{def}(\mathcal{A}) = n. \qquad (4.23)$$

Proof Denote by M the subspace of $\mathbf{X}_n$ such that $\mathbf{X}_n = \text{Ker}(\mathcal{A}) \oplus M$ (see Theorem 3.8, p. 86). Using Theorem 3.9, p. 86, we get $n = \text{def}(\mathcal{A}) + \dim(M)$. Taking into account Theorem 4.3, p. 97, it is enough to prove that the spaces M and $\text{Im}(A)$ are isomorphic. Let us check for this purpose that the operator $\mathcal{A}$ is a bijective map acting from M to $\text{Im}(\mathcal{A})$. In fact, every $x \in \mathbf{X}_n$ can be written in the form $x = x_0 + x_1$, where $x_0 \in \text{Ker}(\mathcal{A})$, $x_1 \in M$. Hence, $\mathcal{A}x = \mathcal{A}x_1$. Therefore, every element of $\text{Im}(\mathcal{A})$ is the image of some element of M. It remains to prove that if $\mathcal{A}x_1 = \mathcal{A}x_2$ for $x_1, x_2 \in M$, then $x_1 = x_2$. Equality $\mathcal{A}(x_1 - x_2) = 0$ means that $x_1 - x_2 \in \text{Ker}(\mathcal{A})$. On the other hand, M is a linear subspace, and thus $x_1 - x_2 \in M$. By Theorem 3.7, p. 85, this implies that $x_1 - x_2 = 0$. $\qquad\square$

4.1.9 The Rank of a Matrix

Let $A(m, n)$ be an $m \times n$ matrix. Let us interpret the set of the matrix columns as a subset of the space $\mathbb{C}^m$. We say that the rank of this set (see Section 2.2.4, p. 63) is the *rank of the matrix* $A(m, n)$ and denote it by $\text{rank}(A)$.

Theorem 4.9 *Suppose* $\mathcal{A} : \mathbf{X}_n \to \mathbf{Y}_m$ *is a linear operator,* $\mathcal{E}_n$ *is a basis for* $\mathbf{X}_n$, $\mathcal{Q}_m$ *is a basis for* $\mathbf{Y}_m$, A_{eq} *is the matrix of the operator* $\mathcal{A}$ *with respect to these bases. Then* $\text{rank}(A_{eq}) = \text{rank}(\mathcal{A})$.

Proof Let $x = \mathcal{E}_n \xi \in \mathbf{X}_n$. Then $\mathcal{A}x = \mathcal{Q}_m \eta$, where $\eta = A_{eq}\xi$ (see Section 4.1.5). Obviously, the vector η belongs to the span of the set of the matrix A_{eq} columns. The rank of the span of the set of the matrix A_{eq} columns is equal to $\text{rank}(A_{eq})$. Since the linear operator $\mathcal{Q}$ is invertible, this span is isomorphic to $\text{Im}(\mathcal{A})$. Therefore, by Theorem 4.3, p. 97, the dimension of $\text{Im}(\mathcal{A})$ is equal to $\text{rank}(A_{eq})$. $\qquad\square$

Consequently, the rank of the matrix of the linear operator $\mathcal{A} : \mathbf{X}_n \to \mathbf{Y}_m$ is an invariant of this operator, i.e., it is the same for all bases in $\mathbf{X}_n$ and $\mathbf{Y}_n$. Hence we could equivalently define the rank of a linear operator as the rank of its matrix.

We can interpret the set of rows of the matrix $A(m, n)$ as a subset of the space $\mathbb{C}^n$. Denote the rank of this set by r_s. The following result is unexpected at first glance.

Theorem 4.10 *For every matrix $A(m, n)$, one has the equality $r_s = \text{rank}(A(m, n))$.*

Proof We can assume without loss of generality that the first r_s rows of the matrix $A(m, n)$ are linearly independent and that each of the other rows is a linear combination of those first rows. Denote by $A(r_s, n)$ the matrix that consists of the first r_s rows of the matrix $A(m, n)$. Let us transform the matrix $A(r_s, n)$ by an algorithm that is in fact equivalent to Gaussian elimination.

Take a nonzero entry in the first row of the matrix $A(r_s, n)$, which is possible because none of the rows of the matrix $A(r_s, n)$ can be equal to zero. Interchange the columns of the matrix $A(r_s, n)$ such that the column that contains this nonzero entry takes the first place. Denote this transformed matrix in the same way. Multiply the first row by $-a_{21}/a_{11}$ and add the result to the second row. Then do the analogous transformations of all other rows of the matrix $A(r_s, n)$. As a result, we obtain a matrix with zeros in the first column below $a_{11} \neq 0$.

The second row of the transformed matrix is a nontrivial linear combination of the first two rows. Therefore, it is not equal to zero. Interchanging the second column of the transformed matrix with one of the following columns as needed, we obtain a matrix that has the entry $a_{22} \neq 0$. Multiply the second row by $-a_{32}/a_{22}$ and add the result to the third row. Do the analogous transformations of all following rows of the matrix $A(r_s, n)$. Continuing these transformations, we finally get a matrix of the block form

$$(\tilde{A}(r_s, r_s), \ B(r_s, n - r_s)), \tag{4.24}$$

where $\tilde{A}(r_s, r_s)$ is an upper triangular matrix that has nonzero entries along the main diagonal.

At each step of the described transformation process we get a row that is a nontrivial linear combination of the previous rows of the matrix $A(r_s, n)$. Therefore, this row is not equal to zero, and the transformations are valid. Clearly, we can assume without loss of generality that the original matrix $A(r_s, n)$ has such first r_s columns that by performing the described transformations without interchanging any columns, we get a matrix of the form (4.24).

Evidently, $\det(\tilde{A}(r_s, r_s)) \neq 0$, and hence the first r_s columns of the original matrix $A(r_s, n)$ are linearly independent. Thus the first r_s columns of the matrix $A(m, n)$ are linearly independent too. Let us check that by uniting this set of columns with any other column of the matrix $A(m, n)$, we get a linearly dependent set.

Let Δ_{r_s} be a *leading principal minor*[1] of degree r_s of the matrix $A(m, n)$. By the previous argumentation, $\Delta_{r_s} \neq 0$. Therefore, the system of linear equations

$$\sum_{j=1}^{r_s} a_{ij} x_j = a_{ik}, \quad i = 1, 2, \ldots, r_s, \tag{4.25}$$

---

[1] The leading principal minor of degree r is the determinant of the submatrix lying in the same set of the first r rows and columns.

has a solution for all $k = 1, 2, \ldots, n$. Since each row of the matrix $A(m, n)$ with a number greater than r_s is a linear combination of the first r_s rows of this matrix, we see that if a vector $(x_1, x_2, \ldots, x_{r_s})$ is the solution of linear system (4.25), then it satisfies the following relations:

$$\sum_{j=1}^{r_s} a_{ij} x_j = a_{ik}, \quad i = r_s + 1, \ldots, m.$$

Consequently, each column of the matrix $A(m, n)$ is a linear combination of the first r_s columns; hence $\text{rank}(A(m, n)) = r_s$. □

It follows immediately from the definition that $\text{rank}(A) \leq \min(m, n)$ for every matrix $A(m, n)$. A matrix $A(m, n)$ is said to have *full rank* if $\text{rank}(A) = \min(m, n)$.

An $n \times n$ matrix is nonsingular if and only if its rank is equal to n. Interchanging any of the matrix's rows or columns evidently does not change the rank of the matrix. Moreover, we have the following result.

Theorem 4.11 *Let $A(m, n)$ be an $m \times n$ matrix. Let $B(m, m)$ and $C(n, n)$ be square nonsingular matrices. Then*

$$\text{rank}(A) = \text{rank}(BA), \tag{4.26}$$

$$\text{rank}(A) = \text{rank}(AC). \tag{4.27}$$

Proof For the justification of equality (4.26), it is enough to check the following statement. If the matrix B is nonsingular, then a necessary and sufficient condition for the linear independence of the columns $Ba^1, \ldots, Ba^p$ is the linear independence of the columns $a^1, \ldots, a^p$ (check it!). If equality (4.26) holds, then equality (4.27) is proved by taking the matrix transpose. □

The reader can easily prove that for all matrices A and B that permit matrix multiplication, the following inequality holds: $\text{rank}(AB) \leq \min\{\text{rank}(A), \text{rank}(B)\}$.

Let us consider two examples.

1. The matrix

$$A = \begin{bmatrix} 3 & 4 & 1 \\ -2 & -3 & 1 \\ 5 & 7 & 0 \end{bmatrix}$$

has rank 2. Indeed, since the first two rows are linearly independent, the rank is at least 2. However, all three rows are linearly dependent, since the first is equal to the sum of the second and third. Thus the rank must be less than 3.

2. The matrix

$$A = \begin{bmatrix} 3 & 3 & 0 & 2 \\ -3 & -3 & 0 & -2 \end{bmatrix}$$

has rank 1: there are nonzero columns, so the rank is positive, but each pair of columns are linearly dependent. Similarly, the transpose of the matrix A,

$$A^T = \begin{bmatrix} 3 & -3 \\ 3 & -3 \\ 0 & 0 \\ 2 & -2 \end{bmatrix},$$

has rank 1. As we have proved above, the rank of a matrix is equal to the rank of its transpose, i.e., $\mathrm{rank}(A) = \mathrm{rank}(A^T)$.

4.1.10 Calculating the Rank of a Matrix Using Determinants

It follows from the proof of Theorem 4.10 that if $\mathrm{rank}(A) = r$, then we can interchange the rows and the columns of the matrix A such that the leading principal minor Δ_r of the transformed matrix will not vanish. This minor is called *basic*.

Let us formulate and prove the converse statement. Namely, let A be a rectangular matrix. The leading principal minor Δ_r of order $r < \min(m, n)$ of the matrix A is bordered by the leading principal minor Δ_{r+1}. We can construct different minors bordering Δ_r by interchanging the rows and the columns of the matrix A whose numbers are greater than r.

Lemma 4.1 *If the leading principal minor Δ_r is nonzero and all minors of order $r + 1$ that border Δ_r are equal to zero, then* $\mathrm{rank}(A) = r$.

Proof Since $\Delta_r \neq 0$, the first r columns of the matrix A are linearly independent. Let us show that each column of A whose number is greater than r is a linear combination of its first r columns. This means that $\mathrm{rank}(A) = r$. Assume the contrary. Then there exists a column of A such that the rank of the matrix that consists of this column and the first r columns of A is equal to $r + 1$. Therefore, this matrix has $r + 1$ linearly independent rows. The first r rows of this matrix are linearly independent, since $\Delta_r \neq 0$. Hence there exists a row whose number is greater than r that is not a linear combination of the first r rows. If we turn this row into the $(r + 1)$th row of the matrix A, then we get $\Delta_{r+1} \neq 0$, but this contradicts the assumption of the lemma. □

Lemma 4.1 gives the following method of calculating the rank of a matrix A.[2]

1. We check all elements of A. If all elements are zero, then $\mathrm{rank}(A) = 0$.

2. If an element of A is not equal to zero, then we interchange the rows and the columns of the matrix A to put this element in the place of a_{11}.

3. We calculate all minors of order two that border $\Delta_1 = |a_{11}|$. If all these minors are equal to zero, then $\mathrm{rank}(A) = 1$.

4. If a minor of order two is not equal to zero, then we interchange the rows and the columns to put this minor in the place of Δ_2, i.e., to put it in the top left corner of the matrix A.

---

[2]Usually, the algorithm, which is described in Section 5.1.1, p. 163, is used for numerical realizations.

5. We calculate all minors of order three that border Δ_2 until we find a nonzero minor, and so on. If at a step of this algorithm we see that the leading principal minor Δ_r is nonzero and all minors of order $r + 1$ that border Δ_r are equal to zero, then $\text{rank}(A) = r$.

Clearly, it is not necessary to interchange the rows and the columns of the matrix at each step of this algorithm. It is enough to calculate all minors of order $r + 1$ that border an arbitrary nonzero minor of order r.

For example, let us calculate the rank of the matrix

$$A = \begin{pmatrix} 2 & -4 & 3 & 1 & 0 \\ 1 & -2 & 1 & -4 & 2 \\ 0 & 1 & -1 & 3 & 1 \\ 4 & -7 & 4 & -4 & 5 \end{pmatrix}.$$

Note that A includes the nonzero minor

$$d = \begin{vmatrix} -4 & 3 \\ -2 & 1 \end{vmatrix}.$$

The minor of order three

$$d' = \begin{vmatrix} 2 & -4 & 3 \\ 1 & -2 & 1 \\ 0 & 1 & -1 \end{vmatrix}$$

borders d and is not equal to zero, but both of the fourth-order minors

$$\begin{vmatrix} 2 & -4 & 3 & 1 \\ 1 & -2 & 1 & -4 \\ 0 & 1 & -1 & 3 \\ 4 & -7 & 4 & -4 \end{vmatrix}, \quad \begin{vmatrix} 2 & -4 & 3 & 0 \\ 1 & -2 & 1 & 2 \\ 0 & 1 & -1 & 1 \\ 4 & -7 & 4 & 5 \end{vmatrix}$$

bordering d' evidently vanish. Thus $\text{rank}(A) = 3$.

4.1.11 The General Solution of a Linear Equation

Let $\mathcal{A}$ be a linear operator mapping a linear space $\mathbf{X}_n$ into a linear space $\mathbf{Y}_m$. Consider the linear equation

$$\mathcal{A}x = y, \tag{4.28}$$

where y is a given element of $\mathbf{Y}_m$ and $x \in \mathbf{X}_n$ is an unknown. In this section we suppose that Eq. (4.28) is solvable and describe the form of all its possible solutions. In other words, we describe the form of the *general solution* of equation (4.28).

Suppose x_1 and x_2 are two solutions of equation (4.28) for the same right-hand side y. Then evidently, $\mathcal{A}(x_1 - x_2) = 0$, i.e., $x_1 - x_2 \in \text{Ker}(\mathcal{A})$. This yields that if system (4.28) possesses a solution x_0 (it is called a *particular solution* of the inhomogeneous equation), then every other solution of equation (4.28) has the form $x = x_0 + \tilde{x}$, where $\tilde{x} \in \text{Ker}(\mathcal{A})$.

Let $\varphi_1, \varphi_2, \ldots, \varphi_p$ be a basis of $\text{Ker}(\mathcal{A})$. Then

$$x = x_0 + \sum_{k=1}^{p} c_k \varphi_k. \tag{4.29}$$

Therefore, the general solution of equation (4.28) has the form (4.29). We can obtain every solution of equation (4.28) by changing the coefficients in (4.29). The vectors $\varphi_1, \varphi_2, \ldots, \varphi_p$ are called the *fundamental set of solutions* of the homogeneous equation

$$\mathcal{A}x = 0. \tag{4.30}$$

The vector

$$\tilde{x} = \sum_{k=1}^{p} c_k \varphi_k$$

is called the *general solution* of the homogeneous equation. Thus the general solution of equation (4.28) is the sum of a particular solution of (4.28) and the general solution of homogeneous equation (4.30).

4.1.12 Systems of Linear Algebraic Equations. Solvability Conditions

For practical construction of the solution of linear equation (4.28) it is necessary to introduce some bases $\mathcal{E}_n = \{e_k\}_{k=1}^{n}$, $\mathcal{Q}_m = \{q_k\}_{k=1}^{m}$ for the spaces $\mathbf{X}_n$, $\mathbf{Y}_m$ and to reduce Eq. (4.28) to a system of linear algebraic equations

$$A_{eq}\xi = \eta. \tag{4.31}$$

The unknown vector $\xi \in \mathbb{C}^n$ is the coordinate representation of x with respect to the basis $\mathcal{E}_n$. The vector $\eta \in \mathbb{C}^m$ is the coordinate representation of y with respect to the basis $\mathcal{Q}_m$. The matrix A_{eq} is the matrix of the linear operator $\mathcal{A}$ (see Section 4.1.5).

Let us write system (4.31) in the components of ξ and η:

$$\sum_{j=1}^{n} a_{ij}^{(eq)} \xi_j = \eta_i, \quad i = 1, 2, \ldots, m. \tag{4.32}$$

Here $a_{ij}^{(eq)}$ (the entries of the matrix A_{eq} of the linear operator $\mathcal{A}$) and η_i are given numbers, and ξ_j are unknowns.

In contrast to the systems of linear algebraic equations that were discussed in Section 1.2.3, p. 27, system (4.32) has, generally speaking, different numbers of equations and unknowns.

Problems (4.28) and (4.31) are equivalent to each other in the sense that if ξ is a solution of (4.31), then $x = \mathcal{E}_n \xi$ is the solution of equation (4.28) with $y = \mathcal{Q}_m \eta$; conversely, if x is a solution of equation (4.28), then the coordinate representations of x and y with respect to the corresponding bases are connected by relationship (4.31).

Let us obtain necessary and sufficient conditions for solvability of the system of linear algebraic equations

$$Ax = b, \tag{4.33}$$

where A is a given $m \times n$ matrix with complex (generally speaking) entries, and $b \in \mathbb{C}^m$ is a given vector. Let us attach the column b to the matrix A and denote this resulting $m \times (n + 1)$ matrix by (A, b). The matrix (A, b) is called the *augmented matrix* of system (4.33).

Theorem 4.12 (Kronecker–Capelli[3] theorem). *A system of linear algebraic equations has a solution if and only if the matrix A has the same rank as the matrix (A, b).*

Proof Evidently, the rank of the augmented matrix (A, b) is greater than or equal to the rank of the matrix A, and $\text{rank}(A) = \text{rank}(A, b)$ if and only if b is a linear combination of the columns of the matrix A. The last condition is equivalent to the statement that there exists a vector $x \in \mathbb{C}^n$ that is a solution of (4.33). $\qquad\square$

Theorem 4.13 (Fredholm's[4] matrix theorem). *A system of linear algebraic equations has a solution if and only if for every solution of the homogeneous system of equations $zA = 0$, the equality $zb = 0$ holds.*

Note here that b is a column and z is a row.

Proof Sufficiency. Let $r = \text{rank}(A)$. We can assume without loss of generality that the first r rows of the matrix A are linearly independent. Clearly, this implies that the first r rows of the matrix (A, b) are linearly independent too. If the kth row of the matrix A is a linear combination of the first r rows of A, then there exists a nonzero vector z such that $zA = 0$. Under the hypothesis of the theorem, $zb = 0$. This implies that the kth row of the matrix (A, b) is a linear combination of the first r rows of (A, b). Thus, $\text{rank}(A) = \text{rank}(A, b)$, and by the Kronecker–Capelli theorem, system (4.33) has a solution.

Necessity. Suppose that system (4.33) has a solution, i.e., that there exists $x \in \mathbb{C}^n$ such that $Ax = b$. Then for every $z \in \mathbb{C}^m$, we have the equality $zAx = zb$. Clearly, if $zA = 0$, then $zb = 0$. $\qquad\square$

---

[3] Alfredo Capelli (1858–1916) was an Italian mathematician.

[4] Erik Ivar Fredholm (1866–1927) was a Swedish mathematician.

Let us give an example of how to apply Fredholm's matrix theorem. Consider the symmetric $n \times n$ matrix

$$
A = \begin{pmatrix}
1 & -1 & 0 & \cdots & \cdots & \cdots & 0 \\
-1 & 2 & -1 & 0 & \cdots & \cdots & 0 \\
\cdots & \cdots & \cdots & \cdots & \cdots & \cdots & \cdots \\
0 & \cdots & -1 & 2 & -1 & \cdots & 0 \\
\cdots & \cdots & \cdots & \cdots & \cdots & \cdots & \cdots \\
0 & \cdots & \cdots & 0 & -1 & 2 & -1 \\
0 & 0 & \cdots & \cdots & 0 & -1 & 1
\end{pmatrix}.
$$

We have to calculate $\operatorname{rank}(A)$ and describe necessary and sufficient solvability conditions for the system of linear algebraic equations

$$
Ax = b. \tag{4.34}
$$

Let us interpret the matrix A as a linear operator acting on the space $\mathbb{R}^n$ and describe its kernel. Consider the homogeneous system of linear algebraic equations

$$
Ax = 0. \tag{4.35}
$$

The ith equation of this system for $i = 2, 3, \ldots, n-1$ can be written in the following form: $-x_{i-1} + 2x_i - x_{i+1} = 0$, or $x_i - x_{i-1} = x_{i+1} - x_i$. Therefore, if x is a solution of system (4.35), then

$$
x_1 = x_2 = \cdots = x_n,
$$

i.e., the kernel of the operator A is a one-dimensional subspace of the space $\mathbb{R}^n$; each vector x of this subspace has the form $x = c(1, \ldots, 1)$, where c is a real number. Hence, using Theorem 4.8, p. 103, we see that $\operatorname{rank}(A) = n - 1$.

Further, since the matrix A is symmetric, using Fredholm's matrix theorem, we see that a necessary and sufficient solvability condition for system (4.34) is $z^T b = 0$, where z is any solution of equation (4.35). Consequently, a solution of system (4.34) exists if and only if $b_1 + b_2 + \cdots + b_n = 0$.

4.1.13   The General Solution of a System of Linear Algebraic Equations

Let us describe an elementary method of calculating the general solution of the system of linear algebraic equations[5]

---

[5]Usually, methods based on the singular value decomposition (see Section 5.1.1, p. 163) are used for numerical approximations of general solutions of systems of linear algebraic equations.

$$Ax = b. \tag{4.36}$$

Our consideration is based on the results of Section 4.1.11. Now we suppose that system (4.36) has a solution, and denote by r the rank of the augmented matrix of system (4.36).

Let us begin with the calculation of a particular solution of system (4.36). Using the method of calculation of the rank of a matrix described in Section 4.1.10, we transform the matrix (A, b) such that the leading principal minor of order r of the transformed matrix is not equal to zero and each row starting from the $(r + 1)$th is a linear combination of the first r rows.

Clearly, the transformed system of linear equations is equivalent to the original one, i.e., every solution of system (4.36) is a solution of the transformed system, and conversely, every solution of the transformed system is the solution of (4.36).

Since the last $m - r$ equations of the transformed system follow from the first r equations of this system, we delete the last $m - r$ equations. In the first r equations we move to the right-hand side all summands with the variables starting from the $(r+1)$th variable. These variables are called *free*.

After that, we assign some values to the free variables (usually there is no reason why we cannot take $x_{r+1} = \cdots = x_n = 0$). As a result, we get a system of r linear equations in r unknowns. The matrix of this system is nonsingular, and we find the values of the variables x_1, x_2, ..., x_r as the unique solution of this system. Thus we have the solution of the original system (4.36) of the form $x = (x_1, x_2, \ldots, x_r, x_{r+1}, \ldots, x_n)$.

For example, let us calculate a particular solution of the following system:

$$x_1 - x_2 + x_3 - x_4 = 4, \tag{4.37}$$
$$x_1 + x_2 + 2x_3 + 3x_4 = 8, \tag{4.38}$$
$$2x_1 + 4x_2 + 5x_3 + 10x_4 = 20. \tag{4.39}$$

The determinant

$$\Delta_2 = \begin{vmatrix} 1 & -1 \\ 1 & 1 \end{vmatrix},$$

which is located in the top left corner of the matrix A of this system, is nonzero. All determinants bordering Δ_2 are equal to zero:

$$\begin{vmatrix} 1 & -1 & 1 \\ 1 & 1 & 2 \\ 2 & 4 & 5 \end{vmatrix} = 0.$$

Hence rank$(A) = 2$, and the rank of the augmented matrix is also two. System (4.37)–(4.39) is solvable, and the last equation follows from the first two equations of this system. Thus, to find a particular solution of system (4.37)–(4.39), it is enough to solve the system of two equations (4.37), (4.38), equating x_3, x_4 to some numbers.

If we take $x_3 = x_4 = 0$ in (4.37), (4.38), then we get $x_1 = 6$, $x_2 = 2$. Therefore, the vector $x = (6, 2, 0, 0)$ is a particular solution of (4.37)–(4.39).

Now we construct the fundamental set of solutions of the homogeneous system of linear equations

$$Ax = 0 \tag{4.40}$$

with an $m \times n$ matrix. Let $\operatorname{rank}(A) = r$. Using Theorem 4.8, we see that it is enough to construct $n - r$ arbitrary linearly independent solutions of system (4.40). Of course, we assume that $n > r$.

Arguing as in the first part of this subsection, we transform system (4.40) to the following form:

$$A(r, r)x(r, 1) + B(r, n - r)y(n - r, 1) = 0. \tag{4.41}$$

Here $A(r, r)$ is a nonsingular matrix, and the column $y((n - r), 1)$ corresponds to the free variables. We take the vectors

$$y_1((n - r), 1), \ y_2((n - r), 1), \ \ldots, y_{n-r}((n - r), 1) \tag{4.42}$$

such that they are linearly independent (the simplest way is to equate them to the vectors of the standard basis in the space $\mathbb{C}^{n-r}$). Using vectors (4.42), we calculate vectors

$$x_1(r, 1), \ x_2(r, 1), \ \ldots, \ x_{n-r}(r, 1)$$

as the solutions of the following systems:

$$A(r, r)x_k(r, 1) + B(r, (n - r))y_k((n - r), 1) = 0, \quad k = 1, 2, \ldots, n - r.$$

Writing together the components of the vectors $x_k(r, 1)$ and $y_k((n - r), 1)$, we construct the following vectors:

$$z_k(n, 1) = (x_k(r, 1), \ y_k((n - r), 1)), \quad k = 1, 2, \ldots, n - r.$$

By construction, $Az_k = 0$ for $k = 1, \ldots, n - r$. Clearly, the vectors z_k, $k = 1, \ldots, n - r$, are linearly independent, since the vectors (4.42) are linearly independent. Thus the vectors z_k, $k = 1, 2, \ldots, n - r$, form a fundamental set of solutions of the homogeneous system of linear equations (4.40).

For example, let us calculate the fundamental set of solutions of the system of linear equations

$$x_1 - x_2 + x_3 - x_4 = 0, \tag{4.43}$$
$$x_1 + x_2 + 2x_3 + 3x_4 = 0, \tag{4.44}$$
$$2x_1 + 4x_2 + 5x_3 + 10x_4 = 0, \tag{4.45}$$

which corresponds to system (4.37)–(4.39). As we know from the previous example, the rank of the matrix of this system is equal to two. Therefore, we have to calculate two linearly independent solutions of system (4.43)–(4.45). As we have seen, the last equation in this system follows from the first two equations. If we put $x_3 = 1$ and $x_4 = 0$ Eqs. (4.43), (4.44), we get

$$x_1 - x_2 + 1 = 0, \tag{4.46}$$

$$x_1 + x_2 + 2 = 0. \tag{4.47}$$

Hence $x_1 = -3/2$, $x_2 = -1/2$. If we put $x_3 = 0$, $x_4 = 1$ in (4.43), (4.44), then we obtain $x_1 = -1$, $x_2 = -2$. Hence $x_1 = (-3/2, -1/2, 1, 0)$ and $x_2 = (-1, -2, 0, 1)$ form a fundamental set of solutions of the homogeneous system of linear equations (4.43)–(4.45). Each vector of the form

$$x = c_1(-3/2, -1/2, 1, 0) + c_2(-1, -2, 0, 1), \tag{4.48}$$

where c_1, c_2 are arbitrary numbers, is a solution of system (4.43)–(4.45), and conversely, each solution of system (4.43)–(4.45) can be represented in the form (4.48) for some c_1, c_2. Thus the general solution of system (4.37)–(4.39) can be represented in the form $x = (6, 2, 0, 0) + c_1(-3/2, -1/2, 1, 0) + c_2(-1, -2, 0, 1)$, where c_1, c_2 are arbitrary numbers.

4.2 Eigenvalues and Eigenvectors of a Linear Operator

4.2.1 Invariant Subspaces

Let $\mathcal{A} : \mathbf{X} \to \mathbf{X}$ be a linear operator. A subspace $L \subset \mathbf{X}$ is said to be *invariant* under $\mathcal{A}$ if $\mathcal{A}x \in L$ for all $x \in L$. The subspaces $L = \{0\}$ and $L = \mathbf{X}$ are invariant for every $\mathcal{A} : \mathbf{X} \to \mathbf{X}$. We refer to these as *trivial* invariant subspaces.

Let the linear space $\mathbf{X}$ be a direct sum of subspaces L and M. Let $\mathcal{P} : \mathbf{X} \to L$ be the projection operator onto the subspace L in parallel with the subspace M. Then $\mathcal{P}x = x$ for every $x \in L$ and $\mathcal{P}x = 0$ for every $x \in M$, i.e., the subspaces L and M are invariant under $\mathcal{P}$.

Let us give an example of an operator that has no nontrivial invariant subspaces. Let $\mathbf{X}_2$ be a two-dimensional real Euclidean space. It is easy to see that if $L \subset \mathbf{X}_2$ is a nontrivial subspace, then L is the set of all vectors having the form $x = \alpha e$, where $e \neq 0$ is a given vector, $\alpha \in \mathbb{R}$. In other words, L is a straight line in the plane containing the origin. Denote by e_1, e_2 an orthonormal basis of the space $\mathbf{X}_2$. Let $\mathcal{Q} : \mathbf{X}_2 \to \mathbf{X}_2$ be the linear operator that maps each vector $x = \xi_1 e_1 + \xi_2 e_2$ into the vector $y = -\xi_2 e_1 + \xi_1 e_2$. The vectors x and y are orthogonal. Hence if L is a nontrivial subspace of $\mathbf{X}_2$, then $\mathcal{Q}x \in L^{\perp}$ for every $x \in L$. Therefore, $\mathcal{Q}x \notin L$ for every $x \neq 0$, i.e., the operator $\mathcal{Q}$ has no nontrivial invariant subspaces.

If a basis of an invariant subspace is known, then the form of the matrix of the linear operator becomes simpler. Namely, let $\{e_k\}_{k=1}^n$ be a basis of the linear space $\mathbf{X}_n$, and let $L \subset \mathbf{X}_n$ be an m-dimensional invariant subspace under the operator $\mathcal{A} : \mathbf{X}_n \to \mathbf{X}_n$. Suppose that the vectors $\{e_k\}_{k=1}^m$ belong to L. Then $\{e_k\}_{k=1}^m$ is a basis for L (prove it!) and

$$\mathcal{A}e_k = \sum_{j=1}^m a_{jk}^{(e)} e_j, \ k = 1, \ldots, m, \quad \mathcal{A}e_k = \sum_{j=1}^n a_{jk}^{(e)} e_j, \ k = m+1, \ldots, n.$$

These equalities show that the entries of the matrix A_e that are located in the intersection of the first m columns and the last $(n - m)$ rows are equal to zero. Therefore, the matrix A_e can be written as the 2×2 block triangular matrix

$$A_e = \begin{pmatrix} A_{11} & A_{12} \\ 0 & A_{22} \end{pmatrix}, \tag{4.49}$$

where A_{11} is a square $m \times m$ matrix, A_{22} is a square $(n - m) \times (n - m)$ matrix, 0 is the zero $(n - m) \times m$ matrix, and A_{12} is an $m \times (n - m)$ matrix.

We get a simpler matrix A_e if the space $\mathbf{X}_n$ is decomposed into a direct sum of invariant subspaces L and M of the operator $\mathcal{A}$, i.e., $\mathbf{X}_n = L \oplus M$ and the basis $\{e_k\}_{k=1}^n$ of the space $\mathbf{X}_n$ is chosen in such way that the vectors $\{e_k\}_{k=1}^m$ form a basis of the subspace L. Then, as is easy to see, the matrix A_{12} in (4.49) is the zero matrix, i.e., the matrix A_e has block diagonal form

$$A_e = \begin{pmatrix} A_{11} & 0 \\ 0 & A_{22} \end{pmatrix}. \tag{4.50}$$

Obviously, the converse statement is also true. Namely, if the matrix of an operator with respect to a basis $\{e_k\}_{k=1}^n$ has the form (4.50), then the space $\mathbf{X}_n$ is the direct sum of two subspaces. The vectors of the set $\{e_k\}_{k=1}^n$ whose numbers are equal to the numbers of the rows of the corresponding blocks form the bases of these subspaces.

If the subspaces L and M are decomposed into direct sums of invariant subspaces of smaller dimensions, then the number of diagonal blocks of the matrix A_e increase and their dimensions decrease.

The simplest case is that in which the space $\mathbf{X}_n$ can be represented as a direct sum of n one-dimensional invariant subspaces of the operator A. Then the matrix A_e is diagonal. However, that is possible only for some special classes of operators.

Lemma 4.2 *Let* $\mathcal{A} : \mathbf{X}_n \to \mathbf{X}_n$ *be a nonsingular operator. Let* $L \subset \mathbf{X}_n$ *be an invariant subspace of the operator* $\mathcal{A}$. *Then for every* $x \in L$ *there exists a unique vector* $y \in L$ *such that* $\mathcal{A}y = x$.[6]

---

[6]Therefore, we can say that a nonsingular operator generates a one-to-one mapping of each of its invariant subspaces onto this subspace.

Proof The subspace L is invariant under the operator $\mathcal{A}$. So we can introduce the operator $\mathcal{A}_L : L \rightarrow L$, assuming $\mathcal{A}_L x = \mathcal{A}x$ for $x \in L$. The operator $\mathcal{A}_L$ is nonsingular, because if $\mathcal{A}_L x = \mathcal{A}x = 0$, then $x = 0$, since $\mathcal{A}$ is nonsingular (see Theorem 4.5, p. 102). This implies that the equation $\mathcal{A}_L y = x$ has a unique solution $y \in L$ for every vector $x \in L$. $\square$

The operator $\mathcal{A}_L$ defined in the proof of Lemma 4.2 is called the *restriction of the operator* $\mathcal{A}$ to its invariant subspace L.

4.2.2 Basic Properties of Eigenvalues and Eigenvectors

A special role of one-dimensional invariant subspaces of operators was shown in the previous section. The concept of one-dimensional invariant subspaces is closely related to the concept of eigenvectors of operators.

A vector $x \in \mathbf{X}$ is called an *eigenvector of the operator* $\mathcal{A} : \mathbf{X} \rightarrow \mathbf{X}$ if x is nonzero and there exists a number λ such that

$$\mathcal{A}x = \lambda x. \tag{4.51}$$

The number λ is called the *eigenvalue of the operator* $\mathcal{A}$. We say that the vector x corresponds to (is associated with) the eigenvalue λ. The pair of the eigenvalue and the eigenvector associated with it is also called the *eigenpair of the operator* $\mathcal{A}$.

Let (x, λ) be an eigenpair of the operator $\mathcal{A}$. Then $\mathcal{A}\alpha x = \lambda \alpha x$ for each $\alpha \in \mathbb{C}$, i.e., the one-dimensional subspace of the space $\mathbf{X}$ spanned by a single eigenvector of the operator $\mathcal{A}$ is invariant under $\mathcal{A}$.

Let λ be an eigenvalue of the operator $\mathcal{A}$. The kernel of the operator $\mathcal{A} - \lambda I$ is called the *eigenspace* of $\mathcal{A}$ corresponding to the eigenvalue λ and is denoted by L_λ. It is clear that $L_\lambda \neq \{0\}$. Every nonzero element of L_λ is an eigenvector of the operator $\mathcal{A}$ corresponding to the eigenvalue λ.

Let us give some examples of operators that have eigenvectors.

1. Every nonzero element of the space $\mathbf{X}_n$ is an eigenvector corresponding to the eigenvalue $\lambda = 0$ of the null operator.
2. Consider the operator αI, where $\alpha \in \mathbb{C}$. Every nonzero element of $\mathbf{X}_n$ is an eigenvector of this operator corresponding to the eigenvalue $\lambda = \alpha$.
3. Let the space $\mathbf{X}$ be a direct sum of subspaces L and M. As usual, we denote by $\mathcal{P} : \mathbf{X} \rightarrow L$ the projection operator onto the subspace L in parallel with the subspace M. Then the following relations hold: $\mathcal{P}x = x$ for every $x \in L$ and $\mathcal{P}x = 0$ for every $x \in M$, i.e., all nonzero elements of L are eigenvectors of the operator $\mathcal{P}$ corresponding to the eigenvalue $\lambda = 1$, and all nonzero elements of M are eigenvectors of $\mathcal{P}$ corresponding to the eigenvalue $\lambda = 0$.

If the linear space $\mathbf{X}_n$ is real, then there exist linear operators $\mathcal{A} : \mathbf{X}_n \rightarrow \mathbf{X}_n$ that do not have any eigenvectors. For example, the linear operator $\mathcal{Q}$ (see Section 4.2.1) does not have any eigenvectors in the real space $\mathbf{X}_2$. This follows immediately from the fact that the operator Q does not have any nontrivial invariant subspaces.

Theorem 4.14 *Every operator A acting in the complex space $\mathbf{X}_n$ has eigenvectors.*

Proof It is enough to prove that there exists a complex number λ such that the linear equation

$$(A - \lambda I)x = 0 \tag{4.52}$$

has a nontrivial solution. Let A_e be the matrix of the operator A with respect to a basis $\mathcal{E}_n$ in the space $\mathbf{X}_n$. Consider the equation

$$\det(A_e - \lambda I) = 0. \tag{4.53}$$

It is easy to see that $\det(A_e - \lambda I)$ is a polynomial in λ of degree n. Thus equation (4.53) has n roots. Every root λ of Eq. (4.53) is an eigenvalue of the operator A. Indeed,

$$(A_e - \lambda I)\xi = 0 \tag{4.54}$$

is a homogeneous system of linear equations with a singular matrix. Hence this system has a nontrivial solution ξ. Then the vector $x = \mathcal{E}_n \xi$ is evidently nonzero, and it is a solution of equation (4.52). □

The proof of the next corollary is left to the reader.

Corollary 4.1 *Suppose that a linear operator A acts on the complex space $\mathbf{X}_n$, and $L \neq \{0\}$ is an invariant subspace of A. Then the operator A has an eigenvector $x \in L$.*

The linear operators A and B acting on the linear space $\mathbf{X}$ are called *permutable* if $AB = BA$.

Lemma 4.3 *Suppose that A, B are permutable transformations of the linear space $\mathbf{X}$, and $L_\lambda \subset \mathbf{X}$ is an eigenspace of the operator A. Then L_λ is an invariant subspace of the operator B.*

Proof Suppose that $x \in L_\lambda$. Then $Ax = \lambda x$. Hence $BAx = \lambda Bx$. By assumption, we have $BA = AB$. Therefore, $ABx = \lambda Bx$, which means that $Bx \in L_\lambda$. □

The polynomial $\det(A - \lambda I)$ is called the *characteristic polynomial of the matrix A*. The equation $\det(A - \lambda I) = 0$ is called the *characteristic equation of the matrix A*. The roots of the characteristic polynomial are called the *characteristic values (eigenvalues) of the matrix A*. The set of all characteristic values of the matrix A is called the *spectrum of A* and is denoted by $\sigma(A)$.

As established in the proof of Theorem 4.14, for every $\lambda \in \sigma(A)$ there exists a nonzero vector $x \in \mathbb{C}^n$ such that

$$Ax = \lambda x.$$

This vector x is called the *eigenvector of the matrix A* corresponding to the characteristic value λ of the matrix.

Theorem 4.15 *Similar matrices have the same characteristic polynomials and therefore the same characteristic values.*

Proof Let T be a nonsingular matrix. By definition of matrix similarity, the matrix $B = T^{-1}AT$ is similar to the matrix A. Then for every $\lambda \in \mathbb{C}$ we have

$$B - \lambda I = T^{-1}AT - \lambda I = T^{-1}(A - \lambda I)T.$$

Since $\det(T^{-1}) = 1/\det(T)$, we see that $\det(B - \lambda I) = \det(A - \lambda I)$. □

The matrices of the same operator $\mathcal{A} : \mathbf{X}_n \to \mathbf{X}_n$ are similar with respect to different bases (see Section 4.1.5, p. 101); hence the characteristic polynomial and its roots do not depend on the choice of basis in $\mathbf{X}_n$. Thus the characteristic polynomial of the matrix of a linear operator is called *the characteristic polynomial of the operator*, and equation (4.53) is called the *characteristic equation of the operator $\mathcal{A}$*.

The characteristic values of the matrix of an operator are called the *characteristic values of the operator*. Therefore, they are invariants of the operator. The set of all characteristic values of the operator $\mathcal{A}$ is called the *spectrum of the operator* and is denoted by $\sigma(\mathcal{A})$.

As follows from the proof of Theorem 4.14, for every operator acting on the complex space $\mathbf{X}_n$, the concepts of characteristic values and eigenvalues are equivalent, and for such operators, the corresponding terms are used as synonyms.

Every operator acting in the space $\mathbf{X}_n$ has no more than n distinct eigenvalues.

Theorem 4.16 (Cayley–Hamilton[7] theorem). *Let*

$$P_n(\lambda) = \lambda^n + a_{n-1}\lambda^{n-1} + \cdots + a_0 \tag{4.55}$$

be the characteristic polynomial of an operator $\mathcal{A}$ acting on the space $\mathbf{X}_n$. Then

$$P_n(\mathcal{A}) = \mathcal{A}^n + a_{n-1}\mathcal{A}^{n-1} + \cdots + a_0 I = 0. \tag{4.56}$$

Proof Let A be the matrix of the operator $\mathcal{A}$ with respect to a basis. Using formula (1.104), p. 40, we get $(A - \lambda I)\widetilde{(A - \lambda I)} = P_n(\lambda)I$ for all $\lambda \in \mathbb{C}$. Obviously, each element of the matrix $\widetilde{(A - \lambda I)}$ is a polynomial of λ of degree no greater than $n - 1$. Therefore, we can write

$$\widetilde{(A - \lambda I)} = \lambda^{n-1}C_{n-1} + \lambda^{n-2}C_{n-2} + \cdots + C_0,$$

where $C_0, C_1, \ldots, C_{n-1}$ are some square matrices of order n, i.e.,

---

[7] Arthur Cayley (1821–1895) was a British mathematician. William Rowan Hamilton (1805–1865) was an Irish physicist and mathematician.

$$P_n(\lambda)I = (A - \lambda I)(\lambda^{n-1}C_{n-1} + \lambda^{n-2}C_{n-2} + \cdots + C_0) \quad \text{for all} \quad \lambda \in \mathbb{C}. \quad (4.57)$$

Equating all coefficients with the same power of λ on both sides of equality (4.57), we obtain

$$\begin{aligned}
AC_0 &= a_0 I, \\
AC_1 - C_0 &= a_1 I, \\
AC_2 - C_1 &= a_2 I, \\
&\cdots\cdots\cdots\cdots\cdots \\
AC_{n-1} - C_{n-2} &= a_{n-1} I, \\
-C_{n-1} &= \quad I.
\end{aligned} \qquad (4.58)$$

Now we multiply the first equality in (4.58) by I, premultiply the second equality by A, premultiply the third equality by A^2, and so on. We premultiply the last equality in (4.58) by A^n. After that, we add together all obtained equalities and get $P_n(A) = 0$, which is equivalent to $P_n(\mathcal{A}) = 0$. $\qquad\square$

The next corollary follows from the Cayley–Hamilton theorem. This corollary plays an important role in applications, for instance in mechanics.

Corollary 4.2 *If the operator $\mathcal{A} : \mathbf{X}_n \to \mathbf{X}_n$ is invertible, then there exists a polynomial Q_{n-1} of degree no greater than $n - 1$ such that $\mathcal{A}^{-1} = Q_{n-1}(\mathcal{A})$.*

The proof of Corollary 4.2 is left to the reader.

Theorem 4.17 *Let $\lambda_1, \lambda_2, \ldots, \lambda_p$ be distinct eigenvalues of the operator $\mathcal{A} : \mathbf{X}_n \to \mathbf{X}_n$. Denote by $x_1, x_2, \ldots, x_p$ the eigenvectors of the operator $\mathcal{A}$ such that $\mathcal{A}x_k = \lambda_k x_k$, $k = 1, 2, \ldots, p$. Then the vectors $x_1, x_2, \ldots, x_p$ are linearly independent.*

Proof Suppose contrary to the assertion of the theorem that the set of vectors x_1, $x_2, \ldots, x_p$ is linearly dependent. Without loss of generality, it can be assumed that the vectors $x_1, x_2, \ldots, x_r, r < p$, form a maximal linearly independent subset of this set. Denote by L_r the subspace of the linear space $\mathbf{X}_n$ spanned by $x_1, x_2, \ldots, x_r$. The subspace L_r is invariant under $\mathcal{A}$ and has dimension r. Let $\mathcal{A}_{L_r}$ be the restriction of the operator $\mathcal{A}$ to L_r. Then $\lambda_1, \lambda_2, \ldots, \lambda_r$ are the eigenvalues of the operator $\mathcal{A}_{L_r}$, and they are distinct. The nonzero vector x_{r+1} belongs to L_r, and we see also that $\mathcal{A}_{L_r}x_{r+1} = \mathcal{A}x_{r+1} = \lambda_{r+1}x_{r+1}$, i.e., λ_{r+1} is an eigenvalue of the operator $\mathcal{A}_{L_r}$, but the operator $\mathcal{A}_{L_r}$ acts on a space of dimension r. Therefore, it cannot have more than r distinct eigenvalues. $\qquad\square$

It follows from the above that if all eigenvalues of the operator $\mathcal{A}$ are distinct, then the corresponding eigenvectors $x_k, k = 1, 2, \ldots, n$, form a basis of the space $\mathbf{X}_n$. We have

$$\mathcal{A}x_k = \lambda_k x_k, \quad k = 1, 2, \ldots, n,$$

and hence the matrix of the operator $\mathcal{A}$ with respect to the basis $\{x_k\}_{k=1}^n$ is diagonal. The eigenvalues $\lambda_k, k = 1, 2, \ldots, n$, form the main diagonal of this matrix.

For example, let us calculate all the eigenvalues and all the eigenvectors of the matrix

$$A = \begin{pmatrix} 4 & -5 & 7 \\ 1 & -4 & 9 \\ -4 & 0 & 5 \end{pmatrix}.$$

The characteristic equation has the form

$$\begin{vmatrix} 4 - \lambda & -5 & 7 \\ 1 & -4 - \lambda & 9 \\ -4 & 0 & 5 - \lambda \end{vmatrix} = 0.$$

Calculating this determinant, we get

$$\lambda^3 - 5\lambda^2 + 17\lambda - 13 = 0. \tag{4.59}$$

Evidently, $\lambda = 1$ is a root of Eq. (4.59). It is easy to see that

$$\lambda^3 - 5\lambda^2 + 17\lambda - 13 = (\lambda - 1)(\lambda^2 - 4\lambda + 13).$$

The equation $\lambda^2 - 4\lambda + 13 = 0$ has two roots: $\lambda = 2 \pm 3i$. Therefore,

$$\lambda_1 = 1, \quad \lambda_2 = 2 + 3i, \quad \lambda_3 = 2 - 3i$$

are all the eigenvalues of the matrix A. The coordinates of the eigenvector corresponding to λ_1 are the solution of the homogeneous system of linear equations

$$3x_1 - 5x_2 + 7x_3 = 0, \tag{4.60}$$
$$x_1 - 5x_2 + 9x_3 = 0, \tag{4.61}$$
$$-4x_1 + 4x_3 = 0. \tag{4.62}$$

We have $\begin{vmatrix} 3 & -5 \\ 1 & -5 \end{vmatrix} \neq 0$. Hence the rank of the matrix of system (4.60)–(4.62) is equal to two, and this system has only one linearly independent solution. Take $x_3 = 1$ and obtain x_1, x_2 as a solution of system (4.60), (4.61). We get $x_1 = 1, x_2 = 2$. Thus the vector $(1, 2, 1)$ is a solution of the system of equations (4.60)–(4.62). Therefore, the set of all eigenvectors corresponding to the eigenvalue $\lambda_1 = 1$ is the set of vectors having the form $c(1, 2, 1)$, where c is an arbitrary nonzero complex number.

The coordinates of the eigenvector corresponding to λ_2 are the solution of the homogeneous system of linear equations

$$(2 - 3i)x_1 - 5x_2 + 7x_3 = 0, \tag{4.63}$$
$$x_1 - (6 + 3i)x_2 + 9x_3 = 0, \tag{4.64}$$
$$-4x_1 + (3 - 3i)x_3 = 0. \tag{4.65}$$

We have $\begin{vmatrix} 2-3i & -5 \\ 1 & -(6+3i) \end{vmatrix} \neq 0$. Hence the coordinates of an eigenvector are the

solution of system (4.63), (4.64) for $x_3 = 1$. We get $x_1 = (3-3i)/4$, $x_2 = (5-3i)/4$. Therefore, the set of all eigenvectors corresponding to the eigenvalue λ_2 is the set of vectors having the form $c(3-3i, 5-3i, 4)$, where c is an arbitrary nonzero complex number. Analogous calculations show that the set of all eigenvectors corresponding to the eigenvalue λ_3 is the set of vectors having the form $c(3+3i, 5+3i, 4)$, where c is an arbitrary nonzero complex number.

In this example all the eigenvalues are distinct and the corresponding eigenvectors form a basis of the space $\mathbb{C}^3$. This can be seen also from the fact that the determinant

$$\begin{vmatrix} 1 & 2 & 1 \\ 3-3i & 5-3i & 4 \\ 3+3i & 5+3i & 4 \end{vmatrix},$$

which is composed of the coordinates of the eigenvectors, is not equal to zero.

If the characteristic polynomial of the operator $\mathcal{A}$ has multiple roots, then the number of corresponding linearly independent eigenvectors can be less than n, and these eigenvectors are not a basis of the space $\mathbf{X}_n$.

Now let us calculate all eigenvalues and all eigenvectors of the matrix

$$A = \begin{pmatrix} 2 & -1 & 2 \\ 5 & -3 & 3 \\ -1 & 0 & -2 \end{pmatrix}.$$

The characteristic equation has the form $\lambda^3 + 3\lambda^2 + 3\lambda + 1 = 0$, and the numbers $\lambda_1 = \lambda_2 = \lambda_3 = -1$ are the roots of this equation. Therefore, we have the following system for the calculation of the coordinates of eigenvectors:

$$3x_1 - x_2 + 2x_3 = 0, \tag{4.66}$$
$$5x_1 - 2x_2 + 3x_3 = 0, \tag{4.67}$$
$$-x_1 - x_3 = 0. \tag{4.68}$$

The determinant $\begin{vmatrix} 3 & -1 \\ 5 & -2 \end{vmatrix}$ is not equal to zero. Hence the rank of the matrix of this system is equal to two, and the linear space of all solutions of system (4.66)–(4.68) is one-dimensional. It is easy to see that the vector $x = (1, 1, -1)$ is a solution of system (4.66)–(4.68). Thus the set of all eigenvectors of the matrix is the set of vectors having the form $c(1, 1, -1)$, where c is an arbitrary nonzero complex number. Clearly, in this example the eigenvectors of the matrix are not a basis of the space $\mathbb{C}^3$.

The dimension of the eigenspace of the operator $\mathcal{A}$ corresponding to the eigenvalue λ of this operator is called the *geometric multiplicity* of the eigenvalue λ. The multiplicity of λ as a root of the characteristic polynomial of the operator $\mathcal{A}$

is called the *algebraic multiplicity* of the eigenvalue λ. In general, these two concepts are different. If the term multiplicity is used without qualification in reference to an eigenvalue, it usually means the algebraic multiplicity. We shall follow this convention.

Theorem 4.18 *For every operator $\mathcal{A}$ acting on the finite-dimensional space $\mathbf{X}_n$ the geometric multiplicity of each eigenvalue is less than or equal to the algebraic multiplicity of that eigenvalue.*

Proof Let L_{λ_0} be the eigenspace of the operator $\mathcal{A}$ corresponding to an eigenvalue λ_0 of that operator and $\dim(L_{\lambda_0}) = m$. Denote by $f_1, f_2, \ldots, f_m$ a basis of the eigenspace L_{λ_0}. Extend this basis to a basis of the space $\mathbf{X}_n$ by some additional vectors $g_{m+1}, g_{m+2}, \ldots, g_n$. Since $\mathcal{A}f_k = \lambda_0 f_k, k = 1, 2, \ldots, m$, it follows that the matrix of the operator $\mathcal{A}$ with respect to this basis of $\mathbf{X}_n$ can be written as a block matrix (see Section 4.2.1):

$$\begin{pmatrix} \Lambda_0 & A_{12} \\ 0 & A_{22} \end{pmatrix}, \tag{4.69}$$

where Λ_0 is a diagonal $m \times m$ matrix all of whose diagonal entries are equal to λ_0. Hence the characteristic polynomial of the operator $\mathcal{A}$ can be written as

$$\det(\mathcal{A} - \lambda I) = (\lambda - \lambda_0)^m Q_{n-m}(\lambda),$$

where $Q_{n-m}(\lambda)$ is a polynomial of degree $n - m$. Evidently, m cannot be greater than the multiplicity of the root λ_0 of the polynomial $\det(\mathcal{A} - \lambda I)$. $\qquad\square$

4.2.3 Diagonalizable Operators

We say that a linear operator $\mathcal{A} : \mathbf{X}_n \to \mathbf{X}_n$ is *diagonalizable* if there is a basis $\mathcal{E}_n$ of the space $\mathbf{X}_n$ consisting entirely of the eigenvectors of $\mathcal{A}$. The matrix of the operator $\mathcal{A}$ with respect to the basis $\mathcal{E}_n$ can be written in the form

$$A_e = \mathrm{diag}(\lambda_1, \ldots, \lambda_1, \lambda_2, \ldots, \lambda_2, \ldots, \lambda_k, \ldots, \lambda_k),$$

where each eigenvalue of the operator $\mathcal{A}$ is repeated according to its geometric multiplicity.

If $\mathcal{A} : \mathbf{X}_n \to \mathbf{X}_n$ is a diagonalizable operator, $\lambda_1, \lambda_2, \ldots, \lambda_k, k \leq n$, are distinct eigenvalues of this operator, and $L_{\lambda_i}, i = 1, 2, \ldots, k$, are corresponding eigenspaces of $\mathcal{A}$, then

$$\mathbf{X}_n = L_{\lambda_1} \oplus L_{\lambda_2} \oplus \cdots \oplus L_{\lambda_k}.$$

For $i = 1, 2, \ldots, k$, denote by $\mathcal{P}_i$ the operator that projects the space $\mathbf{X}_n$ onto the subspace L_{λ_i}. Then it is easy to see that

$$\mathcal{A}x = \lambda_1 \mathcal{P}_1 x + \lambda_2 \mathcal{P}_2 x + \cdots + \lambda_k \mathcal{P}_k x \quad \text{for all} \quad x \in \mathbf{X}_n,$$

i.e.,

$$\mathcal{A} = \lambda_1 \mathcal{P}_1 + \lambda_2 \mathcal{P}_2 + \cdots + \lambda_k \mathcal{P}_k. \tag{4.70}$$

Equality (4.70) is referred to as a *spectral resolution of the operator* $\mathcal{A}$.

Using (4.70) and (4.4), pp. 95, 122, we get $\mathcal{A}^j = \lambda_1^j \mathcal{P}_1 + \lambda_2^j \mathcal{P}_2 + \cdots + \lambda_k^j \mathcal{P}_k$ for every integer $j \geq 0$. Therefore, if Q_m is a polynomial of degree $m \geq 0$, then

$$Q_m(\mathcal{A}) = Q_m(\lambda_1)\mathcal{P}_1 + Q_m(\lambda_2)\mathcal{P}_2 + \cdots + Q_m(\lambda_k)\mathcal{P}_k. \tag{4.71}$$

Since the numbers $\lambda_1, \lambda_2, \ldots, \lambda_k$ are distinct, we can define Lagrange basis functions (see p. 29)

$$\Phi_j(\lambda) = \frac{(\lambda - \lambda_1)(\lambda - \lambda_2) \cdots (\lambda - \lambda_{j-1})(\lambda - \lambda_{j+1}) \cdots (\lambda - \lambda_k)}{(\lambda_j - \lambda_1)(\lambda_j - \lambda_2) \cdots (\lambda_j - \lambda_{j-1})(\lambda_j - \lambda_{j+1}) \cdots (\lambda_j - \lambda_k)}, \quad j = 1, 2, \ldots, k.$$

Then, taking into account (4.71), we obtain

$$\mathcal{P}_j = \Phi_j(\mathcal{A}), \quad j = 1, 2, \ldots, k. \tag{4.72}$$

Equation (4.72) is called *Sylvester's formula*.[8] This shows that for each $j = 1, 2, \ldots, k$, the projection operator $\mathcal{P}_j$ is a polynomial of degree $k - 1$, and the coefficients of this polynomial depend only on the eigenvalues of the operator $\mathcal{A}$.

Theorem 4.19 *A linear operator $\mathcal{A}$ is diagonalizable if and only if the geometric multiplicity of each eigenvalue λ of the operator $\mathcal{A}$ is equal to the algebraic multiplicity of λ.*

The proof of Theorem 4.19 is left to the reader.

Suppose that operators $\mathcal{A}$, $\mathcal{B}$ acting on the finite-dimensional space $\mathbf{X}_n$ are diagonalizable and have the same characteristic polynomial. The reader can easily prove that there exists a nonsingular operator $\mathcal{Q} : \mathbf{X}_n \to \mathbf{X}_n$ such that $\mathcal{B} = \mathcal{Q}\mathcal{A}\mathcal{Q}^{-1}$.

4.2.4 Invariants of an Operator

In this section we use in an essential way the following lemma.

Lemma 4.4 *For every $x \in \mathbb{C}$, the following expansion holds:*

---

[8] James Joseph Sylvester (1814–1897) was an English mathematician.

$$d(x) = \begin{vmatrix} a_{11} + x & a_{12} & \cdots & a_{1n} \\ a_{21} & a_{22} + x & \cdots & a_{2n} \\ \cdots & \cdots & \cdots & \cdots \\ a_{n1} & a_{n2} & \cdots & a_{nn} + x \end{vmatrix}$$

$$= x^n + c_1 x^{n-1} + c_2 x^{n-2} + \cdots + c_{n-1} x + c_n, \tag{4.73}$$

where

$$c_k = \sum_{1 \le p_1 < p_2 < \cdots < p_k \le n} \begin{vmatrix} a_{p_1, p_1} & a_{p_1, p_2} & \cdots & a_{p_1, p_k} \\ a_{p_2, p_1} & a_{p_2, p_2} & \cdots & a_{p_2, p_k} \\ \cdots & \cdots & \cdots & \cdots \\ a_{p_k, p_1} & a_{p_k, p_2} & \cdots & a_{p_k, p_k} \end{vmatrix}, \quad k = 1, 2, \ldots, n. \tag{4.74}$$

For each k, the right-hand side of (4.74) is the sum of all C_n^k determinants of the indicated form. These determinants are called the *principal minors* of order k of the matrix

$$A = \begin{pmatrix} a_{11} & a_{12} & \cdots & a_{1n} \\ a_{21} & a_{22} & \cdots & a_{2n} \\ \cdots\cdots\cdots\cdots\cdots \\ a_{n1} & a_{n2} & \cdots & a_{nn} \end{pmatrix}.$$

Note that $c_1 = a_{11} + a_{22} + \cdots + a_{nn}$, $c_n = \det A$.

Proof of Lemma 4.4. Denote by $a_1, a_2, \ldots, a_n$ the columns of the matrix A. Let us interpret the determinant of the matrix A as a function of its columns, i.e.,

$$\det A = \Delta(a_1, a_2, \ldots, a_n).$$

Then the function $d(x)$ in (4.73) can be represented in the form

$$d(x) = \Delta(a_1 + xi_1, a_2 + xi_2, \ldots, a_n + xi_n),$$

where by $i_1, i_2, \ldots, i_n$ we denote as usual the standard unit vectors in the space $\mathbb{C}^n$. Since the determinant is a linear function of its columns, it follows easily that

$$\begin{aligned} d(x) = {}& \Delta(a_1, a_2, \ldots, a_n) \\ &+ x(\Delta(i_1, a_2, \ldots, a_n) + \Delta(a_1, i_2, \ldots, a_n) + \cdots + \Delta(a_1, a_2, \ldots, a_{n-1}, i_n)) \\ &+ x^2(\Delta(i_1, i_2, a_3 \ldots, a_n) + \cdots + \Delta(a_1, a_2, \ldots, a_{n-2}, i_{n-1}, i_n)) \\ &\qquad\qquad + \cdots + x^n \Delta(i_1, i_2, \ldots, i_n). \end{aligned} \tag{4.75}$$

The multiplier of each x^k in (4.75) is the sum of C_n^k determinants, each of which is obtained from the determinant $\Delta(a_1, a_2, \ldots, a_n)$ by replacing k columns of Δ with the corresponding standard unit vectors. To complete the proof, it is enough to note that $\Delta(i_1, i_2, \ldots, i_n) = 1$ and that each principal minor of order $n - k$ of the matrix A is obtained from $\Delta(a_1, a_2, \ldots, a_n)$ by replacing k corresponding columns with the standard unit vectors with the same numbers. $\qquad\qquad\qquad\qquad\square$

The characteristic polynomial of the matrix A_e of the linear operator $\mathcal{A}$ is equal to $\det(\lambda I - A_e)$ up to sign. Let us expand this determinant as the polynomial in λ:

$$\det(\lambda I - A_e) = P_n(\lambda) = \lambda^n - \mathcal{I}_1\lambda^{n-1} + \mathcal{I}_2\lambda^{n-2} + \cdots + (-1)^n\mathcal{I}_n. \qquad (4.76)$$

As we have noted in Section 4.2.2, the coefficients of the polynomial P_n are *invariants of the operator* $\mathcal{A}$. All of them are functions of entries of the matrix A_e, but they are invariant under every transformation of the basis. In this connection the following notation is used: $\mathcal{I}_k = \mathcal{I}_k(\mathcal{A})$, $k = 1, 2, \ldots, n$. Using (4.73) and (4.74), we get the following representations for the invariants $\mathcal{I}_k(\mathcal{A})$ of the operator $\mathcal{A}$ by entries of the matrix A_e:

$$\mathcal{I}_k(\mathcal{A}) = \sum_{1 \le i_1 < i_2 < \cdots < i_k \le n} \begin{vmatrix} a^e_{i_1,i_1} & a^e_{i_1,i_2} & \cdots & a^e_{i_1,i_k} \\ a^e_{i_2,i_1} & a^e_{i_2,i_2} & \cdots & a^e_{i_2,i_k} \\ \cdots & \cdots & \cdots & \cdots \\ a^e_{i_k,i_1} & a^e_{i_k,i_2} & \cdots & a^e_{i_k,i_k} \end{vmatrix}, \quad k = 1, 2, \ldots, n. \qquad (4.77)$$

In particular,

$$\mathcal{I}_1(\mathcal{A}) = a^e_{11} + a^e_{22} + \cdots + a^e_{nn}, \quad \mathcal{I}_n(\mathcal{A}) = \det A_e. \qquad (4.78)$$

Using Viète's formulas (see Section 1.1.4, p. 14), we have

$$a^e_{11} + a^e_{22} + \cdots + a^e_{nn} = \lambda_1 + \lambda_2 + \cdots + \lambda_n, \quad \det A_e = \lambda_1\lambda_2\cdots\lambda_n, \qquad (4.79)$$

where $\lambda_1, \lambda_2, \ldots, \lambda_n$ are the characteristic values of the operator $\mathcal{A}$. Generally, $\mathcal{I}_k(\mathcal{A})$ is the sum of all products of k distinct characteristic values of the operator $\mathcal{A}$.

Every square matrix $A = \{a_{ij}\}^n_{i,j=1}$ generates a linear operator in $\mathbb{C}^n$ defined by the matrix–vector multiplication rule (4.5), p. 95. Hence it is possible to assign the numbers $\mathcal{I}_k(A)$, $k = 1, 2, \ldots, n$ (calculated by formulas (4.77), where a^e_{ij} are replaced with a_{ij}), to each square matrix. Evidently, these numbers are invariant under similarity transformations, and therefore, they are called *invariants of the matrix A*.

Theorem 4.20 *Let $\mathcal{A}$ be an operator acting on a finite-dimensional space $\mathbf{X}_n$. Then there exists a positive number ε_0 such that if $|\varepsilon| < \varepsilon_0$ and $\varepsilon \ne 0$, then the operator $\mathcal{A} + \varepsilon I$ is invertible.*

The proof of Theorem 4.20 is left to the reader.

The number $\mathcal{I}_1(\mathcal{A}) = a^e_{11} + a^e_{22} + \cdots + a^e_{nn} = \lambda_1 + \lambda_2 + \cdots + \lambda_n$ is called the *trace of the operator* $\mathcal{A}$ and is denoted by $\mathrm{tr}(\mathcal{A})$.

The following equality holds:

$$\text{tr}(\alpha \mathcal{A} + \beta \mathcal{B}) = \alpha \text{tr}(\mathcal{A}) + \beta \text{tr}(\mathcal{B}). \tag{4.80}$$

Here $\mathcal{A}$, $\mathcal{B}$ are linear operators in a finite-dimensional linear space, and α, β are complex numbers. Let $\mathcal{A} : \mathbf{X}_n \to \mathbf{X}_m$, $\mathcal{B} : \mathbf{X}_m \to \mathbf{X}_n$. Then

$$\text{tr}(\mathcal{A}\mathcal{B}) = \text{tr}(\mathcal{B}\mathcal{A}). \tag{4.81}$$

Equality (4.80) follows immediately from the definition of the trace of a linear operator. Equality (4.81) is verified by direct calculation of the sums of diagonal elements of the matrices of operators defined on the left-hand side and right-hand side of (4.81).

Let $\mathcal{A}$, $\mathcal{B}$ be arbitrary linear operators acting on a finite-dimensional vector space. The reader can prove that the characteristic polynomials of the operators $\mathcal{A}\mathcal{B}$ and $\mathcal{B}\mathcal{A}$ coincide. Hint: if the operator $\mathcal{A}$ is nonsingular, the result follows from the similarity of the matrices of the operators $\mathcal{A}\mathcal{B}$ and $\mathcal{B}\mathcal{A}$. In the general case, it is useful to apply Theorem 4.20.

4.2.5 Invariant Subspaces of an Operator in a Real Space

Let $\mathcal{A}$ be a linear operator acting on the real space $\mathbf{X}_n$. Then the matrix A_e of the operator $\mathcal{A}$ is real with respect to every basis $\mathcal{E}_n$. The characteristic equation (4.53) for the matrix A_e is a polynomial equation with real coefficients. This equation has, generally speaking, both real and complex roots.

If λ is a real root of Eq. (4.53), then the system of equations

$$(A_e - \lambda I)\xi = 0 \tag{4.82}$$

has a nontrivial real solution ξ, and for $x = \mathcal{E}_n\xi$, the equality $\mathcal{A}x = \lambda x$ holds, i.e., x is an eigenvector of the operator $\mathcal{A}$. Therefore, all real characteristic values of the matrix A_e are eigenvalues of the operator $\mathcal{A}$.

If equation (4.53) has no real root, then the system of equations (4.82) has no nontrivial real solutions. Hence if all roots of equation (4.53) are complex, then the operator $\mathcal{A}$ does not have eigenvectors. Therefore, there can exist linear operators acting on a real space that have no one-dimensional invariant subspaces.

There is a two-dimensional invariant subspace of the operator $\mathcal{A}$ corresponding to every complex characteristic value of the matrix A_e. In fact, if $\lambda = \alpha + i\beta$ is a complex characteristic value of the matrix A_e, then $\det(A_e - \lambda I) = 0$, and the system of equations

$$(A_e - \lambda I)\xi = 0 \tag{4.83}$$

has a nontrivial complex solution $\xi = \zeta + i\eta$. Here ζ and η belong to $\mathbb{R}^n$. The matrix A_e is real, and thus writing system (4.83) in terms of the complex solution, we get

$$A_e\zeta + iA_e\eta = (\alpha + i\beta)(\zeta + i\eta) = \alpha\zeta - \beta\eta + i(\beta\zeta + \alpha\eta).$$

Equating the real and imaginary parts of the last equation, we obtain

$$A_e\zeta = \alpha\zeta - \beta\eta,$$

$$A_e\eta = \beta\zeta + \alpha\eta.$$

If $x = \mathcal{E}_n\zeta$ and $y = \mathcal{E}_n\eta$, then

$$\mathcal{A}x = \alpha x - \beta y, \tag{4.84}$$

$$\mathcal{A}y = \beta x + \alpha y. \tag{4.85}$$

Denote by L the subspace of the linear space $\mathbf{X}_n$ spanned by x and y. Suppose that $z \in L$, which means that $z = \gamma x + \delta y$ for some $\gamma, \delta \in \mathbb{R}$. Then $\mathcal{A}z \in L$. Indeed,

$$\mathcal{A}z = \gamma\mathcal{A}x + \delta\mathcal{A}y = \gamma(\alpha x - \beta y) + \delta(\beta x + \alpha y) =$$
$$= (\alpha\gamma + \beta\delta)x + (\alpha\delta - \beta\gamma)y \in L.$$

Therefore, L is an invariant subspace of the operator $\mathcal{A}$.

To complete the proof, the reader can show that the vectors x and y that satisfy relationships (4.84), (4.85) are linearly independent, i.e., the subspace L is two-dimensional.

The reader can easily prove that if a linear operator $\mathcal{A}$ acts on the real space $\mathbf{X}_n$ and a subspace $L_m \subset \mathbf{X}_n$ is invariant under $\mathcal{A}$ and has dimension $m \geq 2$, then the operator $\mathcal{A}$ has either a one-dimensional or a two-dimensional invariant subspace of the subspace L_m.

4.2.6 Nilpotent Operators

A linear operator $\mathcal{A}$ acting on a finite-dimensional space $\mathbf{X}_n$ is called *nilpotent* if $\mathcal{A}^q = 0$ for some integer $q \geq 1$. The smallest such q is called the *index of nilpotency* of the operator $\mathcal{A}$. The definition of a square *nilpotent matrix* is similar.

Using (4.16), p. 99, we see that $A_e^q = \mathcal{E}^{-1}\mathcal{A}^q\mathcal{E}$. Therefore, if an operator $\mathcal{A}$ is nilpotent, then its matrix with respect to every basis is nilpotent of the same index. Conversely, if the matrix of an operator is nilpotent, then the operator is nilpotent of the same index.

Theorem 4.21 *An operator $\mathcal{A}$ is nilpotent if and only if all its eigenvalues are equal to zero.*

Proof Let $\mathcal{A}$ be a nilpotent operator of index q and let (λ, x) be an eigenpair of the operator $\mathcal{A}$. Then $\mathcal{A}x = \lambda x$. Therefore, $\mathcal{A}^q x = \lambda^q x$. We have assumed that $\mathcal{A}^q = 0$, and hence $\lambda^q x = 0$, but $x \neq 0$. Thus $\lambda = 0$. Conversely, suppose that all eigenvalues of the operator $\mathcal{A}$ are equal to zero. Then the characteristic equation of the operator $\mathcal{A}$ has the form $\lambda^n = 0$, and by Theorem 4.16, p. 117, we get $\mathcal{A}^n = 0$. $\square$

The following corollary is obvious.

Corollary 4.3 *The index of nilpotency of a nilpotent operator acting on an n-dimensional space is less than or equal to n.*

Let $\mathcal{A} : \mathbf{X}_n \to \mathbf{X}_n$ be a nilpotent operator of index q. Then evidently, there exists a vector $x_0 \in \mathbf{X}_n$ such that $\mathcal{A}^{q-1} x_0 \neq 0$. The reader can easily prove that the vectors $x_0, \mathcal{A}x_0, \ldots, \mathcal{A}^{q-1} x_0$ are linearly independent.

4.2.7 The Triangular Form of the Matrix of an Operator

Theorem 4.22 *For each operator $\mathcal{A}$ acting on the complex space $\mathbf{X}_n$, there exists a basis such that the matrix of $\mathcal{A}$ with respect to this basis is triangular, the eigenvalues of $\mathcal{A}$ forming the main diagonal of this matrix.*

The proof of Theorem 4.22 is based on the following result.

Theorem 4.23 (**Schur's**[9] **theorem**). *Let A be an $n \times n$ matrix. Let $\lambda_1, \lambda_2, \ldots, \lambda_n$ be its characteristic values numbered in arbitrary order. Then there exists a unitary matrix U such that*

$$U^* A U = T, \tag{4.86}$$

where T is an upper triangular matrix of the form

$$T = \begin{pmatrix} \lambda_1 & t_{12} & \cdots & t_{1n} \\ 0 & \lambda_2 & \cdots & t_{2n} \\ \cdots & \cdots & \cdots & t_{n-1,n} \\ 0 & 0 & \cdots & \lambda_n \end{pmatrix}. \tag{4.87}$$

Proof Denote by u_1 an eigenvector of the matrix A corresponding to the eigenvalue λ_1. Every eigenvector is defined up to a scalar multiplier. Hence we can assume that $|u_1| = 1$.[10] Let us construct an orthonormal basis $\{u_k\}_{k=1}^n$ of the space $\mathbb{C}^n$ containing u_1 (see Section 3.2, p. 71). Denote by U_1 the matrix with columns consisting of the elements of the vectors $\{u_k\}_{k=1}^n$. Taking into account that $Au_1 = \lambda_1 u_1$ and $(u_k, u_1) = 0$ for $k = 2, 3, \ldots, n$, we get

---

[9]Issai Schur (1875–1941) was a German mathematician.

[10]In this subsection we use only the standard inner product on $\mathbb{C}^n$.

$$U_1^* A U_1 = \begin{pmatrix} \lambda_1 & * \\ 0 & A_1 \end{pmatrix}. \tag{4.88}$$

The right-hand side of this equality is a block 2×2 matrix. The first diagonal block of this matrix consists of the number λ_1 only. The second diagonal block is a square matrix of order $n - 1$. The block in location $(2, 1)$ is a zero $(n - 1)$-dimensional column. The block in location $(1, 2)$ is an $(n - 1)$-dimensional row with nonzero, generally speaking, elements. Analogous notations will be used in the proof below. The matrix $U_1^* A U_1$ is similar to A; hence (see Theorem 4.15, p. 117)

$$\sigma(U_1^* A U_1) = \sigma(A).$$

Using (4.88) and expanding the determinant $\det(\lambda I - U_1^* A U_1)$ in terms of the first column, we get $\sigma(U_1^* A U_1) = \lambda_1 \cup \sigma(A_1)$. Therefore,

$$\sigma(A_1) = \{\lambda_2, \ldots, \lambda_n\}.$$

By analogy with U_1, we can construct a unitary matrix U_2 such that

$$U_2^* A_1 U_2 = \begin{pmatrix} \lambda_2 & * \\ 0 & A_2 \end{pmatrix}. \tag{4.89}$$

Let

$$V_2 = \begin{pmatrix} 1 & 0 \\ 0 & U_2 \end{pmatrix}.$$

Then V_2 is a unitary matrix of order n. By elementary calculations,

$$V_2^* U_1^* A U_1 V_2 = \begin{pmatrix} \lambda_1 & * & * \\ 0 & \lambda_2 & * \\ 0 & 0 & A_2 \end{pmatrix}.$$

Continuing this process, we can construct unitary matrices $V_3, \ldots, V_{n-1}$ such that the matrix

$$V_{n-1}^* \cdots V_2^* U_1^* A U_1 V_2 \cdots V_{n-1}$$

is an upper triangular matrix with the numbers $\lambda_1, \lambda_2, \ldots, \lambda_n$ on the main diagonal. Let $U = U_1 V_2 \cdots V_{n-1}$. Then U is a unitary matrix, since it is represented as a product of unitary matrices (see Section 1.2.7, p. 48), and $U^* = V_{n-1}^* \cdots V_2^* U_1^*$. Therefore, the matrix $T = U^* A U$ has the form (4.87). $\square$

Arguing as above, we see that there exists a unitary matrix V such that

$$V^* A V = L,$$

where L is a lower triangular matrix, and all characteristic values of A form the main diagonal of L.

Remark 4.1 From the proof of Schur's theorem, we see that if a matrix A is real and all its characteristic values (and hence all eigenvectors) are real, then the matrix U in (4.86) can be chosen as a real unitary matrix, in other words, as an orthogonal matrix.

Proof of Theorem 4.22. Let $\mathcal{A}$ be a linear operator acting on the space $\mathbf{X}_n$, and let $\mathcal{F}_n = \{f_k\}_{k=1}^n$ be an arbitrarily chosen basis in $\mathbf{X}_n$. Then $\mathcal{A}\mathcal{F}_n = \mathcal{F}_n A_f$, where A_f is the matrix of the operator $\mathcal{A}$ with respect to this basis (see (4.11), p. 99). Using Schur's theorem, we see that there exists a unitary matrix U such that $A_f = UTU^*$, where T is a matrix of the form (4.87), $\lambda_1, \lambda_2, \ldots, \lambda_n$ are characteristic values of A_f (i.e., eigenvalues of $\mathcal{A}$). Hence, $\mathcal{A}\mathcal{F}_n = \mathcal{F}_n UTU^*$. Therefore, we get $\mathcal{A}\mathcal{F}_n U = \mathcal{F}_n UT$. Let $\mathcal{E}_n = \mathcal{F}_n U$. Then $\mathcal{A}\mathcal{E}_n = \mathcal{E}_n T$. Thus T is the matrix of the operator $\mathcal{A}$ with respect to the basis $\mathcal{E}_n$. □

Remark 4.2 If the space $\mathbf{X}_n$ is unitary and the basis $\mathcal{F}_n$ is orthonormal, then the basis $\mathcal{E}_n$ is also orthonormal.

The matrix T that appears in Theorem 4.22 is usually called the *Schur form* of the matrix of the operator. The following simplification of the matrix of an operator is often useful.

Theorem 4.24 *Let A be a square matrix of order n, let λ_1, λ_2, ..., λ_k be distinct characteristic values of A having multiplicities $n_1, n_2, \ldots, n_k$, respectively, where $n_1 + n_2 + \cdots + n_k = n$. Then there exists a nonsingular matrix S such that*

$$S^{-1}AS = \begin{pmatrix} T_1 & & & 0 \\ & T_2 & & \\ & & \ddots & \\ 0 & & & T_k \end{pmatrix} \tag{4.90}$$

is a block diagonal matrix, and each diagonal block T_i is an upper triangular matrix of order n_i, $i = 1, 2, \ldots, k$. All diagonal elements of each block T_i are identical and equal to λ_i.

Proof At the first stage, using Schur's theorem, we transform the matrix A to an upper triangular matrix T by a unitary similarity transformation. We can order the characteristic values on the diagonal of the triangular matrix T according to the statement of the theorem, i.e., the first n_1 numbers of the diagonal are equal to λ_1, the next n_2 numbers are equal to λ_2, and so on. To complete the proof, it is enough to transform the matrix T to the form (4.90) by a similarity transformation. We construct this transformation as a result of a sequence of elementary similarity transformations. Let us write the above-mentioned upper triangular matrix T in the block form

$$T = \begin{pmatrix} T_{11} & T_{12} \\ 0 & T_{22} \end{pmatrix}.$$

Here T_{11} is an upper triangular matrix of order n_1 each of whose diagonal elements are equal to λ_1, and T_{22} is an upper triangular matrix of order $n - n_1$ none of whose diagonal elements is equal to λ_1. We consider the upper triangular matrices of the form

$$\begin{pmatrix} I_{n_1} & P \\ 0 & I_{n-n_1} \end{pmatrix}, \quad \begin{pmatrix} I_{n_1} & -P \\ 0 & I_{n-n_1} \end{pmatrix}, \tag{4.91}$$

where I_{n_1}, I_{n-n_1} are the identity matrices of orders n_1, $n - n_1$, respectively. Using elementary calculations, we see that matrices (4.91) are mutually inverse. Now we find the matrix P such that the following equality holds:

$$\begin{pmatrix} I_{n_1} & P \\ 0 & I_{n-n_1} \end{pmatrix} \begin{pmatrix} T_{11} & T_{12} \\ 0 & T_{22} \end{pmatrix} \begin{pmatrix} I_{n_1} & -P \\ 0 & I_{n-n_1} \end{pmatrix} = \begin{pmatrix} T_{11} & 0 \\ 0 & T_{22} \end{pmatrix}. \tag{4.92}$$

Clearly, the equality (4.92) is true if the matrix P is a solution of the following equation [11]:

$$P T_{22} - T_{11} P = -T_{12}. \tag{4.93}$$

Equation (4.93) is a system of linear algebraic equations for the elements of the matrix P. Let us check that the corresponding homogeneous system

$$P T_{22} - T_{11} P = 0 \tag{4.94}$$

has the trivial solution only. Indeed, we can rewrite Eq. (4.94) in the equivalent form $P(T_{22} - \lambda_1 I_{n-n_1}) = (T_{11} - \lambda_1 I_{n_1}) P$. Obviously, the matrix $T_{22} - \lambda_1 I_{n-n_1}$ is nonsingular. Hence, $P = (T_{11} - \lambda_1 I_{n_1}) P (T_{22} - \lambda_1 I_{n-n_1})^{-1}$. Therefore,

$$P = (T_{11} - \lambda_1 I_{n_1})^q P ((T_{22} - \lambda_1 I_{n-n_1})^{-1})^q$$

for all integers $q \geq 1$. By construction, the matrix $T_{11} - \lambda_1 I_{n_1}$ is nilpotent, and there exists an integer $q \geq 1$ such that $(T_{11} - \lambda_1 I_{n_1})^q = 0$. Hence $P = 0$. Thus a transformation of the form (4.92) exists. In the following steps of the proof we construct analogous transformations, which successively reduce the orders of the blocks of the transformed matrix. Arguing as in the proof of Theorem 4.23, as a result we get relationship (4.90).   □

Using Theorem 4.24, the reader can easily prove the following result.

Theorem 4.25 *For each operator $\mathcal{A}$ acting on the space $\mathbf{X}_n$ there exist invariant subspaces M and N such that $\mathbf{X}_n = M \oplus N$, the restriction of the operator $\mathcal{A}$ to the subspace M is a nilpotent operator, and the restriction of the operator $\mathcal{A}$ to the subspace N is an invertible operator.*

---

[11] Equation (4.93) is a *Sylvester equation*; see, for example, [23, p. 170], [59, 66].

Below is a useful example of applications of Schur's theorem.

Theorem 4.26 *Let $A = \{a_{ij}\}_{i,j=1}^n$ be an arbitrary square matrix. For every $\varepsilon > 0$, there exists an invertible diagonalizable matrix $A_m = \{a_{ij}^{(m)}\}_{i,j=1}^n$ such that*

$$\max_{1 \leq i,j \leq n} |a_{ij} - a_{ij}^{(m)}| \leq \varepsilon. \tag{4.95}$$

Proof Using Theorem 4.23, we represent the matrix A in the form $A = UTU^*$, where U is a unitary matrix and T is an upper triangular matrix. Without loss of generality we can assume that the diagonal elements of the matrix T are ordered in the following way:

$$\lambda_1, \lambda_1, \ldots, \lambda_1, \lambda_2, \lambda_2, \ldots, \lambda_2, \ldots, \lambda_k, \lambda_k, \ldots, \lambda_k.$$

Here each characteristic value of the matrix A is repeated according to its multiplicity. Denote by T_m the upper triangular matrix that differs from the matrix T only by the diagonal elements, which are equal to the following numbers:

$$\lambda_1 + 1/m, \ \lambda_1 + 1/2m, \ \ldots, \ \lambda_1 + 1/n_1 m,$$
$$\lambda_2 + 1/m, \ \lambda_2 + 1/2m, \ \ldots, \ \lambda_2 + 1/n_2 m, \ \ldots,$$
$$\lambda_k + 1/m, \ \lambda_k + 1/2m, \ \ldots, \ \lambda_k + 1/n_k m,$$

where n_i is the multiplicity of λ_i, $i = 1, 2, \ldots, k$, $m \geq 1$. Let $A_m = UT_mU^*$. It is easy to see that all diagonal elements of the matrix T_m for large enough m are nonzero and distinct. Therefore, all characteristic values of the matrix A_m are nonzero and distinct. This means that for all large enough m, the matrices A_m are invertible and diagonalizable. Further, $A - A_m = U(T - T_m)U^*$. Hence

$$\max_{1 \leq i,j \leq n} |a_{ij} - a_{ij}^{(m)}| \leq c/m,$$

where c is a constant depending only on n and on the elements of the matrix U. Thus for every $\varepsilon > 0$, we get (4.95) for large enough m. $\square$

We can say that the sequence of the matrices $\{A_m\}_{m=1}^\infty$ *converges* to the matrix A.

4.2.8 The Real Schur Form

Theorem 4.27 *Let A be a real square matrix of order $n \geq 1$. There exists an orthogonal matrix Q such that $A = Q^T TQ$, where T is a block upper triangular matrix. The diagonal blocks of the matrix T are square matrices of order one or two. The set of all characteristic values of the second-order blocks coincides with the set of all complex characteristic values of the matrix A.*

Proof If all characteristic values of the matrix A are real, then this theorem immediately follows from Theorem 4.23 (see Remark 4.1). Therefore, we assume that among the characteristic values of the matrix A there exists a complex number $\lambda = \alpha + i\beta$. As we have seen in Section 4.2.5, p. 126, a two-dimensional invariant subspace of the matrix A in the space $\mathbb{R}^n$ corresponds to this number. Let q_1, q_2 be a basis of this subspace. Suppose that this basis is orthonormal with respect to the standard inner product on the space $\mathbb{R}^n$. Then

$$Aq_1 = \alpha_{11}q_1 + \alpha_{21}q_2, \quad Aq_2 = \alpha_{12}q_1 + \alpha_{22}q_2. \tag{4.96}$$

The matrices

$$\begin{pmatrix} \alpha & \beta \\ -\beta & \alpha \end{pmatrix} \text{ and } T_{11} = \begin{pmatrix} \alpha_{11} & \alpha_{12} \\ \alpha_{21} & \alpha_{22} \end{pmatrix}$$

are similar to the matrices of the same operator with respect to the different bases. Therefore, they have the same characteristic values λ and $\bar{\lambda}$. We can join the vectors q_1, q_2 to some vectors to complete a basis $\{q_k\}_{k=1}^n$ of the space $\mathbb{R}^n$. Denote by Q the matrix whose columns are the vectors of this basis. Using equalities (4.96) and the orthonormality of the vectors q_1 and q_2, we get

$$Q^T A Q = \begin{pmatrix} T_{11} & T_{12} \\ 0 & T_{22} \end{pmatrix}.$$

The completion of the proof is similar to the corresponding argumentation in the proof of Schur's theorem. □

4.3 Operators on Unitary Spaces

4.3.1 *Linear Functionals*

Let $\mathbf{X}$ be a complex linear space. A linear map l from $\mathbf{X}$ into the one-dimensional space $\mathbf{Y} = \mathbb{C}$ is called a *linear functional* (*linear form*) on $\mathbf{X}$. We point out that a complex number $l(x)$ uniquely corresponds to each vector $x \in \mathbf{X}$.

Theorem 4.28 (**Riesz**[12]). *Let $\mathbf{X}_n$ be a finite-dimensional unitary space, and l a linear functional on $\mathbf{X}_n$. Then there exists a unique vector $u \in \mathbf{X}_n$ such that*

$$l(x) = (x, u) \quad \text{for all} \quad x \in \mathbf{X}_n. \tag{4.97}$$

Proof First, we make sure that exactly one vector u is determined by the linear functional l. Suppose that there is an additional vector $u_1 \in \mathbf{X}_n$ such that

---

[12]Frigyes Riesz (1880–1956) was a Hungarian mathematician.

$$l(x) = (x, u_1) \quad \text{for all} \quad x \in \mathbf{X}_n. \tag{4.98}$$

Then subtracting term by term (4.97) from (4.98), we get $(x, u_1 - u) = 0$ for all $x \in \mathbf{X}_n$. If we put $x = u_1 - u$ in the last equality, then $(u_1 - u, u_1 - u) = 0$, i.e., $u_1 = u$. Let us prove the existence of a vector u defined by (4.97). Let $\{e_k\}_{k=1}^n$ be an orthonormal basis for $\mathbf{X}_n$, and let $x = \sum_{k=1}^n \xi_k e_k$. Since the functional l is linear, we see that

$$l(x) = \sum_{k=1}^n \xi_k l(e_k). \tag{4.99}$$

Put $u = \sum_{k=1}^n \overline{l(e_k)} e_k$. Using (3.19), p. 80, we get $l(x) = (x, u)$ for each $x \in \mathbf{X}_n$. $\qquad \square$

4.3.2 The Adjoint Operator

Let $\mathbf{X}_n$, $\mathbf{Y}_m$ be unitary spaces, and let $\mathcal{A} : \mathbf{X}_n \to \mathbf{Y}_m$ be a linear operator. A linear operator $\mathcal{A}^* : \mathbf{Y}_m \to \mathbf{X}_n$ is called the *adjoint* of $\mathcal{A}$ if

$$(\mathcal{A}x, y) = (x, \mathcal{A}^* y) \quad \text{for all } x \in \mathbf{X}_n \text{ and for all } y \in \mathbf{Y}_m, \tag{4.100}$$

where on the left-hand side of (4.100) we use an inner product on the space $\mathbf{Y}_m$, and on the right-hand side of (4.100) we use an inner product on $\mathbf{X}_n$.

For every linear operator $\mathcal{A} : \mathbf{X}_n \to \mathbf{Y}_m$ there exists an adjoint of $\mathcal{A}$. Indeed, for each fixed $y \in \mathbf{Y}_m$, the inner product $(\mathcal{A}x, y)$ is a functional on $\mathbf{X}_n$. This functional is linear, since the operator $\mathcal{A}$ is linear and the inner product in the first argument is linear too. Using Riesz's theorem, we see that there exists a unique vector $g \in \mathbf{X}_n$ such that

$$(\mathcal{A}x, y) = (x, g) \quad \text{for all} \quad x \in \mathbf{X}_n.$$

Therefore a vector $g \in \mathbf{X}_n$ uniquely corresponds to vector $y \in \mathbf{Y}_m$, and thus the map from $\mathbf{Y}_m$ to $\mathbf{X}_n$ is defined. Denote this map by $\mathcal{A}^*$. Then we can write

$$(\mathcal{A}x, y) = (x, \mathcal{A}^* y) \quad \text{for all } x \in \mathbf{X}_n \text{ and for all } y \in \mathbf{Y}_m. \tag{4.101}$$

We assert that the map $\mathcal{A}^*$ is linear. In fact, if $y_1, y_2 \in \mathbf{Y}_m$, $\alpha, \beta \in \mathbb{C}$, then

$$(\mathcal{A}x, \alpha y_1 + \beta y_2) = \bar{\alpha}(\mathcal{A}x, y_1) + \bar{\beta}(\mathcal{A}x, y_2)$$
$$= \bar{\alpha}(x, \mathcal{A}^* y_1) + \bar{\beta}(x, \mathcal{A}^* y_2) = (x, \alpha \mathcal{A}^* y_1 + \beta \mathcal{A}^* y_2). \tag{4.102}$$

On the other hand, by the definition of $\mathcal{A}^*$ we have

$$(\mathcal{A}x, \alpha y_1 + \beta y_2) = (x, \mathcal{A}^*(\alpha y_1 + \beta y_2)). \tag{4.103}$$

In (4.102) and (4.103) the vector $x \in \mathbf{X}_n$ is arbitrary. Therefore, comparing (4.102) and (4.103), we see that

$$\mathcal{A}^*(\alpha y_1 + \beta y_2) = \alpha \mathcal{A}^* y_1 + \beta \mathcal{A}^* y_2.$$

The reader can easily prove that for each linear operator $\mathcal{A}$ there exists exactly one adjoint of $\mathcal{A}$.

By the definition of the adjoint operator we obviously have

$$(\mathcal{A}^*)^* = \mathcal{A}.$$

It is easy to see that

$$(\mathcal{A}\mathcal{B})^* = \mathcal{B}^* \mathcal{A}^* \tag{4.104}$$

and

$$(\alpha \mathcal{A} + \beta \mathcal{B})^* = \bar{\alpha} \mathcal{A}^* + \bar{\beta} \mathcal{B}^* \tag{4.105}$$

for all operators $\mathcal{A}$, $\mathcal{B}$ and α, $\beta \in \mathbb{C}$, and also that if an operator $\mathcal{A}$ is invertible, then the adjoint $\mathcal{A}^*$ is invertible too, and

$$(\mathcal{A}^*)^{-1} = (\mathcal{A}^{-1})^*. \tag{4.106}$$

If the space $\mathbf{Y}_m$ is unitary, then there exists a useful formula for calculating the matrix of an operator $\mathcal{A} : \mathbf{X}_n \rightarrow \mathbf{Y}_m$. Let $\mathcal{E}_n$ be a basis of $\mathbf{X}_n$, $\mathcal{Q}_m$ a basis of $\mathbf{Y}_m$, and $G_q = \{(q_j, q_i)\}_{i,j=1}^m$ the Gram matrix corresponding to the basis $\mathcal{Q}_m$. Consider the matrix

$$G_{\mathcal{A}} = \begin{pmatrix} (\mathcal{A}e_1, q_1) & (\mathcal{A}e_2, q_1) & \dots & (\mathcal{A}e_n, q_1) \\ (\mathcal{A}e_1, q_2) & (\mathcal{A}e_2, q_2) & \dots & (\mathcal{A}e_n, q_2) \\ \dots\dots\dots\dots\dots\dots\dots\dots\dots\dots\dots\dots\dots\dots\dots \\ (\mathcal{A}e_1, q_m) & (\mathcal{A}e_2, q_m) & \dots & (\mathcal{A}e_n, q_m) \end{pmatrix}.$$

Then

$$G_{\mathcal{A}} = G_q A_{eq}. \tag{4.107}$$

Indeed, calculating the inner products of both sides of equation (4.9), p. 98, with q_l, we get

$$(\mathcal{A}e_i, q_l) = \sum_{j=1}^m a_{ji}^{(eq)}(q_j, q_l), \quad i = 1, 2, \dots, n, \ l = 1, 2, \dots, m. \tag{4.108}$$

Formula (4.107) is the matrix form of equality (4.108). The Gram matrix G_q is nonsingular, since Q_m is a basis. Therefore,

$$A_{eq} = G_q^{-1} G_A. \tag{4.109}$$

If the basis Q_m is orthonormal, then $G_q = I$ and

$$A_{eq} = G_A. \tag{4.110}$$

If both spaces $\mathbf{Y}_m$ and $\mathbf{X}_n$ are unitary and $A^* : \mathbf{Y}_m \to \mathbf{X}_n$ is the adjoint of the operator A, then as above,

$$G_{A^*} = G_e A_{qe}^*, \tag{4.111}$$

where G_e is the Gram matrix of the basis $\mathcal{E}_n$, A_{qe}^* is the matrix of the operator A^* with respect to the bases $Q_m, \mathcal{E}_n$, and

$$G_{A^*} = \begin{pmatrix} (A^*q_1, e_1) & (A^*q_2, e_1) & \dots & (A^*q_m, e_1) \\ (A^*q_1, e_2) & (A^*q_2, e_2) & \dots & (A^*q_m, e_2) \\ \dots\dots\dots\dots\dots\dots\dots\dots\dots\dots\dots\dots\dots \\ (A^*q_1, e_n) & (A^*q_2, e_n) & \dots & (A^*q_m, e_n) \end{pmatrix}.$$

Since $(A^*q_i, e_j) = (q_i, Ae_j) = \overline{(Ae_j, q_i)}$, we see that the matrices G_A and G_{A^*} are mutually adjoint. Hence, using (4.107), we get $G_{A^*} = (A_{eq})^* G_q$, and because of (4.111), the following equality holds:

$$A_{qe}^* = G_e^{-1}(A_{eq})^* G_q. \tag{4.112}$$

Formula (4.112) shows the relationship between the matrices of operators A and A^*. In particular, if the bases $\mathcal{E}_n$ and Q_m are orthonormal, then the matrices of the operators A and A^* are mutually adjoint.

4.3.3 Linear Equations in Unitary Spaces

Theorem 4.29 *Let* $\mathbf{X}_n, \mathbf{Y}_m$ *be unitary spaces. Every linear operator* $A : \mathbf{X}_n \to \mathbf{Y}_m$ *determines an orthogonal decomposition*

$$\mathbf{Y}_m = \mathrm{Ker}(A^*) \oplus \mathrm{Im}(A) \tag{4.113}$$

of the space $\mathbf{Y}_m$.

Proof Suppose that $y \in \mathrm{Im}(A)$, $y_1 \in \mathrm{Ker}(A^*)$. Then there exists $x \in \mathbf{X}_n$ such that $y = Ax$; hence

$$(y, y_1) = (\mathcal{A}x, y_1) = (x, \mathcal{A}^* y_1) = 0,$$

i.e., y is orthogonal to $\text{Ker}(\mathcal{A}^*)$. If the vector $y \in \mathbf{Y}_m$ is orthogonal to $\text{Im}(\mathcal{A})$, then $(y, \mathcal{A}x) = 0$ for every $x \in \mathbf{X}_n$, and $(\mathcal{A}^* y, x) = 0$ for every $x \in \mathbf{X}_n$. Therefore, $\mathcal{A}^* y = 0$, i.e., $y \in \text{Ker}(\mathcal{A}^*)$. These arguments show that $\text{Im}(\mathcal{A})$ is the orthogonal complement of $\text{Ker}(\mathcal{A}^*)$. Thus, using Theorem 3.12, p. 91, we see that equality (4.113) holds. $\square$

Obviously, the following decomposition holds too:

$$\mathbf{X}_n = \text{Ker}(\mathcal{A}) \oplus \text{Im}(\mathcal{A}^*). \tag{4.114}$$

Theorem 4.30 *Suppose that a linear operator $\mathcal{A}$ maps a finite-dimensional unitary space $\mathbf{X}_n$ into a finite-dimensional unitary space $\mathbf{Y}_m$. Then*

$$\text{rank}(\mathcal{A}) = \text{rank}(\mathcal{A}^*). \tag{4.115}$$

Proof The operator $\mathcal{A}$ realizes an isomorphism between $\text{Im}(\mathcal{A}^*)$ and $\text{Im}(\mathcal{A})$. Indeed, using (4.114), for every $x \in \mathbf{X}_n$, we get $\mathcal{A}x = \mathcal{A}x_1$, where $x_1 \in \text{Im}(\mathcal{A}^*)$, i.e., each element of $\text{Im}(\mathcal{A})$ is the image of an element of $\text{Im}(\mathcal{A}^*)$. Suppose that $\mathcal{A}x' = \mathcal{A}x''$ for distinct elements x', x'' of $\text{Im}(\mathcal{A}^*)$. Then $\mathcal{A}(x' - x'') = 0$, and $(x' - x'') \in \text{Ker}(\mathcal{A})$. Since $\text{Im}(\mathcal{A}^*)$ is a linear subspace, we see that $(x' - x'') \in \text{Im}(\mathcal{A}^*)$. Using (4.114) again, we have $x' - x'' = 0$. Hence the finite-dimensional spaces $\text{Im}(\mathcal{A})$ and $\text{Im}(\mathcal{A}^*)$ are isomorphic. Thus (see Theorem 4.3, p. 97) they have the same dimension. $\square$

An immediate consequence of Theorem 4.29 is the following.

Theorem 4.31 **(Fredholm's theorem)**. *Let $\mathbf{X}_n$, $\mathbf{Y}_m$ be unitary spaces and $\mathcal{A} : \mathbf{X}_n \to \mathbf{Y}_m$ a linear operator. A linear equation*

$$\mathcal{A}x = y \tag{4.116}$$

has a solution if and only if the vector y is orthogonal to each solution z of the homogeneous equation $\mathcal{A}^* z = 0$.

Note here that Theorems 4.12, p. 109, and 4.13, p. 109, can be proved on the basis of Fredholm's theorem.

Using the decomposition (4.114) and arguing as in the proof of Theorem 4.30, the reader can prove that if the linear equation (4.116) is solvable, then the set of all its solutions contains a unique element x_0 with minimal length. The element x_0 is called the *normal solution* of equation (4.116). It is easy to see that the vector x_0 belongs to $\text{Im}(\mathcal{A}^*)$.

4.3.4 Pseudosolutions. The Tikhonov Regularization Method

Suppose that a linear operator $\mathcal{A}$ maps a unitary space $\mathbf{X}_n$ into a unitary space $\mathbf{Y}_m$. Let y be a fixed element of $\mathbf{Y}_m$ and x an arbitrary element of $\mathbf{X}_n$. Then the vector $\mathcal{A}x - y$ is called the *residual* corresponding to the Eq. (4.116). The real-valued function

$$F(x) = |\mathcal{A}x - y|^2$$

defined on the space $\mathbf{X}_n$ is called the *residual functional*. If $\mathcal{A}x \neq y$, i.e., the vector x is not a solution of equation (4.116), then $F(x) > 0$. It is important to find a vector x that minimizes the residual functional.

A vector $x \in \mathbf{X}_n$ minimizing the residual functional is called a *pseudosolution* of equation (4.116).[13] If equation (4.116) is solvable, then each of its solutions is a pseudosolution.

A pseudosolution of equation (4.116) exists for every $y \in \mathbf{Y}_m$. Indeed, using (4.113), we can write $y = y_1 + y_0$, where $y_1 \in \text{Im}(\mathcal{A})$, $y_0 \in \text{Ker}(\mathcal{A}^*)$. Then for every $x \in \mathbf{X}_n$, the vector $\mathcal{A}x - y_1$ belongs to $\text{Im}(\mathcal{A})$. Hence

$$F(x) = |\mathcal{A}x - y_1|^2 + |y_0|^2.$$

Evidently, the minimum of the function F is equal to $|y_0|^2$ and is achieved at the vector x that is a solution of the equation

$$\mathcal{A}x = y_1. \tag{4.117}$$

Equation (4.117) is solvable, since $y_1 \in \text{Im}(\mathcal{A})$. The normal solution x_0 of (4.117) is called the *normal pseudosolution* of equation (4.116).

We can write $\mathcal{A}x_0 = \mathcal{P}y$, where $\mathcal{P}$ is the operator of the orthogonal projection of $\mathbf{Y}_m$ onto $\text{Im}(\mathcal{A})$. As we have seen in the proof of Theorem 4.30, the operator $\mathcal{A}$ realizes an isomorphism between $\text{Im}(\mathcal{A}^*)$ and $\text{Im}(\mathcal{A})$. Therefore, there exists a linear operator $\mathcal{A}^+ : \mathbf{Y}_m \to \mathbf{X}_n$ such that $x_0 = \mathcal{A}^+ y$, where x_0 is the normal pseudosolution of the equation $\mathcal{A}x = y$ for a given $y \in \mathbf{Y}_m$. The operator $\mathcal{A}^+$ is called the *pseudoinverse* of $\mathcal{A}$. It is easy to see that if the operator $\mathcal{A}$ is invertible, then $\mathcal{A}^+ = \mathcal{A}^{-1}$.

For every $y \in \mathbf{Y}_m$, the equation

$$\mathcal{A}^* \mathcal{A}x = \mathcal{A}^* y \tag{4.118}$$

is solvable, and every solution of (4.118) is a pseudosolution of equation (4.116). Indeed, since $\mathcal{A}^* y_0 = 0$, we see that equation (4.118) is equivalent to the equation

$$\mathcal{A}^*(\mathcal{A}x - y_1) = 0. \tag{4.119}$$

---

[13]The problem of calculating the pseudosolution is often called the *linear least squares problem*.

Equation (4.119) is solvable because each solution of (4.117) is a solution of equation (4.119). Conversely, if x is a solution of equation (4.119), then the vector $Ax - y_1$ belongs to $\mathrm{Ker}(\mathcal{A}^*)$, and by (4.113), it is orthogonal to $\mathrm{Im}(\mathcal{A})$. On the other hand, $Ax - y_1 \in \mathrm{Im}(\mathcal{A})$. Thus $Ax - y_1 = 0$, i.e., x is a solution of equation (4.117).

We say that the original equation (4.116) is reduced to Eq. (4.118) by the *Gauss transformation*. The Gauss transformation of a linear equation leads to a solvable equation.

The Tikhonov[14] regularization method can be used for the practical construction of the normal pseudosolution of equation (4.116). Along with the residual functional, consider the so-called *regularizing functional* (the *Tikhonov functional*):

$$F_\alpha(x) = F(x) + \alpha|x|^2 = |Ax - y|^2 + \alpha|x|^2. \tag{4.120}$$

Here α is a positive number called the *regularization parameter*.

Theorem 4.32 *For every positive α, there exists a unique vector x_α minimizing the functional F_α on the space $\mathbf{X}_n$. The limit of x_α as $\alpha \to 0$ exists and is equal to the normal pseudosolution x_0 of Eq. (4.116).*

Proof Consider the following equation:

$$A^*Ax + \alpha x = A^*y. \tag{4.121}$$

Equation (4.121) has a unique solution $x_\alpha \in \mathbf{X}_n$ for every $y \in \mathbf{Y}_m$. Indeed, if x is a solution of the homogeneous equation corresponding to (4.121), then calculating the inner products of both sides of the corresponding homogeneous equation with x yields $|Ax|^2 + \alpha|x|^2 = 0$. Hence $x=0$, since $\alpha > 0$. Using the equality $A^*y = A^*Ax_\alpha + \alpha x_\alpha$, by elementary calculations we obtain

$$F_\alpha(x) = (\mathcal{B}_\alpha(x - x_\alpha), x - x_\alpha) + (y, y) - (\mathcal{B}_\alpha x_\alpha, x_\alpha),$$

where $\mathcal{B}_\alpha = A^*A + \alpha I$. Since $(\mathcal{B}_\alpha(x - x_\alpha), x - x_\alpha) > 0$ for every $x \neq x_\alpha$, we see that x_α is the unique minimum point of the functional F_α. Therefore,

$$F_\alpha(x_\alpha) = |Ax_\alpha - y_1|^2 + |y_0|^2 + \alpha|x_\alpha|^2 \leq |Ax - y_1|^2 + |y_0|^2 + \alpha|x|^2 \quad \text{for all} \quad x \in \mathbf{X}_n.$$

If we take here $x = x_0$, then

$$|Ax_\alpha - y_1|^2 + \alpha|x_\alpha|^2 \leq \alpha|x_0|^2. \tag{4.122}$$

This implies that $|x_\alpha| \leq |x_0|$, and hence by the Bolzano–Weierstrass theorem (see a calculus textbook), we can find a sequence $\alpha_k \to 0$ and a vector $x_* \in \mathbf{X}_n$ such

---

[14] Andrei Nikolaevich Tikhonov (1906–1993) was a Soviet and Russian mathematician.

that $x_{\alpha_k} \to x_*$ as $\alpha_k \to 0$. From (4.122), it follows that $A x_* = y_1$. The normal pseudosolution is unique. Therefore, $x_* = x_0$. Using the uniqueness of the normal pseudosolution again, we see that $x_\alpha \to x_0$ as α tends to zero in any manner. $\qquad\square$

4.3.5 Self-adjoint and Skew-Hermitian Operators

A linear operator $\mathcal{A} : \mathbf{X}_n \to \mathbf{X}_n$ is called *self-adjoint (Hermitian)* if $\mathcal{A}^* = \mathcal{A}$, in other words, if

$$(\mathcal{A}x, y) = (x, \mathcal{A}y) \quad \text{for all} \quad x, y \in \mathbf{X}_n. \tag{4.123}$$

A linear operator $\mathcal{A} : \mathbf{X}_n \to \mathbf{X}_n$ is called *skew-Hermitian* if $\mathcal{A}^* = -\mathcal{A}$, i.e.,

$$(\mathcal{A}x, y) = -(x, \mathcal{A}y) \quad \text{for all} \quad x, y \in \mathbf{X}_n. \tag{4.124}$$

The reader can easily prove that if an operator $\mathcal{A}$ is self-adjoint, then the inner product $(\mathcal{A}x, x)$ is real for all $x \in \mathbf{X}_n$; if an operator $\mathcal{A}$ is skew-Hermitian, then the inner product $(\mathcal{A}x, x)$ is imaginary for all $x \in \mathbf{X}_n$.

Since the matrices of operators $\mathcal{A}$ and $\mathcal{A}^*$ with respect to any orthonormal bases are mutually adjoint (see Section 4.3.2), we see that the matrix of a self-adjoint operator with respect to an orthonormal basis is Hermitian, and the matrix of a skew-Hermitian operator is skew-Hermitian.

Theorem 4.33 *If the matrix of an operator $\mathcal{A}$ with respect to an orthonormal basis is Hermitian, then the operator $\mathcal{A}$ is self-adjoint; if the matrix of an operator $\mathcal{A}$ with respect to an orthonormal basis is skew-Hermitian, then the operator $\mathcal{A}$ is skew-Hermitian.*

The proof of Theorem 4.33 is left to the reader.

Theorem 4.34 *Each operator of the orthogonal projection[15] is self-adjoint.*

Proof Let $\mathcal{P}$ be an operator of the orthogonal projection of a unitary space $\mathbf{X}$ onto a subspace $L \subset \mathbf{X}$, and let x and y be arbitrary elements of the space $\mathbf{X}$. By definition, $x = \mathcal{P}x + x_2$, $y = \mathcal{P}y + y_2$, where the vectors x_2 and y_2 are orthogonal to L. Hence, $(\mathcal{P}x, y) = (\mathcal{P}x, \mathcal{P}y)$. Similarly, we have $(x, \mathcal{P}y) = (\mathcal{P}x, \mathcal{P}y)$. Therefore, $(\mathcal{P}x, y) = (y, \mathcal{P}x)$. $\qquad\square$

Theorem 4.35 *If an operator $\mathcal{A}$ is self-adjoint and $\mathcal{A}^2 = \mathcal{A}$, then the operator $\mathcal{A}$ is the operator of the orthogonal projection.*

The proof of Theorem 4.35 is left to the reader.

Arguing exactly as in Section 1.2.7, p. 48, it is easy to verify that every operator can be uniquely represented in the form

---

[15]See the definition on p. 94.

$$\mathcal{A} = \mathcal{H}_1 + i\mathcal{H}_2, \tag{4.125}$$

where i is the imaginary unit, and

$$\mathcal{H}_1 = \frac{1}{2}(\mathcal{A} + \mathcal{A}^*), \quad \mathcal{H}_2 = \frac{1}{2i}(\mathcal{A} - \mathcal{A}^*)$$

are self-adjoint operators.

Theorem 4.36 *Let $\mathcal{A}$ be a linear operator acting on the unitary space $\mathbf{X}_n$. If*

$$(\mathcal{A}x, x) = 0 \quad \text{for all} \quad x \in \mathbf{X}_n, \tag{4.126}$$

then $\mathcal{A} = 0$.

Proof Assume first that $\mathcal{A}$ is self-adjoint. Then for all $x, y \in \mathbf{X}_n$, the following equality holds: $(\mathcal{A}(x + y), x + y) = (\mathcal{A}x, x) + (\mathcal{A}y, y) + 2\mathrm{Re}(\mathcal{A}x, y)$. Combining this with (4.126), we get $\mathrm{Re}(\mathcal{A}x, y) = 0$. The last equality holds for every $y \in \mathbf{X}_n$. Hence we can replace y by iy, but $\mathrm{Re}(\mathcal{A}x, iy) = \mathrm{Im}(\mathcal{A}x, y)$. Therefore, $(\mathcal{A}x, y) = 0$ for all $x, y \in \mathbf{X}_n$. If we put $y = \mathcal{A}x$, we obtain $|\mathcal{A}x| = 0$ for every $x \in \mathbf{X}_n$, i.e., $\mathcal{A} = 0$. So the theorem is true for self-adjoint operators. Let now $\mathcal{A}$ be an arbitrary operator. If $(\mathcal{A}x, x) = 0$, then using (4.125) and considering the self-adjointness of $\mathcal{H}_1$, $\mathcal{H}_2$, we get $(\mathcal{H}_1 x, x) = 0$, $(\mathcal{H}_2 x, x) = 0$ for every $x \in \mathbf{X}_n$. Hence, using the self-adjointness of the operators $\mathcal{H}_1$ and $\mathcal{H}_2$ again, we see that $\mathcal{H}_1, \mathcal{H}_2 = 0$. $\square$

Lemma 4.5 *Let $\mathcal{A}$ be a linear operator acting on the unitary space $\mathbf{X}_n$. If the inner product $(\mathcal{A}x, x)$ is real for all $x \in \mathbf{X}_n$, then $\mathcal{A}$ is self-adjoint.*

Proof If $(\mathcal{A}x, x)$ is real, then $(\mathcal{A}^*x, x) = (x, \mathcal{A}x) = (\mathcal{A}x, x)$, and $((\mathcal{A}^* - \mathcal{A})x, x) = 0$ for every $x \in \mathbf{X}_n$. Therefore, using Theorem 4.36, we see that $\mathcal{A}^* - \mathcal{A} = 0$. $\square$

The following lemma is proved similarly.

Lemma 4.6 *Let $\mathcal{A}$ be a linear operator acting on the unitary space $\mathbf{X}_n$. If the inner product $(\mathcal{A}x, x)$ is imaginary for all $x \in \mathbf{X}_n$, then $\mathcal{A}$ is skew-Hermitian.*

Thus we have the following theorem.

Theorem 4.37 *Let $\mathcal{A}$ be a linear operator acting on the unitary space $\mathbf{X}_n$. The operator $\mathcal{A}$ is self-adjoint if and only if the inner product $(\mathcal{A}x, x)$ is real for all vectors $x \in \mathbf{X}_n$; $\mathcal{A}$ is skew-Hermitian if and only if the inner product $(\mathcal{A}x, x)$ is imaginary for all $x \in \mathbf{X}_n$.*

It follows from Theorem 4.37 that all eigenvalues of each self-adjoint operator are real and all eigenvalues of each skew-Hermitian operator are imaginary. Indeed, if (x, λ) is an eigenpair of the operator $\mathcal{A}$, then $(\mathcal{A}x, x) = \lambda(x, x)$.

The reader can easily prove now that the determinant of every self-adjoint operator is real.

4.3.6 Positive Definite and Positive Semidefinite Operators

A self-adjoint operator $\mathcal{A} : \mathbf{X}_n \to \mathbf{X}_n$ is called *positive semidefinite* if

$$(\mathcal{A}x, x) \geq 0 \;\; \text{for all } x \in \mathbf{X}_n. \tag{4.127}$$

A self-adjoint operator $\mathcal{A} : \mathbf{X}_n \to \mathbf{X}_n$ is called *positive definite* if

$$(\mathcal{A}x, x) > 0 \;\; \text{for all nonzero } x \in \mathbf{X}_n. \tag{4.128}$$

A Hermitian matrix A of order n is called *positive semidefinite* if

$$(Ax, x) = \sum_{i,j=1}^{n} a_{ij} x_j \bar{x}_i \geq 0 \;\; \text{for all } x \in \mathbb{C}^n. \tag{4.129}$$

A Hermitian matrix A of order n is called *positive definite* if

$$(Ax, x) = \sum_{i,j=1}^{n} a_{ij} x_j \bar{x}_i > 0 \;\; \text{for all nonzero } x \in \mathbb{C}^n. \tag{4.130}$$

In the rest of this subsection we give without proof some useful properties of positive definite operators and matrices. The proof of the following properties is left to the reader.

1. The equality $(x, y)_{\mathcal{A}} = (\mathcal{A}x, y)$ defines an inner product on the space $\mathbf{X}_n$ for every positive definite operator $\mathcal{A} : \mathbf{X}_n \to \mathbf{X}_n$.
2. For every operator $\mathcal{A} : \mathbf{X}_n \to \mathbf{X}_n$, the operator $\mathcal{A}^*\mathcal{A}$ is self-adjoint and positive semidefinite. If $\mathcal{A}$ is invertible, then $\mathcal{A}^*\mathcal{A}$ is positive definite.
3. Let $\mathcal{A}$ be a linear operator acting on the unitary space $\mathbf{X}_n$. If the operator $\mathcal{A} + \mathcal{A}^*$ is positive definite, then the operator $\mathcal{A}$ is nonsingular.
4. The matrix of a positive definite operator with respect to an orthonormal basis is positive definite.
5. All elements of the main diagonal of a positive definite matrix are positive.
6. The Gram matrix of a set of vectors in a unitary space is positive semidefinite.
7. The Gram matrix of a set of vectors is positive definite if and only if the set of vectors is linearly independent.

4.3.7 Unitary Operators

An operator $\mathcal{A} : \mathbf{X}_n \to \mathbf{X}_n$ is called *unitary* if

$$\mathcal{A}\mathcal{A}^* = \mathcal{A}^*\mathcal{A} = I. \tag{4.131}$$

The proof of the following properties of unitary operators is left to the reader.

1. An operator $\mathcal{A} : \mathbf{X}_n \to \mathbf{X}_n$ is unitary if and only if its matrix with respect to every orthonormal basis of $\mathbf{X}_n$ is unitary (see p. 49).
2. The modulus of the determinant of a unitary operator is equal to one.
3. The product of two unitary operators is a unitary operator.

If an operator $\mathcal{A}$ is unitary, then we have $(\mathcal{A}x, \mathcal{A}y) = (x, \mathcal{A}^*\mathcal{A}y) = (x, y)$ for all x, $y \in \mathbf{X}_n$, i.e., a unitary operator does not change the inner product of vectors. Hence it does not change the length of vectors.

Conversely, if a linear operator does not change the inner product of vectors in $\mathbf{X}_n$, then this operator is unitary. Indeed, taking into account the equality $(\mathcal{A}x, \mathcal{A}y) = (x, y)$, we obtain $(x, \mathcal{A}^*\mathcal{A}y) = (x, y)$. Since the last equality holds for all x, $y \in \mathbf{X}_n$, we see that

$$\mathcal{A}^*\mathcal{A} = I. \tag{4.132}$$

We prove that the equality $\mathcal{A}\mathcal{A}^* = I$ holds too. From (4.132) it follows that the operator $\mathcal{A}$ is invertible. Then, using left multiplication of both sides of equality (4.132), by $\mathcal{A}$ and then right multiplication by $\mathcal{A}^{-1}$, we obtain $\mathcal{A}\mathcal{A}^* = I$.

Now the reader can easily prove that if $|\mathcal{A}x| = |x|$ for all $x \in \mathbf{X}_n$, then the operator $\mathcal{A}$ is unitary.

Thus a linear operator $\mathcal{A} : \mathbf{X}_n \to \mathbf{X}_n$ is unitary if and only if it does not change the length of every vector in the space $\mathbf{X}_n$.

The modulus of every eigenvalue of a unitary operator is equal to one. Indeed, if $\mathcal{A}x = \lambda x, x \neq 0$, then since $|\mathcal{A}x| = |x|$ for every unitary operator (see Section 4.3.7), we get $|\lambda||x| = |\mathcal{A}x| = |x|$, i.e., $|\lambda| = 1$.

Let us point out the following useful corollary. Its proof is obvious.

Corollary 4.4 *All eigenvalues of a Hermitian matrix are real; all eigenvalues of a skew-Hermitian matrix are imaginary; the modulus of every eigenvalue of a unitary matrix is equal to one.*

4.3.8 Normal Operators

A linear operator $\mathcal{A}$ acting on a unitary space $\mathbf{X}_n$ is called *normal* if

$$\mathcal{A}\mathcal{A}^* = \mathcal{A}^*\mathcal{A}.$$

Evidently, self-adjoint operators, skew-Hermitian operators, and unitary operators are normal. An operator is normal if and only if its matrix with respect to every orthonormal basis of the space $\mathbf{X}_n$ is normal (see the definition of a normal matrix on p. 50).

Theorem 4.38 *Let $\mathcal{A} : \mathbf{X}_n \to \mathbf{X}_n$ be a normal operator. Then $\mathrm{Ker}(\mathcal{A}) = \mathrm{Ker}(\mathcal{A}^*)$.*

Proof Suppose that $\mathcal{A}x = 0$. Then

$$0 = (\mathcal{A}x, \mathcal{A}x) = (\mathcal{A}^*\mathcal{A}x, x) = (\mathcal{A}\mathcal{A}^*x, x) = (\mathcal{A}^*x, \mathcal{A}^*x),$$

and hence $\mathcal{A}^*x = 0$. The same calculations show that if $\mathcal{A}^*x = 0$, then $\mathcal{A}x=0$. □

Theorems 4.29 and 4.38, p. 136, immediately imply the following corollary.

Corollary 4.5 *Let* $\mathcal{A} : \mathbf{X}_n \to \mathbf{X}_n$ *be a normal operator. Then*

$$\mathbf{X}_n = \mathrm{Ker}(\mathcal{A}) \oplus \mathrm{Im}(\mathcal{A}) = \mathrm{Ker}(\mathcal{A}^*) \oplus \mathrm{Im}(\mathcal{A}^*), \quad \mathrm{Im}(\mathcal{A}) = \mathrm{Im}(\mathcal{A}^*).$$

Theorem 4.39 *Let* $\mathcal{A} : \mathbf{X}_n \to \mathbf{X}_n$ *be a normal operator and* (x, λ) *an eigenpair of* $\mathcal{A}$, *i.e.,* $\mathcal{A}x = \lambda x$. *Then* $(x, \bar{\lambda})$ *is an eigenpair of the operator* $\mathcal{A}^*$.

Proof It is obvious that if an operator $\mathcal{A}$ is normal, then for each $\lambda \in \mathbb{C}$, the operator $\mathcal{A} - \lambda I$ is normal too, and $(\mathcal{A} - \lambda I)^* = \mathcal{A}^* - \bar{\lambda}I$. If we combine this equality with Theorem 4.38, we get $\mathrm{Ker}(\mathcal{A} - \lambda I) = \mathrm{Ker}(\mathcal{A}^* - \bar{\lambda}I)$. □

Theorem 4.40 *Eigenvectors of a normal operator with distinct eigenvalues are mutually orthogonal.*

Proof Let $\mathcal{A}$ be a normal operator, and let $\mathcal{A}x = \lambda x$, $\mathcal{A}y = \mu y$, where $\lambda \neq \mu$. Then $\lambda(x, y) = (\mathcal{A}x, y) = (x, \mathcal{A}^*y)$. By Theorem 4.39, it follows that $\mathcal{A}^*y = \bar{\mu}y$; hence $(x, \mathcal{A}^*y) = \mu(x, y)$. Thus, $\lambda(x, y) = \mu(x, y)$, and $(x, y) = 0$, since $\lambda \neq \mu$. □

Theorem 4.41 *Let* $\mathcal{A}$ *be a linear operator acting on the space* $\mathbf{X}_n$. *There exists an orthonormal basis* $\{e_k\}_{k=1}^n \subset \mathbf{X}_n$ *such that* $\mathcal{A}e_k = \lambda_k e_k$, $k = 1, 2, \ldots, n$, *if and only if the operator* $\mathcal{A}$ *is normal.*

Proof Necessity The matrices of mutually adjoint operators with respect to an orthonormal basis are mutually adjoint (see Section 4.3.2, p. 133). Hence if

$$A_e = \mathrm{diag}(\lambda_1, \lambda_2, \ldots, \lambda_n)$$

is the matrix of the operator $\mathcal{A}$ with respect to the orthonormal basis $\{e_k\}_{k=1}^n$, then

$$A_e^* = \mathrm{diag}(\bar{\lambda}_1, \bar{\lambda}_2, \ldots, \bar{\lambda}_n)$$

is the matrix of the operator $\mathcal{A}^*$ with respect to the same basis. The matrix of the product of two operators is equal to the product of the matrices of those operators (see Section 4.1.5, p. 98), and diagonal matrices are permutable. Therefore,

$$(A^*A)_e = A_e^* A_e = A_e A_e^* = (AA^*)_e,$$

and thus $\mathcal{A}^*\mathcal{A} = \mathcal{A}\mathcal{A}^*$, i.e., the operator $\mathcal{A}$ is normal.

Sufficiency. Let (e_1, λ_1) be an eigenpair of $\mathcal{A}$. Suppose that $|e_1| = 1$. By Theorem 4.39, it follows that $(e_1, \bar{\lambda}_1)$ is an eigenpair of the operator $\mathcal{A}^*$. Denote by L_{n-1} the subspace of all vectors in $\mathbf{X}_n$ that are orthogonal to e_1. The subspace L_{n-1} is invariant under the operator $\mathcal{A}$. Indeed, if $x \in L_{n-1}$, i.e., $(x, e_1) = 0$, then we get $(\mathcal{A}x, e_1) = (x, \mathcal{A}^* e_1) = \lambda_1(x, e_1) = 0$. Therefore, using Corollary 4.1, p. 116, we see that there exist a *normalized*[16] vector $e_2 \in L_{n-1}$ and a number λ_2 such that $\mathcal{A}e_2 = \lambda_2 e_2$. Let now L_{n-2} be the subspace of all vectors in $\mathbf{X}_n$ that are orthogonal to both vectors e_1 and e_2. Arguing as above, we prove that there exist a normalized vector $e_3 \in L_{n-2}$ and a number λ_3 such that $\mathcal{A}e_3 = \lambda_3 e_3$. Continuing this process, we construct an orthonormal set of vectors $\{e_k\}_{k=1}^n \subset \mathbf{X}_n$ such that $\mathcal{A}e_k = \lambda_k e_k$, where $k = 1, 2, \ldots, n$. $\qquad\square$

Remark 4.3 Theorem 4.41 states that for every normal operator $\mathcal{A}$ there exists an orthonormal basis such that the matrix of $\mathcal{A}$ with respect to this basis is diagonal, and all eigenvalues of $\mathcal{A}$ form the main diagonal of this matrix. Thus every normal operator is diagonalizable (see Section 4.2.3, p. 121).

Remark 4.4 The following equivalent formulation of the last result is often useful. Let $\mathcal{A}$ be a normal operator acting on the space $\mathbf{X}_n$. Denote by $\lambda_1, \lambda_2, \ldots, \lambda_k, k \leq n$, all distinct eigenvalues of $\mathcal{A}$ and by $L_{\lambda_i}, i = 1, 2, \ldots, k$, all corresponding eigenspaces. Then

$$\mathbf{X}_n = L_{\lambda_1} \oplus L_{\lambda_2} \oplus \cdots \oplus L_{\lambda_k}, \tag{4.133}$$

$$\mathcal{A} = \lambda_1 \mathcal{P}_1 + \lambda_2 \mathcal{P}_2 + \cdots + \lambda_k \mathcal{P}_k, \tag{4.134}$$

where the sums in (4.133) are orthogonal and $\mathcal{P}_i$ is the operator of the orthogonal projection of the space $\mathbf{X}_n$ onto the subspace L_{λ_i} for $i = 1, 2, \ldots, k$.

The proof of the following corollary of Theorem 4.41 is left to the reader.

Corollary 4.6 *Let A be a real square matrix of order n such that $A^T A = A A^T$. Then there exist an orthonormal*[17] *set of vectors $\{\xi_k\}_{k=1}^n \subset \mathbb{C}^n$ and numbers $\lambda_1, \ldots, \lambda_n$ such that $A\xi_k = \lambda_k \xi_k, k = 1, 2, \ldots, n$. Moreover, if the number λ_k is real, then we can choose the corresponding real vector ξ_k.*

The proof of the following three propositions is left to the reader.

Proposition 4.1 *If all eigenvalues of a normal operator are real, then the operator is self-adjoint. If all eigenvalues of a normal operator are imaginary, then the operator is skew-Hermitian. If the modulus of each eigenvalue of a normal operator is equal to one, then the operator is unitary.*

Proposition 4.2 *Let $\mathcal{A}$ and $\mathcal{B}$ be normal operators whose characteristic polynomials are equal to each other. Then there exists a unitary operator $\mathcal{Q}$ such that $\mathcal{B} = \mathcal{Q}\mathcal{A}\mathcal{Q}^*$.*

---

[16] As usual, a vector x is called normalized if $|x| = 1$.

[17] With respect to the standard inner product on the space $\mathbb{C}^n$.

Proposition 4.3 *Let A be a normal operator, Q a unitary operator. Then the operator $\tilde{A} = QAQ^*$ is normal, and the following resolution holds:*

$$\tilde{A} = \lambda_1 \tilde{\mathcal{P}}_1 + \lambda_2 \tilde{\mathcal{P}}_2 + \cdots + \lambda_k \tilde{\mathcal{P}}_k. \tag{4.135}$$

Here $\lambda_1, \lambda_2, \ldots, \lambda_k$ are all distinct eigenvalues of A, and $\tilde{\mathcal{P}}_i = Q\mathcal{P}_i Q^*$ is the operator of the orthogonal projection of the space $\mathbf{X}_n$ onto the subspace QL_{λ_i}, where $i = 1, 2, \ldots, k$.

Theorem 4.42 *Normal operators A and B are permutable if and only if they have a common orthonormal basis that consists of their eigenvectors.*

Proof Sufficiency. Let $\{e_j\}_{k=1}^n$ be a common orthonormal basis that consists of the eigenvectors of the operators A and B, i.e., $Ae_k = \lambda_k e_k$ and $Be_k = \mu_k e_k$, where $k = 1, 2, \ldots, n$. Then $BAe_k = \lambda_k \mu_k e_k$, $ABe_k = \lambda_k \mu_k e_k$ for $k = 1, 2, \ldots, n$, i.e., for each vector of the basis, the values of the operators AB and BA coincide. Thus, $AB = BA$.

Necessity. Let us use representation (4.133) of the space $\mathbf{X}_n$ in the form of an orthogonal sum of the eigenspaces of the operator A corresponding to distinct eigenvalues of A. It follows from Lemma 4.3, p. 116, that every subspace L_{λ_i} is invariant under B. Since the operator B is normal, we see that in each L_{λ_i} there exists an orthonormal basis that consists of eigenvectors of the operator B. Clearly, the union of all such bases is a basis of the space $\mathbf{X}_n$, and by construction, all vectors of this basis are eigenvectors of the operator A. $\qquad\square$

4.3.9 The Root of a Positive Semidefinite Self-adjoint Operator

Theorem 4.43 *Let A be a positive semidefinite self-adjoint operator acting on a finite-dimensional unitary space X_n and let $k \geq 2$ be a given integer. Then there exists a unique positive semidefinite self-adjoint operator T such that $T^k = A$.*

The operator T is called the *kth root of the operator* A and is denoted by $A^{1/k}$ or by $\sqrt[k]{A}$.

Proof Since the operator A is self-adjoint, there exists an orthonormal basis $\{e_i\}_{i=1}^n$ consisting entirely of the eigenvectors of A. Let us denote the corresponding eigenvalues by $\lambda_1, \lambda_2, \ldots, \lambda_n$ and define the operator T by the action of this operator on the basis vectors:

$$T e_i = \sqrt[k]{\lambda_i}\, e_i, \quad i = 1, 2, \ldots, n.$$

All eigenvalues of a positive semidefinite operator are nonnegative, and hence we can assume that all numbers $\sqrt[k]{\lambda_i}$, $i = 1, 2, \ldots, n$, are nonnegative. Obviously, the operator T is self-adjoint and positive semidefinite. Moreover, we see that $T^k = A$, i.e., $T = A^{1/k}$. To complete the proof, we shall show that the kth root of the operator A is unique. For this purpose, we first establish that there exists a polynomial P_m of degree $m \leq n - 1$ such that $T = P_m(A)$. Indeed, let $\lambda_1, \lambda_2, \ldots, \lambda_r$, $r \leq n$, be all the distinct eigenvalues of the operator A. Then there exists a polynomial P_{r-1} of degree $r - 1$ such that $P_{r-1}(\lambda_i) = \sqrt[k]{\lambda_i}$, $i = 1, 2, \ldots, r$.[18] Hence

$$P_{r-1}(A)e_i = P_{r-1}(\lambda_i)e_i = \sqrt[k]{\lambda_i}\, e_i, \quad i = 1, 2, \ldots, n,$$

i.e., $P_{r-1}(A) = T$. Let U be an arbitrary positive semidefinite self-adjoint operator such that $U^k = A$. Then

$$TU = P_{r-1}(A)U = P_{r-1}(U^k)U = UP_{r-1}(U^k) = UT,$$

i.e., the operators T and U are permutable. Therefore, by Theorem 4.42, these operators have a common orthonormal basis that consists of their eigenvectors. We also denote this basis by $e_1, e_2, \ldots, e_n$ and write

$$Te_i = \mu_i e_i, \quad Ue_i = \tilde{\mu}_i e_i, \quad \mu_i, \tilde{\mu}_i \geq 0, \quad i = 1, 2, \ldots, n.$$

Hence

$$T^k e_i = \mu_i^k e_i, \quad U^k e_i = \tilde{\mu}_i^k e_i, \quad i = 1, 2, \ldots, n,$$

but $T^k = U^k$. Therefore, $\tilde{\mu}_i^k = \mu_i^k$, and $\tilde{\mu}_i = \mu_i$, $i = 1, \ldots, n$. Thus $U = T$. □

4.3.10   Congruent Hermitian Operators

Hermitian operators $A, B : \mathbf{X}_n \to \mathbf{X}_n$ are said to be *congruent* if there exists a nonsingular operator X such that $B = X^* A X$. Let $n_+(A)$ be the number of positive characteristic values of A, $n_-(A)$ the number of negative characteristic values of A, and $n_0(A)$ the number of zero characteristic values of A, all counting multiplicity. Since all characteristic values of a Hermitian operator are real, we see that $n_+(A) + n_-(A) + n_0(A) = n$. The *inertia* of A is the triple $(n_+(A), n_-(A), n_0(A))$.

Theorem 4.44 (**Sylvester's law of inertia**). *Hermitian operators A, B are congruent if and only if they have the same inertia.*

Proof Sufficiency. Let (n_+, n_-, n_0) be the inertia of the operator A and

$$Ae_k = \lambda_k(A)e_k, \quad k = 1, 2, \ldots, n, \tag{4.136}$$

---

[18]The polynomial P_{r-1} can be written in an explicit form, for example using the Lagrange interpolation formula (see p. 29).

where $e_k, k = 1, 2, \ldots, n$, is the orthonormal set of all eigenvectors of $\mathcal{A}$. We assume that all eigenvalues of the operator $\mathcal{A}$ are ordered by increasing values such that the first n_- eigenvalues are negative, the next n_0 eigenvalues are zero, and finally, the last n_+ eigenvalues are positive. Let us define the Hermitian operator $\mathcal{D}$ by the action of this operator on the basis vectors $\mathcal{E}_n = \{e_k\}_{k=1}^n$:

$$\mathcal{D}e_k = \begin{cases} |\lambda_k(\mathcal{A})|^{-1/2}e_k, & \lambda_k(\mathcal{A}) \neq 0, \\ e_k, & \lambda_k(\mathcal{A}) = 0, \end{cases}$$

where $k = 1, 2, \ldots, n$. Then we can write equality (4.136) in the form

$$\mathcal{D}\mathcal{A}\mathcal{D}\mathcal{E} = \mathcal{E}T_A, \tag{4.137}$$

where T_A is a diagonal matrix. The first n_- elements of its diagonal are equal to -1, the next n_0 elements are zero, and the last n_+ elements are equal to one. Let $\mathcal{Q}_n = \{q^k\}_{k=1}^n$ be an orthonormal basis in $\mathbf{X}_n$. We define the operator $\mathcal{M}$ by the following equality:

$$\mathcal{M}\mathcal{Q} = \mathcal{Q}T_A. \tag{4.138}$$

The bases $\mathcal{E}_n$ and $\mathcal{Q}_n$ are orthonormal; hence there exists a unitary operator $\mathcal{U}$ such that $\mathcal{E} = \mathcal{U}\mathcal{Q}$ (see p. 76), and we can write (4.137) in the form

$$\mathcal{U}^*\mathcal{D}\mathcal{A}\mathcal{D}\mathcal{U}\mathcal{Q} = \mathcal{Q}T_A. \tag{4.139}$$

Comparing the left-hand sides of (4.138) and (4.139), we see that the operators $\mathcal{A}$ and $\mathcal{M}$ are congruent. Thus all operators having the same inertia (n_+, n_-, n_0) are congruent to the operator $\mathcal{M}$. Therefore, all of them are mutually congruent.

Necessity. We denote by L_+, L_-, L_0 the subspaces of the space $\mathbf{X}_n$ spanned by the eigenvectors of the operator $\mathcal{A}$ corresponding to the positive, negative, and zero eigenvalues of the operator $\mathcal{A}$, respectively. Let us decompose the space $\mathbf{X}_n$ into the orthogonal sum $\mathbf{X}_n = L_+ \oplus L_- \oplus L_0$ (see Remark 2, p. 144). Then we see that $\dim(L_+) + \dim(L_-) + \dim(L_0) = n$. Denote by M_+ the subspace of $\mathbf{X}_n$ spanned by all eigenvectors of the operator $\mathcal{B}$ corresponding to all of its positive eigenvalues. For each $x \in M_+$, $x \neq 0$, we have $(\mathcal{B}x, x) = (\mathcal{A}\mathcal{X}x, \mathcal{X}x) = (\mathcal{A}y, y) > 0$, where $y = \mathcal{X}x$. This means that $(\mathcal{A}y, y) > 0$ for each y belonging to the subspace $\tilde{M}_+ = \mathcal{X}M_+$. Since $\mathcal{X}$ is invertible, $\dim(M_+) = \dim(\tilde{M}_+)$. Obviously, $\tilde{M}_+ \cap (L_- \oplus L_0) = \{0\}$; hence $\dim(M_+) + \dim(L_-) + \dim(L_0) \leq n$, and $\dim(M_+) \leq \dim(L_+)$. Arguing similarly, we get the opposite inequality, whence $\dim(M_+) = \dim(L_+)$, or $n_+(\mathcal{A}) = n_+(\mathcal{B})$. For the same reason, we get $n_-(\mathcal{A}) = n_-(\mathcal{B})$, $n_0(\mathcal{A}) = n_0(\mathcal{B})$. □

4.3.11 Variational Properties of Eigenvalues of Self-adjoint Operators

Recall that a linear operator $\mathcal{A} : \mathbf{X}_n \to \mathbf{X}_n$ is self-adjoint if

$$(\mathcal{A}x, y) = (x, \mathcal{A}y) \quad \text{for all} \quad x, y \in \mathbf{X}_n. \tag{4.140}$$

Recall also that all eigenvalues of a self-adjoint operator are real and that there exists an orthonormal basis of the space $\mathbf{X}_n$ consisting of the eigenvectors of the operator $\mathcal{A}$.

Let $\mathcal{A} : \mathbf{X}_n \to \mathbf{X}_n$ be a self-adjoint operator, $\lambda_1, \lambda_2, \ldots, \lambda_n$ the eigenvalues of $\mathcal{A}$, and $\{e_k\}_{k=1}^n$ the orthonormal basis of the corresponding eigenvectors. We assume that the eigenvalues are ordered by increasing values:

$$\lambda_1 \leq \lambda_2 \cdots \leq \lambda_n. \tag{4.141}$$

Let us point out that we consider the characteristic values of the matrix of the operator $\mathcal{A}$ to be the eigenvalues of $\mathcal{A}$, i.e., each multiple eigenvalue is repeated according to its multiplicity. Therefore, generally speaking, the inequalities in (4.141) are not strict.

Let p, q be integers such that $1 \leq p \leq q \leq n$. Denote by L_{pq} the subspace of the space $\mathbf{X}_n$ spanned by the vectors $\{e_k\}_{k=p}^q$. Clearly, $L_{1n} = \mathbf{X}_n$.

Lemma 4.7 *For every $x \in L_{pq}$, the following inequalities hold:*

$$\lambda_p(x, x) \leq (\mathcal{A}x, x) \leq \lambda_q(x, x). \tag{4.142}$$

Moreover,

$$\lambda_p = \min_{x \in L_{pq}, \ x \neq 0} \frac{(\mathcal{A}x, x)}{(x, x)}, \quad \lambda_q = \max_{x \in L_{pq}, \ x \neq 0} \frac{(\mathcal{A}x, x)}{(x, x)}. \tag{4.143}$$

Proof For every $x \in L_{pq}$, we have

$$(\mathcal{A}x, x) = \left(\mathcal{A} \sum_{k=p}^q \xi_k e_k, \sum_{k=p}^q \xi_k e_k \right)$$

$$= \left(\sum_{k=p}^q \lambda_k \xi_k e_k, \sum_{k=p}^q \xi_k e_k \right) = \sum_{k=p}^q \lambda_k |\xi_k|^2. \tag{4.144}$$

Evidently,

$$\lambda_p \sum_{k=p}^q |\xi_k|^2 \leq \sum_{k=p}^q \lambda_k |\xi_k|^2 \leq \lambda_q \sum_{k=p}^q |\xi_k|^2, \quad \sum_{k=p}^q |\xi_k|^2 = (x, x),$$

and hence (4.142) holds, and for every $x \neq 0$ that belongs to L_{pq}, the following inequalities hold:

$$\lambda_p \leq \frac{(\mathcal{A}x, x)}{(x, x)} \leq \lambda_q.$$

We have

$$\frac{(\mathcal{A}e_p, e_p)}{(e_p, e_p)} = \lambda_p, \quad \frac{(\mathcal{A}e_q, e_q)}{(e_q, e_q)} = \lambda_q.$$

Thus equalities (4.143) hold also. $\square$

Obviously, the next theorem follows from Lemma 4.7.

Theorem 4.45 *For each $k = 1, 2, \ldots, n$, the following equalities hold:*

$$\lambda_k = \min_{x \in L_{kn}, \, x \neq 0} \frac{(\mathcal{A}x, x)}{(x, x)}, \quad \lambda_k = \max_{x \in L_{1k}, \, x \neq 0} \frac{(\mathcal{A}x, x)}{(x, x)}. \tag{4.145}$$

Note that $L_{kn} = L_{1,k-1}^{\perp}$, $L_{1k} = L_{k+1,n}^{\perp}$. Therefore, for calculating the kth eigenvalue, we need to know all eigenvectors e_j for $j = 1, 2, \ldots, k - 1$ or for $j = k + 1, \ldots, n$. Thus formulas (4.145) are inconvenient. The next two theorems give descriptions of each eigenvalue of the self-adjoint operator $\mathcal{A}$ without reference to the preceding or succeeding eigenvectors.

Theorem 4.46 *For each $k = 1, 2, \ldots, n$, the following equality holds:*

$$\lambda_k = \max_{R_{n-k+1}} \min_{x \in R_{n-k+1}, \, x \neq 0} \frac{(\mathcal{A}x, x)}{(x, x)}. \tag{4.146}$$

Here R_{n-k+1} is an $(n - k + 1)$-dimensional subspace of the space $\mathbf{X}_n$. The maximum is taken over all subspaces $R_{n-k+1} \subset \mathbf{X}_n$ of dimension $n - k + 1$.

Proof Clearly, $\dim(R_{n-k+1}) + \dim(L_{1k}) = n + 1$. Hence (see Corollary 3.1, p. 87) there exists a vector $x \neq 0$ belonging to $R_{n-k+1} \cap L_{1k}$. Therefore, using (4.145), we see that for each subspace R_{n-k+1}, there exists a vector $x \in R_{n-k+1}$ such that $(\mathcal{A}x, x)/(x, x) \leq \lambda_k$. Thus for each subspace R_{n-k+1}, we get

$$\min_{x \in R_{n-k+1}, \, x \neq 0} \frac{(\mathcal{A}x, x)}{(x, x)} \leq \lambda_k.$$

If we choose now a subspace R_{n-k+1} for which

$$\min_{x \in R_{n-k+1}, \, x \neq 0} \frac{(\mathcal{A}x, x)}{(x, x)} = \lambda_k,$$

then we have proved equality (4.146). It follows from Theorem 4.45 that the desired subspace R_{n-k+1} is L_{kn}. $\square$

Theorem 4.47 *For each $k = 1, 2, \ldots, n$, the following equality holds:*

$$\lambda_k = \min_{R_k} \max_{x \in R_k,\, x \neq 0} \frac{(\mathcal{A}x, x)}{(x, x)}. \tag{4.147}$$

Here R_k is a k-dimensional subspace of the space $\mathbf{X}_n$. The minimum is taken over all subspaces $R_k \subset \mathbf{X}_n$ of dimension k.

Proof Clearly, $\dim(R_k) + \dim(L_{kn}) = n + 1$ for each subspace R_k, and therefore, we see that $R_k \cap L_{kn} \neq \{0\}$. By Theorem 4.45, we have

$$\min_{x \in L_{kn},\, x \neq 0} \frac{(\mathcal{A}x, x)}{(x, x)} = \lambda_k,$$

and hence for each subspace R_k, we get

$$\max_{x \in R_k,\, x \neq 0} \frac{(\mathcal{A}x, x)}{(x, x)} \geq \lambda_k.$$

To conclude the proof, it remains to choose a k-dimensional subspace R_k for which

$$\max_{x \in R_k,\, x \neq 0} \frac{(\mathcal{A}x, x)}{(x, x)} = \lambda_k.$$

Using Theorem 4.45, we see that the desired subspace is L_{1k}.   □

It follows immediately from (4.142) that a self-adjoint operator $\mathcal{A}$ is positive semidefinite (see (4.127), p. 141) if and only if all eigenvalues of $\mathcal{A}$ are nonnegative; a self-adjoint operator $\mathcal{A}$ is positive definite (see (4.128), p. 141) if and only if all eigenvalues of $\mathcal{A}$ are positive. Using the last statement, the reader can easily prove the following proposition.

Proposition 4.4 *If an operator $\mathcal{A}$ is positive definite, then $\det(\mathcal{A}) > 0$.*

Now the reader can easily prove the Cauchy–Schwarz inequality (see Theorem 3.1, p. 72) using the Gram matrix (see (3.7), p. 74) for a set of two vectors x, y in a unitary space.

4.3.12 Examples of Application of Variational Properties of Eigenvalues

Theorem 4.48 *Let $\mathcal{A}, \mathcal{B}, \mathcal{C} : \mathbf{X}_n \to \mathbf{X}_n$ be self-adjoint operators, and let*

$$\lambda_1(\mathcal{A}) \leq \lambda_2(\mathcal{A}) \leq \cdots \leq \lambda_n(\mathcal{A}),$$

$$\lambda_1(\mathcal{B}) \le \lambda_2(\mathcal{B}) \le \cdots \le \lambda_n(\mathcal{B}),$$

$$\lambda_1(\mathcal{C}) \le \lambda_2(\mathcal{C}) \le \cdots \le \lambda_n(\mathcal{C})$$

be eigenvalues of $\mathcal{A}$, $\mathcal{B}$, *and* $\mathcal{C}$, *respectively. Suppose that* $\mathcal{A} = \mathcal{B} + \mathcal{C}$. *Then*

$$\lambda_1(\mathcal{C}) \le \lambda_k(\mathcal{A}) - \lambda_k(\mathcal{B}) \le \lambda_n(\mathcal{C}), \quad k = 1, 2, \ldots, n. \tag{4.148}$$

Proof To prove this statement it is enough to note that for each arbitrarily fixed subspace R_k of the space $\mathbf{X}_n$, we have

$$\frac{(\mathcal{A}x, x)}{(x, x)} = \frac{(\mathcal{B}x, x)}{(x, x)} + \frac{(\mathcal{C}x, x)}{(x, x)} \quad \text{for all} \quad x \in R_k, \ x \ne 0.$$

Since (4.142), we see that

$$\frac{(\mathcal{C}x, x)}{(x, x)} \le \lambda_n(\mathcal{C}) \quad \text{for all} \quad x \in \mathbf{X}_n, \ x \ne 0.$$

Hence

$$\max_{x \in R_k, \ x \ne 0} \frac{(\mathcal{A}x, x)}{(x, x)} \le \max_{x \in R_k, \ x \ne 0} \frac{(\mathcal{B}x, x)}{(x, x)} + \lambda_n(\mathcal{C}),$$

and thus

$$\min_{R_k} \max_{x \in R_k, \ x \ne 0} \frac{(\mathcal{A}x, x)}{(x, x)} \le \min_{R_k} \max_{x \in R_k, \ x \ne 0} \frac{(\mathcal{B}x, x)}{(x, x)} + \lambda_n(\mathcal{C}).$$

By Theorem 4.47, the last inequality is equivalent to the following:

$$\lambda_k(\mathcal{A}) - \lambda_k(\mathcal{B}) \le \lambda_n(\mathcal{C}). \tag{4.149}$$

Note that $\mathcal{B} = \mathcal{A} + (-\mathcal{C})$. The eigenvalues of the operator $-\mathcal{C}$ are equal to $-\lambda_k(\mathcal{C})$, $k = 1, 2, \ldots, n$, and the maximal eigenvalue of $-\mathcal{C}$ is equal to $-\lambda_1(\mathcal{C})$. Therefore, arguing as above, we get

$$\lambda_k(\mathcal{B}) - \lambda_k(\mathcal{A}) \le -\lambda_1(\mathcal{C}). \tag{4.150}$$

Combining (4.149) and (4.150), we obtain (4.148). □

The estimates (4.148) are useful, because they show how the eigenvalues of a self-adjoint operator $\mathcal{B}$ can change if we add to $\mathcal{B}$ a self-adjoint operator $\mathcal{C}$. It is evident that if the eigenvalues of the operator $\mathcal{C}$ are small, then changes of the eigenvalues of $\mathcal{B}$ are small too.

Theorem 4.49 *Let* $A_{n+1} = \{a_{ij}\}_{i,j=1}^{n+1}$ *be an arbitrary Hermitian matrix of order* $n + 1$, *and let* $A_n = \{a_{ij}\}_{i,j=1}^{n}$ *be the matrix corresponding to its leading principal* *minor of order n. Let* $\hat{\lambda}_1 \leq \hat{\lambda}_2 \leq \cdots \leq \hat{\lambda}_{n+1}$ *be the eigenvalues of the matrix* A_{n+1}, *and let* $\lambda_1 \leq \lambda_2 \leq \cdots \leq \lambda_n$ *be the eigenvalues of* A_n. *Then*

$$\hat{\lambda}_1 \leq \lambda_1 \leq \hat{\lambda}_2 \leq \lambda_2 \leq \cdots \leq \lambda_n \leq \hat{\lambda}_{n+1}, \tag{4.151}$$

i.e., the eigenvalues of A_{n+1} *are interlaced with the eigenvalues of* A_n.

Proof In this proof we use the standard inner product on $\mathbb{C}^n$. Let $1 \leq k \leq n$. By Theorem 4.47,

$$\hat{\lambda}_{k+1} = \min_{R_{k+1}} \max_{x \in R_{k+1}, \, x \neq 0} \frac{(A_{n+1}x, x)}{(x, x)}. \tag{4.152}$$

The minimum here is taken over all subspaces R_{k+1} of dimension $k+1$ of the space $\mathbb{C}^{n+1}$. Denote by $R_k \subset \mathbb{C}^n$ the set of all vectors in R_{k+1} such that the $(n+1)$th coordinate with respect to the natural basis is zero. Then

$$\max_{x \in R_{k+1}, \, x \neq 0} \frac{(A_{n+1}x, x)}{(x, x)} \geq \max_{x \in R_k, \, x \neq 0} \frac{(A_n x, x)}{(x, x)}.$$

To justify this inequality, it is enough to note that on the left-hand side, the maximum is taken over a broader set of vectors than on the right-hand side. Therefore, using (4.152), we get

$$\hat{\lambda}_{k+1} = \min_{R_{k+1}} \max_{x \in R_{k+1}, \, x \neq 0} \frac{(A_{n+1}x, x)}{(x, x)} \geq \min_{R_k} \max_{x \in R_k, \, x \neq 0} \frac{(A_n x, x)}{(x, x)},$$

but by Theorem 4.47, the right-hand side of this inequality is equal to λ_k. Thus, $\hat{\lambda}_{k+1} \geq \lambda_k$ for all $k = 1, 2, \ldots, n$.

Let us now use Theorem 4.46. By this theorem,

$$\hat{\lambda}_k = \max_{R_{n+2-k}} \min_{x \in R_{n+2-k}, \, x \neq 0} \frac{(A_{n+1}x, x)}{(x, x)}. \tag{4.153}$$

The maximum here is taken over all subspaces R_{n+2-k} of dimension $n+2-k$ of the space $\mathbb{C}^{n+1}$. If we narrow the set of vectors over which the minimum is taken, then this minimum cannot decrease. Therefore, analogously to the previous case, we can write

$$\hat{\lambda}_k = \max_{R_{n+2-k}} \min_{x \in R_{n+2-k}, \, x \neq 0} \frac{(A_{n+1}x, x)}{(x, x)}$$

$$\leq \max_{R_{n+1-k}} \min_{x \in R_{n+1-k}, \, x \neq 0} \frac{(A_n x, x)}{(x, x)} = \lambda_k. \tag{4.154}$$

Thus inequalities (4.151) hold.   □

In the same way we can prove the following more general result.

Theorem 4.50 *Let A be a Hermitian matrix of order n and A_m be the Hermitian matrix of order $m < n$ corresponding to a principal minor of order m of the matrix A (see Section 4.2.4, p. 121). Let $\lambda_1(A) \leq \lambda_2(A) \leq \cdots \leq \lambda_n(A)$ be the eigenvalues of the matrix A, and let $\lambda_1(A_m) \leq \lambda_2(A_m) \leq \cdots \leq \lambda_m(A_m)$ be the eigenvalues of the matrix A_m. Then*

$$\lambda_k(A) \leq \lambda_k(A_m) \leq \lambda_{k+n-m}(A), \quad k = 1, 2, \ldots, m. \tag{4.155}$$

Remark 4.5 Clearly, Theorem 4.49 is the particular case of Theorem 4.50 when $m = n - 1$ and A_{n-1} corresponds to the leading principal minor of the matrix A of order $n - 1$. Sometimes it is convenient to order the eigenvalues by nonincreasing values. Then, obviously, the estimate (4.155) has the form

$$\lambda_{k+n-m}(A) \leq \lambda_k(A_m) \leq \lambda_k(A), \quad k = 1, 2, \ldots, m. \tag{4.156}$$

Theorem 4.51 (**Sylvester's criterion**). *A Hermitian matrix A is positive definite if and only if all the leading principal minors of A are positive.*

Proof Necessity. Take an integer k, $1 \leq k \leq n$. If in condition (4.130), p. 141, we put $x = (x_1, \ldots, x_k, 0, \ldots, 0) = (y, 0, \ldots, 0)$, where y is an arbitrary vector in $\mathbb{C}^k$, then $(Ax, x) = (A_k y, y)$. Here A_k is the matrix corresponding to the leading principal minor of order k of the matrix A.[19] Evidently, it follows now from condition (4.130) that $(A_k y, y) > 0$ for every nonzero vector $y \in \mathbb{C}^k$, i.e., the matrix A_k is positive definite. Therefore, its determinant (the leading principal minor of order k of the matrix A) is positive (see Proposition 4.4, p. 150).

Sufficiency. Now we prove that if all leading principal minors of the matrix A are positive, then all its eigenvalues are positive. The last condition means that the matrix A is positive definite. Actually, we prove more, namely, that all eigenvalues of all leading principal minors of the matrix A are positive. Obviously, for the minor of order one, i.e., for a_{11}, the assertion is true. Let us assume that all eigenvalues $\lambda_1 \leq \cdots \leq \lambda_k$ of the matrix A_k corresponding to the leading principal minor of order k are positive, and prove that all eigenvalues $\hat{\lambda}_1 \leq \cdots \leq \hat{\lambda}_{k+1}$ of the matrix A_{k+1} are positive too. Using Theorem 4.49, we see that the following inequalities hold:

$$\hat{\lambda}_1 \leq \lambda_1 \leq \hat{\lambda}_2 \leq \lambda_2 \leq \cdots \leq \lambda_k \leq \hat{\lambda}_{k+1}.$$

Therefore, $\hat{\lambda}_2, \ldots, \hat{\lambda}_{k+1} > 0$. Since by hypothesis, $\det(A_{k+1}) > 0$, and by equality (4.79), p. 124, $\det(A_{k+1}) = \hat{\lambda}_1 \hat{\lambda}_2 \cdots \hat{\lambda}_{k+1}$, we get $\hat{\lambda}_1 > 0$. $\qquad \square$

Now we introduce two concepts which will be used below. Let $x, y \in \mathbb{R}^n$. Additionally, we assume that $x_1 \geq x_2 \geq \cdots \geq x_n$, $y_1 \geq y_2 \geq \cdots \geq y_n$. We write $x \prec_w y$ and say that x is *weakly majorized* by y if

---

[19] The matrix A_k is usually called the *leading principal submatrix* of order k of the matrix A.

$$\sum_{i=1}^{k} x_i \le \sum_{i=1}^{k} y_i, \quad k = 1, 2, \ldots, n.$$

We write $x \prec y$ and say that x is *majorized* by y if $x \prec_w y$ and

$$\sum_{i=1}^{n} x_i = \sum_{i=1}^{n} y_i. \tag{4.157}$$

Theorem 4.52 (Schur). *Let A be a Hermitian matrix of order n. Let $\lambda(A) \in \mathbb{R}^n$ be the vector consisting of all the eigenvalues of the matrix A ordered by nonincreasing values, and let $d(A) \in \mathbb{R}^n$ be the vector consisting of all the diagonal entries of the matrix A ordered by nonincreasing values. Then*

$$d(A) \prec \lambda(A). \tag{4.158}$$

Proof Since for every permutation matrix P, the eigenvalues of the matrices A and PAP are equal, without loss of generality we can assume that the matrix A is such that all its diagonal entries are ordered by nonincreasing values, i.e., $a_{11} \ge a_{22} \ge \cdots \ge a_{nn}$. Let A_k be the leading principal submatrix of A of order k. Using equality (4.79), p. 124, and the estimate (4.156), we get

$$\sum_{i=1}^{k} a_{ii} = \sum_{i=1}^{k} \lambda_i(A_k) \le \sum_{i=1}^{k} \lambda_i(A). \tag{4.159}$$

Now using (4.79), p. 124, with respect to the matrix A, we see that for $k = n$, inequality (4.159) transforms to an equality.   □

The next corollary is obvious.

Corollary 4.7 *Let A be a Hermitian matrix and U a unitary matrix. Then*

$$d(U^*AU) \prec \lambda(A).$$

Theorem 4.53 *Let A be a Hermitian matrix of order n. Assume that all the eigenvalues of A are ordered by nonincreasing values. Then*

$$\sum_{i=1}^{k} \lambda_i(A) = \max_{V} \operatorname{tr}(V^*AV), \quad k = 1, 2, \ldots, n.$$

The maximum here is taken over all rectangular unitary matrices $V \in M_{n,k}$.[20]

---

[20] See the definition on p. 50.

Proof Let V be an arbitrary rectangular $n \times k$ unitary matrix. Let $U = (V, W)$ be a square unitary matrix of order n. For every matrix W, the diagonal elements of the matrix $V^* A V$ equal the first k diagonal elements of the matrix $U^* A U$. By Corollary 4.7, their sum is no more than the number $\sum_{i=1}^{k} \lambda_i (U^* A U)$, which is equal to $\sum_{i=1}^{k} \lambda_i (A)$. If the columns of the matrix V are the eigenvectors of the matrix A corresponding to $\lambda_1(A), \lambda_2(A), \ldots, \lambda_k(A)$ and are orthonormal with respect to the standard inner product on the space $\mathbb{C}^n$, then $\operatorname{tr}(V^* A V) = \sum_{i=1}^{k} \lambda_i (A)$. $\square$

Theorem 4.54 (**Fan**[21]). Let A, B be Hermitian matrices of the same order. Then

$$\lambda(A + B) \prec (\lambda(A) + \lambda(B)).$$

This theorem follows immediately from Theorem 4.53 and the fact that the trace of the sum of matrices is equal to the sum of their traces (see (4.80), p. 124).

4.4 Operators on Euclidean Spaces

4.4.1 Overview

Let $\mathbf{X}_n$ be a Euclidean space (i.e., an n-dimensional real inner product space). In this section we consider linear operators $\mathcal{A}$ acting on the Euclidean space $\mathbf{X}_n$ and note some features related to the assumption that $\mathbf{X}_n$ is real.

The matrices of operators $\mathcal{A}$ and $\mathcal{A}^*$ with respect to an orthonormal basis of the space $\mathbf{X}_n$ are transposes of each other.

A linear operator is self-adjoint if and only if the matrix of this operator with respect to every orthonormal basis of the space $\mathbf{X}_n$ is symmetric.

Skew-Hermitian operators acting on a Euclidean space are usually called *skew-symmetric*. A linear operator is skew-symmetric if and only if the matrix of this operator with respect to every orthonormal basis of the space $\mathbf{X}_n$ is skew-symmetric.

Every linear operator $\mathcal{A} : \mathbf{X}_n \to \mathbf{X}_n$ can be uniquely represented in the form

$$\mathcal{A} = \mathcal{A}_1 + \mathcal{A}_2,$$

where $\mathcal{A}_1$ is a self-adjoint operator, $\mathcal{A}_2$ is a skew-symmetric operator, and

$$\mathcal{A}_1 = \frac{1}{2}(\mathcal{A} + \mathcal{A}^*), \quad \mathcal{A}_2 = \frac{1}{2}(\mathcal{A} - \mathcal{A}^*).$$

---

[21] Ky Fan (1914–2010) was an American mathematician.

Similar arguments for matrices can be found on pp. 48, 49.

Theorem 4.55[22] *A linear operator $\mathcal{A}$ acting on the Euclidean space $\mathbf{X}_n$ is skew-symmetric if and only if*

$$(\mathcal{A}x, x) = 0 \quad \text{for all} \quad x \in \mathbf{X}_n. \tag{4.160}$$

Proof If $\mathcal{A} = -\mathcal{A}^*$, then

$$(\mathcal{A}x, x) = (x, \mathcal{A}^*x) = -(x, \mathcal{A}x),$$

i.e., $(\mathcal{A}x, x) = 0$. The sufficiency of condition (4.160) follows from the obvious identity $(\mathcal{A}(x + y), x + y) = (\mathcal{A}x, x) + (\mathcal{A}y, y) + (\mathcal{A}x + \mathcal{A}^*x, y)$. $\qquad\square$

Unitary operators (i.e., operators satisfying the condition $\mathcal{A}\mathcal{A}^* = I$) acting on a Euclidean space are called *orthogonal*. A linear operator is orthogonal if and only if the matrix of this operator with respect to every orthonormal basis of the space $\mathbf{X}_n$ is orthogonal (see Section 1.2.7, p. 48).

An orthogonal operator does not change the lengths of vectors and the angles between vectors, which follows immediately from the definition. The determinant of an orthogonal operator is equal to plus or minus one. Every eigenvalue of an orthogonal operator is equal to plus or minus one.

Recall that a linear operator $\mathcal{A}$ is normal if $\mathcal{A}\mathcal{A}^* = \mathcal{A}^*\mathcal{A}$. Self-adjoint operators, skew-symmetric operators, and orthogonal operators are normal.

If an operator $\mathcal{A} : \mathbf{X}_n \to \mathbf{X}_n$ is normal, then the matrix A_e of the operator $\mathcal{A}$ with respect to every orthonormal basis of the space $\mathbf{X}_n$ is normal, i.e., A_e satisfies the following condition:

$$A_e A_e^T = A_e^T A_e. \tag{4.161}$$

The converse is also true: if there exists an orthonormal basis $\mathcal{E}_n$ of the space $\mathbf{X}_n$ such that the matrix of the operator $\mathcal{A}$ with respect to this basis satisfies condition (4.161), then the operator $\mathcal{A}$ is normal.

4.4.2   The Structure of Normal Operators

In this subsection we consider linear operators acting on the Euclidean space $\mathbf{X}_n$.

Theorem 4.56 *Let $\mathcal{A}$ be a linear operator acting on the Euclidean space $\mathbf{X}_n$. The operator $\mathcal{A}$ is normal if and only if there exists an orthonormal basis $\mathcal{E}_n$ of the space $\mathbf{X}_n$ such that the matrix of the operator $\mathcal{A}$ with respect to this basis is block diagonal:*

---

[22]Compare with Theorem 4.36 p. 140.

$$A_e = \begin{pmatrix} A_1 & & & \\ & A_2 & & \\ & & \ddots & \\ & & & A_k \end{pmatrix}. \tag{4.162}$$

Every diagonal block here is an 1×1 *matrix or a* 2×2 *matrix. Each* 1×1 *block is a real number, while each* 2×2 *block is a matrix of the form*

$$A_p = \begin{pmatrix} \alpha_p & -\beta_p \\ \beta_p & \alpha_p \end{pmatrix}, \tag{4.163}$$

where α_p, β_p *are real numbers.*

Proof Sufficiency. By direct calculations we can easily verify that the matrix A_e as described in the statement of the theorem satisfies condition (4.161).

Necessity. Let A_e be the matrix of the normal operator $\mathcal{A}$ with respect to an arbitrarily chosen orthonormal basis $\mathcal{E}_n$. Then A_e satisfies condition (4.161). Using Corollary 4.6, p. 144, we see that for the matrix A_e, there exists an orthonormal basis $\mathcal{F}_n = \{f_k\}_{k=1}^n$ of the space $\mathbb{C}^n$ such that

$$A_e f_k = \lambda_k f_k, \quad k = 1, 2, \ldots, n, \tag{4.164}$$

where $\lambda_1, \lambda_2, \ldots, \lambda_n$ are the characteristic values of the matrix A_e, and if λ_k is real, then the corresponding vector f_k is real. Let us enumerate the characteristic values of the matrix A_e in the following order: $\lambda_1 = \alpha_1$, $\lambda_2 = \alpha_2$, $\ldots$, $\lambda_m = \alpha_m$, where $0 \le m \le n$ are real, and $\lambda_{m+j} = \alpha_{m+j} + i\beta_{m+j}$, $\bar{\lambda}_{m+j} = \alpha_{m+j} - i\beta_{m+j}$, for $j = 1, 2, \ldots, p$, where $p = (n - m)/2$, are complex. Then the eigenvectors f_k for $k = 1, 2, \ldots, m$ are real, and the other corresponding eigenvectors are complex, i.e., $f_k = g_k + ih_k$, where $g_k, h_k \in \mathbb{R}^n$, $k > m$. The matrix A_e is real, and therefore, if λ_k is a complex characteristic value of A_e and $A_e f_k = \lambda_k f_k$, then $A_e \bar{f}_k = \bar{\lambda}_k \bar{f}_k$. By Theorem 4.40, p. 143, we see that the eigenvectors of the normal operator $\mathcal{A}$ satisfying distinct eigenvalues are orthogonal to each other. Hence $(f_k, \bar{f}_k) = 0$, and $(g_k, g_k) = (h_k, h_k)$, $(g_k, h_k) = 0$. Moreover, we have $(f_k, f_k) = 1$. This easily yields that $(g_k, g_k) = (h_k, h_k) = 1/2$. Let now $f_k, f_l \in \mathcal{F}_n$, $k \ne l$, be complex vectors such that $f_k \ne \bar{f}_l$. Then we have $(f_k, f_l) = 0$ and $(f_k, \bar{f}_l) = 0$, whence by elementary calculations we obtain $(g_k, g_l), (h_k, h_l), (g_k, h_l), (h_k, g_l) = 0$. Recall that (see Section 4.2.5, p. 125) if $A_e f_k = \lambda_k f_k$, where $\lambda_k = \alpha_k + i\beta_k$, $f_k = g_k + ih_k$, then $A_e g_k = \alpha_k g_k - \beta_k h_k$, $A_e h_k = \alpha_k g_k + \beta_k h_k$. We now associate with each real eigenvalue λ_k of the matrix A_e the real eigenvector $f_k \in \mathcal{F}_n$; we associate the pair of real eigenvectors $\tilde{g}_k = \sqrt{2}\, g_k$, $\tilde{h}_k = \sqrt{2}\, h_k$ with each pair of complex-conjugate characteristic values $\lambda_k, \bar{\lambda}_k$ of the matrix A_e. As a result, we obtain the following set of n vectors in the space $\mathbb{R}^n$:

$$\tilde{\mathcal{F}}_n = \{f_1, f_2, \ldots, f_m,\ \tilde{g}_1, \tilde{h}_1, \tilde{g}_2, \tilde{h}_2, \ldots, \tilde{g}_p, \tilde{h}_p\}.$$

We have proved that this set is orthonormal. For the vectors of the set $\widetilde{\mathcal{F}}_n$ we get

$$A_e f_k = \alpha_k f_k, \quad k = 1, 2, \ldots, m, \tag{4.165}$$

$$\begin{aligned}
A_e \tilde{g}_j &= \alpha_j \tilde{g}_j - \beta_j \tilde{h}_j, \\
A_e \tilde{h}_j &= \beta_j \tilde{g}_j + \alpha_j \tilde{h}_j,
\end{aligned} \tag{4.166}$$

where $j = 1, 2, \ldots, p$. Using (4.165) and (4.166), we see that the matrix of the operator $\mathcal{A}$ with respect to the orthonormal basis $\widetilde{\mathcal{E}}_n = \mathcal{E}\widetilde{\mathcal{F}}_n$ of the space $\mathbf{X}_n$ has the form (4.162). The blocks of this matrix consist of the corresponding elements of the matrix A_e. $\square$

Let us discuss two important special cases using Corollary 4.4, p. 142.

1. Self-adjoint operators. The matrix of a self-adjoint operator $\mathcal{A}$ with respect to an orthonormal basis is symmetric. By Corollary 4.4, p. 142, all the characteristic values of this matrix are real. Therefore, all the numbers β_j, $j = 1, 2, \ldots, p$, in equalities (4.166) are equal to zero. Thus there exists an orthonormal basis of the space $\mathbf{X}_n$ such that the matrix of the operator $\mathcal{A}$ with respect to this basis is diagonal.

2. Skew-symmetric operators. The matrix of a skew-symmetric operator $\mathcal{A}$ with respect to an orthonormal basis is skew-symmetric. By Corollary 4.4, p. 142, all the characteristic values of this matrix are imaginary. Therefore all numbers α_j in equalities (4.165), (4.166) are equal to zero. Thus there exists an orthonormal basis of the space $\mathbf{X}_n$ such that the matrix of the operator $\mathcal{A}$ with respect to this basis has the form (4.162), where all the diagonal blocks of order one are equal to zero, and all the blocks of order two are skew-symmetric:

$$A_j = \begin{pmatrix} 0 & -\beta_j \\ \beta_j & 0 \end{pmatrix},$$

where $j = 1, 2, \ldots, p$.

4.4.3 The Structure of Orthogonal Operators

The matrix of an orthogonal operator with respect to an orthonormal basis is orthogonal. By Corollary 4.4, p. 142, the modulus of each characteristic value of this matrix is equal to one. Therefore, all the numbers α_k, $k = 1, 2, \ldots, m$, in equalities (4.165) are equal to plus or minus one; the numbers α_j, β_j for $j = 1, 2, \ldots, p$ in (4.166) satisfy the conditions $\alpha_j^2 + \beta_j^2 = 1$; hence there exist angles $\varphi_j \in [0, 2\pi)$ such that $\alpha_j = \cos\varphi_j$, $\beta_j = \sin\varphi_j$. Thus there exists an orthonormal basis of the space $\mathbf{X}_n$ such that the matrix of the orthogonal operator with respect to this basis has the form (4.162), where each diagonal block of order one is a number equal to plus or minus one, and each diagonal block of order two has the following form:

$$\begin{pmatrix} \cos \varphi_j & -\sin \varphi_j \\ \sin \varphi_j & \cos \varphi_j \end{pmatrix}.$$

Now we can give a clear geometric interpretation of each orthogonal transformation of the Euclidean space $\mathbf{X}_n$.

Let us begin with the two-dimensional case. As follows from the above, for each orthogonal transformation $\mathcal{A}$ of the Euclidean space $\mathbf{X}_2$ there exists an orthonormal basis e_1, e_2 such that the matrix of the transformation with respect to this basis has either the form

$$A_e = \begin{pmatrix} -1 & 0 \\ 0 & 1 \end{pmatrix}$$

or the form

$$A_e = \begin{pmatrix} \cos \varphi & -\sin \varphi \\ \sin \varphi & \cos \varphi \end{pmatrix}.$$

In the first case, the operator $\mathcal{A}$ transforms each vector $x = \xi_1 e_1 + \xi_2 e_2 \in \mathbf{X}_2$ into the vector $\mathcal{A}x = -\xi_1 e_1 + \xi_2 e_2$, i.e., the operator $\mathcal{A}$ performs a specular reflection with respect to the coordinate axis ξ_2.

In the second case, $(\mathcal{A}x, x) = |x||\mathcal{A}x| \cos \varphi$, i.e., the operator $\mathcal{A}$ performs a rotation of each vector $x \in \mathbf{X}_2$ through an angle φ. For $\varphi > 0$, the direction of the rotation coincides with the direction of the shortest rotation from e_1 to e_2.

In the three-dimensional case, every orthogonal operator $\mathcal{A}$ has at least one eigenvalue, since the corresponding characteristic equation is an polynomial equation of degree three with real coefficients. Therefore, the matrix A_e of the operator $\mathcal{A}$ with respect to the orthonormal basis $e_1, e_2, e_3 \in \mathbf{X}_3$ (renumbered if necessary) has one of the following forms:

$$A_e = \begin{pmatrix} 1 & 0 & 0 \\ 0 & \cos \varphi & -\sin \varphi \\ 0 & \sin \varphi & \cos \varphi \end{pmatrix}, \tag{4.167}$$

$$A_e = \begin{pmatrix} -1 & 0 & 0 \\ 0 & \cos \varphi & -\sin \varphi \\ 0 & \sin \varphi & \cos \varphi \end{pmatrix}. \tag{4.168}$$

Observe that if the operator $\mathcal{A}$ has exactly one eigenvalue, then these representations follow immediately from Theorem 4.56. If the operator $\mathcal{A}$ has three eigenvalues, then we obtain representation (4.167) or (4.168) by choosing a special angle φ.

Arguing by analogy with the two-dimensional case, it is easy to verify that an operator $\mathcal{A}$ that has the matrix (4.167) performs a rotation through an angle φ about the coordinate axis ξ_1, while an operator $\mathcal{A}$ that has the matrix (4.168) performs first a rotation through an angle φ about the coordinate axis ξ_1 and then a specular reflection with respect to the $\xi_2 \xi_3$ coordinate plane. In the first case, the determinant of the operator $\mathcal{A}$ is equal to one, while in the second case, it is equal to minus one.

As we know, the determinant of a linear operator does not depend on the choice of the basis of the space. Therefore, we can divide all orthogonal transformations of a three-dimensional space into two classes: proper rotations and improper rotations. A *proper rotation* is a transformation with a positive determinant. It performs a rotation of the space about an axis. An *improper rotation* is a transformation with a negative determinant. It is a combination of a rotation about an axis and a reflection in the plane that is orthogonal to that axis.

Using Theorem 4.56, we can represent the Euclidean space X_n of arbitrary dimension n as an orthogonal sum of some one-dimensional invariant subspaces of the orthogonal operator $\mathcal{A} : X_n \to X_n$ and some two-dimensional invariant subspaces of $\mathcal{A}$. In each two-dimensional invariant subspace, the operator $\mathcal{A}$ performs a rotation through an angle. Generally speaking, these angles can differ for different subspaces. In each one-dimensional invariant subspace, only the direction of a coordinate axis can be transformed.

The proof of the following proposition is left to the reader.

Proposition 4.5 *Every real symmetric matrix A is orthogonally similar to a diagonal matrix, i.e., $Q^T A Q = \Lambda$, where Λ is a diagonal matrix and Q is an orthogonal matrix. The columns of the matrix Q are the eigenvectors of A. The diagonal elements of the matrix Λ are the eigenvalues of A.*

4.4.4   Givens Rotations and Householder Transformations

In this subsection we consider two important types of orthogonal matrices that are often used in applications.

1. Givens[23] rotations. A real matrix $Q_{st}(\varphi) = \{q_{ij}(\varphi)\}_{i,j=1}^n$, $1 \le s < t \le n$, is called a *Givens rotation* if $q_{ss}(\varphi) = q_{tt}(\varphi) = \cos\varphi$, $q_{ii}(\varphi) = 1$ for $i \ne s, t$, $q_{st}(\varphi) = -\sin\varphi$, $q_{ts}(\varphi) = \sin\varphi$, and all other elements of the matrix $Q_{st}(\varphi)$ are equal to zero.

It is easy to see that the matrix $Q = Q_{st}(\varphi)$ is orthogonal. This matrix defines an orthogonal transformation of the Euclidean space $\mathbb{R}^n$ with the standard inner product, and it performs a rotation trough an angle φ in the two-dimensional space (in the plane) spanned by the vectors i_s, i_t of the natural basis in the space $\mathbb{R}^n$. The matrix Q^T is the inverse of Q and performs the opposite rotation in the same plane.

Let x be an arbitrary vector in the space $\mathbb{R}^n$. Obviously, $(Qx)_i = x_i$ for $i \ne s, t$, $(Qx)_s = x_s \cos\varphi - x_t \sin\varphi$, $(Qx)_t = x_s \sin\varphi + x_t \cos\varphi$. Take $\rho = (x_s^2 + x_t^2)^{1/2}$. Suppose that $\varphi = 0$ if $\rho = 0$ and $\cos\varphi = x_s/\rho$, $\sin\varphi = -x_t/\rho$ if $\rho > 0$. Then we get $(Qx)_s = \rho$, $(Qx)_t = 0$.

Now it is perfectly clear that if x is an arbitrary nonzero vector in $\mathbb{R}^n$, then sequentially choosing the angles φ_n, φ_{n-1}, ..., φ_2, we can construct the Givens rotations $Q_{1,n}(\varphi_n)$, $Q_{1,n-1}(\varphi_{n-1})$, ..., $Q_{1,2}(\varphi_2)$ such that $Qx = |x| \, i_1$, where

---

[23] Wallace Givens (1910–1993) was an American mathematician.

$$Q = Q_{1,2}(\varphi_2) \cdots Q_{1,n-1}(\varphi_{n-1}) Q_{1,n}(\varphi_n).$$

Thus, using an orthogonal matrix, we can transform any nonzero vector into a vector whose direction coincides with the direction of the vector i_1 of the natural basis.

Let x, y be two arbitrary nonzero vectors in $\mathbb{R}^n$. As we have just shown, there exist orthogonal matrices Q_x and Q_y such that $Q_x x = |x| i_1$, $Q_y y = |y| i_1$. Therefore, $Qx = (|x|/|y|) y$, where $Q = Q_y^T Q_x$, i.e., for every pair of nonzero vectors there exists an orthogonal matrix that transforms the first vector into a vector whose direction coincides with the direction of the second vector.

2. Householder[24] transformations. Let $w = \{w_i\}_{i=1}^n$ be an arbitrarily chosen vector in $\mathbb{R}^n$ with $|w| = 1$. A matrix

$$R = I - 2ww^T$$

is called a *Householder transformation (or reflection)*. We explain that the vector w is treated here as a column vector. Hence $R = \{\delta_{ij} - 2w_i w_j\}_{i,j=1}^n$.

The matrix R is symmetric. Let us show that this matrix is orthogonal. Indeed,

$$R^T R = R^2 = I - 4ww^T + 4ww^T ww^T = I,$$

because $w^T w = |w|^2 = 1$. Note further that

$$Rw = w - 2ww^T w = -w, \quad Rz = z - 2ww^T z = z, \qquad (4.169)$$

if $w^T z = (w, z) = 0$, i.e., the vectors w and z are orthogonal.[25]

Now let x be an arbitrary vector. By Theorem 3.1.3, p. 91, it can be uniquely represented in the form $x = \alpha w + z$, where α is a real number and z is a vector orthogonal to w. Using equalities (4.169), we see that $Rx = -\alpha w + z$. We can say therefore that the matrix R performs a specular reflection of the vector x with respect to the $(n-1)$-dimensional hyperplane orthogonal to the vector w. This property of the matrix R allows us to call it a Householder reflection.

Consider the following problem. A nonzero vector a and an unit vector e are given. It is necessary to construct a Householder reflection R such that $Ra = \mu e$, where μ is a number (clearly, the equality $|\mu| = |a|$ holds, since R is orthogonal).

It is easy to see (make a drawing!) that the solution of this problem is the Householder reflection defined by the vector

$$w = \frac{a - |a| e}{|a - |a| e|} \qquad (4.170)$$

or by the vector $w = (a + |a| e)/|a + |a| e|$. For minimization of effects of rounding errors in numerical calculations, we should take the vector w that has the larger denominator.

---

[24] Alston Scott Householder (1904–1993) was an American mathematician.

[25] With respect to the standard inner product on the space $\mathbb{R}^n$.

It is useful to note that if a is an arbitrary nonzero vector, then a Householder transformation R can be constructed such that for every vector $x \in \mathbb{R}^n$, the following condition holds:

$$(a, Rx) = |a|x_k, \qquad (4.171)$$

where k is a given integer lying in the range from 1 to n, and x_k is the kth component of the vector x. Evidently, to do this we need to choose $e = i_k$ in formula (4.170).

Chapter 5
Canonical Forms and Factorizations

In this chapter we explore in detail the problem of reducing the matrix of an operator to a simple form using special bases in finite-dimensional spaces. The singular value decomposition of an operator is constructed. The Jordan canonical form of the matrix of a finite-dimensional operator is obtained. A special section is devoted to studying matrix pencils. We obtain their canonical forms and describe applications to investigate the structure of solutions of systems of ordinary linear differential equations.

5.1 The Singular Value Decomposition

5.1.1 Singular Values and Singular Vectors of an Operator

In this section we show that for every linear operator $\mathcal{A}$ mapping a finite-dimensional unitary space $\mathbf{X}_n$ into a finite-dimensional unitary space $\mathbf{Y}_m$, there exist orthonormal bases $\{e_k\}_{k=1}^n \subset \mathbf{X}_n$ and $\{q_k\}_{k=1}^m \subset \mathbf{Y}_m$ such that

$$\mathcal{A}e_k = \begin{cases} \sigma_k q_k, & k \le r, \\ 0 & , \ k > r, \end{cases} \tag{5.1}$$

where $\sigma_k > 0$, $k = 1, 2, \ldots, r$. The numbers $\sigma_1, \sigma_2, \ldots, \sigma_r$ are called the *singular values* of the operator $\mathcal{A}$. Sometimes it is convenient to include $\min(m, n) - r$ zeros in the set of singular values.

Relationships (5.1) show that the numbers $\sigma_1, \sigma_2, \ldots, \sigma_r$ form the main diagonal of the leading principal (basic) minor of the matrix A_{eq} of the operator $\mathcal{A}$ with respect to the bases $\{e_k\}_{k=1}^n$, $\{q_k\}_{k=1}^m$, and all other elements of the matrix A_{eq} are equal to zero.

© Springer International Publishing AG 2017
L. Beilina et al., *Numerical Linear Algebra: Theory and Applications*,
DOI 10.1007/978-3-319-57304-5_5

The vectors $\{e_k\}_{k=1}^n$, $\{q_k\}_{k=1}^m$ are called respectively the left and the right *singular vectors* of the operator $\mathcal{A}$. Let us construct them. The operator $\mathcal{A}^*\mathcal{A}$ is self-adjoint and positive semidefinite (see Property 2, p. 140). Therefore (see Theorem 4.41, p. 142, and Section 4.3.11, p. 150), there exist the orthonormal eigenvectors $\{e_k\}_{k=1}^n$ of the operator $\mathcal{A}^*\mathcal{A}$, and all its eigenvalues are nonnegative. Thus,

$$A^*Ae_k = \sigma_k^2 e_k, \quad k = 1, 2, \ldots, n. \tag{5.2}$$

Here $\sigma_k^2 \geq 0$ are the eigenvalues of the operator $\mathcal{A}^*\mathcal{A}$. Let us enumerate them as follows: $\sigma_1 \geq \sigma_2 \geq \cdots \geq \sigma_r > 0$, $\sigma_{r+1} = \cdots = \sigma_n = 0$. Put $z_k = \mathcal{A}e_k$ for $k = 1, \ldots, r$ and note that $(z_p, z_q) = (\mathcal{A}e_p, \mathcal{A}e_q) = (\mathcal{A}^*\mathcal{A}e_p, e_q) = \sigma_p^2(e_p, e_q)$. Hence

$$(z_p, z_q) = \begin{cases} 0, & p \neq q, \\ \sigma_p^2, & p = q, \end{cases} \tag{5.3}$$

and the vectors

$$q_k = \sigma_k^{-1} \mathcal{A}e_k, \quad k = 1, 2, \ldots, r, \tag{5.4}$$

form an orthonormal set in the space $\mathbf{Y}_m$. If $r < m$, then we join this set with some vectors q_k, $k = r+1, r+2, \ldots, m$, to complete an orthonormal basis of the space $\mathbf{Y}_m$. Relationships (5.1) now follow immediately from the definition of the vectors $\{e_k\}_{k=1}^n$ and $\{q_k\}_{k=1}^m$.

Using (5.1), we see that the vectors $\{q_k\}_{k=1}^r$ form a basis of $\mathrm{Im}(\mathcal{A})$. Hence it follows from Theorem 4.29, p. 136, that the vectors $\{q_k\}_{k=r+1}^m$ form a basis of $\mathrm{Ker}(\mathcal{A}^*)$. Therefore,

$$\mathcal{A}^* q_k = 0 \text{ for } k = r+1, r+2, \ldots, m. \tag{5.5}$$

For $k = 1, 2, \ldots, r$, using (5.4), (5.2), we get

$$A^* q_k = \sigma_k^{-1} A^*Ae_k = \sigma_k e_k. \tag{5.6}$$

Combining (5.6), (5.4), and (5.5), we obtain

$$AA^* q_k = \sigma_k^2 q_k, \ k = 1, 2, \ldots, r, \quad AA^* q_k = 0, \ k = r+1, r+2, \ldots, m. \tag{5.7}$$

It follows from (5.2) and (5.7) that all the nonzero eigenvalues of the operators $\mathcal{A}^*\mathcal{A}$ and $\mathcal{A}\mathcal{A}^*$ coincide, i.e., the spectra of these operators can differ only by the multiplicity of the zero eigenvalue.

Moreover, the next equalities follow from the previous arguments:

$$\mathrm{rank}(\mathcal{A}) = \mathrm{rank}(\mathcal{A}^*\mathcal{A}) = \mathrm{rank}(\mathcal{A}\mathcal{A}^*),$$

$$\mathrm{def}(\mathcal{A}^*\mathcal{A}) = n - \mathrm{rank}(\mathcal{A}), \quad \mathrm{def}(\mathcal{A}\mathcal{A}^*) = m - \mathrm{rank}(\mathcal{A}).$$

Clearly, the rank r of the operator $\mathcal{A}$ is equal to the number of all nonzero singular values of the operator $\mathcal{A}$. This remark gives us a real opportunity to compute the rank of the operator $\mathcal{A}$: we have to solve the eigenvalue problem for the positive semidefinite self-adjoint operator $\mathcal{A}^*\mathcal{A}$ and calculate the number of all nonzero eigenvalues. Precisely this method is typically used in practical computations of the rank. Evidently, the eigenvectors $\{e_i\}_{i=r+1}^n$ of the operator $\mathcal{A}^*\mathcal{A}$ form an orthonormal basis of the kernel of the operator $\mathcal{A}$.

If the singular values and the singular vectors of the operator $\mathcal{A}$ are known, then the pseudosolution (see Section 4.3.4, p. 136) of the equation

$$\mathcal{A}x = y \qquad (5.8)$$

can be easily constructed. Indeed, in Section 4.3.4 we proved that every solution of the equation

$$\mathcal{A}^*\mathcal{A}x = \mathcal{A}^*y \qquad (5.9)$$

is a pseudosolution of (5.8). Substituting in (5.9) the expansions $x = \displaystyle\sum_{k=1}^n \xi_k e_k$ and $y = \displaystyle\sum_{k=1}^m \eta_k q_k$ with respect to the bases $\{e_k\}_{k=1}^n \subset \mathbf{X}_n$ and $\{q_k\}_{k=1}^m \subset \mathbf{Y}_m$ for x and y and using after that (5.2), (5.5), (5.6), we get

$$\sum_{k=1}^r (\sigma_k^2 \xi_k - \sigma_k \eta_k) e_k = 0. \qquad (5.10)$$

Therefore, $\xi_k = \eta_k/\sigma_k$ for $k = 1, 2, \ldots, r$. Thus every vector

$$x = \sum_{k=1}^r \frac{\eta_k}{\sigma_k} e_k + \sum_{k=r+1}^n \xi_k e_k, \qquad (5.11)$$

where $\xi_{r+1}, \ldots, \xi_n$ are arbitrary numbers, is a pseudosolution of Eq. (5.8).

If $y \in \operatorname{Im}(\mathcal{A})$, i.e., Eq. (5.8) is solvable, then formula (5.11) gives the general solution (see Section 4.1.11, p. 107) of Eq. (5.8). Indeed, in this case, the vector $x_0 = \displaystyle\sum_{k=1}^r (\eta_k/\sigma_k) e_k$ is a particular solution of Eq. (5.8), and $\displaystyle\sum_{k=r+1}^n \xi_k e_k$ is the general solution of the corresponding homogeneous equation.

For each pseudosolution x of Eq. (5.8), we have

$$|x|^2 = \sum_{k=1}^r \frac{|\eta_k|^2}{\sigma_k^2} + \sum_{k=r+1}^n |\xi_k|^2.$$

If we take $\xi_{r+1}, \ldots, \xi_n = 0$, then we get the pseudosolution of minimal length. This pseudosolution is *normal*. Obviously, it is orthogonal to the kernel of the operator $\mathcal{A}$.

The proof of the following four propositions is left to the reader.

Proposition 5.1 *The absolute value of the determinant of every operator acting on a finite-dimensional space is equal to the product of all its singular values.*

Proposition 5.2 *Let $A \in M_{m,n}$ be an arbitrary rectangular matrix of rank r. Then there exist unitary matrices U and V (of order m and n, respectively) such that*

$$A = U \Sigma V, \tag{5.12}$$

where

$$\Sigma = \begin{pmatrix} S & O_{1,2} \\ O_{2,1} & O_{2,2} \end{pmatrix}$$

is a block 2×2 matrix, $S = \text{diag}(\sigma_1, \sigma_2, \ldots, \sigma_r)$, all elements of the diagonal S are positive, and all elements of the matrices $O_{1,2}$, $O_{2,1}$, $O_{2,2}$ are equal to zero. Formula (5.12) determines the so-called singular value decomposition *of a rectangular matrix.*

Proposition 5.3 *Let $A \in M_{m,n}$ be an arbitrary matrix, and U, V arbitrary unitary matrices of orders m and n, respectively. The singular values of the matrices A and $U A V$ coincide (therefore, we say that the singular values of a matrix are invariant under unitary transformations).*

Proposition 5.4 *Let $A \in M_{m,n}$ be an arbitrary matrix, $\sigma_1, \sigma_2, \ldots, \sigma_r$ its singular values. Then*

$$\max_{1 \leq k \leq r} \sigma_k \leq \left(\sum_{i,j=1}^{m,n} |a_{ij}|^2 \right)^{1/2}. \tag{5.13}$$

The singular values of an operator characterize the sensitivity of the solution of a linear equation with respect to changes in its right-hand side. Let $\mathcal{A}$ be a nonsingular operator acting on a finite-dimensional unitary space $\mathbf{X}_n$. Consider two equations:

$$\mathcal{A}x = y \tag{5.14}$$

and

$$\mathcal{A}x = \tilde{y}. \tag{5.15}$$

Since the operator $\mathcal{A}$ is nonsingular, these equations are uniquely solvable. Denote by x the solution of Eq. (5.14) and by $\tilde{x}$ the solution of Eq. (5.15). The number $\delta_x = |x - \tilde{x}|/|x|$ is the *relative change in the solution* with respect to the change in the right-hand side. Let us clarify the dependence of δ_x on the *relative change in the right-hand side* $\delta_y = |y - \tilde{y}|/|y|$. If we represent the vectors y and $\tilde{y}$ in the form of expansions $y = \sum_{k=1}^{n} \eta_k q_k$ and $\tilde{y} = \sum_{k=1}^{n} \tilde{\eta}_k q_k$, then using (5.1), we obtain

$$x = \mathcal{A}^{-1} y = \sum_{k=1}^{n} \frac{\eta_k}{\sigma_k} e_k, \quad \tilde{x} = \mathcal{A}^{-1} \tilde{y} = \sum_{k=1}^{n} \frac{\tilde{\eta}_k}{\sigma_k} e_k.$$

Here, as usual, $\sigma_k, k = 1, 2, \ldots, n$, are the singular values, and $\{e_k\}_{k=1}^{n}$ and $\{q_k\}_{k=1}^{m}$ are the singular vectors of $\mathcal{A}$. Therefore, using the inequalities $\sigma_1 \geq \sigma_2 \geq \cdots \geq \sigma_n > 0$, we get

$$\delta_x^2 = \frac{\sum_{k=1}^{n} \frac{|\eta_k - \tilde{\eta}_k|^2}{\sigma_k^2}}{\sum_{k=1}^{n} \frac{|\eta_k|^2}{\sigma_k^2}} \leq \frac{\sigma_1^2}{\sigma_n^2} \frac{\sum_{k=1}^{n} |\eta_k - \tilde{\eta}_k|^2}{\sum_{k=1}^{n} |\eta_k|^2} = \frac{\sigma_1^2}{\sigma_n^2} \delta_y^2. \tag{5.16}$$

Thus,

$$\delta_x \leq \frac{\sigma_1}{\sigma_n} \delta_y. \tag{5.17}$$

The number σ_1/σ_n, which characterizes the stability of the solution of Eq. (5.14) with respect to changes in its right-hand side, is called the *condition number* of the operator $\mathcal{A}$ and is denoted by cond $(\mathcal{A})$. Evidently, cond $(\mathcal{A}) \geq 1$ for every operator $\mathcal{A}$.

The reader can easily check the following properties of the condition number.

1. There exist vectors y and $\tilde{y}$ such that the two sides in (5.17) are equal. It this sense, estimate (5.17) cannot be improved.
2. There exist operators whose condition numbers are equal to one (give some examples!).

5.1.2 The Polar Decomposition

Theorem 5.1 *Let $\mathcal{A} : \mathbf{X}_n \rightarrow \mathbf{Y}_m$ be an arbitrary operator. There exist operators $\mathcal{U} : \mathbf{X}_n \rightarrow \mathbf{Y}_m$, $\mathcal{S} : \mathbf{X}_n \rightarrow \mathbf{X}_n$, and $\mathcal{T} : \mathbf{Y}_m \rightarrow \mathbf{Y}_m$ such that*

$$\mathcal{U}^* \mathcal{U} = I \;\; \text{if } n \leq m, \quad \mathcal{U} \mathcal{U}^* = I \;\; \text{if } n \geq m. \tag{5.18}$$

The operators $\mathcal{S}$, $\mathcal{T}$ are self-adjoint and positive semidefinite, and

$$A = \mathcal{US} = \mathcal{TU}. \tag{5.19}$$

Proof Let $\{e^k\}_{k=1}^n$, $\{q^k\}_{k=1}^m$ be the singular vectors of the operator $\mathcal{A}$ (which form orthonormal bases in the spaces $\mathbf{X}_n$, $\mathbf{Y}_m$; see Section 5.1.1). Let $\sigma_1, \sigma_2, \ldots, \sigma_r$ be the singular values of $\mathcal{A}$. If $n \leq m$, we define the operator $\mathcal{U}$ by the relationships

$$\mathcal{U}e_k = q_k, \quad k = 1, 2, \ldots, n. \tag{5.20}$$

If $n \geq m$, we put

$$\mathcal{U}e_k = q_k, \ k = 1, 2, \ldots, m, \quad \mathcal{U}e_k = 0, \ k = m+1, m+2, \ldots, n. \tag{5.21}$$

We define the operators $\mathcal{S}$, $\mathcal{T}$ by the following equalities:

$$\mathcal{S}e_k = \sigma_k e_k, \ k = 1, 2, \ldots, r, \quad \mathcal{S}e_k = 0, \ k = r+1, r+2, \ldots, n,$$

$$\mathcal{T}q_k = \sigma_k q_k, \ k = 1, 2, \ldots, r, \quad \mathcal{T}q_k = 0, \ k = r+1, r+2, \ldots, m.$$

The operators $\mathcal{T}$ and $\mathcal{S}$ are self-adjoint and positive semidefinite, since, as is easy to see, the numbers $(\mathcal{S}x, x)$, $(\mathcal{T}y, y)$ are nonnegative for all $x \in \mathbf{X}_n$ and for all $y \in \mathbf{Y}_m$. Obviously,

$$\mathcal{US}e_k = \mathcal{A}e_k, \quad \mathcal{TU}e_k = \mathcal{A}e_k, \quad k = 1, 2, \ldots, n,$$

i.e., the relationships (5.19) are true. By direct calculations we verify that the operator $\mathcal{U}^*$ can be defined by the relationships

$$\mathcal{U}^*q_k = e_k, \ k = 1, 2, \ldots, m, \ \text{if} \ m \leq n, \tag{5.22}$$

$$\mathcal{U}^*q_k = e_k, \ k = 1, 2, \ldots, n, \ \mathcal{U}^*q_k = 0, \ k = n+1, n+2, \ldots, m, \ \text{if} \ m \geq n. \tag{5.23}$$

Clearly, equalities (5.18) follow from (5.20)–(5.23). $\square$

Formulas (5.19) define the *polar decomposition* of the operator $\mathcal{A}$.

Now let us dwell on the case in which an operator $\mathcal{A}$ acts on a space $\mathbf{X}_n$. The relationships (5.18) show that in this case, the operator $\mathcal{U}$ is unitary. It follows from equalities (5.19) that every linear transformation of a finite-dimensional space $\mathbf{X}_n$ is the result of the sequential execution of a unitary transformation, which does not change the length of vectors, and a positive semidefinite self-adjoint transformation, which stretches the space $\mathbf{X}_n$ in n mutually orthogonal directions.

It immediately follows from (5.19) that $\mathcal{A}^*\mathcal{A} = \mathcal{S}^2$, $\mathcal{A}\mathcal{A}^* = \mathcal{T}^2$. Since the operators $\mathcal{S}$ and $\mathcal{T}$ are self-adjoint and positive semidefinite, the last two equalities show that S and T are uniquely determined by the operator A, namely (see Theorem 4.43, p. 145),

$$S = \sqrt{A^*A}, \quad T = \sqrt{AA^*}. \tag{5.24}$$

If the operator A is nonsingular, then the operator A^*A is nonsingular. Hence the operator S is also nonsingular. Therefore, in this case, the operator $U = AS^{-1}$ is also uniquely determined.

The next theorem readily follows from formulas (5.19), (5.24).

Theorem 5.2 *An operator A is normal if and only if the operators T and S in the factorization* (5.19) *coincide, in other words, if and only if the operators U and S commute.*

Note that if the space $\mathbf{X}_n$ is real, then there is a valid polar decomposition, but the operator U in (5.19) is orthogonal, and the operators T and S are symmetric and positive semidefinite.

5.1.3 Basic Properties of the Pseudoinverse Operator

Let $\mathbf{X}_n$, $\mathbf{Y}_m$ be finite-dimensional unitary spaces, $A : \mathbf{X}_n \to \mathbf{Y}_m$ a linear operator, $\{e_k\}_{k=1}^n \subset \mathbf{X}_n$, $\{q_k\}_{k=1}^m \subset \mathbf{Y}_m$ its singular vectors, $\sigma_1, \sigma_1, \ldots, \sigma_r$ the singular values of A, where $r = \operatorname{rank}(A)$, $r \leq \min(m, n)$.

As we have seen in Section 5.1.1, the formula

$$x_0 = \sum_{k=1}^r \frac{\eta_k}{\sigma_k} e_k,$$

where $\eta_k = (q_k, y)$, $k = 1, 2, \ldots, m$, are the coordinates of the vector y with respect to the basis $\{q_k\}_{k=1}^m$, defines the normal pseudosolution x_0 of the equation $Ax = y$. Thus the pseudoinverse of the operator A (see Section 4.3.4, p. 109) can be represented in the form

$$A^+ y = \sum_{k=1}^r \frac{(q_k, y)}{\sigma_k} e_k. \tag{5.25}$$

Here are the basic properties of the pseudoinverse operator:

1. $(A^*)^+ = (A^+)^*$,
2. $(A^+)^+ = A$,
3. $(AA^+)^* = AA^+$, $(AA^+)^2 = AA^+$,
4. $(A^+A)^* = A^+A$, $(A^+A)^2 = A^+A$,
5. $AA^+A = A$,

6. $\mathcal{A}^+\mathcal{A}\mathcal{A}^+ = \mathcal{A}^+$,
7. if rank $\mathcal{A} = n$, then $\mathcal{A}^+\mathcal{A} = I$.

We prove only the first and third equalities. The reader can easily prove all the other properties.

1. Let $x = \sum_{k=1}^{n} \xi_k e_k$, $y = \sum_{k=1}^{m} \eta_k q_k$. Solving the equation $\mathcal{A}\mathcal{A}^* y = \mathcal{A}x$ in the same way as (5.9), we see that $(\mathcal{A}^*)^+ x = \sum_{k=1}^{r} (\xi_k/\sigma_k) q_k$. By elementary calculations, we get $((\mathcal{A}^*)^+ x, y) = \sum_{k=1}^{r} \xi_k \bar{\eta}_k/\sigma_k$. Using (5.25), we obtain $(x, \mathcal{A}^+ y) = \sum_{k=1}^{r} \xi_k \bar{\eta}_k/\sigma_k$. This means that $(\mathcal{A}^*)^+ = (\mathcal{A}^+)^*$.

3. It follows from (5.25) that $\mathcal{A}^+ q_k = \sigma_k^{-1} e_k$ for $k = 1, 2, \ldots, r$ and $\mathcal{A}^+ q_k = 0$ for $k = r+1, r+2, \ldots, m$. Hence, $\mathcal{A}\mathcal{A}^+ q_k = q_k$ for $k = 1, 2, \ldots, r$ and $\mathcal{A}\mathcal{A}^+ q_k = 0$ for $k = r+1, r+2, \ldots, m$. Therefore, $(\mathcal{A}\mathcal{A}^+)^2 = \mathcal{A}\mathcal{A}^+$. By elementary calculations, we get $(\mathcal{A}\mathcal{A}^+ y, y) = \sum_{k=1}^{r} |\eta_k|^2 \geq 0$ for each $y \in \mathbf{Y}_m$. Thus we obtain $(\mathcal{A}\mathcal{A}^+)^* = \mathcal{A}\mathcal{A}^+$ (see Lemma 4.5, p. 140).

It follows from Properties 3 and 4 that the operators $\mathcal{A}\mathcal{A}^+$ and $\mathcal{A}^+\mathcal{A}$ are orthogonal projection operators (see Theorem 4.35, p. 140).

5.1.4   Elements of the Theory of Majorization

A real-valued function f of a real variable is called *convex* on an interval (a, b) if for all points x_1, x_2 in this interval and for every $t \in [0, 1]$, the following inequality holds:

$$f(tx_1 + (1 - t)x_2) \leq tf(x_1) + (1 - t)f(x_2). \tag{5.26}$$

Geometrically, this means that every point on the graph of the function f on the closed interval $[x_1, x_2]$ lies below or on the chord subtending the points $(x_1, f(x_1))$ and $(x_2, f(x_2))$.

Theorem 5.3 (Jensen's[1] inequality). *If a function f is convex on an interval (a, b), then for all points $x_1, x_2, \ldots, x_m$ that belong to (a, b) and for all nonnegative numbers $\alpha_1, \alpha_2, \ldots, \alpha_m$ such that $\alpha_1 + \alpha_2 + \cdots + \alpha_m = 1$, the following inequality is valid:*

$$f\left(\sum_{i=1}^{m} \alpha_i x_i\right) \leq \sum_{i=1}^{m} \alpha_i f(x_i). \tag{5.27}$$

---

[1]Johan Ludvig William Valdemar Jensen (1859–1925) was a Danish mathematician and engineer.

Proof We easily get (5.27) by induction on m using the obvious identities

$$\sum_{i=1}^{m} \alpha_i x_i = \alpha_m x_m + (1 - \alpha_m) \sum_{i=1}^{m-1} \frac{\alpha_i}{(1 - \alpha_m)} x_i, \quad \sum_{i=1}^{m-1} \frac{\alpha_i}{(1 - \alpha_m)} = 1.$$

□

Theorem 5.4 *Suppose that a function f is differentiable on an interval (a, b) and the derivative of f is nondecreasing on (a, b). Then the function f is convex on the interval (a, b).*

Proof It is enough to prove that for all $x_1, x_2 \in (a, b)$, $x_1 < x_2$, the function

$$\varphi(t) = f((1 - t)x_1 + tx_2) - (1 - t)f(x_1) - tf(x_2)$$

is nonpositive for all $t \in [0, 1]$. It is easy to see that $\varphi(0) = 0$, $\varphi(1) = 0$, and $\varphi'(t)$ is nondecreasing on the segment $[0, 1]$. Using the Lagrange finite-increments formula, we see that $\varphi(t) = \varphi(t) - \varphi(0) = t\varphi'(t_1)$, where t_1 is a point in the interval $(0, t)$. Similarly, $\varphi(t) = (t - 1)\varphi'(t_2)$, where t_2 is a point in the interval $(t, 1)$. Hence it is evident that $\varphi(t) = t(t - 1)(\varphi'(t_2) - \varphi'(t_1)) \leq 0$. □

Below we will use the following definitions. A real matrix is called *nonnegative* if all its elements are nonnegative. A nonnegative square matrix is called *stochastic* if it is nonnegative and the sum of all the elements of each of its rows is equal to one. A stochastic matrix is called *doubly stochastic* if the sum of all the elements of each of its columns is also equal to one.

Theorem 5.5 *Let $x, y \in \mathbb{R}^n$, $x_1 \geq x_2 \geq \cdots \geq x_n$, $y_1 \geq y_2 \geq \cdots \geq y_n$, and $x \prec y$.[2] Then there exists a doubly stochastic matrix S such that $x = Sy$.*

Proof We prove the theorem by induction on n. For $n = 1$, the theorem is trivial. Now we assume that the assertion is true for all vectors of length $n - 1$ and prove that it holds for all vectors of length n. We easily check that if vectors x and y satisfy all conditions of the theorem, then $x_1 \geq y_n$.[3] Therefore, there exist an integer k, $1 \leq k \leq n - 1$, and a real number $\tau \in [0, 1]$ such that

$$x_1 = \tau y_k + (1 - \tau)y_{k+1}. \tag{5.28}$$

Let us consider the two following vectors of length $n - 1$:

$$\tilde{x} = (x_2, x_3, \ldots, x_n) \text{ and } \tilde{y} = (y_1, y_2, \ldots, y_{k-1}, y_k + y_{k+1} - x_1, y_{k+2}, \ldots, y_n).$$

---

[2] We use the notation defined on p. 153.

[3] Otherwise, equality 4.157, p. 153, is impossible.

It is easy to see that the components of these vectors are monotonically nonincreasing and $\tilde{x} \prec \tilde{y}$. Therefore, by the induction hypothesis, there exists a doubly stochastic matrix $\tilde{S}$ of order $n - 1$ such that

$$\tilde{x} = \tilde{S}\tilde{y}. \tag{5.29}$$

Writing (5.28), (5.29) in the form of a single matrix equality, we get $x = Sy$, where S is a doubly stochastic matrix.   □

Theorem 5.6 *Let $x, y \in \mathbb{R}^n$, $x_1 \geq x_2 \geq \cdots \geq x_n$, $y_1 \geq y_2 \geq \cdots \geq y_n$, and $x \prec_w y$. Let f be a function that is nondecreasing and convex on the whole real axis. Then*

$$\sum_{i=1}^{n} f(x_i) \leq \sum_{i=1}^{n} f(y_i). \tag{5.30}$$

Proof By assumption, $\alpha = \sum\limits_{i=1}^{n} y_i - \sum\limits_{i=1}^{n} x_i \geq 0$. We take numbers x_{n+1}, y_{n+1} such that the following conditions hold: $x_{n+1} \leq x_n$, $y_{n+1} \leq y_n$, and $x_{n+1} - y_{n+1} = \alpha$. Then for the vectors $(x_1, x_2, \ldots, x_{n+1})$ and $(y_1, y_2, \ldots, y_{n+1})$, all conditions of Theorem 5.5 hold. Hence there exists a doubly stochastic matrix $S = \{s_{ij}\}_{i,j=1}^{n+1}$ such that

$$x_i = \sum_{j=1}^{n+1} s_{ij} y_j, \quad i = 1, 2, \ldots, n + 1,$$

whence, using Jensen's inequality, we get

$$f(x_i) \leq \sum_{j=1}^{n+1} s_{ij} f(y_j), \quad i = 1, 2, \ldots, n + 1. \tag{5.31}$$

Summing all inequalities (5.31), we see that

$$\sum_{i=1}^{n+1} f(x_i) \leq \sum_{j=1}^{n+1} \sum_{i=1}^{n+1} s_{ij} f(y_j) = \sum_{i=1}^{n+1} f(y_i). \tag{5.32}$$

By construction, $x_{n+1} \geq y_{n+1}$. By assumption, the function f is nondecreasing. Therefore, $f(x_{n+1}) \geq f(y_{n+1})$. Thus (5.30) follows from (5.32).   □

Remark 5.1 Obviously, if all conditions of Theorem 5.6 hold and additionally $x \prec y$, then the proof is simpler. In this case, we can omit the condition that f is a nondecreasing function.

Corollary 5.1 *Suppose that $x, y \in \mathbb{R}^n$, $x_1 \geq x_2, \geq \cdots \geq x_n$, $y_1 \geq y_2, \geq \cdots \geq y_n$, and $x \prec_w y$. Additionally, we assume that all components of the vectors x and y are nonnegative. Then for every $p > 1$, the following inequality holds:*

$$\sum_{i=1}^{n} x_i^p \le \sum_{i=1}^{n} y_i^p. \tag{5.33}$$

Proof It follows immediately from Theorem 5.4 that for every $p > 1$, the function $f(t) = t^p$ is convex on the positive semiaxis. Extending $f(t)$ to the whole real axis such that $f(t) = 0$ if $t < 0$, we get (5.33). $\qquad\square$

Corollary 5.2 *Suppose that*

$$x_1 \ge x_2 \ge \cdots \ge x_n \ge 0,$$

$$y_1 \ge y_2 \ge \cdots \ge y_n \ge 0,$$

and

$$\prod_{i=1}^{k} x_i \le \prod_{i=1}^{k} y_i, \quad k = 1, 2, \ldots, n. \tag{5.34}$$

Then

$$x \prec_w y. \tag{5.35}$$

Proof If $x = 0$, then the assertion is obvious. Let $x_{p+1} = x_{p+2} = \cdots = x_n = 0$, where $p \ge 1$, and let all other components of the vector x be positive. Then, using conditions (5.34), we see that the first q, $p \le q \le n$, components of the vector y are also positive (and all other components are zero). In this case, conditions (5.34) have the form

$$\prod_{i=1}^{k} x_i \le \prod_{i=1}^{k} y_i, \quad k = 1, 2, \ldots, q. \tag{5.36}$$

It is easy to see that there exists a positive δ such that for all $\varepsilon \in (0, \delta)$ and for

$$\tilde{x}_i = \begin{cases} x_i, & i = 1, 2, \ldots, p, \\ \varepsilon, & i = p+1, p+2, \ldots, q, \end{cases}$$

as a consequence of (5.36), we have

$$\prod_{i=1}^{k} \tilde{x}_i \le \prod_{i=1}^{k} y_i, \quad k = 1, 2, \ldots, q. \tag{5.37}$$

Taking the logarithm of all inequalities (5.37), we obtain

$$\sum_{i=1}^{k} \log \tilde{x}_i \leq \sum_{i=1}^{k} \log y_i, \quad k = 1, 2, \ldots, q. \tag{5.38}$$

If we now put $f(t) = e^t, t \in \mathbb{R}$, then using Theorem 5.6, we get

$$\sum_{i=1}^{k} \tilde{x}_i \leq \sum_{i=1}^{k} y_i, \quad k = 1, 2, \ldots, q. \tag{5.39}$$

Taking the limit as $\delta \to 0$ in these inequalities, we obtain (5.35).   □

5.1.5 Some Estimates of Eigenvalues and Singular Values

It readily follows from the definition that the singular values of a matrix A are calculated by the formulas $\sigma_k = \sqrt{\lambda_k(A^*A)}$, where $\lambda_k(A^*A)$, $k = 1, 2, \ldots, r$, are the nonzero eigenvalues of the matrix A^*A. The next lemma gives another (sometimes more useful) representation of the singular values using the eigenvalues of a Hermitian matrix.

Lemma 5.1 *Let $A \in M_{m,n}$ be an arbitrary matrix, $\sigma_1, \sigma_2, \ldots, \sigma_r$ its singular values, and $\{e_k\}_{k=1}^{n} \subset \mathbb{C}^n$, $\{q_k\}_{k=1}^{m} \subset \mathbb{C}^m$ its singular vectors (see Section 5.1.1, p. 163). Let $\tilde{A}$ be the Hermitian matrix of order $m + n$ of the form*

$$\tilde{A} = \begin{pmatrix} 0 & A \\ A^* & 0 \end{pmatrix}.$$

Then the vectors $u_k = (q_k, e_k) \in \mathbb{C}^{m+n}$, $k = 1, 2, \ldots, r$; $u_{r+k} = (q_k, -e_k)$, $k = 1, 2, \ldots, r$; $u_{2r+k} = (q_k, 0)$, $k = r + 1, r + 2, \ldots, m$; $u_{r+m+k} = (0, e_k)$, $k = r + 1, r + 2, \ldots, n$, form a complete orthogonal set of eigenvectors of $\tilde{A}$. The corresponding eigenvalues are $\pm\sigma_1, \pm\sigma_2, \ldots, \pm\sigma_r$, and $m + n - 2r$ zeros.

The reader can prove this lemma by multiplying the vectors u_k by the matrix $\tilde{A}$ for $k = 1, 2, \ldots, m + n$.

Using Lemma 5.1, we can easily prove, for example, a theorem that is analogous to Theorem 4.54, p. 154. Let us introduce the necessary notation. Let $A \in M_{m,n}$ be an arbitrary matrix, and $\sigma_1, \sigma_2, \ldots, \sigma_r$ its singular values. We denote by $\sigma(A)$ the vector of length $\min(m, n)$ that consists of the singular values of the matrix A completed by zeros if $r < \min(m, n)$ and ordered by nonincreasing values of all its elements.

Theorem 5.7 *Let A and B be arbitrary m × n matrices. Then*

$$\sigma(A + B) \prec_w (\sigma(A) + \sigma(B)).$$

The next theorem is useful for the estimation of the singular values of the product of matrices.

Theorem 5.8 *Let A and B be arbitrary square matrices of order n. Then*

$$\sigma_i(AB) \leq \sigma_1(A)\sigma_i(B), \quad i = 1, 2, \dots, n. \tag{5.40}$$

Proof Let $M = \sigma_1^2(A)I - A^*A$. Obviously, the matrix M is Hermitian and positive semidefinite, and $\sigma_1^2(A)B^*B = B^*(A^*A + M)B = (AB)^*(AB) + B^*MB$. The matrix B^*MB is also Hermitian and positive semidefinite. Therefore, inequalities (5.40) are valid. □

The next corollary is obvious but useful.

Corollary 5.3 *For all $k = 1, 2, \dots, n$ and $p > 1$, the following inequalities hold:*

$$\left(\sum_{i=1}^{k} \sigma_i^p(AB)\right)^{1/p} \leq \left(\sum_{i=1}^{k} \sigma_i^p(A)\right)^{1/p} \left(\sum_{i=1}^{k} \sigma_i^p(B)\right)^{1/p}. \tag{5.41}$$

Theorem 5.9 *Let $A \in M_{m,n}$ be given, and let A_r denote a submatrix of A obtained by deleting a total of r columns and (or) rows from A. Then*

$$\sigma_{k+r}(A) \leq \sigma_k(A_r) \leq \sigma_k(A), \quad k = 1, 2, \dots, \min(m, n), \tag{5.42}$$

where for $X \in M_{p,q}$, we set $\sigma_j(X) = 0$ if $j > \min(p, q)$.

Proof It suffices to note that if columns from the matrix A are deleted (replaced by zeros), then the nonzero rows and columns of the matrix $A_r^*A_r$ form a submatrix corresponding to the principal minor of the matrix A^*A of the corresponding order.[4] After that, to conclude the proof, we use Theorem 4.50, p. 152. □

Lemma 5.2 *Let $A \in M_{m,n}$, $V_k \in M_{m,k}$, $W_k \in M_{n,k}$, $k \leq \min(m, n)$. We assume that the columns of the matrices V_k, W_k are orthonormal with respect to the standard inner products on the spaces $\mathbb{C}^m$, $\mathbb{C}^n$, respectively. Then*

$$\sigma_i(V_k^*A W_k) \leq \sigma_i(A), \quad i = 1, 2, \dots, k. \tag{5.43}$$

Proof Let matrices $V = (V_k, V_{m-k}) \in M_m$, $W = (W_k, W_{n-k}) \in M_n$ be unitary. It is easy to see that the matrix $V_k^*A W_k$ is the leading principal submatrix of order k of the matrix $V^*A W$. Therefore, using Theorem 5.9 and the fact that the singular values

---

[4]Similarly, if rows from the matrix A are deleted, then we get a submatrix of AA^* of the same order.

of every matrix are invariant under its unitary transformations, we get the following estimates:

$$\sigma_i(V_k^* A \, W_k) \le \sigma_i(V^* A \, W) = \sigma_i(A), \quad i = 1, 2, \ldots, k.$$

□

Theorem 5.10 (**Weyl**[5]). *Let $A \in M_n$ have singular values $\sigma_1(A) \ge \ldots \ge \sigma_n(A) \ge 0$ and eigenvalues $\lambda_1(A), \ldots, \lambda_n(A)$ ordered so that $|\lambda_1(A)| \ge \ldots \ge |\lambda_n(A)|$. Then*

$$|\lambda_1(A)\lambda_2(A) \cdots \lambda_k(A)| \le \sigma_1(A)\sigma_2(A) \cdots \sigma_k(A), \quad k = 1, 2, \ldots, n, \qquad (5.44)$$

with equality for $k = n$.

Proof By Schur's theorem, Theorem 4.23, p. 127, there exists a unitary matrix U such that $U^* A U = T$, where T is upper triangular and the numbers $\lambda_1(A), \lambda_2(A), \ldots, \lambda_n(A)$ form the main diagonal of T. Let $U_k \in M_{n,k}$ be the matrix that consists of the first k columns of U. By elementary calculations, we get

$$U^* A U = (U_k, U_{n-k})^* A (U_k, U_{n-k}) = \begin{pmatrix} U_k^* A U_k & T_{12} \\ T_{21} & T_{22} \end{pmatrix} = T,$$

where $U_k^* A U_k$ is an upper triangular matrix, and $\lambda_1(A), \lambda_2(A), \ldots, \lambda_k(A)$ form its main diagonal. Evidently,

$$|\lambda_1(A)\lambda_2(A) \cdots \lambda_k(A)| = |\det(U_k^* A U_k)|$$
$$= \sigma_1(U_k^* A U_k)\sigma_2(U_k^* A U_k) \cdots \sigma_k(U_k^* A U_k).$$

Hence (5.44) follows from Lemma 5.2. Equality in (5.44) for $k = n$ holds, since, as we know, for every square matrix A of order n the following equalities hold: $\det(A) = \lambda_1(A)\lambda_2(A) \cdots \lambda_n(A)$, $|\det(A)| = \sigma_1(A)\sigma_2(A) \cdots \sigma_n(A)$.   □

Now it follows from Corollary 5.2, p. 172, that for every matrix $A \in M_n$ we have

$$\sum_{i=1}^{k} |\lambda_i(A)| \le \sum_{i=1}^{k} \sigma_i(A), \quad k = 1, 2 \ldots, n. \qquad (5.45)$$

Theorem 5.11 *Let $A \in M_{m,p}$, $B \in M_{p,n}$ be arbitrary matrices, $q = \min(m, n, p)$. Then*

$$\prod_{i=1}^{k} \sigma_i(AB) \le \prod_{i=1}^{k} \sigma_i(A)\sigma_i(B), \quad k = 1, 2, \ldots, q. \qquad (5.46)$$

If $m = n = p$, then equality holds in (5.46) for $k = n$.

---

[5]Hermann Klaus Hugo Weyl (1885–1955) was a German mathematician.

Proof Let $AB = UDV$ be the singular value decomposition of the product AB. Then $D = U^*ABV^*$. Denote by U_k, V_k^* the matrices consisting of the first k columns of the matrices U, V^*, respectively. Then $U_k^* ABV_k^* = \mathrm{diag}(\sigma_1(AB), \sigma_2(AB), \ldots, \sigma_k(AB))$, because it is the leading principal submatrix of order k of the matrix D. By assumption, $p \geq k$. Therefore, by Theorem 5.1, there exists a polar decomposition $BV_k^* = X_k Q_k$, where $X_k^* X_k = I_k$, and $Q_k \in M_k$ is a non-negative semidefinite Hermitian matrix, and I_k is the identity matrix of order k. By elementary calculations, we get the equality $Q_k^2 = (BV_k^*)^* BV_k^*$. Therefore, by Lemma 5.2, we obtain

$$\det(Q_k^2) = \det(V_k B^* BV_k^*) \leq \sigma_1(B^*B)\sigma_2(B^*B)\cdots\sigma_k(B^*B) = \sigma_1^2(B)\sigma_2^2(B)\cdots\sigma_k^2(B).$$

Using Lemma 5.2 one more time, we see that

$$\begin{aligned}
\sigma_1(AB)\sigma_2(AB)\cdots\sigma_k(AB) &= |\det(U_k^*ABV_k^*)| \\
&= |\det(U_k^*AX_kQ_k)| = |\det(U_k^*AX_k)\det(Q_k)| \\
&\leq \sigma_1(A)\sigma_2(A)\cdots\sigma_k(A)\sigma_1(B)\sigma_2(B)\cdots\sigma_k(B).
\end{aligned}$$

Finally, if $m = n = p$, then

$$\begin{aligned}
\sigma_1(AB)\sigma_2(AB)\cdots\sigma_n(AB) &= |\det(AB)| \\
&= |\det(A)||\det(B)| = \sigma_1(A)\sigma_2(A)\cdots\sigma_n(A)\sigma_1(B)\sigma_2(B)\cdots\sigma_n(B).
\end{aligned}$$

$\square$

Using Corollary 5.2, p. 172, we see that under the assumptions of Theorem 5.11, the following inequalities hold:

$$\sum_{i=1}^{k}\sigma_i(AB) \leq \sum_{i=1}^{k}\sigma_i(A)\sigma_i(B), \quad k = 1, 2, \ldots, n. \tag{5.47}$$

Sums of singular values of a matrix have useful variational characterizations. Let us introduce the following concept, which we will use below in the formulations of corresponding results. A matrix $C \in M_{m,n}$ is called a *partial isometry* of rank k if it has rank k and all its (nonzero) singular values are equal to one. If $m = n = k$, we get the set of all unitary matrices of order n.

Theorem 5.12 *Let $A \in M_{m,n}$, $q = \min(m, n)$. Then for each $k = 1, 2, \ldots, q$, the following equalities hold:*

$$\sum_{i=1}^{k}\sigma_i(A) = \max_{X,Y}|\mathrm{tr}X^*AY|, \tag{5.48}$$

$$\sum_{i=1}^{k} \sigma_i(A) = \max_{C} |\text{tr}AC|. \tag{5.49}$$

In the first case, the maximum is taken over all matrices

$$X \in M_{m,k}, \ Y \in M_{n,k} \text{ such that } X^*X = I, \ Y^*Y = I. \tag{5.50}$$

In the second case, the maximum is taken over all matrices $C \in M_{n,m}$ that are partial isometries of rank k.

Proof First we show that formulations (5.48) and (5.49) are equivalent. Using formula (4.81), p. 125, we get $\text{tr}X^*AY = \text{tr}AYX^* = \text{tr}AC$, where $C = YX^* \in M_{n,m}$. Therefore, $C^*C = XX^*$. As we have seen on p. 164, all the nonzero eigenvalues of the self-adjoint matrices XX^* and X^*X coincide. Hence all their singular values coincide too, but X^*X is the identity matrix of order k. Thus the matrix C has exactly k singular values, and all of them are equal to one, i.e., C is a partial isometry of rank k. Conversely, if $C \in M_{n,m}$ is a partial isometry of rank k, then by definition, the following singular value decomposition of C holds:

$$C = (Y_k, Y_{n-k}) \begin{pmatrix} I_k & 0 \\ 0 & 0 \end{pmatrix} \begin{pmatrix} X_k^* \\ X_{m-k}^* \end{pmatrix}, \tag{5.51}$$

where I_k is the identity matrix of order k, and the matrices

$$Y = (Y_k, Y_{n-k}) \in M_n, \quad X^* = \begin{pmatrix} X_k^* \\ X_{m-k}^* \end{pmatrix} \in M_m$$

are unitary. Using equality (5.51), by elementary calculations, we see that $C = Y_k X_k^*$, and the matrices Y_k, X_k^* satisfy conditions (5.50). Thus the equivalence of formulations (5.48) and (5.49) is established.

Now successively using (5.45) and (5.47), we can write that if C is an arbitrary partial isometry of rank k, then

$$|\text{tr}(AC)| \le \sum_{i=1}^{m} |\lambda_i(AC)| \le \sum_{i=1}^{m} \sigma_i(AC) \le \sum_{i=1}^{q} \sigma_i(A)\sigma_i(C) = \sum_{i=1}^{k} \sigma_i(A). \tag{5.52}$$

To conclude the proof, it is enough to find a partial isometry C of rank k such that inequality (5.52) transforms to an equality. Let $A = UDV$ be the singular value decomposition of the matrix A. Put $C = V^*PU^*$, where

$$P = \begin{pmatrix} I_k & 0 \\ 0 & 0 \end{pmatrix} \in M_{n,m}.$$

By construction, $C \in M_{n,m}$, and it is a partial isometry of rank k. Moreover,

$$AC = UDVV^*PU^* = UDPU^*,$$

and therefore, $\mathrm{tr}(AC) = \mathrm{tr}(DP)$, and using elementary calculations, we obtain the equality $\mathrm{tr}(DP) = \sum_{i=1}^{k} \sigma_i(A)$. □

5.2 The Jordan Canonical Form

In this section we show that for every linear operator acting on a complex finite-dimensional space $\mathbf{X}_n$, there exists a basis such that the matrix of the operator with respect to this basis has a very simple form. It is bidiagonal. The elements of the main diagonal of this matrix form the set of all eigenvalues of the operator. Each element of the diagonal above the main diagonal is either one or zero. A matrix of this form is called a *Jordan matrix*.[6] To obtain the *Jordan canonical form* of the operator it is necessary to take its matrix in an arbitrarily chosen basis and after that to reduce this matrix to the Jordan canonical form by a similarity transformation. This plan is realized in this section.

The following question naturally arises: is it possible to reduce every matrix to diagonal form by a similarity transformation? Simple examples show that it is impossible. For instance, if we require that the matrix SAS^{-1} be diagonal, where

$$A = \begin{pmatrix} 0 & 1 \\ 0 & 0 \end{pmatrix}$$

and S is a nonsingular matrix, then we get contradictory equalities.

5.2.1 Existence and Uniqueness of the Jordan Canonical Form

We begin with a definition. A *Jordan block* $J_k(\lambda)$ is a $k \times k$ upper triangular matrix of the form

---

[6]Marie Ennemond Camille Jordan (1838–1922) was a French mathematician.

$$J_k(\lambda) = \begin{pmatrix} \lambda & 1 & & & 0 \\ & \lambda & 1 & & \\ & & \ddots & \ddots & \\ & & & \lambda & 1 \\ 0 & & & & \lambda \end{pmatrix}. \tag{5.53}$$

We explain that all k elements of the main diagonal of the matrix $J_k(\lambda)$ are equal to λ, all $k-1$ elements of the diagonal above the main diagonal are equal to one, and all other entries of this matrix are zero.

It is useful to note that if the matrix of the operator $\mathcal{A} \colon \mathbf{X}_k \to \mathbf{X}_k$ with respect to a basis $\{e_i\}_{i=1}^k$ is the Jordan block $J_k(0)$, then, evidently, the vectors of this basis are connected to each other by the following relationships:

$$\mathcal{A}e_1 = 0, \quad \mathcal{A}e_2 = e_1, \quad \ldots, \quad \mathcal{A}e_k = e_{k-1}.$$

If we denote the vector e_k by f, then we see that the basis $\{e_i\}_{i=1}^k$ consists of the vectors $f, \mathcal{A}f, \mathcal{A}^2 f, \ldots, \mathcal{A}^{k-1} f,$[7] and moreover, $\mathcal{A}^k f = 0$.

Let us formulate now the main result of this section.

Theorem 5.13 *Let A be a given complex matrix of order n. There is a nonsingular matrix S such that*

$$S^{-1}AS = J, \tag{5.54}$$

where

$$J = \begin{pmatrix} J_{n_1}(\lambda_1) & & & 0 \\ & J_{n_2}(\lambda_2) & & \\ & & \ddots & \\ 0 & & & J_{n_k}(\lambda_k) \end{pmatrix} \tag{5.55}$$

and $n_1 + n_2 + \cdots + n_k = n$. The numbers λ_i, $i = 1, 2, \ldots, k$ (which are not necessarily distinct), form the set of all characteristic values of the matrix A (according to their multiplicities).

The matrix (5.55) is called the *Jordan canonical form* of the matrix A. Obviously, Theorem 5.13 is equivalent to the following statement. For every operator $\mathcal{A}$ acting on a finite-dimensional complex space, there exists a basis $\mathcal{E}_n$ such that the matrix of the operator $\mathcal{A}$ with respect to this basis has the form (5.55), i.e.,

$$\mathcal{A}\mathcal{E}_n = \mathcal{E}_n J. \tag{5.56}$$

The basis $\mathcal{E}_n$ is called a *Jordan basis*.

---

[7]Which are listed in reverse order.

The easiest proof of the existence of a Jordan basis is for a nilpotent operator. Using Theorem 4.21, p. 126, and Schur's theorem, p. 127, we see that an operator is nilpotent if and only if there exists a basis such that the matrix of the operator with respect to this basis is upper triangular and all elements of the main diagonal of this matrix are zero.

Theorem 5.14 *Let* $\mathcal{A} : \mathbf{X}_n \to \mathbf{X}_n$ *be a nilpotent operator acting on a complex vector space* $\mathbf{X}_n$. *Then there exists a basis of the space* $\mathbf{X}_n$ *such that the matrix of the operator* $\mathcal{A}$ *with respect to this basis has the following Jordan canonical form:*

$$
\begin{pmatrix}
J_{n_1}(0) & & & 0 \\
& J_{n_2}(0) & & \\
& & \ddots & \\
0 & & & J_{n_m}(0)
\end{pmatrix}. \tag{5.57}
$$

Here $n_1 + n_2 + \cdots + n_m = n$.

Proof Taking into account the remark in the last paragraph before Theorem 5.13, it is easy to see that the assertion is equivalent to the following one: for every nilpotent operator $\mathcal{A} : \mathbf{X}_n \to \mathbf{X}_n$, there exist vectors $f_1, f_2, \ldots f_m$ such that the vectors

$$
f_1, \ \mathcal{A}f_1, \ \mathcal{A}^2 f_1, \ \ldots, \ \mathcal{A}^{n_1-1} f_1, \ f_2, \ \mathcal{A}f_2, \ \mathcal{A}^2 f_2, \ \ldots, \ \mathcal{A}^{n_2-1} f_2, \ldots,
$$
$$
f_m, \ \mathcal{A}f_m, \ \mathcal{A}^2 f_m, \ \ldots, \ \mathcal{A}^{n_m-1} f_m \tag{5.58}
$$

form a basis in the space $\mathbf{X}_n$, and

$$
\mathcal{A}^{n_1} f_1 = \mathcal{A}^{n_2} f_2 = \cdots = \mathcal{A}^{n_m} f_m = 0. \tag{5.59}
$$

We prove the existence of the required basis by induction on the dimension of the space. In the case of a nilpotent operator acting on a one-dimensional space, the assertion is obviously true. Now we suppose that the assertion is true for every space whose dimension is less than n, and prove that then the statement is true for every n-dimensional space.

The operator $\mathcal{A}$ is nilpotent. Therefore, $\mathrm{def}(\mathcal{A}) \geq 1$, and hence $\mathrm{rank}(\mathcal{A}) < n$ (see equality (4.23), p. 102). Evidently, the subspace $\mathrm{Im}(\mathcal{A})$ is invariant under $\mathcal{A}$, whence by the induction hypothesis, we conclude that there exist vectors $u_1, u_2, \ldots, u_k$ such that the vectors

$$
u_1, \ \mathcal{A}u_1, \ \mathcal{A}^2 u_1, \ \ldots, \ \mathcal{A}^{p_1-1} u_1, \ u_2, \ \mathcal{A}u_2, \ \mathcal{A}^2 u_2, \ \ldots, \ \mathcal{A}^{p_2-1} u_2, \ldots,
$$
$$
u_k, \ \mathcal{A}u_k, \ \mathcal{A}^2 u_k, \ \ldots, \ \mathcal{A}^{p_k-1} u_k \tag{5.60}
$$

form a basis of the subspace $\mathrm{Im}(\mathcal{A})$, and

$$\mathcal{A}^{p_1} u_1 = \mathcal{A}^{p_2} u_2 = \cdots = \mathcal{A}^{p_k} u_k = 0. \tag{5.61}$$

For $i = 1, 2, \ldots, k$, the vectors u_i belong to $\mathrm{Im}(\mathcal{A})$. Hence there exist vectors $v_i \in \mathbf{X}_n$ such that

$$u_i = \mathcal{A} v_i. \tag{5.62}$$

The vectors

$$\mathcal{A}^{p_i - 1} u_i, \ i = 1, 2, \ldots, k, \tag{5.63}$$

belong to the basis (5.60). Hence they are linearly independent. Relationships (5.61) show that these vectors belong to $\mathrm{Ker}(\mathcal{A})$. Thus we can join vectors (5.63) with some vectors $w_1, w_2, \ldots, w_l$ to complete the basis of the subspace $\mathrm{Ker}(\mathcal{A})$.

If we prove now that the vectors

$$v_1, \ \mathcal{A} v_1, \ \ldots, \ \mathcal{A}^{p_1} v_1, \ v_2, \ \mathcal{A} v_2, \ \ldots, \ \mathcal{A}^{p_2} v_2, \ \ldots, \ v_k, \ \mathcal{A} v_k, \ \ldots, \ \mathcal{A}^{p_k} v_k,$$
$$w_1, \ w_2, \ \ldots, \ w_l \tag{5.64}$$

form a basis of the space $\mathbf{X}_n$, then obviously, this basis is the required Jordan basis of the operator $\mathcal{A}$. The set (5.64) consists of n vectors. Indeed, this set consists of $p_1 + \cdots + p_k + k + l$ elements, and moreover, $p_1 + \cdots + p_k = \mathrm{rank}(\mathcal{A})$, $k + l = \mathrm{def}(\mathcal{A})$, but $\mathrm{rank}(\mathcal{A}) + \mathrm{def}(\mathcal{A}) = n$ for every operator $\mathcal{A}$. Further, put

$$\alpha_{1,0} v_1 + \alpha_{1,1} \mathcal{A} v_1 + \cdots + \alpha_{1,p_1} \mathcal{A}^{p_1} v_1 + \alpha_{2,0} v_2 + \alpha_{2,1} \mathcal{A} v_2 + \cdots + \alpha_{2,p_2} \mathcal{A}^{p_2} v_2$$
$$+ \cdots + \alpha_{k,0} v_k + \alpha_{k,1} \mathcal{A} v_k + \cdots + \alpha_{k,p_k} \mathcal{A}^{p_k} v_k$$
$$+ \beta_1 w_1 + \beta_2 w_2 + \cdots + \beta_l w_l = 0. \tag{5.65}$$

Acting on both sides of the last equality by the operator $\mathcal{A}$, using relationships (5.61), (5.62), and also using the fact that $w_1, w_2, \ldots, w_l \in \mathrm{Ker}(\mathcal{A})$, we get

$$\alpha_{1,0} u_1 + \alpha_{1,1} \mathcal{A} u_1 + \cdots + \alpha_{1,p_1 - 1} \mathcal{A}^{p_1 - 1} u_1$$
$$+ \alpha_{2,0} u_2 + \alpha_{2,1} \mathcal{A} u_2 + \cdots + \alpha_{2,p_2 - 1} \mathcal{A}^{p_2 - 1} u_2$$
$$+ \cdots + \alpha_{k,0} u_k + \alpha_{k,1} \mathcal{A} u_k + \cdots + \alpha_{k,p_k - 1} \mathcal{A}^{p_k - 1} u_k = 0. \tag{5.66}$$

The vectors (5.60) are linearly independent. Therefore, all the coefficients in the linear combination on the left-hand side of (5.66) are zero, and (5.65) has the form

$$\alpha_{1,p_1} \mathcal{A}^{p_1} v_1 + \alpha_{2,p_2} \mathcal{A}^{p_2} v_2 + \cdots + \alpha_{k,p_k} \mathcal{A}^{p_k} v_k$$
$$+ \beta_1 w_1 + \beta_2 w_2 + \cdots + \beta_l w_l = 0. \tag{5.67}$$

The left-hand side of (5.67) is a linear combination of the vectors of the basis of the subspace $\mathrm{Ker}(\mathcal{A})$. Therefore, all the coefficients in this linear combination are equal to zero. Thus we have proved that all the coefficients in the linear combination on the left-hand side of equality (5.65) must all be zero, i.e., the set of vectors (5.64) is linearly independent and consists of n vectors, and hence it is a basis of the space $\mathbf{X}_n$. $\qquad\square$

The next theorem is an immediate generalization of Theorem 5.14.

Theorem 5.15 *Suppose that an operator $\mathcal{A}$ acting on a complex space $\mathbf{X}_n$ has the form $\mathcal{A} = \mathcal{A}_0 + \lambda I$, where $\mathcal{A}_0$ is a nilpotent operator, and λ is an arbitrary number. Then the matrix of the operator $\mathcal{A}$ with respect to the Jordan basis of the operator $\mathcal{A}_0$ has the following Jordan canonical form:*

$$
\begin{pmatrix}
J_{n_1}(\lambda) & & & 0 \\
& J_{n_2}(\lambda) & & \\
& & \ddots & \\
0 & & & J_{n_m}(\lambda)
\end{pmatrix}. \tag{5.68}
$$

This statement follows immediately from the facts that linear operations with their matrices correspond to linear operations with operators and that the matrix of the identity operator with respect to every basis is the identity matrix.

Proof of Theorem 5.13. Representation (5.54) is the result of the sequential realization of the following steps.

1. Using Schur's theorem, p. 127, we construct an upper triangular matrix T that is unitarily similar to the matrix A.

2. Using Theorem 4.24, p. 129, we reduce the matrix T to block diagonal form. Each block here is an upper triangular matrix. All diagonal elements of this matrix are equal and coincide with a characteristic value of the matrix A.

3. Using Theorems 5.14 and 5.15, we independently reduce each block constructed in the second step to the form (5.68). $\qquad\square$

The next lemma is useful for the investigation of the uniqueness of the Jordan canonical form.

Lemma 5.3 *The following relationships hold for a Jordan block $J_k(0)$:*

$$
(J_k(0))^k = 0, \tag{5.69}
$$

$$
(J_k(0))^j \neq 0, \quad j = 1, 2, \ldots, k - 1. \tag{5.70}
$$

Proof The relationship (5.69) follows immediately from Theorem 4.21, p. 126, and Corollary 4.3, p. 126. The relationships (5.70) are easily verified by direct calculation. It is important to note that while the orders of the matrices $J_k(0)$ increases sequentially with k, the nonzero columns of $J_k(0)$ are displaced to the right. $\qquad\square$

Theorem 5.16 *The Jordan matrix* (5.55) *is uniquely determined by the matrix A* (*up to permutation of the diagonal Jordan blocks*).

Proof Two possible Jordan canonical forms of the matrix A are similar to the matrix A. Therefore, they have the same set of characteristic values (according to their multiplicities). Hence to complete the proof, it is enough to show the coincidence of the orders of the Jordan blocks that correspond to a given characteristic value of the matrix A.

This problem can be formulated as follows: prove the coincidence of the orders of the Jordan blocks of two possible Jordan canonical forms of a matrix that has a unique characteristic value. Moreover, arguing as in the proof of Theorem 5.15, it is easy to see that it is sufficient to consider a matrix A_0 with a unique zero characteristic value.

Thus, let

$$
J(0) = \begin{pmatrix} J_{n_1}(0) & & 0 \\ & \ddots & \\ 0 & & J_{n_k}(0) \end{pmatrix}, \quad \tilde{J}(0) = \begin{pmatrix} J_{m_1}(0) & & 0 \\ & \ddots & \\ 0 & & J_{m_r}(0) \end{pmatrix}
$$

be two possible Jordan canonical forms of the matrix A_0. We assume that the Jordan blocks are sorted according to nondecreasing values of their orders (this may be achieved by the corresponding numeration of the Jordan bases) such that

$$
n_1 \geq n_2 \geq \cdots \geq n_k, \quad n_1 + n_2 + \cdots + n_k = n,
$$

$$
m_1 \geq m_2 \geq \cdots \geq m_r, \quad m_1 + m_2 + \cdots + m_r = n,
$$

where n is the order of the matrix A_0. Suppose that the first $l - 1, l \geq 1$, Jordan blocks of the matrices $J(0)$ and $\tilde{J}(0)$ coincide. By assumption, there exists a nonsingular matrix S such that

$$
J(0) = S\tilde{J}(0)S^{-1}. \tag{5.71}
$$

As a result of the assumption of the coincidence of the first Jordan blocks, the matrix S has the following form:

$$
S = \begin{pmatrix} I_p & 0 \\ 0 & S_{n-p} \end{pmatrix},
$$

where I_p is the identity matrix of order $p = n_1 + \cdots + n_{l-1}$. This gives us the opportunity to consider only the matrices $J(0)$ and $\tilde{J}(0)$ such that their first blocks do not coincide, i.e., $J_{n_1}(0) \neq J_{m_1}(0)$. If we prove that this is impossible, then we will have concluded the proof of the theorem. To be definite, assume that $n_1 > m_1$. Raising both sides of equality (5.71) to the power m_1, we get

$$(J(0))^{m_1} = S(\tilde{J}(0))^{m_1} S^{-1}. \tag{5.72}$$

Using Lemma 5.3, we see that $(\tilde{J}(0))^{m_1} = 0$ and also that $(J(0))^{m_1} \neq 0$. This contradiction concludes the proof of the theorem. $\qquad\square$

5.2.2 Root and Cyclic Subspaces

A Jordan matrix is block diagonal. Hence a space $\mathbf{X}_n$ can be represented as a direct sum of invariant subspaces of the operator $\mathcal{A}$ corresponding to the blocks of the Jordan matrix (see Section 4.2.1, p. 114). The subspace corresponding to the block $J_{n_j}(\lambda_j)$ in representation (5.54) is called a *cyclic* subspace. The direct sum of all cyclic subspaces corresponding to the same eigenvalue λ of the operator $\mathcal{A}$ is called the *root* subspace.

Let us investigate the structure of cyclic and root subspaces. Suppose that an m-dimensional cyclic subspace corresponds to the eigenvalue λ of the operator $\mathcal{A}$. For the sake of being definite, assume that the vectors $\{e_k\}_{k=1}^m$ of the basis $\mathcal{E}_n$ belong to this subspace. Using (5.56), we see that

$$\mathcal{A}e_1 = \lambda e_1, \quad \mathcal{A}e_2 = \lambda e_2 + e_1, \quad \dots, \quad \mathcal{A}e_m = \lambda e_m + e_{m-1}. \tag{5.73}$$

This immediately implies that e_1 is an eigenvector of the operator $\mathcal{A}$. Clearly, the vectors $e_1, e_2, \dots, e_{m-1}$ are nonzero, and therefore no other vectors $e_2, e_3, \dots, e_m$ are eigenvectors of the operator $\mathcal{A}$.

Every cyclic subspace includes exactly one eigenvector of the operator $\mathcal{A}$. Indeed, if we assume that $x = \xi_1 e_1 + \xi_2 e_2 + \cdots + \xi_m e_m$ is an eigenvector of the operator $\mathcal{A}$, then $J_m(\lambda)\xi = \lambda\xi$, where $\xi = (\xi_1, \xi_2, \dots, \xi_m)^T$. The last equality is equivalent to $J_m(0)\xi = 0$. The rank of the matrix $J_m(0)$ is $m - 1$. Indeed, $\det J_m(0) = 0$, and the minor obtained from the determinant $\det J_m(0)$ by deleting the first column and the last row is equal to one. Thus the dimension of the kernel of the matrix $J_m(0)$ is one.

Clearly, if the root subspace corresponding to an eigenvalue λ of the operator $\mathcal{A}$ is the direct sum of k cyclic subspaces, then it contains exactly k linearly independent eigenvectors of the operator $\mathcal{A}$ corresponding to the eigenvalue λ. Therefore, the number of cyclic subspaces of a given root subspace is equal to the geometric multiplicity of the corresponding eigenvalue λ.

The sum of dimensions of all cyclic subspaces corresponding to the eigenvalue λ is equal the multiplicity of λ as the root of the characteristic equation, i.e., it is equal to the algebraic multiplicity of the eigenvalue λ.

It follows immediately from (5.73) that

$$(\mathcal{A} - \lambda I)^j e_j = 0, \quad j = 1, 2, \dots, m. \tag{5.74}$$

Also it is easy to see that $(A - \lambda I)^p e_j \neq 0$ for $p < j$. For this reason, the integer j is called the *height* of the *cyclic vector* e_j. Particularly, an eigenvector is a cyclic vector of height one.

It is easy to guess that if l is the dimension of the root subspace corresponding to the eigenvalue λ of the operator $\mathcal{A}$, then for every vector x of this subspace, the following equality holds:

$$(A - \lambda I)^l x = 0. \tag{5.75}$$

Remark 5.2 Obviously, a Jordan basis is not uniquely determined by the operator $\mathcal{A}$. Moreover, if we have a Jordan basis, then using it, we can easily construct another Jordan basis. For example, if in the basis $\mathcal{E}_n$ we replace the vector e_2 by the vector $\tilde{e}_2 = e_2 + \alpha e_1$, where α is an arbitrary number, then for this new basis, the equalities (5.73) hold, i.e., it is also a Jordan basis of the operator $\mathcal{A}$. However, since the Jordan matrix is uniquely determined by the operator $\mathcal{A}$ (up to permutation of the diagonal Jordan blocks), all Jordan bases have the structure described above.

5.2.3   The Real Jordan Canonical Form

Theorem 5.17 *Let A be a real square matrix of order $n \geq 1$. There exists a nonsingular real matrix S such that $S^{-1} A S = J$, where*

$$J = diag(J_{m_1}(\lambda_1), J_{m_2}(\lambda_2), \ldots, J_{m_l}(\lambda_l))$$

and $\lambda_1, \lambda_2,\ldots, \lambda_l$ are the characteristic values of the matrix A. If λ_k is real, then the Jordan block $J_{m_k}(\lambda_k)$ that corresponds to λ_k is of exactly the same form as (5.55) in Theorem 5.13. The diagonal block of the form

$$\begin{pmatrix}
\alpha & \beta & 1 & 0 & 0 & 0 & \ldots & 0 & 0 \\
-\beta & \alpha & 0 & 1 & 0 & 0 & \ldots & 0 & 0 \\
0 & 0 & \alpha & \beta & 1 & 0 & \ldots & 0 & 0 \\
0 & 0 & -\beta & \alpha & 0 & 1 & \ldots & 0 & 0 \\
\ldots & \ldots & \ldots & \ldots & \ldots & \ldots & \ldots & \ldots & \ldots \\
\ldots & \ldots & \ldots & \ldots & \ldots & \ldots & \ldots & \ldots & \ldots \\
\ldots & \ldots & \ldots & \ldots & \ldots & \ldots & \ldots & 0 & 1 \\
0 & 0 & 0 & 0 & 0 & 0 & \ldots & \alpha & \beta \\
0 & 0 & 0 & 0 & 0 & 0 & \ldots & -\beta & \alpha
\end{pmatrix} \tag{5.76}$$

in the matrix J corresponds to each pair of the complex conjugate characteristic values $\lambda = \alpha + i\beta$, $\bar{\lambda} = \alpha - i\beta$ of the matrix A. The dimension of block (5.76) is twice the dimension of the corresponding Jordan block of the eigenvalue λ in the matrix A in representation (5.54), p. 180.

Proof Let $\{e_k\}_{k=1}^n$ be the Jordan basis of the matrix A constructed in the proof of Theorem 5.13. Suppose that the vectors $\{e_k\}_{k=1}^m$, $m \le n$, correspond to the real characteristic values of the matrix A. Taking into account the relationships of the form (5.73), it is easy to see that these vectors can be real. Suppose that the vectors $\{e_k\}_{k=m+1}^{m+p}$ correspond to the Jordan block $J_p(\lambda)$ of the complex characteristic value $\lambda = \alpha + i\beta$ of the matrix A. Then we can assume that the vectors $\{\bar{e}_k\}_{k=m+1}^{m+p}$ of the same Jordan basis correspond to the block $J_p(\bar{\lambda})$. Put $x_k = \operatorname{Re} e_k$, $y_k = \operatorname{Im} e_k$, where $k = m + 1, m + 2, \ldots, m + p$. It is easy to see that the vectors

$$e_1, e_2, \ldots, e_m, x_{m+1}, y_{m+1}, \ldots, x_{m+p}, y_{m+p}, e_{m+2p+1}, \ldots, e_n \quad (5.77)$$

are linearly independent by writing the change of basis matrix from the Jordan basis $\{e_k\}_{k=1}^n$ to (5.77). Now let us move from the Jordan basis to the basis (5.77). Note that by the definition of the Jordan basis,

$$Ae_{m+1} = \lambda e_{m+1}, \ Ae_{m+2} = \lambda e_{m+2} + e_{m+1}, \ \ldots, \ Ae_{m+p} = \lambda e_{m+p} + e_{m+p-1}.$$

Equating the real and imaginary parts of these equalities,[8] we see that the block of the form (5.76) corresponds to the vectors $x_{m+1}, y_{m+1}, \ldots, x_{m+p}, y_{m+p}$ in the basis (5.77). To conclude the proof, we apply this process to all other pairs of the complex conjugate characteristic values of the matrix A. □

Let us give some examples of applications of the Jordan canonical form.

Theorem 5.18 *A square matrix A is similar to its transpose A^T.*

Proof If we represent A in Jordan canonical form and write $A = SJS^{-1}$, then $A^T = (S^{-1})^T J^T S^T$. If there exists a nonsingular matrix P such that

$$J^T = P^{-1}JP, \quad (5.78)$$

then the assertion is true. Indeed, in this case the matrix A^T is similar to the matrix J, and J is similar to A. Obviously, it is enough to check the equality of the form (5.78) only for an arbitrary Jordan block. Moreover, since every Jordan block is equal to $\lambda I + J(0)$, it is sufficient to specify a matrix P such that $(J(0))^T = P^{-1}J(0)P$. By elementary calculations, we see that the permutation matrix

$$P = \begin{pmatrix} 0 & & & 1 \\ & & \cdot^{\cdot^{\cdot}} & \\ & 1 & & \\ 1 & & & 0 \end{pmatrix} \quad (5.79)$$

satisfies the required condition. □

---

[8] Analogous calculations were done in Section 4.2.5, p. 125.

Theorem 5.19 *Every real square matrix A can be represented as the product of two real symmetric matrices, one of which can be nonsingular.*

Proof By Theorem 5.17, the matrix A is represented in the real Jordan canonical form: $A = SJS^{-1}$. Using elementary calculations, we see that the matrix JP, where the matrix P is defined by equality (5.79), is symmetric (it is convenient to perform calculations for each Jordan block of the matrix J separately). Now we write the evident equalities $A = SJPPS^{-1} = SJPS^T(S^{-1})^T PS^{-1}$. Clearly, the matrix $SJPS^T$ is symmetric, and the matrix $(S^{-1})^T PS^{-1}$ is symmetric and nonsingular. $\square$

5.2.4 Power Series of Matrices

Let us recall that by M_n we denote the set of all square matrices of order n with complex, generally speaking, elements. An infinite series of the form

$$a_0 I + a_1 A + a_2 A^2 + \cdots + a_k A^k + \cdots , \tag{5.80}$$

where $a_0, a_1, \ldots,$ are complex numbers, $A \in M_n$, $n \geq 1$, is called a *power series of matrices*. We say that series (5.80) *converges* if there is a matrix B such that

$$\lim_{m \to \infty} \sum_{k=0}^{m} a_k A^k = B.$$

We connect power series of matrices (5.80) with the following power series:

$$\sum_{k=0}^{\infty} a_k \lambda^k, \quad \lambda \in \mathbb{C}. \tag{5.81}$$

Let us recall some results from calculus. The series (5.81) is connected with the series of nonnegative terms: $\sum_{k=0}^{\infty} |a_k| t^k$, $t \geq 0$. The set of all numbers $t \geq 0$ such that $\sum_{k=0}^{\infty} |a_k| t^k < \infty$ forms an interval on the positive semiaxis. This interval includes the point $t = 0$, can be open or closed on the right, finite or infinite. The length of this interval (denote it by r) is called the radius of convergence of the power series (5.81). The series (5.81) converges absolutely for all $|\lambda| < r$. For $|\lambda| > r$, the series (5.81) diverges.

Now we clarify the conditions of convergence of the series (5.80).

Let $\lambda_1, \lambda_2, \ldots, \lambda_n$ be all the eigenvalues of the matrix A. The *spectral radius* of A is the nonnegative number

$$\rho(A) = \max_{1 \leq j \leq n} |\lambda_j|. \tag{5.82}$$

This is just the radius of the smallest closed disk centered at the origin in the complex plane that includes all eigenvalues of A.

Let λ be an eigenvalue of the matrix A. Denote by n_λ the maximal order of the Jordan blocks corresponding to λ.

Theorem 5.20 *(1) If $\rho(A) < r$, then the series (5.80) converges. (2) If $\rho(A) > r$, then the series (5.80) diverges. (3) If $\rho(A) = r$, then the series (5.80) converges if and only if for each characteristic value λ of the matrix A such that $|\lambda| = \rho(A)$, the following series (which are obtained by differentiation of the series (5.81)) converge:*

$$\sum_{k=1}^{\infty} a_k k(k-1)\cdots(k-j+1)\lambda^{k-j}, \ j = 0, 1, \ldots, n_\lambda - 1. \tag{5.83}$$

Proof Suppose that the matrix S reduces the matrix A to Jordan canonical form, i.e., $A = SJS^{-1}$, where the matrix J is defined by equality (5.55), p. 180. Then for each $m \geq 0$, we have

$$\sum_{k=0}^{m} a_k A^k = S\left(\sum_{k=0}^{m} a_k J^k\right) S^{-1}.$$

Therefore, the series (5.80) converges if and only if the series

$$\sum_{k=0}^{\infty} a_k J^k \tag{5.84}$$

converges. The series (5.84) converges if and only if each of the following series converges: $\sum_{k=0}^{\infty} a_k J_l^k(\lambda)$, where λ is the characteristic value of the matrix A, $J_l(\lambda)$ is the Jordan block corresponding to λ, and l is the order of the matrix $J_l(\lambda)$. By definition, $1 \leq l \leq n_\lambda$. Now we note that $J_l(\lambda) = \lambda I_l + J_l(0)$. By Lemma 5.3, p. 183, it follows that $(J_l(0))^l = 0$, $(J_l(0))^j \neq 0$, $j = 1, 2, \ldots, l-1$. Therefore, for each $k > l - 1$, we get

$$(J_l(\lambda))^k = \lambda^k I + C_1^k \lambda^{k-1} J_l(0) + \cdots + C_{l-1}^k \lambda^{k-l+1} (J_l(0))^{l-1}. \tag{5.85}$$

Thus the investigation of convergence of the series (5.84) is reduced to the investigation of convergence of l power series of the form (5.83). If $\rho(A) > r$, then there exists a characteristic value λ of the matrix A such that $|\lambda| > r$. In this case, the series $\sum_{k=l}^{\infty} a_k \lambda^k$ corresponding to $j = 0$ diverges, and hence the condition $\rho(A) \leq r$ is necessary for the convergence of the series (5.80). Let now $\rho(A) < r$. Then for

each characteristic value λ of the matrix A, the following inequality holds: $|\lambda| < r$.[9]
Therefore, for each $j = 1, 2, \ldots, l - 1$ and for all large enough k, we get

$$|a_k k(k - 1) \cdots (k - j + 1)\lambda^{k-j}|$$

$$= |a_k|k(k - 1) \cdots (k - j + 1) \left(\frac{|\lambda|}{r}\right)^{k-j} r^{k-j} \leq r^j |a_k| r^k,$$

Thus all series (5.83) converge, and the series (5.80) also converges. Finally, if $\rho(A) = r$, and for each characteristic value λ of the matrix A such that $|\lambda| = \rho(A)$ all series (5.83) converge, then as follows from the previous arguments, the series (5.80) also converges; if at least one of them diverges, then the series (5.80) also diverges.[10]   □

Let us give some examples of power series of matrices that arise in different applications.

1. A *Neumann series* (or geometric progression) is a power series of matrices of the form

$$I + A + A^2 + \cdots + A^k + \cdots . \tag{5.86}$$

The power series $\sum\limits_{k=1}^{\infty} \lambda^k$, which diverges only for $|\lambda| < 1$, corresponds to the series (5.86). Therefore, the series (5.86) converges if and only if $\rho(A) < 1$. If this condition holds, then as a consequence, we get

$$A^k \to 0 \quad \text{as} \quad k \to \infty. \tag{5.87}$$

A matrix that satisfies condition (5.87) is called *convergent*. Using (5.85), we see that if $\rho(A) \geq 1$, then condition (5.87) does not hold. Thus a matrix A is convergent if and only if $\rho(A) < 1$.

Theorem 5.21 *Let A be a convergent matrix. Then the inverse matrix of $I - A$ exists and can be expressed as a power series:*

$$(I - A)^{-1} = I + A + A^2 + \cdots \tag{5.88}$$

Proof Clearly, if λ is an eigenvalue of the matrix $I - A$, then $1 - \lambda$ is an eigenvalue of A. Since $\rho(A) < 1$, none of the eigenvalues of A are equal to one; hence none of the eigenvalues of the matrix $I - A$ are equal to zero. For each integer $k \geq 1$, we obviously have $(I - A)(I + A + \cdots + A^k) = I - A^{k+1}$. Therefore,

---

[9]We can assume that $r > 0$, since otherwise, the matrix A is nilpotent and the series (5.80) consists of a finite number of terms.

[10]Here we take into account the structure of the powers of the matrix $J_l(0)$); see the proof of Lemma 5.3, p. 183.

$$\sum_{i=0}^{k} A^i = (I - A)^{-1} - (I - A)^{-1} A^{k+1}.$$

Since A is convergent, the limit as $k \to \infty$ on the right-hand side of the last equality exists and is equal to $(I - A)^{-1}$. Hence the limit on the left-hand side of the last equality exists too, and the relationship (5.88) holds. □

2. A *matrix exponential* is a power series of matrices of the following form:

$$e^A = I + A + \frac{1}{2!}A^2 + \cdots + \frac{1}{k!}A^k + \cdots . \qquad (5.89)$$

The power series $\sum_{k=1}^{\infty} \frac{1}{k!}\lambda^k$ corresponding to (5.89) has infinite radius of convergence. Therefore, the series (5.89) converges for every $A \in M_n, n \geq 1$.

In the rest of this section we give without proof some useful properties of the matrix exponential. The proof of these properties is left to the reader.

1. For all $A \in M_n$, the following equality holds:

$$e^A = \lim_{k \to \infty} \left(I + \frac{1}{k}A\right)^k.$$

Hint: note that for each $m \geq 1$, we have

$$e^A - \left(I + \frac{1}{m}A\right)^m = \sum_{k=1}^{\infty} \left(\frac{1}{k!} - \frac{C_m^k}{m^k}\right) A^k,$$

and the coefficients of A^k are nonnegative.
2. If $A, B \in M_n$ commute, then $e^{A+B} = e^A e^B$. Hint: use the well-known equality $e^{x+y} = e^x e^y$, which holds for all $x, y \in \mathbb{C}$.
3. For all $A \in M_n, t \in \mathbb{R}$, the following equality holds:

$$\frac{d\, e^{tA}}{d\,t} = A e^{tA}. \qquad (5.90)$$

5.3 Matrix Pencils

5.3.1 Definitions and Basic Properties

In this section, all vectors are elements of the space $\mathbb{C}^n$, all matrices are, generally speaking, complex, and we use only the standard inner product on $\mathbb{C}^n$.

Let A, B be rectangular $m \times n$ matrices. The function that assigns to each $\lambda \in \mathbb{C}$ the matrix $A + \lambda B$ is called a *matrix pencil*. Since the pencil is uniquely defined by the ordered pair of matrices A, B, we usually denote the pencil by (A, B). If $m = n$, i.e., A and B are square matrices of order n, then the polynomial $\det(A - \lambda B)$ is called the *characteristic polynomial of the pencil* (A, B). Arguing as in the derivation of formula (4.75), p. 123, we get

$$
\begin{aligned}
\det(A - \lambda B) = {}& \Delta(a_1, a_2, \ldots, a_n) \\
& - \lambda \left(\Delta(b_1, a_2, \ldots, a_n) + \Delta(a_1, b_2, \ldots, a_n) + \cdots + \Delta(a_1, a_2, \ldots, b_n) \right) \\
& + \lambda^2 (\Delta(b_1, b_2, \ldots, a_n) + \cdots + \Delta(a_1, a_2, \ldots, b_{n-1}, b_n)) - \cdots \\
& \pm \lambda^n \Delta(b_1, b_2, \ldots, b_n).
\end{aligned}
$$
$$(5.91)$$

Here $a_1, a_2, \ldots, a_n$ are the columns of the matrix A; $b_1, b_2, \ldots, b_n$ are the columns of the matrix B. All other symbols are the same as in formula (4.75), p. 123.

It immediately follows from (5.91) that the degree of the characteristic polynomial of the pencil (A, B) is less than or equal to $\operatorname{rank}(B)$.

If $\det(A - \lambda B) = 0$ for all $\lambda \in \mathbb{C}$, then the pencil (A, B) is called *singular*. If $m \neq n$, then the pencil is also called singular. In all other cases, the pencil is called *regular*. Thus a regular pencil is a pencil of the square matrices A and B such that there exists $\lambda \in \mathbb{C}$ for which $\det(A - \lambda B) \neq 0$.

Theorem 5.22 *If* $\operatorname{Ker}(A) \cap \operatorname{Ker}(B) \neq \{0\}$*, then the pencil* (A, B) *of square matrices is singular.*

Proof Indeed, if $x \neq 0$ and $Ax = 0$, $Bx = 0$, then $Ax - \lambda Bx = 0$ for all $\lambda \in \mathbb{C}$. Therefore, $\det(A - \lambda B) = 0$ for all $\lambda \in \mathbb{C}$. □

Corollary 5.4 *If the pencil* (A, B) *is regular, then* $\operatorname{Ker}(A) \cap \operatorname{Ker}(B) = \{0\}$.

Let us give some examples.

1. $A = I_3 = \begin{pmatrix} 1 & 0 & 0 \\ 0 & 1 & 0 \\ 0 & 0 & 1 \end{pmatrix}$, $B = \begin{pmatrix} 1 & 1 & 1 \\ 0 & 1 & 1 \\ 0 & 0 & 1 \end{pmatrix}$, $\operatorname{rank}(A) = \operatorname{rank}(B) = 3$,

   $p(\lambda) = \det(A - \lambda B) = (1 - \lambda)^3$, $\deg(p) = 3$.[11]

2. $A = I_3$, $B = \begin{pmatrix} 1 & 1 & 1 \\ 0 & 1 & 1 \\ 0 & 0 & 0 \end{pmatrix}$, $\operatorname{rank}(B) = 2$,

   $p(\lambda) = \det(A - \lambda B) = (1 - \lambda)^2$, $\deg(p) = 2$.

---

[11] We denote degree of the polynomial p by $\deg(p)$.

3. $A = I_3$, $B = \begin{pmatrix} 0 & 1 & 1 \\ 0 & 0 & 1 \\ 0 & 0 & 1 \end{pmatrix}$, $\mathrm{rank}(B) = 2$,

   $p(\lambda) = \det(A - \lambda B) = (1 - \lambda)$, $\deg(p) = 1$.

4. $A = I_3$, $B = \begin{pmatrix} 0 & 1 & 1 \\ 0 & 0 & 1 \\ 0 & 0 & 0 \end{pmatrix}$, $\mathrm{rank}(B) = 2$,

   $p(\lambda) = \det(A - \lambda B) = 1$, $\deg(p) = 0$.

5. $A = \begin{pmatrix} 1 & 0 & 0 \\ 0 & 1 & 0 \\ 0 & 0 & 0 \end{pmatrix}$, $B = \begin{pmatrix} 0 & 1 & 1 \\ 0 & 0 & 1 \\ 0 & 0 & 0 \end{pmatrix}$, $\mathrm{rank}(A) = \mathrm{rank}(B) = 2$,

   $\mathrm{Ker}(A) \cap \mathrm{Ker}(B) = \{0\}$, $p(\lambda) = \det(A - \lambda B) \equiv 0$; the pencil (A, B) is singular.

A number $\lambda \in \mathbb{C}$ is called a *characteristic value* of a regular pencil (A, B) if $\det(A - \lambda B) = 0$. Let λ be a characteristic value of the pencil (A, B). Then a vector $x \neq 0$ is called an *eigenvector* corresponding to λ if $Ax = \lambda Bx$.

Two pencils (A, B) and (A_1, B_1) are called *equivalent* if there exist nonsingular matrices U, V such that $A_1 = UAV$, $B_1 = UBV$.[12] It is useful to note that the conversion to an equivalent pencil is in fact a change of bases in the spaces $\mathbb{C}^n$, $\mathbb{C}^m$ (see Section 4.1.5, p. 97).

Theorem 5.23 *The characteristic polynomials of two equivalent pencils coincide up to a constant nonzero factor.*

Proof Indeed, $\det(UAV - \lambda UBV) = \det(U) \det(A - \lambda B) \det(V)$. □

We also note that if x is an eigenvector of the pencil (A, B) corresponding to a characteristic value λ, then the vector $y = V^{-1}x$ is an eigenvector of the equivalent pencil (UAV, UBV) corresponding to the same characteristic value. Indeed, the vector y is nonzero, and if $Ax = \lambda Bx$, then $AVy = \lambda BVy$. Therefore, $UAVy = \lambda UBVy$.

Theorem 5.24 (Generalized Schur's theorem). *If the pencil (A, B) is regular, then there exist unitary matrices U, V such that the matrices $A_1 = UAV$ and $B_1 = UBV$ are upper triangular.*

Proof First let us prove that there exist unitary matrices U_1 and V_1 such that all entries of the first columns of the matrices $\tilde{A} = U_1 A V_1$ and $\tilde{B} = U_1 B V_1$ that are located below the main diagonal are zero. By assumption, the pencil (A, B) is regular. Therefore, two cases are possible: (1) the characteristic polynomial of the pencil has a root (denote it by λ_1); (2) the characteristic polynomial of the pencil is identically equal to $\det(A) \neq 0$ (see (5.91)).

---

[12]The matrices A, B can be rectangular.

Let us consider the first case. Let v_1 be a normalized eigenvector of the pencil corresponding to λ_1, i.e.,

$$Av_1 = \lambda_1 Bv_1. \tag{5.92}$$

We join the vector v_1 with some vectors to complete the orthonormal basis $\{v_k\}_{k=1}^{n}$ of the space $\mathbb{C}^n$. Let V be a unitary matrix whose columns are formed by the vectors $\{v_k\}_{k=1}^{n}$. Now we note that the vector Bv_1 is nonzero, since otherwise, using (5.92), we see that Av_1 would also equal zero, but that contradicts the assumption on the regularity of the pencil (see Corollary 5.4). We take $u_1 = Bv_1/|Bv_1|$, join the vector u_1 with some vectors to complete the orthonormal basis $\{u_k\}_{k=1}^{n}$ of the space $\mathbb{C}^n$, and construct the matrix U_1 whose rows are $\bar{u}_1, \bar{u}_2, \ldots, \bar{u}_n$. Using elementary calculations, we see that the elements of the first column of the matrix $\tilde{B} = U_1 B V_1$ are computed by the formulas $\tilde{b}_{j,1} = |Bv_1|(u_1, u_j)$, $j = 1, 2, \ldots, n$. Therefore, $\tilde{b}_{1,1} = |Bv_1| > 0$, $\tilde{b}_{j,1} = 0$ for $j = 2, 3, \ldots, n$. The elements of the first column of the matrix $\tilde{A} = U_1 A V_1$ are analogously determined. As a result, we get $\tilde{a}_{11} = (Av_1, u_1) = \lambda_1 |Bv_1|$, $\tilde{a}_{1j} = (Av_1, u_j) = 0$ for $j = 2, 3, \ldots, n$. Now let us turn to the case that the characteristic polynomial of the pencil (A, B) has no roots. Then $\det(B) = 0$, but $\det(A) \neq 0$, and hence there exists a normalized vector v_1 such that $Bv_1 = 0$ and $Av_1 \neq 0$. As in the first case, we construct orthonormal bases $\{v_k\}_{k=1}^{n}$ and $u_1 = v_1/|Av_1|$, $u_2, \ldots, u_n$ of the space $\mathbb{C}^n$, and using them, we form the unitary matrices U, V. Using elementary calculations, we see that the first column of the matrix $\tilde{B} = U_1 B V_1$ is zero, the diagonal element of the first column of the matrix $\tilde{A} = U_1 A V_1$ is equal to $|Av_1| > 0$, and all other elements of this column are zero. The further arguments are based on decreasing the order of the considered matrices and are completely analogous to the corresponding arguments in the proof of Theorem 4.23, p. 127.   □

It is useful to note that if the triangular matrices A_1 and B_1 that appear in Theorem 5.24 are constructed, then the characteristic equation of the pencil (A, B) can be written in the form

$$\prod_{i=1}^{n}(a_{ii}^{(1)} - \lambda b_{ii}^{(1)}) = 0,$$

where $a_{ii}^{(1)}$ and $b_{ii}^{(1)}$, $i = 1, 2, \ldots, n$, are the diagonal elements of the matrices A_1 and B_1, respectively. Therefore, if the characteristic polynomial of a pencil has degree k, then the characteristic values of the pencil are calculated by the formulas $\lambda_i = a_{ii}^{(1)}/b_{ii}^{(1)}$ for $i = 1, 2, \ldots, k$. Obviously, $b_{ii}^{(1)} = 0$ for $i > k$. Hence if the polynomial of a pencil has degree $k < n$, then we say that the pencil has an infinite characteristic value of multiplicity $n - k$.

5.3.2 The Quasidiagonal Form of a Regular Pencil

Theorem 5.25 *Every regular pencil* (A, B) *can be reduced to an equivalent* quasi-diagonal form, *namely, to the pencil* (A_1, B_1), *where*

$$A_1 = \begin{pmatrix} A_{11} & 0 \\ 0 & I_{n-k} \end{pmatrix}, \quad B_1 = \begin{pmatrix} I_k & 0 \\ 0 & B_{22} \end{pmatrix}, \tag{5.93}$$

the matrix A_{11} *is upper triangular, all characteristic values* $\lambda_1, \lambda_2, \ldots, \lambda_k$ *of the pencil* (A, B) *form the main diagonal of this matrix;* B_{22} *is an upper triangular matrix with a zero main diagonal;* I_k, I_{n-k} *are identity matrices of the corresponding orders.*

Proof As we proved in Theorem 5.24, the pencil (A, B) is equivalent to the pencil

$$\left(\begin{pmatrix} A_{11} & A_{12} \\ 0 & A_{22} \end{pmatrix}, \begin{pmatrix} B_{11} & B_{12} \\ 0 & B_{22} \end{pmatrix} \right). \tag{5.94}$$

Here A_{11} is an upper triangular matrix of order k whose main diagonal is formed by the numbers $b_{11}\lambda_1, b_{22}\lambda_2, \ldots, b_{kk}\lambda_k$; B_{11} is an upper triangular matrix of order k whose main diagonal is formed by the numbers $b_{11}, b_{22}, \ldots, b_{kk}$, all of which are nonzero; A_{22} is an upper triangular matrix of order $n - k$ whose diagonal elements $a_{k+1,k+1}, a_{k+2,k+2}, \ldots, a_{nn}$ are nonzero; B_{22} is an upper triangular matrix with zero main diagonal. Multiplying both matrices of the pencil (5.94) by the block diagonal matrix $\mathrm{diag}(B_{11}^{-1}, A_{22}^{-1})$, we move to the equivalent pencil

$$\left(\begin{pmatrix} \tilde{A}_{11} & \tilde{A}_{12} \\ 0 & I_{n-k} \end{pmatrix}, \begin{pmatrix} I_k & \tilde{B}_{12} \\ 0 & \tilde{B}_{22} \end{pmatrix} \right).$$

Here $\tilde{A}_{11}$ is an upper triangular matrix whose main diagonal is formed by the numbers $\lambda_1, \lambda_2, \ldots, \lambda_k$; the main diagonal of the upper triangular matrix $\tilde{B}_{22}$ is zero. We shall complete the proof if we can construct $k \times (n - k)$ rectangular matrices P and Q such that

$$\begin{pmatrix} I_k & Q \\ 0 & I_{n-k} \end{pmatrix} \begin{pmatrix} \tilde{A}_{11} & \tilde{A}_{12} \\ 0 & I_{n-k} \end{pmatrix} \begin{pmatrix} I_k & P \\ 0 & I_{n-k} \end{pmatrix} = \begin{pmatrix} \tilde{A}_{11} & 0 \\ 0 & I_{n-k} \end{pmatrix},$$

$$\begin{pmatrix} I_k & Q \\ 0 & I_{n-k} \end{pmatrix} \begin{pmatrix} I_k & \tilde{B}_{12} \\ 0 & \tilde{B}_{22} \end{pmatrix} \begin{pmatrix} I_k & P \\ 0 & I_{n-k} \end{pmatrix} = \begin{pmatrix} I_k & 0 \\ 0 & \tilde{B}_{22} \end{pmatrix}.$$

By elementary calculations we get the following equations for the determination of the matrices P and Q:

$$\tilde{A}_{11}P + Q = -\tilde{A}_{12}, \quad P + Q\tilde{B}_{22} = -\tilde{B}_{12}. \tag{5.95}$$

We can consider the system of Eq. (5.95) as a system of linear algebraic equations for the elements of the matrices P and Q. To prove its solvability for every $\tilde{A}_{12}$ and $\tilde{B}_{12}$, it is enough to check that the corresponding homogeneous system

$$\tilde{A}_{11}P + Q = 0, \quad P + Q\tilde{B}_{22} = 0 \tag{5.96}$$

has the trivial solution only. If the matrices P and Q satisfy (5.96), then $P = \tilde{A}_{11}P\tilde{B}_{22}$ and $Q = \tilde{A}_{11}Q\tilde{B}_{22}$. The matrix $\tilde{B}_{22}$ is nilpotent. Therefore, arguing as in the proof of Theorem 4.24, p. 157, we see that $P = 0$ and $Q = 0$. $\square$

5.3.3   Weierstrass Canonical Form

In this subsection we show that every regular matrix pencil is equivalent to a pencil of bidiagonal matrices that is analogous to the matrix Jordan canonical form.

Theorem 5.26 (Weierstrass[13]**).** *Let (A,B) be a regular pencil of matrices of order n, and let $\lambda_1, \lambda_2, ..., \lambda_k, k \leq n$, be all its characteristic values. Then there exist nonsingular matrices U and V such that $UAV = diag(J, I_{n-k})$ and $UBV = diag(I_k, H)$. Here I_{n-k} and I_k are the identity matrices of order $n - k$ and k, respectively; J is a Jordan matrix whose main diagonal is formed by the numbers $\lambda_1, \lambda_2, ..., \lambda_k$; and H is a nilpotent Jordan matrix.*

Proof Let S_k and S_{n-k} be nonsingular matrices that reduce the matrices A_{11} and B_{22} of the equivalent quasidiagonal form (5.93) of the pencil (A, B) to Jordan canonical form. Then we reduce the pencil (A_1, B_1) to the required form by the following similarity transformation:

$$diag(S_k, S_{n-k})A_1 diag(S_k^{-1}, S_{n-k}^{-1}), \quad diag(S_k, S_{n-k})B_1 diag(S_k^{-1}, S_{n-k}^{-1}).$$

$\square$

The pencil of matrices $(diag(J, I_{n-k}), diag(I_k, H))$ that appears in Theorem 5.26 is called the *Weierstrass canonical form*.

Theorem 5.27 *The Jordan matrices J and H in the Weierstrass canonical form are uniquely determined by the matrices of the original pencil (A, B) up to permutation of the diagonal Jordan blocks.*

Proof Let $(diag(J, I_{n-k}), diag(I_k, H))$ and $(diag(J_1, I_{n-k}), diag(I_k, H_1))$ be two different Weierstrass canonical forms of the same pencil of matrices (A, B). Then there exist nonsingular matrices U and V such that

$$\begin{pmatrix} U_{11} & U_{12} \\ U_{21} & U_{22} \end{pmatrix} \begin{pmatrix} J & 0 \\ 0 & I \end{pmatrix} = \begin{pmatrix} J_1 & 0 \\ 0 & I \end{pmatrix} \begin{pmatrix} V_{11} & V_{12} \\ V_{21} & V_{22} \end{pmatrix}, \tag{5.97}$$

---

[13] Karl Theodor Wilhelm Weierstrass (1815–1897) was a German mathematician.

$$\begin{pmatrix} U_{11} & U_{12} \\ U_{21} & U_{22} \end{pmatrix} \begin{pmatrix} I & 0 \\ 0 & H \end{pmatrix} = \begin{pmatrix} I & 0 \\ 0 & H_1 \end{pmatrix} \begin{pmatrix} V_{11} & V_{12} \\ V_{21} & V_{22} \end{pmatrix}. \tag{5.98}$$

In equalities (5.97), (5.98), we use the block representations of the matrices U and V that correspond to the orders of the blocks J and H. Here we do not write the indices that indicate the orders of the identity blocks. Using elementary calculations, as a consequence of (5.97), (5.98) we get $U_{11} = V_{11}$ and $U_{22} = V_{22}$,

$$U_{11}J = J_1 U_{11}, \quad U_{22}H = H_1 U_{22}, \tag{5.99}$$

$$V_{21} = U_{21}J, \quad U_{21} = H_1 V_{21}, \tag{5.100}$$

$$U_{12} = J_1 V_{12}, \quad V_{12} = U_{12}H. \tag{5.101}$$

It follows from (5.100) and (5.101) that $U_{21} = H_1 U_{21} J$ and $U_{12} = J_1 U_{12} H$. The matrices H and H_1 are nilpotent, and hence (see the final part of the proof of Theorem 5.93) $U_{12} = 0$ and $U_{21} = 0$, and therefore, $V_{12} = 0$ and $V_{21} = 0$. The matrix U is nonsingular, whence the matrices U_{11} and U_{22} are nonsingular. Thus (see (5.99)) the matrix J is similar to J_1, the matrix H is similar to H_1, and the assertion follows now from Theorem 5.16, p. 184. □

5.3.4 Hermitian and Definite Pencils

If $\det(B) \neq 0$, then the matrix pencil (A, B) is equivalent to the pencils $(B^{-1}A, I)$ and (AB^{-1}, I). The set of all characteristic values of the pencil (A, B) coincides with the spectrum of the matrix $B^{-1}A$. Obviously, the eigenvectors of the matrix $B^{-1}A$ are also connected with the eigenvectors of the pencil (A, B). These facts are useful for theoretical investigations, but spectral problems for matrix pencils with $\det(B) \neq 0$ are not solved numerically as eigenvalue problems for the matrix $B^{-1}A$, since usually, such important properties of the matrices A and B as symmetry, sparseness, and so on, are lost.

The pencil (A, B) is called *Hermitian* if $A = A^*$, $B = B^*$. All characteristic values of a Hermitian pencil are real. Indeed, if $x \neq 0$, $Ax = \lambda Bx$, then $(Ax, x) = \lambda(Bx, x)$. The numbers (Ax, x) and (Bx, x) are real (see Theorem 4.37, p. 140).

It is important to note that as a consequence of Theorem 5.19, p. 187, we see that the problem of calculating eigenvalues and eigenvectors of an arbitrary real matrix is equivalent to the problem of characteristic values and eigenvectors of a pencil of symmetric real matrices with $\det(B) \neq 0$. Thus the spectral problem for a pencil of symmetric real matrices with $\det(B) \neq 0$ in the general case is as difficult as the spectral problem for an arbitrary real matrix. The situation improves if we narrow the class of the allowable matrices B.

A pencil (A, B) is called *definite* if it is Hermitian and the matrix B is positive definite. The next theorem shows that every definite pencil can be reduced to diagonal form by a similarity transformation.

Theorem 5.28 *If a pencil (A, B) is definite, then there exists a nonsingular matrix U such that $U^* B U = I$ and $U^* A U = \Lambda$, where $\Lambda = diag(\lambda_1, \lambda_2, \ldots, \lambda_n)$.*[14]

Proof Let us define a new inner product on $\mathbb{C}^n$ by the formula $(x, y)_B = (Bx, y)$. The operator $C = B^{-1} A : \mathbb{C}^n \to \mathbb{C}^n$ is self-adjoint with respect to this inner product, since for all $x, y \in \mathbb{C}^n$, we have

$$(Cx, y)_B = (B B^{-1} Ax, y) = (x, Ay) = (Bx, B^{-1} Ay) = (x, Cy)_B.$$

Therefore, by Theorem 4.41, p. 143, there exist vectors $e_1, e_2, \ldots, e_n$ and numbers $\lambda_1, \lambda_2, \ldots, \lambda_n$ such that

$$C e_k = \lambda_k e_k, \ k = 1, 2, \ldots, n, \quad (e_k, e_l)_B = \delta_{kl}, \ k, l = 1, 2, \ldots, n. \qquad (5.102)$$

Let us construct the matrix U whose columns are the vectors $e_1, e_2, \ldots, e_n$. It is easy to see that the matrix U is nonsingular. We can write relationships (5.102) in the form $B^{-1} A U = U \Lambda$ and $U^* B U = I$, where $\Lambda = diag(\lambda_1, \lambda_2, \ldots, \lambda_n)$. Evidently, we have the equality $U^* A U = \Lambda$. $\qquad \square$

5.3.5 Singular Pencils. The Theorem on Reduction

In this subsection we show that every singular pencil is equivalent to a quasidiagonal pencil of 2×2 block matrices of a special form. Let us denote by r the maximal order of the minors of the matrix $A + \lambda B$ that are not identically zero as functions of $\lambda \in \mathbb{C}$. The number r is called the *rank* of the pencil (A, B). We assume that the pencil (A, B) is singular; hence we have one of the following inequalities: either $r < m$ or $r < n$. Here, as usual, m is the number of rows of the pencil, and n is the number of columns of the pencil. To be definite, assume that $r < n$. Since the pencil (A, B) is singular, for every $\lambda \in \mathbb{C}$, there exists a nonzero vector $x(\lambda) \in \mathbb{C}^n$ such that

$$(A + \lambda B)x(\lambda) = 0 \quad \text{for all} \ \ \lambda \in \mathbb{C}. \qquad (5.103)$$

Solving the homogeneous system (5.103) by the method described in Section 4.1.13, p. 110, we see that the components $x_k(\lambda)$ of the vector $x(\lambda)$ are calculated by the formulas $P_{l_k}(\lambda)/Q_{m_k}(\lambda)$, where P_{l_k} and Q_{m_k} are some polynomials, $k = 1, 2, \ldots, n$. Multiplying the vector $x(\lambda)$ by an arbitrary function of the

---

[14]Clearly, $\lambda_1, \lambda_2, \ldots, \lambda_n$ are the characteristic values of the pencil (A, B).

variable λ, we also get a solution of system (5.103). Therefore, we can assume that the vector $x(\lambda)$ is a polynomial of degree ε:

$$x(\lambda) = x_0 + \lambda x_1 + \cdots + \lambda^\varepsilon x_\varepsilon, \quad \varepsilon \geq 0. \tag{5.104}$$

We choose the polynomial of minimal degree among all polynomials satisfying (5.103). The corresponding integer ε is called the *minimal index* of the singular pencil (A, B).

Lemma 5.4 *The minimal indices of equivalent pencils are equal.*

Proof If $x(\lambda)$ is a polynomial of the minimal degree satisfying (5.103), then for all nonsingular matrices U and V, we have

$$U(A - \lambda B)VV^{-1}x(\lambda) = 0 \quad \text{for all} \quad \lambda \in \mathbb{C}^n.$$

Obviously, the degrees of the polynomials $V^{-1}x(\lambda)$ and $x(\lambda)$ are equal. $\qquad \square$

Substituting the polynomial (5.104) into Eq. (5.103), collecting all coefficients with the same power of λ, and equating them to zero, we get a homogeneous system of linear equations for vectors $x_0, x_1, \ldots, x_\varepsilon$, which is equivalent to (5.103):

$$Ax_0 = 0, \ Bx_0 + Ax_1 = 0, \ \ldots, \ Bx_{\varepsilon-1} + Ax_\varepsilon = 0, \ Bx_\varepsilon = 0. \tag{5.105}$$

(Clearly, the matrix of system (5.105) is block bidiagonal.) If $\varepsilon > 0$ is the minimal index of the pencil (A, B), then the system

$$Ax_0 = 0, \ Bx_0 + Ax_1 = 0, \ \ldots, \ Bx_{k-1} + Ax_k = 0, \ Bx_k = 0 \tag{5.106}$$

has only the trivial solution for each $k < \varepsilon$. Thus we have the following lemma.

Lemma 5.5 *If $\varepsilon > 0$ is the minimal index of the pencil (A, B), then the columns of the matrix of system (5.106) are linearly independent for each $k < \varepsilon$.*

Lemma 5.6 *Let ε be the minimal index of the pencil (A, B) and let $x_0, x_1, \ldots, x_\varepsilon$ be the solutions of system (5.105). Then these vectors are linearly independent. The vectors $Ax_1, Ax_2, \ldots, Ax_\varepsilon$ are also linearly independent.*

Proof First let us prove that none of the vectors $\{x_i\}_{i=0}^\varepsilon$ are equal to zero. Indeed, if $x_0 = 0$, then the polynomial $\lambda^{-1}x(\lambda)$ has degree $\varepsilon - 1$ and satisfies relationship (5.103), which contradicts the assumption of the minimality of the index ε. If $x_j = 0$ for every $j \geq 1$, then $Ax_j = 0$. Not all vectors $x_0, x_1, \ldots, x_{j-1}$ are zero. It follows from (5.105) that these vectors satisfy system (5.106) for $k = j - 1$, which contradicts the assumption of the minimality of the index ε. In the same way, we prove that none of the vectors $\{Ax_i\}_{i=1}^\varepsilon$ are equal to zero. Let us prove that the vectors $\{Ax_i\}_{i=1}^\varepsilon$ are linearly independent. If we assume the contrary, then there exist an integer $h \in [1, \varepsilon]$ and numbers $\alpha_1, \alpha_2, \ldots, \alpha_{h-1}$, not all zero, such that

$$Ax_h = \alpha_1 Ax_{h-1} + \alpha_2 Ax_{h-2} + \cdots + \alpha_{h-1} Ax_1. \tag{5.107}$$

Let

$$y_0 = x_0, \quad y_1 = x_1 - \alpha_1 x_0, \quad y_2 = x_2 - \alpha_1 x_1 - \alpha_2 x_0, \ldots,$$

$$y_{h-1} = x_{h-1} - \alpha_1 x_{h-2} - \cdots - \alpha_{h-1} x_0.$$

Using Eqs. (5.105) and (5.107), it is easy to see that

$$Ay_0 = Ax_0 = 0, \quad Ay_1 = Ax_1 - \alpha_1 Ax_0 = -Bx_0 = -By_0,$$

$$Ay_2 = Ax_2 - \alpha_1 Ax_1 - \alpha_2 Ax_0 = -Bx_1 + \alpha_1 Bx_0 = -B(x_1 - \alpha_1 x_0) = -By_1, \ldots,$$

$$Ay_{h-1} = -By_{h-2}, \quad By_{h-1} = 0.$$

Therefore, the polynomial $y(\lambda) = y_0 + \lambda y_1 + \cdots + \lambda^{h-1} y_{h-1}$ has degree $h - 1 < \varepsilon$ and satisfies a relationship of the form (5.103), but this fact contradicts the assumption of the minimality of the index ε. It remains to prove that the vectors $\{x_i\}_{i=0}^{\varepsilon}$ are linearly independent. Now we assume that $\alpha_0 x_0 + \alpha_1 x_1 + \cdots + \alpha_\varepsilon x_\varepsilon = 0$ for some numbers $\alpha_0, \alpha_1, \ldots, \alpha_\varepsilon$. Then $\alpha_1 Ax_1 + \alpha_2 Ax_2 + \cdots + \alpha_\varepsilon Ax_\varepsilon = 0$. As a consequence of the linear independence of the vectors $\{Ax_i\}_{i=1}^{\varepsilon}$, we get $\alpha_1 = \alpha_2 = \cdots = \alpha_\varepsilon = 0$. Hence $\alpha_0 x_0 = 0$, but $x_0 \neq 0$, and therefore, $\alpha_0 = 0$. $\qquad\square$

Lemma 5.7 *If a pencil (A, B) has minimal index $\varepsilon > 0$, then it is equivalent to the quasitriangular pencil (A_1, B_1), where*

$$A_1 = \begin{pmatrix} L_\varepsilon^{(0)} & D \\ 0 & \hat{A} \end{pmatrix}, \quad B_1 = \begin{pmatrix} L_\varepsilon^{(1)} & F \\ 0 & \hat{B} \end{pmatrix}, \tag{5.108}$$

$$L_\varepsilon^{(0)} = (0, I_\varepsilon), \quad L_\varepsilon^{(1)} = (I_\varepsilon, 0), \tag{5.109}$$

and the minimal index of the pencil $(\hat{A}, \hat{B})$ is greater than or equal to ε.

Proof Using Lemma 5.6, we see that we can introduce a basis in the space $\mathbb{C}^n$ such that the first $\varepsilon + 1$ vectors of this basis are $\{(-1)^i x_i\}_{i=0}^{\varepsilon}$. Analogously, we can introduce a basis in the space $\mathbb{C}^m$ whose first vectors are $\{(-1)^i Ax_i\}_{i=1}^{\varepsilon}$.[15] Equalities (5.105) show that conversion to the specified bases leads to a pencil of matrices (5.108). Let us prove the second part of the theorem. Note that there is no polynomial $y(\lambda)$ of degree less than ε that satisfies the identity

$$(L_\varepsilon^{(0)} + \lambda L_\varepsilon^{(1)}) y(\lambda) = 0 \quad \text{for all} \quad \lambda \in \mathbb{C}. \tag{5.110}$$

---

[15]The alternation of signs in these bases will be convenient for some formulas below.

Indeed, in this case, the polynomial $x(\lambda) = (y(\lambda), 0)$ of degree less than ε satisfies the identity

$$(A_1 + B_1)x(\lambda) = 0 \quad \text{for all} \quad \lambda \in \mathbb{C}, \tag{5.111}$$

but that is impossible, since the minimal indices of the equivalent pencils (A, B) and (A_1, B_1) must be equal. Now we suppose that contrary to the assertion of the theorem, there exists a polynomial $z(\lambda)$ of degree less than ε that satisfies the identity

$$(\hat{A} + \hat{B})z(\lambda) = 0 \quad \text{for all} \quad \lambda \in \mathbb{C}. \tag{5.112}$$

If we construct a polynomial $v(\lambda)$ such that

$$(L_\varepsilon^{(0)} + \lambda L_\varepsilon^{(1)})v(\lambda) + (D + \lambda F)z(\lambda) = 0 \quad \text{for all} \quad \lambda \in \mathbb{C}, \tag{5.113}$$

then $x(\lambda) = (v(\lambda), z(\lambda)) \in \mathbb{C}^n$ satisfies (5.111). Let $z(\lambda) = z_0 + \lambda z_1 + \cdots + \lambda^{\varepsilon-1}z_{\varepsilon-1}$, and we let $v(\lambda) = v_0 + \lambda v_1 + \cdots + \lambda^{\varepsilon-1}v_{\varepsilon-1}$. Substituting these representations into (5.113), collecting all coefficients with the same power of λ, and equating them to zero, we get the system of equations

$$L_\varepsilon^{(0)}v_0 = -g_0, \ L_\varepsilon^{(1)}v_0 + L_\varepsilon^{(0)}v_1 = -g_1, \ldots, \ L_\varepsilon^{(1)}v_{\varepsilon-2} + L_\varepsilon^{(0)}v_{\varepsilon-1} = -g_{\varepsilon-1},$$
$$L_\varepsilon^{(1)}v_{\varepsilon-1} = -g_\varepsilon, \quad (5.114)$$

where $g_0 = Dz_0$, $g_1 = Fz_0 + Dz_1, \ldots, g_{\varepsilon-1} = Fz_{\varepsilon-2} + Dz_{\varepsilon-1}, g_\varepsilon = Fz_{\varepsilon-1}$. It has been proved that the minimal index of the pencil $(L_\varepsilon^{(0)}, L_\varepsilon^{(1)})$ is equal to ε. Therefore, by Lemma 5.5, it follows that the rank of the matrix of the system (5.114) is equal to $\varepsilon(\varepsilon + 1)$ (the number of its columns). It is easy to compute that the number of equations in the system (5.114) is also equal to $\varepsilon(\varepsilon + 1)$. Therefore, the system (5.114) is uniquely solvable for every right-hand side. Thus the polynomial $x(\lambda) = (v(\lambda), z(\lambda))$ has degree $\varepsilon - 1$ and satisfies identity (5.111). This contradiction concludes the proof of the lemma. $\qquad\Box$

Theorem 5.29 (on reduction). *If the minimal index of the pencil (A, B) is equal to $\varepsilon > 0$ and the rank of the pencil is less than n, then the pencil (A, B) is equivalent to the pencil*

$$\left(\begin{pmatrix} L_\varepsilon^{(0)} & 0 \\ 0 & \hat{A} \end{pmatrix}, \ \begin{pmatrix} L_\varepsilon^{(1)} & 0 \\ 0 & \hat{B} \end{pmatrix} \right), \tag{5.115}$$

where as above, $L_\varepsilon^{(0)} = (0, I_\varepsilon)$, $L_\varepsilon^{(1)} = (I_\varepsilon, 0)$, and the minimal index of the pencil $(\hat{A}, \hat{B})$ is greater than or equal to ε.

Proof Let (A_1, B_1) be the pencil that was constructed in the proof of Lemma 5.7 and is equivalent to the pencil (A, B). Using elementary calculations, we see that

$$\begin{pmatrix} I_\varepsilon & Q \\ 0 & I_{m-\varepsilon} \end{pmatrix} A_1 \begin{pmatrix} I_{\varepsilon+1} & -P \\ 0 & I_{n-\varepsilon-1} \end{pmatrix} = \begin{pmatrix} L_\varepsilon^{(0)} & R \\ 0 & \hat{A} \end{pmatrix},$$

$$\begin{pmatrix} I_\varepsilon & Q \\ 0 & I_{m-\varepsilon} \end{pmatrix} B_1 \begin{pmatrix} I_{\varepsilon+1} & -P \\ 0 & I_{n-\varepsilon-1} \end{pmatrix} = \begin{pmatrix} L_\varepsilon^{(1)} & S \\ 0 & \hat{B} \end{pmatrix}.$$

Here P and Q are some yet unknown rectangular matrices of corresponding orders,

$$R = D + Q\hat{A} - L_\varepsilon^{(0)} P, \quad S = F + Q\hat{B} - L_\varepsilon^{(1)} P.$$

We shall complete the proof of the theorem if we show that the matrices P and Q can be chosen such that $R = 0$ and $S = 0$. Let $p_1, p_2, \ldots, p_{\varepsilon+1}$ be the rows of the matrix P. It is easy to see that the matrix $L_\varepsilon^{(0)} P$ consists of the rows $p_2, p_3, \ldots, p_{\varepsilon+1}$, and the matrix $L_\varepsilon^{(1)} P$ consists of the rows $p_1, p_2, \ldots, p_\varepsilon$. Therefore, if $R = 0$ and $S = 0$, then

$$q_j \hat{A} + q_{j+1} \hat{B} + f_{j+1} + d_j = 0, \quad j = 1, 2, \ldots, \varepsilon - 1. \tag{5.116}$$

Here the lowercase letters with indices denote the rows of the corresponding matrices. Let us consider (5.116) as a system of equations for the elements of the rows $(-1)^j q_j$, $j = 1, 2, \ldots, \varepsilon$. Obviously, the matrix of this system has the same form as the matrix of the system (5.106) for $k = \varepsilon - 2$. Since by Lemma 5.7, the minimal index of the pencil $(\hat{A}, \hat{B})$ is greater than or equal to ε, the rank of this matrix is equal to $(\varepsilon - 1)(n - \varepsilon - 1)$, i.e., it is equal to the number of its columns. It is easy to calculate that the number of equalities in system (5.116) is also equal to $(\varepsilon - 1)(n - \varepsilon - 1)$. Therefore, system (5.116) is solvable for every D and F. If we find the matrix Q, then the matrix P is easily found as the solution of the system of equations $L_\varepsilon^{(0)} P = D + Q\hat{A}$ and $L_\varepsilon^{(1)} P = F + Q\hat{B}$.   □

Remark 5.3 Let us recall that we have assumed above that the rank of the pencil (A, B) is less than n. If the rank of the pencil (A, B) is less than m, then it is easy to establish by reduction to the pencil (A^T, B^T) that (A, B) is equal to the pencil

$$\left(\begin{pmatrix} (L_\eta^{(0)})^T & 0 \\ 0 & \tilde{A} \end{pmatrix}, \begin{pmatrix} (L_\eta^{(1)})^T & 0 \\ 0 & \tilde{B} \end{pmatrix} \right),$$

where η is the minimal index of the pencil (A^T, B^T), and the minimal index of the pencil $(\tilde{A}^T, \tilde{B}^T)$ is greater than or equal to η. The number ε is called the *right minimal index* of the pencil (A, B), and the number η is called the *left minimal index*.

5.3.6 Kronecker Canonical Form

In this subsection we show that every singular pencil (A, B) is equivalent to a quasidiagonal pencil such that each of the diagonal blocks is a bidiagonal matrix.

First we assume that the right minimal index of the pencil (A, B) is equal to zero. This means that there is a nonzero vector $x \in \mathbb{C}^n$ such that $Ax = 0$ and $Bx = 0$. In other words, the defect of the 2×1 block matrix

$$M = \begin{pmatrix} A \\ B \end{pmatrix} \tag{5.117}$$

is positive. Denote it by h_r. Evidently, choosing the vectors of a basis of $\mathrm{Ker}(M)$ as the first h_r vectors for the basis in the space $\mathbb{C}^n$, we reduce the pencil (A, B) to the pencil $((0(m, h_r), A_0), (0(m, h_r), B_0))$, where $0(m, h_r)$ is the $m \times h_r$ zero matrix, and the right minimal index of the pencil (A_0, B_0) is positive.

Now we assume that the left minimal index of the pencil (A_0, B_0) is equal to zero. Then, arguing as above, we reduce the original pencil (A, B) to the quasidiagonal pencil $(\mathrm{diag}(0(h_l, h_r), A_1), \mathrm{diag}(0(h_l, h_r), B_1))$, where h_l is the dimension of the kernel of the 1×2 block matrix $M_0 = (A_0, B_0)$. Clearly, in this case the left and right minimal indices of the pencil (A_1, B_1) are positive. To be definite, assume that the rank of the pencil (A_1, B_1) is less than the number of its columns. Then by Theorem 5.29, it follows that the pencil (A_1, B_1) is equivalent to the pencil $(\mathrm{diag}(L_{\varepsilon_1}^{(0)}, \hat{A}_1), \mathrm{diag}(L_{\varepsilon_1}^{(1)}, \hat{B}_1))$, where $\varepsilon_1 > 0$, and the right minimal index of the pencil $(\hat{A}_1, \hat{B}_1)$ is greater than or equal to ε_1. Continuing this process, we get the pencil

$$(\mathrm{diag}(0(h_l, h_r), L_{\varepsilon_1}^{(0)}, L_{\varepsilon_2}^{(0)} \ldots, L_{\varepsilon_p}^{(0)}, \hat{A}_p), \mathrm{diag}(0(h_l, h_r)), L_{\varepsilon_1}^{(1)}, L_{\varepsilon_2}^{(1)} \ldots, L_{\varepsilon_p}^{(1)}, \hat{B}_q),$$

where $0 < \varepsilon_1 \leq \varepsilon_1 \cdots \leq \varepsilon_p$, and the rank of the pencil $(\hat{A}_p, \hat{B}_p)$ is equal to the number of its columns.

Suppose that the number of rows of the pencil $(\hat{A}_p, \hat{B}_p)$ is more than its rank (in the contrary case, this pencil is regular). It is easy to see that since the left minimal index of the pencil (A_1, B_1) is positive, the left minimal index of the pencil $(\hat{A}_p, \hat{B}_p)$ is also positive. Consistently applying Theorem 5.29 again (see also Remark 5.3), we reduce the pencil $(\hat{A}_p, \hat{B}_p)$ to the pencil

$$((L_{\eta_1}^{(0)})^T, (L_{\eta_2}^{(0)})^T \ldots, (L_{\eta_q}^{(0)})^T, \hat{A}_q), (L_{\eta_1}^{(1)})^T, (L_{\eta_2}^{(1)})^T \ldots, (L_{\eta_q}^{(1)})^T, \hat{B}_q),$$

where $0 < \eta_1 \leq \eta_1 \cdots \leq \eta_q$. Here the pencil $(\hat{A}_q, \hat{B}_q)$ is regular, and therefore it can be reduced to Weierstrass canonical form (see Theorem 5.26, p. 196).

Thus we have proved that every arbitrary singular pencil can be reduced to an equivalent pencil

$$(\text{diag}(0(h_l, h_r), L_{\varepsilon_1}^{(0)}, L_{\varepsilon_2}^{(0)} \ldots, L_{\varepsilon_p}^{(0)}, (L_{\eta_1}^{(0)})^T, (L_{\eta_2}^{(0)})^T \ldots, (L_{\eta_q}^{(0)})^T, J, I_{n-k})),$$

$$\text{diag}(0(h_l, h_r), L_{\varepsilon_1}^{(1)}, L_{\varepsilon_2}^{(1)} \ldots, L_{\varepsilon_p}^{(1)}, (L_{\eta_1}^{(1)})^T, (L_{\eta_2}^{(1)})^T \ldots, (L_{\eta_q}^{(1)})^T, I_k, H)).$$

$$(5.118)$$

Here n is the order of the pencil $(\hat{A}_q, \hat{B}_q)$, k is the number of its characteristic values, J is the corresponding Jordan matrix, and H is a nilpotent Jordan matrix (details see in Theorem 5.26, p. 191).

The pair of matrices (5.118) is so called the *Kronecker canonical form* of the singular pencil in the most general case. Clearly, in specific particular situations both numbers h_r and h_l or one of them can be equal to zero, and therefore, every group of diagonal blocks of the pencil (5.118) can be omitted.

5.3.7 Applications to Systems of Linear Differential Equations

The Jordan, Weierstrass, and Kronecker canonical forms have many applications to the study of systems of linear differential equations.

1. Let us begin with the Cauchy problem for a system of ordinary linear differential equations with constant coefficients:

$$\dot{x}(t) = Ax(t) + f(t), \tag{5.119}$$

$$x(0) = x_0. \tag{5.120}$$

Here A is a given square matrix of order n, f is a given continuous vector-valued function of the variable $t \in \mathbb{R}$ with values in the space $\mathbb{C}^n$, the vector $x_0 \in \mathbb{C}^n$ is given, and x is an unknown vector-valued function.

Let us recall (see (5.90), p. 191) that

$$\frac{\mathrm{d}e^{tA}}{\mathrm{d}t} = Ae^{tA}, \tag{5.121}$$

$$e^{tA}|_{t=0} = I. \tag{5.122}$$

Using relationships (5.121), (5.122), it is easy to check by direct substitution that the solution of problem (5.119), (5.120) is given by the formula

$$x(t) = e^{tA}x_0 + \int_0^t e^{(t-\tau)A} f(\tau)\mathrm{d}\tau. \tag{5.123}$$

In the simplest case of $n = 1$, i.e., if $A = \lambda \in \mathbb{C}$ and x_0, $f(t) \in \mathbb{C}$, the solution of problem (5.119), (5.120) is calculated as follows:

$$x(t) = x_0 e^{\lambda t} + \int_0^t f(\tau) e^{\lambda(t-\tau)} d\tau. \tag{5.124}$$

Let S be a nonsingular matrix that reduces the matrix A to Jordan canonical form (i.e., $A = SJS^{-1}$, $J = \mathrm{diag}(J_{n_1}(\lambda_1), J_{n_2}(\lambda_2), \dots, J_{n_k}(\lambda_k))$; see Section 5.2.1, p. 180). Then problem (5.119), (5.120) is reduced to the problem

$$\dot{y}(t) = J y(t) + g(t), \tag{5.125}$$

$$y(0) = y_0, \tag{5.126}$$

where $y = S^{-1}x$, $g = S^{-1}f$, $y_0 = S^{-1}y(0)$. Clearly, problem (5.125), (5.126) splits into the following system of independent equations:

$$\dot{y}_{n_i}(t) = J_{n_i}(\lambda_{n_i}) y_{n_i}(t) + g_{n_i}(t), \tag{5.127}$$

$$y_{n_i}(0) = y_{0,n_i}, \tag{5.128}$$

where $i = 1, 2, \dots, k$. Thus it is enough to solve the problem

$$\dot{y}(t) = J(\lambda) y(t) + g(t), \tag{5.129}$$

$$y(0) = y_0, \tag{5.130}$$

where $J(\lambda)$ is the Jordan block whose order we denote by m. Let us write the homogeneous system corresponding to (5.129) in detail:

$$\dot{y}_1(t) = \lambda y_1(t) + y_2(t),$$
$$\dot{y}_2(t) = \lambda y_2(t) + y_3(t),$$
$$\dots\dots\dots\dots\dots\dots\dots\dots$$
$$\dot{y}_m(t) = \lambda y_m(t).$$

It follows from the last equation that

$$y_m(t) = y_m(0) e^{\lambda t}.$$

Therefore,

$$\dot{y}_{m-1}(t) = \lambda y_{m-1}(t) + y_m(0) e^{\lambda t},$$

whence by formula (5.124), we get

$$y_{m-1}(t) = (y_m(0)t + y_{m-1}(0)) e^{\lambda t}.$$

Analogously,

$$y_{m-2}(t) = \left(y_m(0)\frac{t^2}{2} + y_{m-1}(0)t + y_{m-2}(0) \right) e^{\lambda t}, \dots .$$

Finally,

$$y_1(t) = \left(y_m(0)\frac{t^{m-1}}{(m-1)!} + y_{m-1}(0)\frac{t^{m-2}}{(m-2)!} + \cdots + y_2(0)t + y_1(0) \right) e^{\lambda t}.$$

Writing the obtained relationships in matrix form, we see that

$$y(t) = E_{J(\lambda)}(t)y(0),$$

where

$$E_{J(\lambda)}(t) = e^{\lambda t} \begin{pmatrix} 1 & t & t^2/2 & \dots & t^{m-1}/(m-1)! \\ 0 & 1 & t & \dots & t^{m-2}/(m-2)! \\ 0 & 0 & 1 & \dots & t^{m-3}/(m-3)! \\ \dots & \dots & \dots & \dots & \dots \\ 0 & 0 & \dots & \dots & 1 \end{pmatrix}.$$

It is easy to see that relationships (5.121), (5.122), where $A = J(\lambda)$, hold for the matrix-valued function $E_{J(\lambda)}(t)$. Thus,

$$E_{J(\lambda)}(t) = e^{tJ(\lambda)}, \tag{5.131}$$

and the solution of problem (5.125), (5.126) can be calculated by the formula

$$y(t) = E_{J(\lambda)}(t)y_0 + \int_0^t E_{J(\lambda)}(t-\tau)g(\tau)d\tau,$$

which is often used in the theory of differential equations for the analysis of systems of the form (5.119).

Note that the reader can easily prove formula (5.131) using Property 2, p. 191. Hint: use the equality $J(\lambda) = \lambda I + J(0)$.

2. Now we consider a system of linear differential equations of the form:

$$B\dot{x}(t) = Ax(t) + f(t). \tag{5.132}$$

First we assume that A, $B \in M_n$ and that the pencil (A, B) is regular. For all non-singular matrices U, $V \in M_n$, we have

$$UBVV^{-1}\dot{x}(t) = UAVV^{-1}x(t) + Uf(t). \tag{5.133}$$

Suppose that the matrices U, V reduce the pencil (A, B) to Weierstrass canonical form. Put $y = V^{-1}x$, $g = Uf(t)$. Then (see Theorem 5.26, p. 196)

$$\text{diag}(I_k, H)\dot{y} = \text{diag}(J, I_{n-k})y + g. \tag{5.134}$$

System (5.134) splits into independent subsystems of the form

$$\dot{y} = J(\lambda)y + g, \tag{5.135}$$

where $J(\lambda)$ is the Jordan block corresponding to a characteristic value λ of the pencil (A, B), and

$$J(0)\dot{y} = y + g, \tag{5.136}$$

where $J(0)$ is a nilpotent Jordan block. We denote its order by m and write system (5.136) in detail as follows:

$$\dot{y}_2 = y_1 + g_1,$$
$$\dot{y}_3 = y_2 + g_2,$$
$$\dots\dots\dots\dots\dots$$
$$\dot{y}_m = y_{m-1} + g_{m-1},$$
$$y_m + g_m = 0.$$

Starting from the last equation, we get

$$y_m = -g_m,$$
$$y_{m-1} = -g_{m-1} - \dot{g}_m, \quad \dots.$$

Generally,

$$y_i = -\sum_{k=i}^{m} \frac{d^{k-i}}{dt^{k-i}} g_k, \quad i = 1, 2, \dots, m. \tag{5.137}$$

The method of solving the system (5.135) has been described above.

Thus, if the pencil (A, B) is regular, then the Cauchy problem for system (5.132) is solvable. It is important to note that as follows from relationship (5.137), the solution of (5.132) can include the derivatives of the function f. Therefore, if the function f is insufficiently smooth, then the solution of (5.132) can be a discontinuous function.

If the pencil (A, B) is singular, then the system (5.132), as we will show below, is solvable only if some conditions on the vector-valued function $f(t)$ hold. When using some matrices U, V (see (5.133)), we reduce the pencil (A, B) to Kronecker canonical form (5.118), we get a system of differential equations that splits into independent subsystems of the investigated forms and the following forms:

$$0(h_l, h_r)\dot{y}(t) = 0(h_l, h_r)y + g, \tag{5.138}$$

$$L_\varepsilon^{(0)}\dot{y}(t) = L_\varepsilon^{(1)}y + g, \tag{5.139}$$

$$(L_\eta^{(0)})^T \dot{y}(t) = (L_\eta^{(1)})^T y + g. \tag{5.140}$$

First of all, using (5.138), we see that for solvability of the system (5.132) it is necessary that the first h_l components of the vector Uf be equal to zero. Note that the number of equations in system (5.139) is less than the number of unknowns, i.e., this system is underdetermined. System (5.140) is overdetermined: the number of equations in (5.140) is greater than the number of unknowns. Writing the system (5.139) in detail, we get (see (5.109), p. 200)

$$\dot{y}_2 = y_1 + g_1, \quad \dot{y}_3 = y_2 + g_2, \quad \ldots, \quad \dot{y}_{\varepsilon+1} = y_\varepsilon + g_\varepsilon. \tag{5.141}$$

Obviously, we can equate the function y_1 with any function that is integrable on every finite interval. After that, using (5.141), we can sequentially define all other components of the vector-valued function y.

The system (5.140) in detail looks as follows:

$$0 = y_1 + g_1, \quad \dot{y}_1 = y_2 + g_2, \ldots, \quad \dot{y}_{\varepsilon-1} = y_\varepsilon + g_\varepsilon, \quad \dot{y}_\varepsilon = g_{\varepsilon+1}.$$

Hence, $y_1 = -g_1, y_2 = -g_2 - \dot{g}_1, \ldots, y_\varepsilon = -g_\varepsilon - \dot{g}_{\varepsilon-1} - \cdots - \dfrac{d^{\varepsilon-1}}{dt^{\varepsilon-1}}g_1,$

$$g_{\varepsilon+1} = -\frac{d}{dt}\left(g_\varepsilon + \dot{g}_{\varepsilon-1} + \cdots + \frac{d^{\varepsilon-1}}{dt^{\varepsilon-1}}g_1\right). \tag{5.142}$$

Equality (5.142) is a necessary condition of solvability of the system (5.140). Obviously, this equality defines some additional conditions for the components of the vector-valued function $f(t)$ that are necessary for the solvability of the original system (5.132).

Chapter 6
Vector and Matrix Norms

In this chapter the concept of a norm on the vector space $\mathbb{C}^n$ is introduced. We investigate relationships between different norms. We give the definition of a norm on the space of complex rectangular matrices and study its properties in detail, particularly with regard to estimates of eigenvalues and singular values of operators.

6.1 Basic Inequalities

Using Theorem 5.4, p. 171, it is easy to see that the function $-\ln(x)$ is convex on the interval $(0, \infty)$. Therefore, for arbitrary positive numbers a, b and $p, q > 1$ such that $1/p + 1/q = 1$, we have $\ln(a^p/p + b^q/q) \geq \ln(a^p)/p + \ln(b^q)/q = \ln(ab)$; hence we get $ab \leq a^p/p + b^q/q$. Clearly, the last inequality holds also for $ab = 0$. Further, since $|ab| = |a||b|$, we get

$$|ab| \leq \frac{|a|^p}{p} + \frac{|b|^q}{q} \tag{6.1}$$

for all, generally speaking, complex numbers a, b and for all $p, q > 1$ such that $1/p + 1/q = 1$. Inequality (6.1) is called *Young's inequality*.[1]

Theorem 6.1 (Hölder's[2] inequality). *Let* $x, y \in \mathbb{C}^n$, $p > 1$, $1/p + 1/q = 1$. *Then*

$$\left| \sum_{k=1}^{n} x_k y_k \right| \leq \left(\sum_{k=1}^{n} |x_k|^p \right)^{1/p} \left(\sum_{k=1}^{n} |y_k|^q \right)^{1/q}. \tag{6.2}$$

---

[1] William Henry Young (1863–1942) was an English mathematician.
[2] Otto Ludwig Hölder (1859–1937) was a German mathematician.

© Springer International Publishing AG 2017
L. Beilina et al., *Numerical Linear Algebra: Theory and Applications*,
DOI 10.1007/978-3-319-57304-5_6

Proof If at least one of the vectors x, y is equal to zero, then inequality (6.2) holds. Assume that the vectors x, y are nonzero. Then using Young's inequality, we get

$$\frac{|x_l|}{\left(\sum\limits_{k=1}^{n}|x_k|^p\right)^{1/p}}\frac{|y_l|}{\left(\sum\limits_{k=1}^{n}|y_k|^q\right)^{1/q}} \leq \frac{|x_l|^p}{p\sum\limits_{k=1}^{n}|x_k|^p} + \frac{|y_l|^q}{q\sum\limits_{k=1}^{n}|y_k|^q}, \quad l = 1, 2, \ldots, n.$$

Summing these inequalities over all l, we obtain

$$\sum_{k=1}^{n}|x_k||y_k| \leq \left(\sum_{k=1}^{n}|x_k|^p\right)^{1/p}\left(\sum_{k=1}^{n}|y_k|^q\right)^{1/q}.$$

Hence inequality (6.2) holds.   □

In the special case of $p = 2$, inequality (6.2) is called the *Cauchy–Schwarz inequality*.

Theorem 6.2 (Minkowski's inequality). *Let x, $y \in \mathbb{C}^n$, $p > 1$. Then*

$$\left(\sum_{k=1}^{n}|x_k + y_k|^p\right)^{1/p} \leq \left(\sum_{k=1}^{n}|x_k|^p\right)^{1/p} + \left(\sum_{k=1}^{n}|y_k|^p\right)^{1/p}. \tag{6.3}$$

Proof If the left-hand side of (6.3) is equal to zero, then inequality (6.3) evidently holds. We therefore assume that the left-hand side of (6.3) is positive. Clearly,

$$\sum_{k=1}^{n}|x_k + y_k|^p = \sum_{k=1}^{n}|x_k + y_k|^{p-1}|x_k + y_k|$$

$$\leq \sum_{k=1}^{n}|x_k + y_k|^{p-1}|x_k| + \sum_{k=1}^{n}|x_k + y_k|^{p-1}|y_k|. \tag{6.4}$$

Let us estimate the sums on the right-hand side of the last inequality using Hölder's inequality:

$$\sum_{k=1}^{n}|x_k + y_k|^{p-1}|x_k| \leq \left(\sum_{k=1}^{n}|x_k + y_k|^{(p-1)q}\right)^{1/q}\left(\sum_{k=1}^{n}|x_k|^p\right)^{1/p}, \tag{6.5}$$

$$\sum_{k=1}^{n}|x_k + y_k|^{p-1}|y_k| \leq \left(\sum_{k=1}^{n}|x_k + y_k|^{(p-1)q}\right)^{1/q}\left(\sum_{k=1}^{n}|y_k|^p\right)^{1/p}, \tag{6.6}$$

where $1/p + 1/q = 1$ and $(p-1)q = p$. Thus, combining (6.4)–(6.6), we obtain

$$\sum_{k=1}^{n} |x_k + y_k|^p \le \left(\sum_{k=1}^{n} |x_k + y_k|^p \right)^{1/q} \left(\left(\sum_{k=1}^{n} |x_k|^p \right)^{1/p} + \left(\sum_{k=1}^{n} |y_k|^p \right)^{1/p} \right),$$

and taking into account that $1 - 1/q = 1/p$, we get (6.3). $\square$

6.2 Norms on the Space $\mathbb{C}^n$

In this section we consider the concept of a vector norm on the space $\mathbb{C}^n$. This concept generalizes the notion of the length of a vector $x \in \mathbb{C}^n$ and in many cases is more convenient.

We say that a *norm* is introduced on the space $\mathbb{C}^n$ if for each $x \in \mathbb{C}^n$, there exists a unique real number $\|x\|$ (the norm of x) such that the following conditions (*the axioms of a norm*) are satisfied:

1. $\|x\| \ge 0$ for all $x \in \mathbb{C}^n$; $\|x\| = 0$ if and only if $x = 0$;
2. $\|\alpha x\| = |\alpha| \|x\|$ for all $x \in \mathbb{C}^n$ and for all $\alpha \in \mathbb{C}$;
3. $\|x + y\| \le \|x\| + \|y\|$ for all $x, y \in \mathbb{C}^n$.

Condition 3 is usually called the *triangle inequality*. Also, the following inequality holds:

4. $\big| \|x\| - \|y\| \big| \le \|x - y\|$ for all $x, y \in \mathbb{C}^n$.

Inequality 4 follows from Axiom 3. Indeed,

$$\|x\| = \|x - y + y\| \le \|x - y\| + \|y\|.$$

Similarly,

$$\|y\| \le \|x - y\| + \|x\|.$$

Combining the last two inequalities, we obtain inequality 4.

Examples of norms on the space $\mathbb{C}^n$.

1. Let $p \ge 1$. The equality $\|x\|_p = \left(\sum_{k=1}^{n} |x_k|^p \right)^{1/p}$ defines a norm. Indeed, Axioms 1 and 2 evidently hold; inequality 3 for $p = 1$ immediately follows from the properties of the modulus of a complex number, and for $p > 1$, it coincides with Minkowski's inequality. Note that $\|x\|_2 = |x| = (x, x)^{1/2}$ for all $x \in \mathbb{C}^n$. Here and further in this section, by $(\cdot, \cdot)$ we denote the standard inner product on the space $\mathbb{C}^n$.

2. Let $\|x\|_\infty = \max_{1 \le k \le n} |x_k|$. It is easy to verify that this equality defines a norm.

3. The function $\|x\|_A = (Ax, x)^{1/2}$ is a norm on the space $\mathbb{C}^n$ if A is a Hermitian positive definite matrix. To substantiate this fact it is enough to recall that the

relationship $(x, y)_A = (Ax, y)$ defines an inner product on the space $\mathbb{C}^n$ (see Property 1, p. 141, and also Section 3.2.2, p. 73).

A vector norm is continuous on the entire space $\mathbb{C}^n$. Indeed, let x and y be arbitrary vectors in $\mathbb{C}^n$. Expanding these vectors into the finite series $x = \sum_{k=1}^{n} x_k i_k$ and $y = \sum_{k=1}^{n} y_k i_k$ with respect to the natural basis of $\mathbb{C}^n$ and using the triangle inequality, we obtain $\|x - y\| \leq \sum_{k=1}^{n} \|i_k\| |x_k - y_k|$. Obviously, this inequality implies that if x tends to y, then $\|x - y\|$ tends to zero.

We say that a sequence $\{x_k\} \subset \mathbb{C}^n$ *converges* to a vector $x \in \mathbb{C}^n$ *in norm* if $\lim_{k \to \infty} \|x - x_k\| = 0$. As we have seen in the previous paragraph, the convergence of a sequence of vectors in any norm introduced on the space $\mathbb{C}^n$ follows from *componentwise convergence*. Below we will prove that the converse statement is also true.

We say that two norms $\| \cdot \|_{(1)}$ and $\| \cdot \|_{(2)}$ are *equivalent* if there exist two positive constants c_1 and c_2 such that

$$c_1 \|x\|_{(1)} \leq \|x\|_{(2)} \leq c_2 \|x\|_{(1)} \quad \text{for all } x \in \mathbb{C}^n. \tag{6.7}$$

Theorem 6.3 *Any two norms on the space $\mathbb{C}^n$ are equivalent.*

Proof Clearly, the relation of the equivalence of norms is transitive. Therefore, it is sufficient to prove that every norm $\| \cdot \|$ is equivalent to the norm $\| \cdot \|_2 = | \cdot |$, i.e., that there exist two positive constants c_1 and c_2 such that

$$c_1 |x| \leq \|x\| \leq c_2 |x| \quad \text{for all } x \in \mathbb{C}^n. \tag{6.8}$$

Let $S_1(0)$ be the set of all vectors in the space $\mathbb{C}^n$ such that $|x| = 1$ ($S_1(0)$ is the unit sphere centered at the origin). This set is closed and bounded in the space $\mathbb{C}^n$. The function $\varphi(x_1, x_2 \ldots, x_n) = \|x\|$ is continuous on the whole space $\mathbb{C}^n$. Therefore, by Weierstrass's theorem (see a calculus textbook), it follows that there exist points $x_1, x_2 \in S_1(0)$ such that $\|x_1\| = \min_{x \in S_1(0)} \|x\|$ and $\|x_2\| = \max_{x \in S_1(0)} \|x\|$. Let $c_1 = \|x_1\|$, $c_2 = \|x_2\|$. Clearly, $0 \leq c_1 \leq c_2$, and we see also that c_1 is not equal to zero, since otherwise, $x_1 = 0$, but $x_1 \in S_1(0)$, whence $|x_1| = 1$ and $x_1 \neq 0$. Thus, $0 < c_1 \leq \|x\| \leq c_2$ for all $x \in S_1(0)$. Let now x be an arbitrary nonzero vector in the space $\mathbb{C}^n$. Then evidently, the vector $(1/|x|)x$ belongs to $S_1(0)$, and therefore, $c_1 \leq \|(1/|x|)x\| \leq c_2$, whence it follows that the vector x satisfies inequalities (6.8). Obviously, if x is equal to zero, then inequalities (6.8) hold.   □

It follows from Theorem 6.3 that every norm on the space $\mathbb{C}^n$ is equivalent to the norm $\| \cdot \|_\infty$. Therefore, the componentwise convergence of a sequence of vectors follows from the convergence of this sequence in any norm.

It is important to note that generally, the constants c_1, c_2 depend on n, i.e., they depend on the dimension of the space $\mathbb{C}^n$. For example, the following estimates hold:

$$\|x\|_\infty \leq \|x\|_p \quad \text{for all } x \in \mathbb{C}^n \text{ and for all } p \geq 1; \tag{6.9}$$

$$\|x\|_p \leq \|x\|_q \quad \text{for all } x \in \mathbb{C}^n \text{ if } p \geq q \geq 1; \tag{6.10}$$

$$\|x\|_p \leq n^{1/p-1/q}\|x\|_q \quad \text{for all } x \in \mathbb{C}^n \text{ if } q > p \geq 1; \tag{6.11}$$

$$\|x\|_p \leq n^{1/p}\|x\|_\infty \quad \text{for all } x \in \mathbb{C}^n \text{ and for all } p \geq 1. \tag{6.12}$$

Before proving these estimates, we note that they are the best possible, i.e., for each of them there exists a nonzero vector x such that the inequality becomes an equality. In particular, if $x = (1, 0, \ldots, 0)$, then the first two inequalities become equalities; if $x = (1, 1, \ldots, 1)$, then the last two inequalities become equalities.

Now we present appropriate proofs.

1. Let $\|x\|_\infty = \max\limits_{1 \leq k \leq n} |x_k| = |x_i|$. Evidently,

$$|x_i| = (|x_i|^p)^{1/p} \leq \left(\sum_{k=1}^n |x_k|^p\right)^{1/p} = \|x\|_p.$$

2. Doing the obvious calculations, we obtain

$$\|x\|_p = \left(\sum_{k=1}^n |x_k|^q |x_k|^{p-q}\right)^{1/p} \leq \|x\|_\infty^{(p-q)/p}\|x\|_q^{q/p};$$

hence using (6.9), we get (6.10).

3. Writing $|x_k|^p$ in the form $|x_k|^p \cdot 1$ and then using Hölder's inequality with $t = q/p > 1$, $r = t/(t-1) = q/(q-p)$ for the estimation of $\|x\|_p$, we get

$$\|x\|_p = \left(\sum_{k=1}^n |x_k|^p\right)^{1/p} \leq \left(\sum_{k=1}^n |x_k|^q\right)^{1/q} \left(\sum_{k=1}^n 1\right)^{(q-p)/(pq)} = n^{1/p-1/q}\|x\|_q.$$

The reader can easily prove inequality (6.12). Then it can be proved that

$$\|x\|_\infty = \lim_{p\to\infty} \|x\|_p \quad \text{for all } x \in \mathbb{C}^n.$$

The proof is also left to the reader.

A vector norm is called *absolute* if it depends only on the absolute values of the components of the vector. For example, the norm $\|\cdot\|_p$ for any $p \geq 1$ is absolute; the norm $\|x\| = (|x_1|^2 + |x_2|^2 - \text{Re}(x_1\overline{x}_2))^{1/2}$ on the space $\mathbb{C}^2$ is not absolute.

Let $D = \text{diag}(d_1, d_2, \ldots, d_n), 0 \leq d_i \leq 1, i = 1, 2, \ldots, n, x \in \mathbb{C}^n$. Then for any absolute norm we get $\|Dx\| \leq \|x\|$. Evidently, it is sufficient to verify this inequality for $D = \text{diag}(1, \ldots, 1, d_k, 1, \ldots, 1), d_k \in [0, 1]$. We have

$$Dx = \frac{1}{2}(1 - d_k)(x_1, x_2, \ldots, -x_k, \ldots, x_n) + \frac{1}{2}(1 - d_k)x + d_k x;$$

therefore, $\|Dx\| \leq \frac{1}{2}(1 - d_k)\|x\| + \frac{1}{2}(1 - d_k)\|x\| + d_k\|x\| = \|x\|$.

A vector norm on $\mathbb{C}^n$ is called *monotone* if the inequality $\|x\| \leq \|y\|$ follows from the inequalities $|x_k| \leq |y_k|, k = 1, 2, \ldots, n$. Every monotone norm is absolute. Indeed, if a norm is monotone, then for each vector x, the following inequalities hold:

$$\|(|x_1|, |x_2|, \ldots, |x_n|)\| \leq \|(x_1, x_2, \ldots, x_n)\| \leq \|(|x_1|, |x_2|, \ldots, |x_n|)\|.$$

Conversely, every absolute norm is monotone. Indeed, if for vectors x and y we have $|x_k| \leq |y_k|, k = 1, 2, \ldots, n$, then there exists a matrix[3]

$$D = \text{diag}(d_1 e^{i\varphi_1}, d_2 e^{i\varphi_2}, \ldots, d_n e^{i\varphi_n}), \quad 0 \leq d_k \leq 1, \quad k = 1, 2, \ldots, n,$$

such that $x = Dy$. Using now the definition of an absolute norm and the inequality $\|Dy\| \leq \|y\|$, we get $\|x\| \leq \|y\|$.

6.3 The Hahn–Banach Theorem. Dual Norms

Let us recall that a *linear functional* f is defined on the space $\mathbb{C}^n$ if a complex number $f(x)$ uniquely corresponds to each vector $x \in \mathbb{C}^n$ and this map is linear, i.e.,

$$f(\alpha x + \beta y) = \alpha f(x) + \beta f(y) \quad \text{for all} \quad x, y \in \mathbb{C}^n \text{ and for all } \alpha, \beta \in \mathbb{C}. \quad (6.13)$$

We say that a *real linear functional* f is defined on the space $\mathbb{C}^n$ if a real number $f(x)$ uniquely corresponds to each vector $x \in \mathbb{C}^n$ and

$$f(\alpha x + \beta y) = \alpha f(x) + \beta f(y) \quad \text{for all} \quad x, y \in \mathbb{C}^n \text{ and for all } \alpha, \beta \in \mathbb{R}. \quad (6.14)$$

If a norm $\|\cdot\|$ is defined on the space $\mathbb{C}^n$, then for every linear functional (real or complex), we can define its norm by the formula

$$\|f\| = \sup_{x \in \mathbb{C}^n, \, x \neq 0} \frac{|f(x)|}{\|x\|} = \sup_{x \in \mathbb{C}^n, \, \|x\|=1} |f(x)|. \quad (6.15)$$

---

[3]Let us recall that by definition, $e^{i\varphi} = \cos\varphi + i \sin\varphi$.

For every linear functional we have

$$\|f\| < \infty. \tag{6.16}$$

Let us prove inequality (6.16) for real functionals. For complex functionals the proof is analogous and easier. Let $z = (z_1, z_2, \ldots, z_n) \in \mathbb{C}^n$, $\|z\| = 1$. If we assume now that $z_k = x_k + iy_k$, $x_k, y_k \in \mathbb{R}$, $k = 1, 2, \ldots, n$, then we get

$$f(z) = f\left(\sum_{k=1}^{n}(x_k + iy_k)i_k\right) = \sum_{k=1}^{n}(x_k f(i_k) + y_k f(ii_k)).$$

Hence $|f(z)| \leq \max(\max_{1 \leq k \leq n}|f(i_k)|, \max_{1 \leq k \leq n}|f(ii_k)|)\sum_{k=1}^{n}|z_k|$. Since all norms on $\mathbb{C}^n$ are equivalent, using the last inequality, we conclude that $|f(z)| \leq c\|z\| = c$, where c is a constant that depends only on n. This means that (6.16) is valid.

Theorem 6.4   (Hahn–Banach[4]). *Let L be a subspace of the space $\mathbb{C}^n$ and f a linear functional defined on L,*

$$\|f\| = \sup_{x \in L,\, \|x\|=1}|f(x)|. \tag{6.17}$$

Then there exists a linear functional F defined on $\mathbb{C}^n$ such that $F(x) = f(x)$ for all $x \in L$ and[5]

$$\|F\| = \sup_{x \in \mathbb{C}^n,\, \|x\|=1}|F(x)| = \|f\|. \tag{6.18}$$

Proof First we assume that f is a real linear functional. Naturally, we suppose that f is not identically zero. Therefore, without loss of generality, we can assume that $\|f\| = 1$. We do not consider the trivial case $L = \mathbb{C}^n$. Let $u \notin L$ and let $L_1 \supset L$ be the set of all vectors of the form $x + tu$, where $x \in L$, $t \in \mathbb{R}$. Using the triangle inequality, we see that

$$f(x) - f(y) \leq \|x - y\| \leq \|x + u\| + \|y + u\|$$

for all $x, y \in L$. Hence, $f(x) - \|x + u\| \leq f(y) + \|y + u\|$. Therefore, there exists a number a such that

$$\sup_{x \in L}(f(x) - \|x + u\|) \leq a \leq \inf_{x \in L}(f(x) + \|x + u\|). \tag{6.19}$$

Let us define a functional f_1 on L_1 by the formula $f_1(x + tu) = f(x) - at$ (check that f_1 is a real linear functional!). It follows from inequalities (6.19) that

---

[4]Hans Hahn (1879–1934) was an Austrian mathematician. Stefan Banach (1892–1945) was a Polish mathematician.

[5]One says, F is the norm-preserving extension of the functional f onto the entire space $\mathbb{C}^n$.

$$|f(x) - a| \le \|x + u\| \quad \text{for all} \quad x \in L,$$

and $|f_1(x+u)| \le \|x+u\|$ for all $x \in L$. For $t \ne 0$ we get $f_1(x+tu) = tf_1(t^{-1}x+u)$; hence

$$|f_1(x + tu)| = |t||f_1(t^{-1}x + u)| \le |t|\|t^{-1}x + u)\| = \|x + tu\|,$$

or $|f_1(x)| \le \|x\|$ for all $x \in L_1$. Arguing as above, we construct a real linear functional f_2 defined on the set $L_2 \supset L_1$ of all vectors of the form $x + t(iu)$, where $x \in L_1, t \in \mathbb{R}$, such that $|f_2(x)| \le \|x\|$ for all $x \in L_2$. It is easy to see that the set L_2 coincides with the subspace of the space $\mathbb{C}^n$ spanned by a basis of the subspace L and the vector u. Thus we have constructed an extension of the real linear functional f defined on L onto the wider subspace. Increasing sequentially the dimension of the subspaces, we can construct a real linear functional F defined on the entire space $\mathbb{C}^n$ such that $F(x) = f(x)$ for all $x \in L$ and $|F(x)| \le \|x\|$ for all $x \in \mathbb{C}^n$. It follows from the last estimate and (6.17) that $\|F\| = \|f\|$.

Let now f be a (complex) linear functional defined on L. We represent it in the form $f(x) = g(x) + ih(x)$ for all $x \in L$, where g and h are real linear functionals on L. Since the functional f is linear, we get $f(ix) = g(ix) + ih(ix) = if(x) = ig(x) - h(x)$. Hence $h(x) = -g(ix)$, and $f(x) = g(x) - ig(ix)$. By assumption, $\|f\| = 1$, and consequently, we have $\|g\| \le 1$. Using the construction described in the previous part of the proof, we construct a real linear functional $G(x)$ defined on the entire space $\mathbb{C}^n$ such that

$$G(x) = g(x) \quad \text{for all} \quad x \in L, \quad \text{and} \quad |G(x)| \le \|x\| \quad \text{for all} \quad x \in \mathbb{C}^n.$$

Further, let $F(x) = G(x) - iG(ix)$ for all $x \in \mathbb{C}^n$. Clearly, $F(x) = f(x)$ for all $x \in L$. Now we prove that the functional F is linear. For this purpose, it is enough to show additionally that $F(ix) = iF(x)$ for all $x \in \mathbb{C}^n$. This fact follows directly from the definition. Indeed, $F(ix) = G(ix)+iG(x) = i(G(x)-iG(ix))$. To complete the proof, we check equality (6.18). Let $x \in \mathbb{C}^n$ be a given vector. Take a real number θ such that $F(x)e^{i\theta}$ is nonnegative. Then

$$|F(x)| = F(e^{i\theta}x) = G(e^{i\theta}x) \le \|e^{i\theta}x\| = \|x\|.$$

Combining (6.17) with the last inequality, we get (6.18). □

Corollary 6.1 *Let $x_0 \in \mathbb{C}^n$ be a given vector. There exists a linear functional F defined on $\mathbb{C}^n$ such that $F(x_0) = \|x_0\|$ and $\|F\| = 1$.*

Proof Let us consider the subspace $L \subset \mathbb{C}^n$ of all vectors of the form αx_0, where $\alpha \in \mathbb{C}$, and let us define on this subspace a linear functional f by the formula $f(\alpha x_0) = \alpha\|x_0\|$. Then obviously, $f(x_0) = \|x_0\|$ and $\|f\| = 1$. To conclude the proof, using the Hahn–Banach theorem, we construct a norm-preserving extension of the functional f onto the entire space $\mathbb{C}^n$. □

We can consider the space $\mathbb{C}^n$ to be a unitary space if we define an inner product (for example, standard) on it. Using Riesz's theorem (see p. 132), we see that for each linear functional f on $\mathbb{C}^n$ there exists one and only one vector $y \in \mathbb{C}^n$ such that $f(x) = (x, y)$ for all $x \in \mathbb{C}^n$, and conversely, every vector $y \in \mathbb{C}^n$ generates a linear functional: $f(x) = (x, y)$ for all $x \in \mathbb{C}^n$. Let $\|\cdot\|$ be a norm on the space $\mathbb{C}^n$. For each vector $y \in \mathbb{C}^n$, we put

$$\|y\|_* = \|f\| = \sup_{x \in \mathbb{C}^n,\, x \neq 0} \frac{|(x, y)|}{\|x\|} = \sup_{x \in \mathbb{C}^n,\, \|x\| = 1} |(x, y)|. \tag{6.20}$$

The reader can easily prove that the relationship (6.20) defines a norm on the space $\mathbb{C}^n$. This norm is called *dual* to the original norm. The next theorem shows that the concept of the duality of norms is reciprocal.

Theorem 6.5 *Let $\|\cdot\|$ be a norm on the space $\mathbb{C}^n$ and let $\|\cdot\|_*$ be its dual norm. Then*

$$\|x\| = \sup_{y \in \mathbb{C}^n,\, \|y\|_* = 1} |(x, y)|. \tag{6.21}$$

Proof It follows immediately from the definition of the dual norm that for every nonzero $y \in \mathbb{C}^n$, the following inequality holds: $\|x\| \geq |(x, y)|/\|y\|_*$. Using Corollary 6.1, we see that there exists a vector y such that $\|x\| = |(x, y)|/\|y\|_*$. These arguments show that equality (6.21) is valid. $\square$

In the proof of Theorem 6.5 we have established the following result.

Corollary 6.2 *For all $x, y \in \mathbb{C}^n$, the following inequality holds:*

$$|(x, y)| \leq \|x\|\|y\|_*. \tag{6.22}$$

Inequality (6.22) is called the *generalized Cauchy–Schwarz inequality*.

For example, the norms $\|\cdot\|_p$, $\|\cdot\|_q$ for $p > 1$, $1/p + 1/q = 1$ are dual to each other with respect to the standard inner product on $\mathbb{C}^n$. Indeed, for all $x, y \in \mathbb{C}^n$, by Hölder's inequality (see (6.2)), we have $|(x, y)| \leq \|x\|_p \|y\|_q$. Let $x_k = \rho_k e^{i\varphi_k}$, where $k = 1, 2, \ldots, n$. Put $y_k = \rho_k^{p-1} e^{i\varphi_k}$, $k = 1, 2, \ldots, n$. By elementary calculations, we see that $|(x, y)| = \|x\|_p \|y\|_q$. Therefore,

$$\|x\|_p = \sup_{y \in \mathbb{C}^n,\, y \neq 0} \frac{|(x, y)|}{\|y\|_q}.$$

Now the reader can easily prove that the norms $\|\cdot\|_1$ and $\|\cdot\|_\infty$ are dual to each other with respect to the standard inner product on the space $\mathbb{C}^n$.

6.4   Norms on the Space of Matrices

As above, we denote by $M_{m,n}$ the set of all rectangular matrices with m rows, n columns, and (generally speaking) complex elements. If $m = n$, we write M_n. If we define on the set $M_{m,n}$ the operations of matrix addition and multiplication of a matrix by a scalar in the usual way, then this set becomes a complex linear space of dimension mn. On this linear space we introduce a norm, i.e., we associate with each $A \in M_{m,n}$ a number $\|A\|$ such that the following axioms hold:

1. $\|A\| \geq 0$ for all $A \in M_{m,n}$; $\|A\| = 0$ if and only if $A = 0$;
2. $\|\alpha A\| = |\alpha| \|A\|$ for all $A \in M_{m,n}$ and for all $\alpha \in \mathbb{C}$;
3. $\|A + B\| \leq \|A\| + \|B\|$ for all $A, B \in M_{m,n}$.

We say in this case that a *vector norm* has been introduced on the space $M_{m,n}$. Clearly, this norm has all the properties that were investigated in the last section for the norms of vectors.

So-called *consistent norms* are used often on spaces of matrices. For consistent norms, in addition to axioms 1–3, the following axiom must be satisfied:

4. $\|AB\|_{(mp)} \leq \|A\|_{(mn)} \|B\|_{(np)}$ for all matrices $A \in M_{m,n}$, $B \in M_{n,p}$.

Here subscripts indicate norms on the corresponding spaces of matrices.

Not all vector norms on spaces of matrices are consistent. For example, we may put

$$\|A\| = \max_{1 \leq i, j \leq n} |a_{ij}| \tag{6.23}$$

for $A \in M_n$. Obviously, this is a vector norm, but it is not a consistent norm on M_n. Indeed, if

$$A = \begin{pmatrix} 1 & 1 \\ 1 & 1 \end{pmatrix}, \quad \text{then} \quad AA = \begin{pmatrix} 2 & 2 \\ 2 & 2 \end{pmatrix},$$

and $\|A\| = 1$, $\|AA\| = 2$, then the inequality $\|AA\| \leq \|A\| \|A\|$ is not satisfied.

Let $\| \cdot \|$ be a consistent norm on M_n and let $S \in M_n$ be an arbitrary nonsingular matrix. Then, as the reader can easily prove, the formula

$$\|A\|_{(s)} = \|SAS^{-1}\| \quad \text{for all} \quad A \in M_n$$

defines a consistent norm on M_n.

Here are important examples of consistent matrix norms.

1. Let $\|A\|_{l_1} = \sum_{i,j=1}^{n} |a_{ij}|$ for $A \in M_n$. Evidently, the first three axioms hold. Let us verify Axiom 4. By definition, for $A, B \in M_n$ we have

$$\|AB\|_{l_1} = \sum_{i,j=1}^{n} \left| \sum_{k=1}^{n} a_{ik} b_{kj} \right|.$$

Therefore,

$$\|AB\|_{l_1} \leq \sum_{i,j,k=1}^{n} |a_{ik}||b_{kj}|.$$

Adding nonnegative items to the right-hand side of the last inequality, we get

$$\|AB\|_{l_1} \leq \sum_{i,j,k,m=1}^{n} |a_{ik}||b_{mj}|.$$

It remains to note that

$$\sum_{i,j,k,m=1}^{n} |a_{ik}||b_{mj}| = \sum_{i,k}^{n} |a_{ik}| \sum_{j,m}^{n} |b_{mj}| = \|A\|_{l_1}\|B\|_{l_1}.$$

2. Let $\|A\|_E = \left(\sum_{i,j=1}^{m,n} |a_{ij}|^2 \right)^{1/2}$ for $A \in M_{m,n}$. This norm is generated by the

standard inner product on the space $\mathbb{C}^{mn}$. Hence the first three axioms for this norm hold. Usually, the norm $\|A\|_E$ is called the *Euclidean* norm or the *Frobenius*[6] norm. Using the Cauchy–Schwarz inequality (see p. 210), we verify Axiom 4. Let $A \in M_{m,n}$ and $B \in M_{n,p}$. Then

$$\|AB\|_E^2 = \sum_{i,j=1}^{m,p} \left| \sum_{k=1}^{n} a_{ik}b_{kj} \right|^2 \leq \sum_{i,j=1}^{m,p} \sum_{k=1}^{n} |a_{ik}|^2 \sum_{k=1}^{n} |b_{kj}|^2$$

$$= \sum_{i,k=1}^{m,n} |a_{ik}|^2 \sum_{k,j=1}^{n,p} |b_{kj}|^2 = \|A\|_E^2 \|B\|_E^2.$$

3. The reader can easily prove that the norm $\|A\| = n \max_{1 \leq i,j \leq n} |a_{ij}|$ is consistent on the space $M_{m,n}$.

Let $A \in M_{m,n}$ and let $\|\cdot\|_{(m)}$, $\|\cdot\|_{(n)}$ be some norms on the spaces $\mathbb{C}^m$, $\mathbb{C}^n$, respectively. Then there exists a nonnegative number N_A such that

$$\|Ax\|_{(m)} \leq N_A \|x\|_{(n)} \quad \text{for all} \quad x \in \mathbb{C}^n. \tag{6.24}$$

Indeed, since every norm on $\mathbb{C}^n$ is equivalent to the norm $\|\cdot\|_\infty$, i.e.,

$$c_1 \|x\|_\infty \leq \|x\|_{(n)} \quad \text{for all} \quad x \in \mathbb{C}^n,$$

$$\|x\|_{(m)} \leq c_2 \|x\|_\infty \quad \text{for all} \quad x \in \mathbb{C}^m,$$

---

[6]Ferdinand Georg Frobenius (1849–1917) was a German mathematician.

where c_1, c_2 are positive constants independent of x, we see that the following chain of inequalities holds:

$$\|Ax\|_{(m)} \leq c_2 \|Ax\|_\infty = c_2 \max_{1 \leq i \leq m} \left| \sum_{j=1}^{n} a_{ij} x_j \right| \leq c_2 \|x\|_\infty \max_{1 \leq i \leq m} \sum_{j=1}^{n} |a_{ij}|$$

$$\leq \frac{c_2}{c_1} \max_{1 \leq i \leq m} \sum_{j=1}^{n} |a_{ij}| \|x\|_{(n)}.$$

Denote by $\nu(A)$ the infimum of the set of all numbers N_A that satisfy (6.24). Evidently, we can define the function ν on the space $M_{m,n}$ in the following equivalent way:

$$\nu(A) = \sup_{x \in \mathbb{C}^n, \, x \neq 0} \frac{\|Ax\|_{(m)}}{\|x\|_{(n)}} = \sup_{x \in \mathbb{C}^n, \, \|x\|_{(n)}=1} \|Ax\|_{(m)}. \qquad (6.25)$$

Clearly,

$$\|Ax\|_{(m)} \leq \nu(A) \|x\|_{(n)} \quad \text{for all} \quad x \in \mathbb{C}^n.$$

The reader can easily prove that all axioms of a consistent matrix norm hold for the function ν. The matrix norm (6.25) is called *subordinate* or *induced* or an *operator norm*.

For each definition of a norm on the spaces $\mathbb{C}^m$, $\mathbb{C}^n$, there exists a vector $x_0 \in \mathbb{C}^n$ such that $\|x_0\|_{(n)} = 1$ and

$$\|Ax_0\|_{(m)} = \sup_{x \in \mathbb{C}^n, \, \|x\|_{(n)}=1} \|Ax\|_{(m)},$$

i.e., we can replace "sup" by "max" in the definition (6.25). The proof of this statement is left to the reader.

It is easy to see that for each definition of a norm on $\mathbb{C}^n$ the subordinate norm of the identity matrix (of order n) is equal to one.

Not every norm defined on M_n is induced by a vector norm. For example, the Frobenius norm is not induced by any vector norm, since $\|I\|_E = \sqrt{n}$. The norm (6.23) is also not an operator norm, since this norm is not consistent on M_n.

Examples of calculations of subordinate matrix norms.

1. Suppose that a norm on the space $\mathbb{C}^n$ is defined by the following equality (see the first example on p. 211 for $p = 1$): $\|x\|_1 = \sum_{k=1}^{n} |x_k|$. Then the induced matrix norm is

$$\|A\|_1 = \max_{x \in \mathbb{C}^n, \, \|x\|_1=1} \|Ax\|_1.$$

It is easy to see that for every $x \in \mathbb{C}^n$, $\|x\|_1 = 1$, we have

$$\|Ax\|_1 = \sum_{i=1}^{n} \left| \sum_{j=1}^{n} a_{ij} x_j \right| \le \sum_{i=1}^{n} \sum_{j=1}^{n} |a_{ij}||x_j| = \sum_{j=1}^{n} |x_j| \sum_{i=1}^{n} |a_{ij}|$$

$$\le \max_{1 \le j \le n} \sum_{i=1}^{n} |a_{ij}| \sum_{j=1}^{n} |x_j| = \max_{1 \le j \le n} \sum_{i=1}^{n} |a_{ij}|.$$

Suppose that $\max_{1 \le j \le n} \sum_{i=1}^{n} |a_{ij}| = \sum_{i=1}^{n} |a_{ik}|$. Let $\tilde{x}$ be a vector in the space $\mathbb{C}^n$ such that $\tilde{x}_k = 1$ and all other coordinates of the vector $\tilde{x}$ are equal to zero. Then clearly, $\|\tilde{x}\|_1 = 1$ and $\|A\tilde{x}\|_1 = \sum_{i=1}^{n} |a_{ik}|$. Therefore,

$$\|A\|_1 = \max_{x \in \mathbb{C}^n, \|x\|_1 = 1} \|Ax\|_1 = \max_{1 \le j \le n} \sum_{i=1}^{n} |a_{ij}|, \tag{6.26}$$

and so $\|A\|_1$ is called the *maximum absolute column sum* norm of the matrix A.

2. If a norm on the space $\mathbb{C}^n$ is defined by the equality $\|x\|_\infty = \max_{k \le 1 \le n} |x_k|$, then for every $x \in \mathbb{C}^n$ such that $\|x\|_\infty = 1$, we have

$$\|Ax\|_\infty = \max_{1 \le i \le n} \left| \sum_{j=1}^{n} a_{ij} x_j \right| \le \max_{1 \le i \le n} \sum_{j=1}^{n} |a_{ij}||x_j|$$

$$\le \max_{1 \le j \le n} |x_j| \max_{1 \le i \le n} \sum_{j=1}^{n} |a_{ij}| = \max_{1 \le i \le n} \sum_{j=1}^{n} |a_{ij}|.$$

Suppose that $\max_{1 \le i \le 1} \sum_{j=1}^{n} |a_{ij}| = \sum_{j=1}^{n} |a_{kj}|$. Let $\tilde{x} \in \mathbb{C}^n$ be the vector with components

$$\tilde{x}_j = \begin{cases} \bar{a}_{kj}/|a_{kj}|, & a_{kj} \ne 0, \\ 1, & a_{kj} = 0, \end{cases}$$

where $j = 1, 2, \ldots, n$, and as usual, the overline is the symbol of complex conjugation. Clearly, $\|\tilde{x}\|_\infty = 1$, and by elementary calculations, we see that for all $i = 1, 2, \ldots, n$, we have the inequalities

$$\left| \sum_{j=1}^{n} a_{ij} \tilde{x}_j \right| \le \sum_{j=1}^{n} |a_{ij}| \le \sum_{j=1}^{n} |a_{kj}|,$$

and for $i = k$, we have

$$\left| \sum_{j=1}^{n} a_{ij} \tilde{x}_j \right| = \sum_{j=1}^{n} |a_{kj}|,$$

i.e., $\|A\tilde{x}\|_\infty = \max\limits_{1 \le i \le 1} \sum\limits_{j=1}^{n} |a_{ij}|$. Therefore,

$$\|A\|_\infty = \max_{x \in \mathbb{C}^n,\, \|x\|_\infty = 1} \|Ax\|_\infty = \max_{1 \le i \le n} \sum_{j=1}^{n} |a_{ij}|,$$

and so $\|A\|_\infty$ is called the *maximum absolute row sum* norm of the matrix A.

3. Now we introduce norms on the spaces $\mathbb{C}^m$ and $\mathbb{C}^n$ induced by the standard inner product, i.e., we set $\|x\|_2 = |x|$. For every $x \in \mathbb{C}^n$, we have

$$\|Ax\|_2^2 = (Ax, Ax) = (A^*Ax, x).$$

The matrix A^*A is Hermitian and positive semidefinite. Hence there exists an orthonormal basis $\{e_k\}_{k=1}^n$ such that $A^*Ae_k = \sigma_k^2 e_k$, where σ_k^2, $k = 1, 2, \ldots, n$, are the eigenvalues of the matrix A^*A. They all are nonnegative. Expanding x into a finite series $x = \sum\limits_{k=1}^{n} \xi_k e_k$ with respect to the basis $\{e_k\}_{k=1}^n$ and assuming that $\|x\|_2 = 1$, we get $\sum\limits_{k=1}^{n} |\xi_k|^2 = 1$, $\|Ax\|_2^2 = \sum\limits_{k=1}^{n} \sigma_k^2 |\xi_k|^2 \le \max\limits_{1 \le k \le n} \sigma_k^2$. Put now $\sigma_j = \max\limits_{1 \le k \le n} \sigma_k$ and $\tilde{x} = e_j$. Then $\|A\tilde{x}\|_2^2 = \sigma_j^2$. Thus we see that $\max\limits_{x \in \mathbb{C}^n,\, \|x\|_2 = 1} \|Ax\|_2 = \max\limits_{1 \le k \le n} \sigma_k$, i.e.,

$$\|A\|_2 = \max_{1 \le k \le n} \sigma_k. \tag{6.27}$$

The next special case is of interest in many applications. Let $A \in M_n$ be a Hermitian matrix, i.e., $A = A^*$. Then evidently, $\sigma_k = |\lambda_k(A)|$, where $k = 1, 2, \ldots, n$, and $\lambda_k(A)$ is the eigenvalue of the matrix A. Therefore, for every Hermitian matrix, we get

$$\|A\|_2 = \max_{1 \le k \le n} |\lambda_k(A)| = \max_{x \in \mathbb{C}^n,\, x \ne 0} \frac{|(Ax, x)|}{(x, x)} = \rho(A),$$

where $\rho(A)$ is the spectral radius of A (see p. 188). In this connection, the norm $\|A\|_2$ is usually called *spectral*.

The proof of the following propositions is left to the reader.

Proposition 6.1 *For every matrix A we have (see p. 192)*

$$\|A^+\|_2 = \frac{1}{\sigma_r}. \tag{6.28}$$

Here σ_r is the minimal singular value of the matrix A.

Proposition 6.2 *If the matrix A is invertible, then (see p. 167)*

$$\text{cond}\,(A) = \|A\|_2 \|A^{-1}\|_2.$$

Therefore, the notation $\text{cond}\,(A) = \text{cond}\,_2(A)$ is often used.

The calculation of eigenvalues of a matrix, generally speaking, is a complicated problem. Hence it is useful to estimate the norm $\|A\|_2$ using an explicit function of elements of A. Let us prove that for every matrix $A \in M_{mn}$, we have the estimate $\|A\|_2 \leq \|A\|_E$. Indeed, by elementary calculations, we get[7] $\text{tr}(A^*A) = \sum\limits_{i,j=1}^{m,n} |a_{ij}|^2$.

On the other hand, $\text{tr}(A^*A) = \sum\limits_{k=1}^{n} \sigma_k^2 \geq \max\limits_{1 \leq k \leq n} \sigma_k^2$. Hence

$$\|A\|_2 = \max_{1 \leq k \leq n} \sigma_k \leq \left(\sum_{i,j=1}^{m,n} |a_{ij}|^2 \right)^{1/2} = \|A\|_E. \tag{6.29}$$

The reader can easily check the following properties.

1. $\|A\|_2 = \|UAV\|_2$ and $\|A\|_E = \|UAV\|_E$ for every matrix $A \in M_n$ and all unitary matrices U and V.
2. $\|A\|_2 = \|A^*\|_2$ for every matrix $A \in M_n$.

The value of a consistent matrix norm is useful particularly for estimating the spectral radius of a matrix. Namely, for every square matrix A, the following inequality holds:

$$\rho(A) \leq \|A\|, \tag{6.30}$$

where $\|A\|$ is any consistent norm of the matrix A. Indeed, let λ, x be an eigenpair of the matrix A, and let X be a square matrix such that all columns of X are equal to each other and equal to the vector x. Then $AX = \lambda X$. Hence,

$$|\lambda| \|X\| = \|AX\| \leq \|A\| \|X\|$$

for every consistent matrix norm. We also see that $\|X\| \neq 0$, since x is an eigenvector that is not equal to zero by definition. Thus for each eigenvalue λ of the matrix A, the following inequality holds: $|\lambda| \leq \|A\|$. This last inequality is equivalent to (6.30).

Clearly, the next corollary follows from estimate (6.30).

Corollary 6.3 *If a consistent matrix norm of the matrix $A \in M_n$ is less than one, then A is a convergent matrix.*

---

[7] Here the trace of the matrix A^*A is calculated as the sum of all elements on the leading diagonal; see p. 124.

Theorem 6.6 *For every consistent matrix norm introduced on the space M_n and for every matrix $A \in M_n$, the following equality holds:*

$$\rho(A) = \lim_{k \to \infty} \|A^k\|^{1/k}. \tag{6.31}$$

Proof If λ is an eigenvalue of the matrix A, then for every integer $k > 0$, the number λ^k is an eigenvalue of the matrix A^k. Therefore, using inequality (6.30), we get $(\rho(A))^k = \rho(A^k) \leq \|A^k\|$, and $\rho(A) \leq \|A^k\|^{1/k}$ for every integer $k > 0$. Further, let ε be a positive number. Then the matrix $(\rho(A) + \varepsilon)^{-1}A$ is convergent, since the modulus of each eigenvalue of this matrix is less than one. Therefore, $(\rho(A) + \varepsilon)^{-k}A^k \to 0$ as $k \to \infty$. Since every norm on M_n is a continuous function (see pp. 211, 217), we get $\|(\rho(A) + \varepsilon)^{-k}A^k\| \to 0$ as $k \to \infty$. Hence there exists $N > 0$ such that for all $k \geq N$, the following inequalities hold: $\|(\rho(A) + \varepsilon)^{-k}A^k\| \leq 1$ and $\|A^k\|^{1/k} \leq \rho(A) + \varepsilon$. Thus for every $\varepsilon > 0$ and sufficiently large k, we have the following estimates: $\rho(A) \leq \|A^k\|^{1/k} \leq \rho(A) + \varepsilon$. This statement is equivalent to (6.31). $\square$

Using (6.31), the reader can easily prove that

$$\rho(A + B) \leq \rho(A) + \rho(B), \quad \rho(AB) \leq \rho(A)\rho(B)$$

for all permutable matrices A and B.

Theorem 6.7 *For every matrix $A \in M_n$, we have*

$$\rho(A) = \inf_{S \in M_n, \det(S) \neq 0} \|SAS^{-1}\|_1 = \inf_{S \in M_n, \det(S) \neq 0} \|SAS^{-1}\|_\infty. \tag{6.32}$$

Proof Let us prove the theorem for the norm $\|\cdot\|_1$. For the norm $\|\cdot\|_\infty$, all the arguments are repeated verbatim. Matrices A and SAS^{-1} are similar. Therefore, they have the same spectrum, and $\rho(A) = \rho(SAS^{-1})$. Using this and (6.30), we get

$$\rho(A) \leq \|SAS^{-1}\|_1 \quad \text{for all} \quad S \in M_n, \ \det(S) \neq 0. \tag{6.33}$$

By Schur's theorem, there is a unitary matrix U such that

$$U^*AU = T, \tag{6.34}$$

where T is an upper triangular matrix, and all eigenvalues $\lambda_1, \lambda_2, \ldots, \lambda_n$ of the matrix A form the main diagonal of T. Let $D = \operatorname{diag}(d, d^2, \ldots, d^n)$, where d is a positive number. Set

$$Q = DTD^{-1} \tag{6.35}$$

and compute

$$
Q = \begin{pmatrix}
\lambda_1 & d^{-1}t_{12} & d^{-2}t_{13} & \dots & d^{-(n-2)}t_{1,n-1} & d^{-(n-1)}t_{1,n} \\
0 & \lambda_2 & d^{-1}t_{23} & \dots & d^{-(n-3)}t_{2,n-1} & d^{-(n-2)}t_{2,n} \\
0 & 0 & \lambda_3 & \dots & d^{-(n-4)}t_{3,n-1} & d^{-(n-3)}t_{3,n} \\
& & & \dots & & \\
0 & 0 & 0 & \dots & \lambda_{n-1} & d^{-1}t_{n-1,n} \\
0 & 0 & 0 & \dots & 0 & \lambda_n
\end{pmatrix}. \tag{6.36}
$$

Let now ε be a given positive number. Then for d large enough, we can be certain that the sum of all the absolute values of the entries of each column of the matrix Q is less than or equal to $\rho(A) + \varepsilon$. Using (6.34), (6.35), we get $SAS^{-1} = Q$, where $S = DU^{-1}$, and $\|SAS^{-1}\|_1 = \|Q\|_1 \leq \rho(A) + \varepsilon$. Since the last inequality can be achieved by choosing the number d for any positive ε, this together with (6.33) provides the first equality in (6.32). $\qquad\square$

Using Theorem 6.7, the reader can easily prove that for every matrix $A \in M_n$, the following equality holds:

$$
\rho(A) = \inf_{\|\cdot\|} \|A\|, \tag{6.37}
$$

where the infimum is taken over all consistent matrix norms on M_n, but in general, "inf" cannot be replaced by "min" in (6.37).

At the end of this section we consider one important kind of norm on the space of matrices. These are the so-called *Ky Fan norms*.

Theorem 6.8 *Let $A \in M_{m,n}$, $\sigma_1(A) \geq \sigma_2(A) \geq \cdots \geq \sigma_q(A) \geq 0$, $q = \min(m, n)$, be the singular values of the matrix A (zeros are also included). Then*

$$
\|A\|_{k,p} = \left(\sum_{j=1}^{k} \sigma_j^p(A) \right)^{1/p},
$$

where $1 \leq k \leq q$, $p \geq 1$ are given numbers, is a norm on the space $M_{m,n}$. For $m = n$, the norm $\|A\|_{k,p}$ is consistent.

Proof Obviously, in the considered case, Axioms 1 and 2, p. 218, are satisfied. Therefore, we check only the following inequalities:

$$
\|A + B\|_{k,p} \leq \|A\|_{k,p} + \|B\|_{k,p} \quad \text{for all } A, B \in M_{m,n}, \text{ and } 1 \leq k \leq q, \ p \geq 1, \tag{6.38}
$$

$$
\|AB\|_{k,p} \leq \|A\|_{k,p}\|B\|_{k,p} \quad \text{for all } A, B \in M_n, \text{ and } 1 \leq k \leq n, \ p \geq 1. \tag{6.39}
$$

Using Theorem 5.7, p. 174, Corollary 5.1, p. 173, and Minkowski's inequality, p. 210, we get (6.38). Inequality (6.39) follows immediately from Corollary 5.3, p. 175. $\square$

The norm $\|A\|_{k,p}$ is called the *Ky Fan norm*.

Remark 6.1 Since for every matrix A and all unitary matrices U, V of corresponding dimensions, all the singular values of the matrices A and UAV coincide, we can say that the norm $\|A\|_{k,p}$ is unitarily invariant. If $k = 1$, we get the spectral norm. If $k = q$ and $p = 2$, we get the Frobenius norm. For every $k > 1$, the Ky Fan norm of the identity matrix is greater than one. Therefore, it is not an operator norm. By Theorem 5.12, p. 177, the norms $\|A\|_{k,1} = \sum_{j=1}^{k} \sigma_j(A)$, $1 \le k \le q$, can be calculated using formulas (5.48) and (5.49), p. 177.

6.5   The Gap Between Two Subspaces of $\mathbb{C}^n$

In this section, a norm on the space $\mathbb{C}^n$ is induced by an inner product. The norm on M_n is the corresponding subordinate matrix norm.

Let L, M be subspaces of the space $\mathbb{C}^n$ and let P_L, P_M be the operators of the orthogonal projection of $\mathbb{C}^n$ onto L and M, respectively (see p. 94). The number

$$\vartheta(L, M) = \|P_L - P_M\|$$

is called the *gap* between the subspaces L and M. It follows immediately from the definition that the function ϑ satisfies the properties of a distance (a metric):

1. $\vartheta(L, M) \ge 0$, $\vartheta(L, M) = 0$ if and only if $L = M$,
2. $\vartheta(L, M) = \vartheta(M, L)$,
3. $\vartheta(L, M) \le \vartheta(L, N) + \vartheta(N, M)$ for all subspaces L, M, N of $\mathbb{C}^n$.

The following numbers will be useful in our future considerations:

$$d_{L,M} = \max_{x \in L, \|x\|=1} \|x - P_M x\|, \quad \sigma_{L,M} = \min_{x \in L, \|x\|=1} \|P_M x\|.$$

These are connected by the equality

$$d_{L,M}^2 = 1 - \sigma_{L,M}^2. \tag{6.40}$$

Indeed, using the relationships $P_M = P_M^*$, $P_M = P_M^2$ (see pp. 94, 139), we can write

$$\|x - P_M x\|^2 = (x, x) + (P_M x, P_M x) - (x, P_M x) - (P_M x, x)$$
$$= (x, x) + (P_M x, P_M x) - 2(P_M x, P_M x) = \|x\|^2 - \|P_M x\|^2.$$

Therefore, $\max_{x \in L, \|x\|=1} \|x - P_M x\|^2 = 1 - \min_{x \in L, \|x\|=1} \|P_M x\|^2$.

The number $d_{L,M}$ can be calculated by the formula $d_{L,M} = \|(I - P_M)P_L\|$. Indeed, by definition,

$$\|(I - P_M)P_L\| = \sup_{x \in \mathbb{C}^n, \, x \neq 0} \frac{\|(I - P_M)P_L x\|}{\|x\|}.$$

Evidently,

$$\frac{\|(I - P_M)P_L x\|}{\|x\|} \leq \frac{\|(I - P_M)P_L x\|}{\|P_L x\|}$$

if $P_L x \neq 0$. Therefore,

$$\|(I - P_M)P_L\| \leq \sup_{y \in L, \, y \neq 0} \frac{\|(I - P_M)y\|}{\|y\|} = d_{L,M}.$$

On the other hand, for $x \in L$, $x \neq 0$, we have

$$\frac{\|(I - P_M)x\|}{\|x\|} = \frac{\|(I - P_M)P_L x\|}{\|x\|}.$$

Thus,

$$d_{L,M} = \sup_{x \in L, \, x \neq 0} \frac{\|(I - P_M)P_L x\|}{\|x\|} \leq \sup_{x \in \mathbb{C}^n, \, x \neq 0} \frac{\|(I - P_M)P_L x\|}{\|x\|} = \|(I - P_M)P_L\|.$$

Theorem 6.9 *For subspaces L, M of the space $\mathbb{C}^n$, we have the following equality:*

$$\|P_L - P_M\| = \max(d_{L,M}, d_{M,L}).$$

Proof Obviously, $P_L - P_M = P_L(I - P_M) - (I - P_L)P_M$. Hence, using the equality $P_L(I - P_L) = 0$ (see p. 94), for all $x \in \mathbb{C}^n$ we get

$$
\begin{aligned}
\|(P_L - P_M)x\|^2 &= \|P_L(I - P_M)x\|^2 + \|(I - P_L)P_M x\|^2 \\
&= \|P_L(I - P_M)(I - P_M)x\|^2 + \|(I - P_L)P_M P_M x\|^2 \\
&\leq \|P_L(I - P_M)\|^2 \|(I - P_M)x\|^2 + \|(I - P_L)P_M\|^2 \|P_M x\|^2 \\
&\leq \max(\|P_L(I - P_M)\|^2, \|(I - P_L)P_M\|^2)(\|(I - P_M)x\|^2 + \|P_M x\|^2) \\
&= \max(\|P_L(I - P_M)\|^2, \|(I - P_L)P_M\|^2)\|x\|^2.
\end{aligned}
$$

Note that $\|P_L(I - P_M)\| = \|(P_L(I - P_M))^*\| = \|(I - P_M)P_L\|$ (see Property 2, p. 167). Therefore, $\|P_L - P_M\| \leq \max(d_{L,M}, d_{M,L})$. The converse inequality also holds. Indeed,

$$(I - P_L)P_M = P_M - P_L P_M = P_M^2 - P_L P_M = (P_M - P_L)P_M,$$

and $\|(I - P_L)P_M\| \leq \|P_L - P_M\|$. Analogously, $\|(I - P_M)P_L\| \leq \|P_L - P_M\|$. $\square$

Clearly, we have the following corollary.

Corollary 6.4 *For all subspaces* $L, M \in \mathbb{C}^n$, *we have*

$$0 \le \vartheta(L, M) \le 1.$$

Theorem 6.10 *If* $\vartheta(L, M) < 1$, *then* $\dim L = \dim M$.

Proof Using Theorem 4.3, p. 97, we see that it is enough to show that P_L is a bijective map from M to L. By assumption, $\|P_L - P_M\| < 1$. Therefore, the operator $I - (P_L - P_M)$ is invertible (see Theorem 5.21, p. 191, and Corollary 6.3, p. 223), and $\text{Im}(I - (P_L - P_M)) = \mathbb{C}^n$. In other words, $(I - (P_L - P_M))\mathbb{C}^n = \mathbb{C}^n$. Acting on both sides of the last equality by the operator P_L, we get $P_L P_M \mathbb{C}^n = L$. Hence, $P_L M = L$, i.e., the operator P_L maps M onto the entire subspace L. To complete the proof, we show that this map is bijective. If we assume the contrary, then there exists a nonzero vector $x_0 \in M$ such that $P_L x_0 = 0$. Then $x_0 = P_M x_0 - P_L x_0$, and therefore, $\|x_0\| \le \|P_M - P_L\|\|x_0\| < \|x_0\|$, but this inequality cannot be satisfied. $\square$

Theorem 6.11 *If* $\dim L = \dim M$, *then* $d_{L,M} = d_{M,L}$.

Proof First we assume that $\sigma_{L,M} = 0$, i.e., that there exists $x \in L$, $\|x\| = 1$, such that $P_M x = 0$ (in other words, $x \in M^\perp$). Now we show that $\sigma_{M,L} = 0$, i.e., that there exists a vector $y \in M$, $\|y\| = 1$, such that $P_L y = 0$. Denote by $L_x^\perp$ the orthogonal complement in L of the one-dimensional subspace spanned by x. Clearly, $\dim L_x^\perp = \dim L - 1$. Let $(L_x^\perp)^\perp$ be the orthogonal complement in the space $\mathbb{C}^n$ of the subspace $L_x^\perp$. Then $\dim(L_x^\perp)^\perp = n - \dim L + 1$, and hence $\dim(L_x^\perp)^\perp + \dim M = n + 1$. Therefore, there exists a vector y, $\|y\| = 1$, that belongs to $(L_x^\perp)^\perp \cap M$. Since $x \in M^\perp$, we see that y is orthogonal to x, i.e., $y \in L^\perp$. Hence $P_L y = 0$. Now we note that if $\sigma_{L,M} = \sigma_{M,L} = 0$, then $d_{L,M} = d_{M,L} = 1$ (see (6.40)). Thus we can assume that $\sigma_{L,M} > 0$. By the definition of $\sigma_{L,M}$, we see that there exists a vector $x \in L$, $\|x\| = 1$, such that $\|P_M x\|^2 = \sigma_{L,M}^2$. Let us show that $P_M x - \sigma_{L,M}^2 x \in L^\perp$. To do this, using the definition of $\sigma_{L,M}$, we write

$$(P_M(x + v), x + v) \ge \sigma_{L,M}^2(x + v, x + v) \quad \text{for all} \quad v \in L.$$

By elementary calculations, we therefore obtain

$$(1 - \sigma_{L,M}^2)(v, v) + (P_M x - \sigma_{L,M}^2 x, v) + (v, P_M x - \sigma_{L,M}^2 x) \ge 0 \quad \text{for all} \quad v \in L.$$

If we replace here v by tv, $t \in \mathbb{R}$, we get

$$t^2(1 - \sigma_{L,M}^2)(v, v) + t(P_M x - \sigma_{L,M}^2 x, v) + t(v, P_M x - \sigma_{L,M}^2 x) \ge 0 \quad \text{for all} \quad t \in \mathbb{R}. \tag{6.41}$$

It follows from (6.41) that $\text{Re}(P_M x - \sigma_{L,M}^2 x, v) = 0$. Replacing v by iv in (6.41), we get $\text{Im}(P_M x - \sigma_{L,M}^2 x, v) = 0$. Therefore, $(P_M x - \sigma_{L,M}^2 x, v) = 0$ for all $v \in L$. In other words, $P_M x - \sigma_{L,M}^2 x \in L^\perp$. Hence, $P_L P_M x - \sigma_{L,M}^2 P_L x = P_L P_M x - \sigma_{L,M}^2 x = 0$. Now let $y = \sigma_{L,M}^{-1} P_M x$. Then $y \in M$, $\|y\| = 1$, $P_L y = \sigma_{L,M} x$, $\|P_L y\| = \sigma_{L,M}$, and hence $\sigma_{M,L} \le \sigma_{L,M}$. Analogously, $\sigma_{L,M} \le \sigma_{M,L}$, i.e., $\sigma_{L,M} = \sigma_{M,L}$. $\square$

Corollary 6.5 *If* $\dim L = \dim M$, *then*

$$\vartheta(L, M) = \|P_L - P_M\| = \|(I - P_M)P_L\| = \|(I - P_L)P_M\|.$$

As a conclusion, the reader is invited to give a geometric interpretation of the function ϑ and all statements of this section for subspaces of three-dimensional Euclidean space $\mathbf{V}_3$.

Chapter 7
Elements of Perturbation Theory

In this chapter we study the influence of perturbations of matrices on the solutions of such basic problems of linear algebra as calculating eigenvalues and singular values of operators, constructing the inverse matrix, solving systems of linear algebraic equations, and solving the linear least squares problem.

7.1 Perturbations in the Symmetric Eigenvalue Problem

Let A and B be Hermitian matrices of order n. Writing the obvious equality

$$A = B + (A - B),$$

using inequalities (4.148), p. 150, and inequality (6.30), p. 223, we see that

$$\max_{1 \leq k \leq n} |\lambda_k(A) - \lambda_k(B)| \leq \max_{1 \leq k \leq n} |\lambda_k(A - B)|, \tag{7.1}$$

$$\max_{1 \leq k \leq n} |\lambda_k(A) - \lambda_k(B)| \leq \|A - B\|, \tag{7.2}$$

where $\| \cdot \|$ is any consistent matrix norm. Using, for example, the Frobenius norm (see p. 218), we get

$$\max_{1 \leq k \leq n} |\lambda_k(A) - \lambda_k(B)| \leq \left(\sum_{i,j=1}^{n} |a_{ij} - b_{ij}|^2 \right)^{1/2}. \tag{7.3}$$

Inequalities (7.1)–(7.3) are usually called *Weyl's inequalities*.

© Springer International Publishing AG 2017

L. Beilina et al., *Numerical Linear Algebra: Theory and Applications*,
DOI 10.1007/978-3-319-57304-5_7

If we put $|a_{ij} - b_{ij}| \leq \varepsilon$, then $\max_{1 \leq k \leq n} |\lambda_k(A) - \lambda_k(B)| \leq n\varepsilon$. It is easy to see that if $A = I$ and all the elements of the matrix E are equal to $\varepsilon > 0$, then

$$\max_{1 \leq k \leq n} |\lambda_k(A) - \lambda_k(A + E)| = n\varepsilon,$$

i.e., the estimate (7.3) is the best possible for the set of all Hermitian matrices.

A special perturbation of a Hermitian matrix is considered in the next theorem.

Theorem 7.1 (**"relative" Weyl's theorem**). *Let* $\lambda_1 \geq \lambda_2 \geq \cdots \geq \lambda_n$ *be the eigenvalues of a Hermitian matrix* $A \in M_n$, *and let* $\tilde{\lambda}_1 \geq \tilde{\lambda}_2 \geq \cdots \geq \tilde{\lambda}_n$ *be the eigenvalues of the matrix* $X^* A X$, *where* X *is an arbitrary nonsingular matrix. Then*

$$|\tilde{\lambda}_i - \lambda_i| \leq \lambda_i \|I - X^* X\|, \quad i = 1, 2, \ldots, n, \tag{7.4}$$

where $\| \cdot \|$ *is any consistent matrix norm.*

Proof Let us take an integer $i \in [1, n]$ and write the obvious equality

$$X^*(A - \lambda_i I)X = H + F,$$

where $H = X^* A X - \lambda_i I$, $F = \lambda_i(I - X^* X)$. It is easy to see that the ith eigenvalue of the matrix $A - \lambda_i I$ is zero. Using Sylvester's law of inertia, p. 146, we can easily check that the ith eigenvalue of the matrix $X^*(A - \lambda_i I)X$ is also zero. The ith eigenvalue of the matrix H is $\tilde{\lambda}_i - \lambda_i$. Hence using inequality (7.2), we get (7.4). $\square$

Theorem 7.1 shows that if we replace the matrix A by $X^* A X$, where X is a nonsingular matrix, then all zero eigenvalues are preserved, and for all nonzero eigenvalues, we have the relative error estimate

$$\frac{|\tilde{\lambda}_i - \lambda_i|}{|\lambda_i|} \leq \|I - X^* X\|, \quad i = 1, 2, \ldots, n.$$

In the rest of this section we assume that the standard inner product is specified on the space $\mathbb{C}^n$. The following theorem describes how perturbations of a Hermitian matrix influence its eigenspaces.

Theorem 7.2 *Let* A, B *be Hermitian matrices of order* n, *and let*

$$\lambda_1(A) \geq \lambda_2(A) \geq \cdots \geq \lambda_n(A), \quad \lambda_1(B) \geq \lambda_2(B) \geq \cdots \geq \lambda_n(B)$$

be their eigenvalues. Let $k \in [1, n]$ *be a given integer and let* $\lambda_k(A)$ *have multiplicity* r *such that* $\lambda_{k-1}(A) > \lambda_k(A) > \lambda_{k+r}(A)$.[1] *Let* L_k *be the eigenspace (of dimension* r) *of the matrix* A *corresponding to* $\lambda_k(A)$, *and let* M_k *be the subspace of dimension* r *in the space* $\mathbb{C}^n$ *spanned by the orthogonal eigenvectors of the matrix* B *corresponding to its eigenvalues* $\lambda_k(B)$, $\lambda_{k+1}(B)$, ..., $\lambda_{k+r-1}(B)$. *Let*

---

[1]For $k = 1$ and $k = n$, these inequalities are modified in the obvious way.

$$\mathrm{gap}_k(A) = \min(\lambda_{k-1}(A) - \lambda_k(A), \lambda_k(A) - \lambda_{k+r}(A)),$$

$$\|A - B\|_2 < \frac{\mathrm{gap}_k(A)}{2}. \tag{7.5}$$

Then

$$\vartheta(L_k, M_k) \leq \frac{\|A - B\|_2}{\mathrm{gap}_k(A) - \|A - B\|_2} < 1. \tag{7.6}$$

Proof Let $x \in L_k$, $\|x\|_2 = 1$. If we write x in the form of the orthogonal decomposition $x = P_{M_k} x + y$, where $y \in M_k^\perp$, then $\|x - P_{M_k} x\|_2 = \|y\|_2$. Evidently,

$$|((A - B)x, y)| \leq \|A - B\|_2 \|y\|_2. \tag{7.7}$$

On the other hand, by the definition of the vector x, we have

$$((A - B)x, y) = \lambda_k(A)(x, y) - (Bx, y).$$

Note that $(x, y) = (y, y)$, $(Bx, y) = (x, By) = (y, By)$. We have used that $B = B^*$ and that $M_k^\perp$ is an invariant subspace of the operator B. Therefore,

$$((A - B)x, y) = \lambda_k(A)(y, y) - (By, y). \tag{7.8}$$

For $y \neq 0$, we get

$$\lambda_k(A)(y, y) - (By, y) = \left(\lambda_k(A) - \frac{(By, y)}{(y, y)}\right) \|y\|_2^2. \tag{7.9}$$

By the definition of the subspace $M_k^\perp$,

$$\frac{(By, y)}{(y, y)} \geq \lambda_{k-1}(B) \quad \text{or} \quad \frac{(By, y)}{(y, y)} \leq \lambda_{k+r}(B)$$

(see Lemma 4.7, p. 148). Now using inequality (7.2) and condition (7.5), we obtain

$$\left|\lambda_k(A) - \frac{(By, y)}{(y, y)}\right| \geq \mathrm{gap}_k(A) - \|A - B\|_2. \tag{7.10}$$

Obviously, it follows from (7.5), (7.7)–(7.10) that

$$\|y\|_2 \leq \frac{\|A - B\|_2}{\mathrm{gap}_k(A) - \|A - B\|_2} < 1. \tag{7.11}$$

Thus (see also Theorem 6.11, p. 217) we have the inequalities (7.6).   □

Sometimes the next corollary gives a more useful estimate.

Corollary 7.1 *Let all the conditions of Theorem 7.2 hold, and let the multiplicity of the eigenvalue $\lambda_k(A)$ be one. Then*

$$\vartheta(L_k, M_k)\sqrt{1 - \vartheta^2(L_k, M_k)} \le \frac{\|A - B\|_2}{\mathrm{gap}_k(A)}. \tag{7.12}$$

Proof Let $x \in M_k$, $\|x\|_2 = 1$. Writing x in the form of the orthogonal decomposition $x = \tilde{x} + y$, where $\tilde{x} = P_{L_k}x$, $y \in L_k^{\perp}$, and using estimate (7.11), we get

$$\|\tilde{x}\|_2 = \sqrt{1 - \|y\|_2^2} > 0. \tag{7.13}$$

If we put $B = A + E$, then we can write

$$(A + E)(\tilde{x} + y) = \lambda_k(B)(\tilde{x} + y). \tag{7.14}$$

Obviously, $A\tilde{x} = \lambda_k(A)\tilde{x}$. Subtracting these equalities term by term, by elementary calculations we obtain

$$(A - \lambda_k(A)I)y = (\eta I - E)x, \tag{7.15}$$

where $\eta = \lambda_k(B) - \lambda_k(A)$. Now we calculate the inner product of both sides of equality (7.14) with $\tilde{x}$. We note that $(A\tilde{x}, \tilde{x}) = \lambda_k(A)(\tilde{x}, \tilde{x})$ and also that $(y, \tilde{x}) = 0$ and $(Ay, \tilde{x}) = 0$, since $y, Ay \in L_k^{\perp}$. As a result we get

$$\eta = \frac{(Ex, \tilde{x})}{\|\tilde{x}\|_2^2}. \tag{7.16}$$

Computing the inner product of both sides of equality (7.15) with y and using (7.16), by elementary calculations we get

$$((A - \lambda_k(A)I)y, y) = \left(Ex, \frac{(y, y)}{\|\tilde{x}\|_2^2}\tilde{x} - y\right). \tag{7.17}$$

As we have seen in the proof of Theorem 7.2,

$$|((A - \lambda_k(A)I)y, y)| \ge \mathrm{gap}_k(A)\|y\|_2^2. \tag{7.18}$$

The vectors $\tilde{x}$ and y are orthogonal; hence

$$\left\|\frac{(y, y)}{\|\tilde{x}\|_2^2}\tilde{x} - y\right\|_2 = \left(\frac{|(y, y)|^2}{\|\tilde{x}\|_2^2} + \|y\|_2^2\right)^{1/2} = \left(\frac{\|y\|_2^2}{\|\tilde{x}\|_2^2} + 1\right)^{1/2}\|y\|_2 = \frac{\|y\|_2}{\|\tilde{x}\|_2}. \tag{7.19}$$

Combining (7.17)–(7.19) and (7.13), we finally get (7.12). $\square$

Remark 7.1 For one-dimensional subspaces L_k and M_k of a real space, we obtain $\vartheta(L_k, M_k) = \sin\alpha$, where α is the angle between L_k and M_k (prove it!). Therefore, estimate (7.12) is usually written as follows:

$$\frac{1}{2}\sin 2\alpha \leq \frac{\|A - B\|_2}{\text{gap}_k(A)}.$$

We may consider all estimates obtained in this section as a priori estimates. Using them, we can estimate the perturbations of eigenvalues and eigenvectors by known perturbations of the original matrix. In some situations, so-called a posteriori estimates are useful. They give information about errors using results of calculations already performed. Let now a normalized vector x be an approximation of an eigenvector of the matrix A, and let α be an approximation of the corresponding eigenvalue. Then the accuracy of calculations can be characterized by the residual vector $r(x, \alpha) = Ax - \alpha x$. If the matrix A is Hermitian, then it is easy to see that

$$\min_{\alpha \in \mathbb{R}} \|r(x, \alpha)\|_2 = \|Ax - \rho(x)x\|_2,$$

where $\rho(x) = (Ax, x)$.[2] This means that the number (Ax, x) is in a certain sense the best approximation of the eigenvalue of the Hermitian matrix A if we know the approximation x of the corresponding eigenvector.

The next theorem shows that the residual $r(x) = Ax - \rho(x)x$ can actually be used to estimate the accuracy of solutions of spectral problems.

Theorem 7.3 *Let $A \in M_n$ be a Hermitian matrix, $\lambda = \lambda_i$ a simple eigenvalue of the matrix A, and u the corresponding normalized eigenvector. Let $x \in \mathbb{C}^n$, $\|x\|_2 = 1$, $\rho = (Ax, x) \neq \lambda$, $|\rho - \lambda| < \text{gap}_k(A)$, $r = Ax - \rho x$. Let L and M be the one-dimensional subspaces in $\mathbb{C}^n$ spanned by u and x, respectively,*

$$\gamma = \min_{\mu \in \sigma(A),\ \mu \neq \lambda} |\rho - \mu|.$$

Then

$$\vartheta(L, M) \leq \frac{\|r\|_2}{\gamma}, \quad |\lambda - \rho| \leq \frac{\|r\|_2^2}{\gamma}. \tag{7.20}$$

Proof Write the vector x in the form of an orthogonal decomposition $x = \tilde{x} + y$, where $\tilde{x} \in L$, $y \in L^\perp$. Then $r = (\lambda - \rho)\tilde{x} + Ay - \rho y$. Since the vector Ay belongs to the subspace $L^\perp$, we have

$$\|r\|_2^2 = (\lambda - \rho)^2\|\tilde{x}\|_2^2 + \|Ay - \rho y\|_2^2. \tag{7.21}$$

It is easy to see that

$$\|Ay - \rho y\|_2^2 \geq \gamma^2\|y\|_2^2. \tag{7.22}$$

---

[2]Hint: for a given x, write $\|r(x, \alpha)\|_2^2$ as a quadratic trinomial in α.

It follows immediately from (7.21), (7.22) that the first estimate in (7.20) is valid. Further, using the definition of r, we obtain $(r, x) = 0$, or in detail,

$$(\lambda - \rho)\|\tilde{x}\|_2^2 + ((A - \rho I)w, w)\|y\|_2^2 = 0, \tag{7.23}$$

where $w = \|y\|_2^{-1}y$, $\|w\|_2 = 1$. Using the equality $\|\tilde{x}\|_2^2 = 1 - \|y\|_2^2$ and (7.23), by elementary calculations we get

$$\|y\|_2^2 = \frac{\rho - \lambda}{((A - \lambda)w, w)}. \tag{7.24}$$

It follows from (7.23) that $\|\tilde{x}\|_2^2 = ((A - \rho I)w, w)\|y\|_2^2/(\rho - \lambda)$. Combining this with (7.21), after elementary calculations we see that

$$\|r\|_2^2 = ((A - \rho I)w, (A - \lambda I)w)\|y\|_2^2. \tag{7.25}$$

Equalities (7.25) and (7.24) show that

$$\|r\|_2^2 = |\rho - \lambda|\frac{|((A - \rho I)w, (A - \lambda I)w)|}{|((A - \lambda I)w, w)|}. \tag{7.26}$$

If we represent here w in the form of an expansion with respect to the orthonormal set of eigenvectors of the matrix A, then we obtain

$$\|r\|_2^2 = |\rho - \lambda|\frac{\left|\sum_{j \neq i}(\lambda_j - \rho)(\lambda_j - \lambda_i)|c_j|^2\right|}{\left|\sum_{j \neq i}(\lambda_j - \lambda_i)|c_j|^2\right|}, \tag{7.27}$$

where c_j, $j = 1, 2, \ldots, n$, $j \neq i$, are the coefficients of the expansion. Obviously, it follows from the assumptions of the theorem that for $j \neq i$, all the numbers $(\lambda_j - \rho)(\lambda_j - \lambda_i)$ are positive. Therefore,

$$\left|\sum_{j \neq i}(\lambda_j - \rho)(\lambda_j - \lambda_i)|c_j|^2\right| \geq \gamma\left|\sum_{j \neq i}(\lambda_j - \lambda_i)|c_j|^2\right|,$$

i.e., the second estimate in (7.20) is also valid.   □

7.2   Perturbations of Singular Values and Singular Vectors

The next theorem follows immediately from Lemma 5.1, p. 174, estimate (7.1), p. 231, and inequality (5.13), p. 166.

Theorem 7.4 *Let* A, $B \in M_{m,n}$ *be arbitrary matrices,* $q = \min(m, n)$, *and let* $\sigma_1(A)$, $\sigma_2(A)$, ..., $\sigma_q(A)$, $\sigma_1(B)$, $\sigma_2(B)$, ..., $\sigma_q(B)$ *be their singular values (here we also include zeros for uniformity of notation). Then*

$$\max_{1 \leq k \leq q} |\sigma_k(A) - \sigma_k(B)| \leq \max_{1 \leq k \leq q} \sigma_k(A - B), \tag{7.28}$$

$$\max_{1 \leq k \leq n} |\sigma_k(A) - \sigma_k(B)| \leq \left(\sum_{i,j=1}^{m,n} |a_{ij} - b_{ij}|^2 \right)^{1/2}. \tag{7.29}$$

The following theorem is analogous to Theorem 7.2, p. 232.

Theorem 7.5 *Let* A, $B \in M_{m,n}$ *be arbitrary matrices,* $q = \min(m, n)$, *and let* $\sigma_1(A)$, $\sigma_2(A)$, ..., $\sigma_q(A)$, $\sigma_1(B)$, $\sigma_2(B)$, ..., $\sigma_q(B)$ *be their singular values ordered by non-increasing values. Let* $\sigma_k(A)$ *be a positive singular value of the matrix* A *of multiplicity* r. *Let* $L_{u,k}$ *be the subspace of* $\mathbb{C}^n$ *spanned by the right singular vectors of the matrix* A *corresponding to* $\sigma_k(A)$ *and let* $L_{v,k}$ *be the subspace spanned by the left singular vectors of* A *corresponding to* $\sigma_k(A)$. *Denote by* $M_{u,k}$ *the subspace spanned by the right singular vectors of the matrix* B *corresponding to the singular values* $\sigma_k(B)$, $\sigma_{k+1}(B)$, ..., $\sigma_{k+r-1}(B)$, *and by* $M_{v,k}$ *the subspace spanned by the left singular vectors of* B *corresponding to the same singular values.[3] Let*

$$\mathrm{gap}_k(A) = \min(\sigma_{k-1}(A) - \sigma_k(A), \sigma_k(A) - \sigma_{k+r}(A)),$$

$$\|A - B\|_2 < \frac{\mathrm{gap}_k(A)}{2}.$$

Then

$$\max(\vartheta(L_{u,k}, M_{u,k}), \vartheta(L_{v,k}, M_{v,k})) \leq \frac{\|A - B\|_2}{\mathrm{gap}_k(A) - \|A - B\|_2} < 1.$$

The proof of this theorem is left to the reader.

7.3 Perturbations of Characteristic Values of Arbitrary Matrices

Let $A = \{a_{ij}\}_{i,j=1}^n$ be a square matrix, and let

$$R_i(A) = \sum_{1 \leq j \leq n, \ j \neq i} |a_{ij}| \quad \text{for all} \quad i = 1, 2, \ldots, n,$$

---

[3] See the footnote on p. 232.

$$C_j(A) = \sum_{1 \le i \le n, \ i \ne j} |a_{ij}| \quad \text{for all} \quad j = 1, 2, \dots, n.$$

Theorem 7.6 (Gershgorin[4]). *Let A be an arbitrary matrix of order n. Then all the characteristic values of A are located in the union of n disks*

$$G_i^R = \{z \in \mathbb{C} : |z - a_{ii}| \le R_i(A)\}, \quad i = 1, 2, \dots, n. \tag{7.30}$$

Proof Let (λ, x) be an eigenpair of the matrix A, and let x_i be the element of x that has the largest absolute value. Evidently, $x_i \ne 0$. Using the definition of an eigenpair, we get

$$(a_{ii} - \lambda)x_i = - \sum_{1 \le j \le n, \ j \ne i} a_{ij} x_j.$$

Therefore, $|a_{ii} - \lambda||x_i| \le R_i(A)|x_i|$, and $|a_{ii} - \lambda| \le R_i(A)$. Thus every characteristic value of the matrix A belongs to one of the disks $G_i^R, i = 1, 2, \dots, n$. $\square$

This theorem is often called the *Gershgorin disk theorem*. Since A and A^T have the same eigenvalues, they all are located in the union of n disks

$$G_i^C = \{z \in \mathbb{C} : |z - a_{ii}| \le C_i(A)\}, \quad i = 1, 2, \dots, n. \tag{7.31}$$

This is the so-called the *column sum version* of the Gershgorin disk theorem.

Theorem 7.6 can be interpreted as a theorem on perturbations of a diagonal matrix $D = \mathrm{diag}(a_{11}, a_{22}, \dots, a_{nn})$. It shows that if the nondiagonal elements of the matrix A are small, then its characteristic values are not very different from the characteristic values of the matrix D.

The next two theorems are called the *Bauer–Fike* theorems.[5]

Theorem 7.7 *Suppose that for the square matrix $A = \{a_{ij}\}_{i,j=1}^n$ there exists a non-singular matrix V such that*

$$V^{-1}AV = \Lambda = \mathrm{diag}(\lambda_1, \lambda_2, \dots, \lambda_n). \tag{7.32}$$

Let $B = \{b_{ij}\}_{i,j=1}^n$ be an arbitrary square matrix. Then all characteristic values of the matrix $A + B$ are located in the union of n disks

$$G_i = \{z \in \mathbb{C} : |z - \lambda_i| \le \|B\|\|V\|\|V^{-1}\|\}, \quad i = 1, 2, \dots, n. \tag{7.33}$$

Here $\| \cdot \|$ is a matrix norm that is induced by any absolute vector norm.

---

[4]Semyon Aronovich Gershgorin (1901–1933) was a Soviet mathematician.

[5]Friedrich Ludwig Bauer (1924–2015) was a German mathematician. Charles Theodore Fike (born 1933) is an American mathematician.

Proof If (λ, x) is an eigenpair of $A + B$, then $(\lambda I - \Lambda)V^{-1}x = V^{-1}BVV^{-1}x$, whence (see p. 213) we get $\min_{1 \le i \le n} |\lambda - \lambda_i| \|V^{-1}x\| \le \|B\| \|V^{-1}\| \|V\| \|V^{-1}x\|$, but we have $V^{-1}x \ne 0$. Therefore, $\min_{1 \le i \le n} |\lambda - \lambda_i| \le \|B\| \|V^{-1}\| \|V\|$. Thus, $\lambda \in \bigcup_{i=1}^{n} G_i$. $\qquad\square$

Theorem 7.8 *Suppose that the conditions of Theorem 7.7 hold. Then all the characteristic values of the matrix $A + B$ are located in the union of n disks*

$$G_i = \{z \in \mathbb{C} : |z - \lambda_i| \le n s_i \|B\|_2\}, \quad i = 1, 2, \ldots, n, \qquad (7.34)$$

where $s_i = \|u_i\|_2 \|v_i\|_2 / |(u_i, v_i)|$, v_i is the ith column of the matrix V, u_i is the ith column of the matrix $U = (V^{-1})^*$, and the inner product $(\cdot, \cdot)$ is the standard inner product on the space $\mathbb{C}^n$.

Remark 7.2 It is obvious that (λ_i, v_i), $i = 1, 2, \ldots, n$, are the eigenpairs of the matrix A, and $(\bar{\lambda}_i, u_i)$, $i = 1, 2, \ldots, n$, are the eigenpairs of the matrix A^*. Each number s_i for $i = 1, 2, \ldots, n$ is greater than or equal to one. The number s_i is called the *coefficient of skewness* of the corresponding eigenvector v_i of the matrix A. If the algebraic multiplicity of the characteristic value λ_i of the matrix A is equal to one, then evidently, the algebraic multiplicity of the characteristic value $\bar{\lambda}_i$ of the matrix A^* is equal to one, too. The eigenspaces corresponding to these eigenvalues are one-dimensional, and hence the corresponding coefficient of skewness s_i is uniquely determined.

Proof of Theorem 7.8. The matrices $A + B$ and $\Lambda + V^{-1}BV = \Lambda + \widetilde{B}$, where we set $\widetilde{B} = U^*BV$, have the same characteristic values. Using the column sum version of the Gershgorin disk theorem, we see that all the characteristic values of the matrix $\Lambda + \widetilde{B}$ are located in the union of n disks

$$G_i' = \{z \in \mathbb{C} : |z - \lambda_i - \tilde{b}_{ii}| \le C_i(\widetilde{B})\}, \quad i = 1, 2, \ldots, n.$$

Note that $|z - \lambda_i - \tilde{b}_{ii}| \ge |z - \lambda_i| - |\tilde{b}_{ii}|$, $C_i(\widetilde{B}) + |\tilde{b}_{ii}| = \|\tilde{b}_i\|_1$, where as usual, by $\tilde{b}_i$ we denote the ith column of the matrix $\widetilde{B}$. Therefore, all the characteristic values of the matrix $A + B$ are located in the union of n disks

$$G_k'' = \{z \in \mathbb{C} : |z - \lambda_k| \le \|\tilde{b}_k\|_1\}, \quad k = 1, 2, \ldots, n.$$

Let us estimate $\|\tilde{b}_k\|_1$. Consider the vectors $t_k \in \mathbb{C}^n$ with the following components[6]:

$$t_{jk} = \begin{cases} \tilde{b}_{jk}/|\tilde{b}_{jk}|, & \tilde{b}_{jk} \ne 0, \\ 0, & \tilde{b}_{jk} = 0. \end{cases}$$

Trivially, $\|\tilde{b}_k\|_1 = (\widetilde{B}i_k, t_k)$, where i_k is a column of the identity matrix. Using the Cauchy–Schwarz inequality, we then obtain

---

[6] By $\tilde{b}_{jk}$ we denote the jth element of the column $\tilde{b}_k$.

$$\|\tilde{b}_k\|_1 = (BVi_k, Ut_k) \leq \|B\|_2 \|U\|_2 \|v_k\|_2 \|t_k\|_2. \tag{7.35}$$

It is easy to check that $\|t_k\|_2 \leq \sqrt{n}$. Further, using estimate (6.29), p. 223, we obtain $\|U\|_2 \leq \left(\sum_{k=1}^{n} \|u_k\|_2^2\right)^{1/2}$. Obviously, each column of the matrix U is uniquely determined up to a nonzero scalar factor. Therefore, we can normalize them to get $\|u_k\|_2 = 1$ for all $k = 1, 2, \ldots, n$. Then evidently, the columns of the matrix V must be normalized so that $(v_k, u_k) = 1$ for all $k = 1, 2, \ldots, n$. In this case, we see that $\|v_k\|_2 = \|v_k\|_2 \|u_k\|_2 / |(u_k, v_k)| = s_k$. Thus, using (7.35), we finally obtain $\|\tilde{b}_k\|_1 \leq n s_k \|B\|_2$.   □

The next theorem helps to compare estimate (7.33) with (7.34).

Theorem 7.9 *For every normalization of the columns of the matrix V, the following inequality holds:*

$$\|V\|_2 \|V^{-1}\|_2 \geq \max_{1 \leq k \leq n} s_k. \tag{7.36}$$

The columns of the matrix V can be normalized so that

$$\|V\|_2 \|V^{-1}\|_2 \leq \sum_{k=1}^{n} s_k. \tag{7.37}$$

Proof Clearly, we have $Vi_k = v_k$, $k = 1, 2, \ldots, n$, and $\|V\|_2 = \sup_{\|x\|_2=1} \|Vx\|_2 \geq \|v_k\|_2$. Similarly, we see that $\|V^{-1}\|_2 = \|U\|_2 \geq \|u_k\|_2$. Therefore, inequality (7.36) holds. Now we normalize the columns of the matrix V so that $\|v_k\|_2 = s_k^{1/2}$. Then, using the equality $(v_k, u_k) = 1$, we get $\|u_k\|_2 = s_k^{1/2}$, $k = 1, 2, \ldots, n$. Obviously, this implies that $\|V^{-1}\|_E = \|V\|_E = \left(\sum_{k=1}^{n} s_k\right)^{1/2}$. Using inequality (6.29), p. 223, we obtain esstimate (7.37).   □

Remark 7.3 The matrix V that has columns that form a basis of the space $\mathbb{C}^n$ that consists of the eigenvectors of the matrix A is not uniquely determined. For every matrix V, we have $\|V\| \|V^{-1}\| \geq 1$. This inequality becomes an equality if, for example, the matrix V is unitary and the norm of V is spectral. By Theorem 4.41, p. 142, it follows that a matrix is unitarily similar to a diagonal matrix if and only if that matrix is normal. Thus, if A is a normal matrix, λ_i, $i = 1, 2, \ldots, n$, are all the characteristic values of A, and B is an arbitrary square matrix, then all the characteristic values of the matrix $A + B$ are located in the union of n disks $G_i = \{z \in \mathbb{C} : |z - \lambda_i| \leq \|B\|_2\}$, $i = 1, 2, \ldots, n$.

7.4 Perturbations and the Invertibility of a Matrix

Let $A \in M_n$ be an invertible matrix, i.e., $\det A \neq 0$, and let $B \in M_n$. The following question arises: what are sufficient conditions on the matrix B so that the matrix $A + B$ has an inverse? Since $A + B = A(I + A^{-1}B)$, we see that the matrix $A + B$ is invertible if and only if the spectrum of the matrix $A^{-1}B$ does not include -1. Therefore, we have the following practically important sufficient conditions of the invertibility of the matrix $A + B$.

1. The matrix $A + B$ is invertible if $A^{-1}B$ is convergent, i.e., $\rho(A^{-1}B) < 1$.
2. The matrix $A + B$ is invertible if $\|A^{-1}B\| < 1$.
3. The matrix $A + B$ is invertible if $\|A^{-1}\|\|B\| < 1$.

Here and below in this section, the norm of a matrix is any consistent norm. The third condition usually is written in the form

$$\mathrm{cond}(A)\frac{\|B\|}{\|A\|} < 1, \qquad (7.38)$$

where $\mathrm{cond}(A) = \|A^{-1}\|\|A\|$. This number is called the *condition number* of the matrix A (compare it with the definition in Section 5.1.1, p. 167). We can interpret condition (7.38) in the following way: the matrix $A + B$ is invertible if the relative perturbation of the matrix A, i.e., $\|B\|/\|A\|$, is small compared with its condition number.

Example 7.1 Let $A = \{a_{ij}\}_{i,j=1}^{n}$ be a square matrix. The matrix A is said to be *row diagonally dominant* if[7]

$$|a_{ii}| > R_i(A) \quad \text{for all} \quad i = 1, 2, \ldots, n. \qquad (7.39)$$

It is said to be *column diagonally dominant* if

$$|a_{ii}| > C_i(A) \quad \text{for all} \quad i = 1, 2, \ldots, n. \qquad (7.40)$$

Let us prove that if a matrix A is row diagonally dominant, then it is nonsingular. Put $D = \mathrm{diag}(a_{11}, a_{22}, \ldots, a_{nn})$. Using condition (7.39), we see that the matrix D is nonsingular. Writing A in the form $A = D + (A - D)$ and using condition (7.39) one more time, we get $\|D^{-1}(A - D)\|_\infty < 1$. Therefore, condition 2 holds, and the matrix A is nonsingular. Since $\det(A) = \det(A^T)$, we see that every column diagonally dominant matrix is nonsingular as well.

The reader can easily prove that if condition (7.39) or (7.40) holds, then all the leading principal minors of the matrix A are nonzero.

---

[7]See the notation in Section 7.3.

Using Example 7.1, the reader can easily prove the Gershgorin disk theorem, and conversely, using the Gershgorin disk theorem, the reader can prove that if a matrix A is row diagonally dominant, then it is nonsingular.

Theorem 7.10 *Let matrices A and $\tilde{A} = A + B$ be invertible. Then*

$$\frac{\|A^{-1} - \tilde{A}^{-1}\|}{\|\tilde{A}^{-1}\|} \leq \|A^{-1}B\|. \tag{7.41}$$

If $\|A^{-1}B\| < 1$, then

$$\|\tilde{A}^{-1}\| \leq \frac{\|A^{-1}\|}{1 - \|A^{-1}B\|}, \tag{7.42}$$

$$\frac{\|A^{-1} - \tilde{A}^{-1}\|}{\|A^{-1}\|} \leq \frac{\|A^{-1}B\|}{1 - \|A^{-1}B\|}. \tag{7.43}$$

Proof By assumption, $I = (A + B)\tilde{A}^{-1}$. Therefore, $A^{-1} = (I + A^{-1}B)\tilde{A}^{-1}$. This implies that $A^{-1} - \tilde{A}^{-1} = A^{-1}B\tilde{A}^{-1}$, whence obviously, we get (7.41). Further, we have $\tilde{A}^{-1} = A^{-1} - A^{-1}B\tilde{A}^{-1}$, and $\|\tilde{A}^{-1}\| \leq \|A^{-1}\| + \|A^{-1}B\|\|\tilde{A}^{-1}\|$. Hence we have estimate (7.42). Finally, estimate (7.43) is an obvious consequence of estimates (7.41), (7.42). $\qquad\square$

The next corollary follows immediately from Theorem 7.10.

Corollary 7.2 *Let matrices A and $\tilde{A} = A + B$ be invertible. Then*

$$\frac{\|A^{-1} - \tilde{A}^{-1}\|}{\|\tilde{A}^{-1}\|} \leq \text{cond}(A)\frac{\|B\|}{\|A\|}. \tag{7.44}$$

If $\text{cond}(A)(\|B\|/\|A\|) < 1$, then

$$\|\tilde{A}^{-1}\| \leq \frac{\|A^{-1}\|}{1 - \text{cond}(A)(\|B\|/\|A\|)}, \tag{7.45}$$

$$\frac{\|A^{-1} - \tilde{A}^{-1}\|}{\|A^{-1}\|} \leq \frac{\text{cond}(A)(\|B\|/\|A\|)}{1 - \text{cond}(A)(\|B\|/\|A\|)}. \tag{7.46}$$

The following theorem shows that the "distance" between a nonsingular matrix A and the "nearest" singular matrix is characterized by the number $1/\text{cond}(A)$.

Theorem 7.11 *Let A be an invertible matrix and $A + B$ a singular matrix. Then*

$$\|B\|/\|A\| \geq 1/\text{cond}(A). \tag{7.47}$$

If the matrix norm is induced by a vector norm, then we can find a matrix B such that

$$\|B\|/\|A\| = 1/\mathrm{cond}(A) \tag{7.48}$$

and the matrix $A + B$ is singular.

Proof As we have seen, if a matrix A is invertible and a matrix $A + B$ is singular, then the spectrum of the matrix $A^{-1}B$ contains the number -1. Therefore, $\rho(A^{-1}B) \geq 1$, but $\rho(A^{-1}B) \leq \|A^{-1}B\| \leq \|A^{-1}\|\|B\|$, i.e., we have $\|B\| \geq 1/\|A^{-1}\|$. The last inequality is equivalent to (7.47). Now we prove the second part of the theorem. It follows from the definition of the induced matrix norm that there exists a vector x such that $\|x\| = 1$, $\|A^{-1}x\| = \|A^{-1}\|$. Put $y = \|A^{-1}\|^{-1}A^{-1}x$. Then $\|y\| = 1$, $Ay = \|A^{-1}\|^{-1}x$. By Corollary 6.1, p. 216, there exists a linear functional f on the space $\mathbb{C}^n$ such that $\|f\| = \sup_{v \in \mathbb{C}^n,\, \|v\|=1} |f(v)| = 1$. We define the matrix B by the action of this matrix on vectors, using the following relationship: $Bv = -(f(v)/\|A^{-1}\|)x$ for all $v \in \mathbb{C}^n$. Clearly, $By = -\|A^{-1}\|^{-1}x$, whence $(A + B)y = 0$, and therefore, $\det(A + B) = 0$. Moreover,

$$\|B\| = \sup_{v \in \mathbb{C}^n,\, \|v\|=1} \|Bv\| = \|A^{-1}\|^{-1} \sup_{v \in \mathbb{C}^n,\, \|v\|=1} |f(v)| = \|A^{-1}\|^{-1}.$$

The last equality is equivalent to (7.48). $\square$

7.5 The Stability of Systems of Linear Equations

In this section we assume that matrix norms are consistent with vector norms. The next theorem establishes the connection between the relative perturbations of the matrix and the right-hand side of the system with the relative perturbations of its solution. The main role in the estimates obtained below is played by the condition number of the matrix of the system.

Theorem 7.12 *Let A be an invertible matrix and let B be a matrix such that $\|A^{-1}B\| < 1$. Let x be the solution of the system of equations*

$$Ax = y, \tag{7.49}$$

and let $\tilde{x}$ be the solution of the system of equations

$$\tilde{A}\tilde{x} = y + b, \quad \tilde{A} = A + B. \tag{7.50}$$

Then

$$\frac{\|x - \tilde{x}\|}{\|x\|} \leq \frac{\mathrm{cond}(A)}{1 - \|A^{-1}B\|} \left(\frac{\|b\|}{\|y\|} + \frac{\|B\|}{\|A\|} \right). \tag{7.51}$$

If we assume additionally that $\|A^{-1}\|\|B\| < 1$, then

$$\frac{\|x - \tilde{x}\|}{\|x\|} \le \frac{\text{cond}(A)}{1 - \text{cond}(A)(\|B\|/\|A\|)} \left(\frac{\|b\|}{\|y\|} + \frac{\|B\|}{\|A\|} \right). \tag{7.52}$$

Proof By assumption, the inverse matrices A^{-1} and $\tilde{A}^{-1}$ exist. Therefore, $x = A^{-1}y$ and $\tilde{x} = \tilde{A}^{-1}(y + b)$. Hence $\tilde{x} - x = \tilde{A}^{-1}b + (\tilde{A}^{-1} - A^{-1})y$, and

$$\|x - \tilde{x}\| \le \|\tilde{A}^{-1}\| \|b\| + \|\tilde{A}^{-1} - A^{-1}\| \|y\|.$$

Therefore, using (7.42), (7.43), and the inequality $\|y\| \le \|A\| \|x\|$, by elementary calculations we get (7.51). We note that estimate (7.52) is an obvious consequence of (7.51). $\qquad \square$

In many situations, an error estimate based on the residual of the approximate solution is especially useful. Now we introduce a number that we shall use for this estimate. Let A be an invertible matrix and let $x \ne 0$, $Ax = y$. Put $\eta = \|A\| \|x\|/\|y\|$. Obviously, $\eta \ge 1$, and since $\|x\| \le \|A^{-1}\| \|y\|$, we see that $\eta \le \|A\| \|A^{-1}\| = \text{cond}(A)$. For a vector $\tilde{x} \in \mathbb{C}^n$, we put $r = A\tilde{x} - y$. Then $\tilde{x} - x = A^{-1}r$, and

$$\|x - \tilde{x}\| \le \|A^{-1}\| \|r\|. \tag{7.53}$$

Therefore,

$$\frac{\|x - \tilde{x}\|}{\|x\|} \le \frac{\text{cond}(A)}{\eta} \frac{\|r\|}{\|y\|}, \tag{7.54}$$

and as a consequence we get

$$\frac{\|x - \tilde{x}\|}{\|x\|} \le \text{cond}(A) \frac{\|r\|}{\|y\|}.$$

Estimate (7.54) shows that the relative error is estimated better by the relative residual of the approximate solution as the number η better approximates $\text{cond}(A)$.

Let $\tilde{x}$ be an approximate solution of the system of equations $Ax = y$. In some cases, for example in backward error analysis, it is useful to represent the vector $\tilde{x}$ in the form of the exact solution of the system with a perturbed matrix:

$$(A + B)\tilde{x} = y. \tag{7.55}$$

It is natural to seek a matrix B with minimal norm (induced by a norm of vectors). The possibility of such a choice of the matrix B is justified by the next theorem.

Theorem 7.13 (Rigal–Gaches). *Let $\tilde{x} \in \mathbb{C}^n$, $\tilde{x} \ne 0$, $r = A\tilde{x} - y$. There exists a matrix B such that Eq. (7.55) holds and $\|B\| = \|r\|/\|\tilde{x}\|$. If we assume additionally that there exists a matrix $D \ne B$ such that $(A + D)\tilde{x} = y$, then $\|D\| \ge \|B\|$.*

Proof To justify the last assertion it is enough to note that $D\tilde{x} = -r$, and therefore, $\|D\tilde{x}\| = \|r\|$; hence $\|D\| \ge \|r\|/\|\tilde{x}\|$. Let us define the matrix B by the relationship

$Bv = -(f(v)/\|\tilde{x}\|)r$ for all $v \in \mathbb{C}^n$, where f is a linear functional on the space $\mathbb{C}^n$ that satisfies the following conditions: $f(\tilde{x}) = \|\tilde{x}\|$ and $\|f\| = 1$.[8] Then we see that Eq. (7.55) holds, and $\|B\| = \|f\|\|r\|/\|\tilde{x}\| = \|r\|/\|\tilde{x}\|$. $\qquad\square$

In all previous estimates we assumed that the values of perturbations of the matrix and the right-hand side of the system were known in the sense of some norms. However, often it is more natural to define componentwise perturbations. Namely, now we assume that for a given $\varepsilon > 0$, we have

$$|B| \leq \varepsilon|A|, \quad |b| \leq \varepsilon|y|. \tag{7.56}$$

Here and below in this section the symbol $|\cdot|$ indicates a matrix or a vector that consists of the absolute values of its components. Inequalities (7.56) are to be understood componentwise. Thus estimates (7.56) mean that the relative perturbation of each element of the matrix and of the right-hand side of the system is less than or equal to ε.

Theorem 7.14 (Bauer–Skeel[9]). *Let x be the solution of system (7.49) and let $\tilde{x}$ be the solution of system (7.50). We assume that the matrix A is nonsingular, conditions (7.56) hold, and*

$$\varepsilon\||A^{-1}\|\|A\|| < 1. \tag{7.57}$$

Then

$$\|x - \tilde{x}\| \leq \varepsilon \frac{\||A^{-1}|(|A||x| + |y|)\|}{1 - \varepsilon\||A^{-1}\|\|A\||}. \tag{7.58}$$

Here the vector norm is any monotone norm, and the matrix norm is consistent with the vector norm.

Proof Using Eqs. (7.49), (7.50), we get $\tilde{x} - x = A^{-1}(Bx + b + B(\tilde{x} - x))$. Therefore, $|\tilde{x} - x| \leq |A^{-1}|(|B||x| + |b| + |B||\tilde{x} - x|)$, hence, using (7.56) and taking into account the assumed agreements for the vector and matrix norms, we obtain

$$\|\tilde{x} - x\| \leq \varepsilon\||A^{-1}|(|A||x| + |y|)\| + \varepsilon\||A^{-1}\|\|A\||\|\tilde{x} - x\|.$$

Combining the last inequality with (7.57), we get (7.58). $\qquad\square$

It is useful to note that if the right-hand side of the system is known exactly, i.e., $b = 0$, then instead of (7.58), we get

$$\frac{\|\tilde{x} - x\|}{\|x\|} \leq \frac{\varepsilon\||A^{-1}\|\|A\||}{1 - \varepsilon\||A^{-1}\|\|A\||}.$$

This estimate shows the dependence of the relative error of the solution on the relative perturbation of the matrix of the system. For this reason, the number

---

[8]See the proof of Theorem 7.11.

[9]Robert D. Skeel (born 1947) is an American mathematician.

$$\kappa_{BS}(A) = \||A^{-1}||A|\|$$

is called the *relative condition number* of the matrix A or the *Bauer–Skeel condition number*.

It is easy to see that under the assumed agreements for vector and matrix norms, $\kappa_{BS}(A) \geq 1$ for every matrix A. For every diagonal matrix, $\kappa_{BS} = 1$. Thus diagonal systems of equations are ideally conditioned with respect to the perturbations of the matrix.

7.6 Perturbations in the Linear Least Squares Problem

In this section we investigate the stability of a pseudosolution with respect to perturbations of a matrix and the right-hand side of the system. The vector norm in this section is Euclidean. The matrix norm is induced by this norm, i.e., we use the spectral matrix norm.

Lemma 7.1 *Let* $A, B \in M_{m,n}$, $\mathrm{rank}(A) = \mathrm{rank}(B) = r$, $\eta = \|A^+\|\|A - B\| < 1$.[10] *Then*

$$\|B^+\| \leq \frac{1}{1 - \eta}\|A^+\|. \tag{7.59}$$

Proof It follows from (7.28), p. 236, and (6.27), p. 221, that

$$\sigma_r(B) - \sigma_r(A) \geq -\|A - B\|.$$

Using (6.28), p. 223, we can write the last inequality in the form

$$\frac{1}{\|B^+\|} - \frac{1}{\|A^+\|} \geq -\|A - B\|,$$

whence, by elementary calculations, we get (7.59). □

Lemma 7.2 *Let* $A, B \in M_{m,n}$, $\mathrm{rank}(A) = \mathrm{rank}(B)$, $P_A = AA^+$, $P_B = BB^+$. *Then*

$$\|P_A(I - P_B)\| = \|P_B(I - P_A)\| \leq \|A - B\| \min(\|A^+\|, \|B^+\|). \tag{7.60}$$

Proof By construction, P_A, P_B are the orthogonal projectors defined on $\mathbb{C}^m$ (see Section 5.1.3, p. 169). Therefore, the equality $\|P_A(I - P_B)\| = \|P_B(I - P_A)\|$ follows from the results of Section 6.5, p. 226. In Section 6.5, we have seen also that the following equality holds: $\|P_B(I - P_A)\| = \|(I - P_A)P_B\|$. Now we note that

$$(I - P_A)P_B = (I - AA^+)P_B = (I - AA^+)(A + B - A)B^+,$$

---

[10]Let us recall that A^+ is the pseudoinverse of A (see Section 4.3.4, p. 137, and Section 5.1.3, p. 168).

but $(I - AA^+)A = 0$ (see Property 5 on p. 169). Hence

$$\|P_B(I - P_A)\| = \|(I - P_A)P_B\| = \|(I - P_A)(B - A)B^+\|$$
$$\leq \|(B - A)B^+\| \leq \|(B - A)\|\|B^+\|. \tag{7.61}$$

Analogously, $\|P_A(I - P_B)\| \leq \|(B - A)\|\|A^+\|$. □

Theorem 7.15 (Wedin[11]). *Let $A, \tilde{A} \in M_{m,n}$, $m \geq n$, be matrices of full rank. Let x be the normal pseudosolution of the system of equations (7.49), and let $\tilde{x}$ be the normal pseudosolution of system (7.50). Let $r = y - Ax$ and let $\tilde{r} = \tilde{y} - \tilde{A}\tilde{x}$ be the corresponding residuals. Suppose that $\|A^+\|\|B\| < 1$. Then*

$$\frac{\|x - \tilde{x}\|}{\|x\|} \leq \frac{\kappa_2(A)}{1 - \kappa_2(A)(\|B\|/\|A\|)}$$
$$\times \left(\frac{\|B\|}{\|A\|} \left(1 + \kappa_2(A) \frac{\|r\|}{\|A\|\|x\|} \right) + \frac{\|b\|}{\|y\|} \left(1 + \frac{\|r\|}{\|A\|\|x\|} \right) \right), \tag{7.62}$$

$$\frac{\|r - \tilde{r}\|}{\|y\|} \leq \left(\frac{\|b\|}{\|y\|} + 2\kappa_2(A) \frac{\|B\|}{\|A\|} \right), \tag{7.63}$$

where $\kappa_2(A) = \|A^+\|\|A\|$.

Proof By the definition of the pseudoinverse operator, we have

$$\tilde{x} - x = \tilde{A}^+(y + b) - x = \tilde{A}^+(r + Ax + b) - x = \tilde{A}^+(r + \tilde{A}x + b - Bx) - x.$$

Since by hypothesis, $\text{rank}(\tilde{A}) = n$, we get $\tilde{A}^+\tilde{A} = I$ (see Property 7 of the pseudoinverse operator, p. 169). Therefore,

$$\tilde{x} - x = \tilde{A}^+(r + b - Bx). \tag{7.64}$$

Now we note that by Property 6, p. 169, we have $\tilde{A}^+r = \tilde{A}^+\tilde{A}\tilde{A}^+r = \tilde{A}^+P_{\tilde{A}}r$. We note also that

$$P_A r = AA^+(y - Ax) = A(A^+y) - AA^+Ax = Ax - Ax = 0, \tag{7.65}$$

i.e., $\tilde{A}^+r = \tilde{A}^+P_{\tilde{A}}(I - P_A)r$, whence by Lemmas 7.1 and 7.2, we see that

$$\|\tilde{A}^+r\| \leq \frac{\|A^+\|^2}{1 - \|A^+\|\|B\|}\|B\|\|r\| = \frac{\|A^+\|^2\|A\|^2}{1 - \|A^+\|\|A\|(\|B\|/\|A\|)} \frac{\|B\|\|r\|}{\|A\|^2\|x\|}\|x\|. \tag{7.66}$$

---

[11]Per-Åke Wedin (born 1938) is a Swedish mathematician.

Analogously, using the evident inequality $\|y\| \le \|r\| + \|A\|\|x\|$, we get

$$\|\tilde{A}^+(b - Bx)\| \le \|\tilde{A}^+\|(\|b\| + \|B\|\|x\|) \le \frac{\|A^+\|}{1 - \|A^+\|\|B\|}\left(\frac{\|b\|}{\|x\|} + \|B\|\right)\|x\|$$

$$\le \frac{\|A^+\|\|A\|}{1 - \|A^+\|\|A\|(\|B\|/\|A\|)}\left(\frac{\|b\|\|r\|}{\|y\|\|A\|\|x\|} + \frac{\|b\|}{\|y\|} + \frac{\|B\|}{\|A\|}\right)\|x\|. \tag{7.67}$$

Combining (7.66), (7.67), and (7.64), we obtain (7.62). Let us estimate $r - \tilde{r}$. Using the definitions of r and $\tilde{r}$, we get

$$\tilde{r} - r = y + b - (A + B)\tilde{x} - y + Ax = b + \tilde{A}(x - \tilde{x}) - Bx.$$

Moreover, we have the inequality $\tilde{A}(x - \tilde{x}) = -\tilde{A}\tilde{A}^+(r - Bx + b)$. Indeed,

$$\tilde{A}\tilde{A}^+(r - Bx + b) = \tilde{A}\tilde{A}^+(y + b - \tilde{A}x) = \tilde{A}\tilde{A}^+\tilde{y} - \tilde{A}\tilde{A}^+\tilde{A}x = \tilde{A}\tilde{x} - \tilde{A}x.$$

Therefore, $\tilde{r} - r = (I - \tilde{A}\tilde{A}^+)(b - Bx) - \tilde{A}\tilde{A}^+r$. Since $I - \tilde{A}\tilde{A}^+$ is a projector, we get $\|r - \tilde{r}\| \le \|b - Bx\| + \|\tilde{A}\tilde{A}^+r\|$. Recall that $r = r - P_A r$. Hence, we have $\|\tilde{A}\tilde{A}^+r\| \le \|P_{\tilde{A}}(I - P_A)\|\|r\|$, whence, using Lemma 7.2, we obtain the estimate $\|\tilde{A}\tilde{A}^+r\| \le \|A^+\|\|B\|\|r\|$. Thus, $\|r - \tilde{r}\| \le \|b\| + \|B\|\|x\| + \|A^+\|\|B\|\|r\|$. Now we note that $x = A^+y$, $\|r\| = \min_{v \in \mathbb{C}^n}\|y - Av\| \le \|y\|$. Finally, using elementary calculations, we get (7.63).   □

Remark 7.4 Inequalities (7.62), (7.63) show that in the estimates of the perturbations of the linear least squares problem, the number $\kappa_2(A)$ plays an important role. It is called the *condition number of the linear least squares problem*. Note also that if the system of equations (7.49) is solvable, then $r = 0$, and the estimate (7.62) is transformed to an estimate of the form (7.52). Clearly, if A is a square nonsingular matrix, then we get $\kappa_2(A) = \mathrm{cond}_2(A)$.

Chapter 8
Solving Systems of Linear Equations

In this chapter we present algorithms and error analysis of numerical methods for solving linear systems $Ax = b$ with nonsingular square matrices. Here we present only direct methods. They are called direct because in the absence of rounding errors they would give the exact solution of $Ax = b$ after a finite number of steps. Section 8.1 presents Gaussian elimination algorithms. Section 8.2 analyzes their rounding errors and presents practical error bounds. Section 8.3 shows how to improve the accuracy of the computed solution of $Ax = b$ through an iterative refinement procedure. In Section 8.4 we discuss the basic special systems of linear equations (with symmetric positive definite matrices, symmetric indefinite matrices, band matrices) and numerical methods of their solution.

8.1 Algorithms for Gaussian Elimination

In this section we consider numerical algorithms for solving systems of linear equations with nonsingular matrices based on Gaussian elimination. Gaussian elimination was already discussed in Section 1.2.5, p. 42.

8.1.1 LU Factorization with Pivoting

As we saw in Section 1.2.5, p. 42, for every nonsingular matrix A of order n, by Gaussian elimination with pivoting we can construct unit[1] elementary lower triangular matrices L_k, $k = 1, 2, \ldots, n$, permutation matrices P_k, $k = 1, 2, \ldots, n$, and an

---

The original version of this chapter was revised. An Erratum to this book can be found at https://doi.org/10.1007/978-3-319-57304-5_8

[1] All the diagonal entries of a unit triangular matrix are equal to one.

© Springer International Publishing AG 2017
L. Beilina et al., *Numerical Linear Algebra: Theory and Applications*,
DOI 10.1007/978-3-319-57304-5_8

upper triangular matrix U such that

$$A = P_1 L_1^{-1} P_2 L_2^{-1} \cdots P_n L_n^{-1} U. \tag{8.1}$$

If we have obtained representation (8.1), then we can solve the system of linear equations

$$Ax = b \tag{8.2}$$

for any right-hand side b by computing the vector

$$f = L_n P_n \cdots L_1 P_1 b \tag{8.3}$$

and solving the system

$$Ux = f \tag{8.4}$$

with the triangular matrix U. The cost of computing the vector f and solving system (8.4) is approximately $2n^2$ arithmetic operations, which is much cheaper than constructing representation (8.1) (see Section 1.2.5, p. 42).

Calculating the vector f can be performed by solving a system of linear equations with a triangular nonsingular matrix. To show this, let us analyze the matrix $P_1 L_1^{-1} P_2 L_2^{-1} \cdots P_n L_n^{-1}$. The matrix P_2 differs from the identity matrix by a permutation of the second and ith columns, $i \geq 2$ (see the description of Gaussian elimination in Section 1.2.5, p. 42). Therefore, the matrix $L_1^{-1} P_2$ differs from L_1^{-1} by a permutation of the second and ith columns. Hence $L_1^{-1} P_2 = P_2 \hat{L}_1^{-1}$, where the matrix $\hat{L}_1^{-1}$ differs from L_1^{-1} by a permutation of the second element and the ith element in the first column. Thus, we obviously get

$$P_1 L_1^{-1} P_2 L_2^{-1} \cdots P_n L_n^{-1} = P_1 P_2 \cdots P_n \tilde{L}_1^{-1} \tilde{L}_2^{-1} \cdots \tilde{L}_n^{-1},$$

where each matrix $\tilde{L}_i^{-1}$, $i = 1, 2, \ldots, n$, can differ from the matrix L_i^{-1} only by permutations of elements in column i.

Now we can write $A = PLU$, where $P = P_1 P_2 \cdots P_n$, and $L = \tilde{L}_1^{-1} \tilde{L}_2^{-1} \cdots \tilde{L}_n^{-1}$ is a unit lower triangular matrix. This factorization of A is called an *LU factorization with pivoting*. If the matrices P, L, and U have been constructed, then we can solve system (8.2) in the following way:

1. Permute elements of b to get $\tilde{b} = P^{-1} b = P_n P_{n-1} \cdots P_1 b$.
2. Solve $Ly = \tilde{b}$ with a lower triangular matrix.
3. Solve $Ux = y$ with an upper triangular matrix.

A method for constructing the matrices P, L, U was actually described in Section 1.2.5, p. 42. It can be realized by the following algorithm.

Algorithm 8.1 LU factorization with pivoting: calculating a permutation matrix P, a unit lower triangular matrix L, and a nonsingular upper triangular matrix U such that $LU = PA$ for a given nonsingular A.

let $P = I, L = I, U = A$
for $i = 1$ to $n - 1$
    find m such that $|U(m, i)|$ is the largest entry in $|U(i : n, i)|$
    if $m \neq i$
        swap rows m and i in P
        swap rows m and i in U
        if $i \geq 2$ swap elements $L(m, 1 : i - 1)$ and $L(i, 1 : i - 1)$
    end if
    $L(i + 1 : n, i) = U(i + 1 : n, i)/U(i, i)$
    $U(i + 1 : n, i + 1 : n) = U(i + 1 : n, i + 1 : n) - L(i + 1 : n, i)\, U(i, i + 1 : n)$
    $U(i + 1 : n, i) = 0$
end for

Obviously, this algorithm can be improved. For example, in Algorithm 8.1 we observe that when column i of A is used to compute the elements of column i of L, then that column is not used again. Also, when row i of A is used to compute row i of U, that row is not used again. This observation allows us to organize the storage arrangement in Algorithm 8.1, overwriting L and U on A (see Question 8.2, p. 284). Further, to save information on permutations, we can use only one vector with the numbers $m_i, i = 1, 2, \ldots, n$, and so on.

The lu function in MATLAB® expresses a matrix A as the product of two triangular matrices, one of them a permutation of a lower triangular matrix and the other an upper triangular matrix. The function [L,U] = lu(A) returns an upper triangular matrix in U and a permuted lower triangular matrix in L such that $A = LU$. The return value L is a product of lower triangular and permutation matrices. The function [L,U,P] = lu(A) returns an upper triangular matrix in U, a lower triangular matrix L with unit main diagonal, and a permutation matrix P such that $PA = LU$.

The next algorithm is called *forward substitution*. We use it to easily solve a given system $Lx = b$ with a unit lower triangular matrix L.

Algorithm 8.2 Forward substitution: solving $Lx = b$ with a unit lower triangular matrix L.

$x(1) = b(1)$
for $i = 2$ to n
    $x(i) = b(i) - L(i, 1 : (i - 1))\, x(1 : (i - 1))$
end for

The last algorithm is called *backward substitution*.[2] Using this algorithm, we can easily solve a given system $Ux = b$ with an upper triangular matrix U.

Algorithm 8.3 Backward substitution: solving $Ux = b$ with a nonsingular upper triangular matrix U.

$x(n) = b(n)/U(n, n)$
for $i = n - 1$ to 1

---

[2]See also (1.119), (1.120), p. 45.

$$x(i) = (b(i) - U(i, (i + 1) : n) \, x((i + 1) : n))/U(i, i)$$
end for

Note that in Algorithm 8.1 we apply permutations on the rows of the matrix A. This process is called *Gaussian elimination with partial pivoting* (GEPP): swap rows with numbers m_i and i of the matrix A such that $|A(m_i, i)|$ will be the largest entry in $|A(i : n, i)|$. In the case of *Gaussian elimination with complete pivoting* (GECP), we swap rows m_i and i as well as columns k_i and i in the matrix A such that $|A(m_i, k_i)|$ will be the largest entry in $|A(i : n, i : n)|$. GEPP is the most common way to implement Gaussian elimination in practice. GECP is more expensive. It is almost never used in practice.

8.1.2 The Need for Pivoting

First of all, let us describe a class of matrices for which LU factorization can be done without pivoting. In other words, in this case all the matrices $P_i, i = 1, 2, \ldots, n$, can be equal to the identity matrix.

Theorem 8.1 *Let A be a given square matrix of order n. There exist a unique unit lower triangular matrix L and a unique nonsingular upper triangular matrix U such that $A = LU$ if and only if the all leading principal submatrices of A are nonsingular.*

Proof Necessity. The decomposition $A = LU$ may also be written in terms of block matrices as

$$\begin{pmatrix} A_{11} & A_{12} \\ A_{21} & A_{22} \end{pmatrix} = \begin{pmatrix} L_{11} & 0 \\ L_{21} & L_{22} \end{pmatrix} \begin{pmatrix} U_{11} & U_{12} \\ 0 & U_{22} \end{pmatrix} = \begin{pmatrix} L_{11}U_{11} & L_{11}U_{12} \\ L_{21}U_{11} & L_{21}U_{12} + L_{22}U_{22} \end{pmatrix},$$

where A_{11} is a leading principal submatrix of order $j, 1 \leq j \leq n$, as are L_{11} and U_{11}. Therefore, $\det A_{11} = \det(L_{11}U_{11}) = \det L_{11} \det U_{11} = \prod_{k=1}^{j}(U_{11})_{kk} \neq 0$, since L is unit triangular and U is nonsingular.

Sufficiency. The proof is by induction on the order n of the matrix A. For all matrices of order one, we have the obvious decomposition $a_{11} = l_{11}u_{11} = 1a_{11}$. To prove that there exists a decomposition for the matrix $\tilde{A}$ of order n, we need to find unique triangular matrices L and U of order $(n - 1)$, unique $(n - 1) \times 1$ vectors l and u, and a unique nonzero number η such that the following decomposition holds:

$$\tilde{A} = \begin{pmatrix} A & b \\ c^T & \delta \end{pmatrix} = \begin{pmatrix} L & 0 \\ l^T & 1 \end{pmatrix} \begin{pmatrix} U & u \\ 0 & \eta \end{pmatrix} = \begin{pmatrix} LU & Lu \\ l^T U & l^T u + \eta \end{pmatrix}. \tag{8.5}$$

By the induction hypothesis, there exist unique matrices L and U of order $(n - 1)$ such that $A = LU$. Comparing the left- and right-hand sides of (8.5), we get

$$u = L^{-1}b, \quad l^T = c^T U^{-1}, \quad \eta = \delta - l^T u. \tag{8.6}$$

It follows from (8.6) that u, l, η are unique. By the induction hypothesis, the diagonal entries of the matrix U are nonzero, since U is nonsingular. Using (8.5), we get

$$0 \neq \det \tilde{A} = \det \begin{pmatrix} L & 0 \\ l^T & 1 \end{pmatrix} \det \begin{pmatrix} U & u \\ 0 & \eta \end{pmatrix} = \eta \det(U).$$

Thus, $\eta \neq 0$. □

From Theorem 8.1 we conclude that there are classes of matrices important for applications for which pivoting is not necessary. For example, all the leading principal minors of the following matrices are nonzero:

1. Positive definite Hermitian matrices (see Sylvester's criterion, p. 153).
2. Row diagonally dominant and column diagonally dominant matrices (see p. 242).

Theorem 8.1 says also that LU decomposition of a matrix A without pivoting can fail even on well-conditioned nonsingular matrices A. This is because $j \times j$ leading principal minors of these matrices can be singular. For example, the permutation matrix $P = \begin{pmatrix} 0 & 1 \\ 1 & 0 \end{pmatrix}$ is orthogonal, but the first element in the first column of this matrix is zero. Thus, LU decomposition without pivoting will fail on this matrix.

Now we consider an example showing that pivoting in Gaussian elimination can significantly reduce the influence of rounding errors. At the same time, the result of declining to pivot can be catastrophic.

Example 8.1 Let us consider a system of two equations in two unknowns

$$Ax = b, \tag{8.7}$$

where

$$A = \begin{pmatrix} \alpha & 1 \\ 1 & 1 \end{pmatrix},$$

α is a given positive small real number, $b = (1, 2)^T$. The solution of this system is

$$x_1 = 1/(1 - \alpha) \approx 1, \quad x_2 = (1 - 2\alpha)/(1 - \alpha) \approx 1. \tag{8.8}$$

By elementary calculations, we get

$$A^{-1} = \begin{pmatrix} -1/(1 - \alpha), & 1/(1 - \alpha) \\ 1/(1 - \alpha), & -\alpha/(1 - \alpha) \end{pmatrix}.$$

Therefore, it is easy to see that for a small α, we have the inequality

$$\text{cond}_\infty(A) = \|A\|_\infty \|A^{-1}\|_\infty \approx 4,$$

i.e., the matrix A is very well conditioned, and the impact of rounding errors in storage of its elements and of the right-hand side on the solution of system (8.7) must be insignificant.

First, we solve system (8.7) by Gaussian elimination without pivoting. We obtain

$$L = \begin{pmatrix} 1 & 0 \\ 1/\alpha & 1 \end{pmatrix}, \quad U = \begin{pmatrix} \alpha & 1 \\ 0 & 1 - 1/\alpha \end{pmatrix}.$$

The solution of the system $Ly = b$ is $y_1 = b_1 = 1$, $y_2 = b_2 - l_{2,1}y_1 = 2 - 1/\alpha$. The solution of $Ux = y$ is $x_2 = y_2/u_{2,2}$, $x_1 = (y_1 - y_2 u_{1,2})/u_{1,1}$.

Let us put $\alpha = 10^{-14}$ and calculate the matrices L, U and the vectors y, x in MATLAB®, using double precision, by the above formulas:

$$L = \begin{pmatrix} 1.000000000000000e + 000 & 0 \\ 1.000000000000000e + 014 & 1.000000000000000e + 000 \end{pmatrix},$$

$$U = \begin{pmatrix} 1.000000000000000e - 014 & 1.000000000000000e + 000 \\ 0 & -9.999999999999900e + 013 \end{pmatrix},$$

$$y_1 = 1, \quad y_2 = -9.999999999999800e + 013,$$

$$x_2 = 9.999999999999900e - 001 \quad x_1 = 9.992007221626409e - 001, \quad (8.9)$$

$$LU = \begin{pmatrix} 1.000000000000000e - 014 & 1.000000000000000e + 000 \\ 1.000000000000000e + 000 & 1.000000000000000e + 000 \end{pmatrix}.$$

If we calculate the solution of system (8.7) directly by formulas (8.8), we obtain

$$x_2 = 9.999999999999900e - 001, \quad x_1 = 1.000000000000001e + 000. \quad (8.10)$$

Comparing (8.9) and (8.10), we observe that the impact of the rounding errors on the solution is significant. If we set $\alpha = 10^{-16}$, then we get

$$y_1 = 1, \quad y_2 = -9.999999999999998e + 015,$$

$$x_2 = 9.999999999999998e - 001, \quad x_1 = 2.220446049250313e + 0,$$

$$LU = \begin{pmatrix} 1.000000000000000e - 016 & 1.000000000000000e + 000 \\ 1.000000000000000e + 000 & 0 \end{pmatrix},$$

i.e., the influence of the rounding errors is catastrophic. In the considered example, this fact is explained in the following way. When we calculate $y_2 = 2 - 1/\alpha$ for a small α, the impact of the first term is lost because of the rounding errors.

Now we use Gaussian elimination with pivoting. For system (8.7), this means that we have to permute the equations in (8.7) and write

$$A = \begin{pmatrix} 1 & 1 \\ \alpha & 1 \end{pmatrix},$$

$b = (2, 1)^T$. Then

$$L = \begin{pmatrix} 1 & 0 \\ \alpha & 1 \end{pmatrix}, \quad U = \begin{pmatrix} 1 & 1 \\ 0 & 1 - \alpha \end{pmatrix}.$$

In this case, for $\alpha = 10^{-14}$ we get

$$x_2 = 9.999999999999900e - 001, \quad x_1 = 1.000000000000010e + 000,$$

$$LU = \begin{pmatrix} 1.000000000000000e + 000 & 1.000000000000000e + 000 \\ 1.000000000000000e - 014 & 1.000000000000000e + 000 \end{pmatrix}.$$

For $\alpha = 10^{-16}$, we obtain

$$x_2 = 9.999999999999999e - 001, \quad x_1 = 1,$$

$$LU = \begin{pmatrix} 1.000000000000000e + 000 & 1.000000000000000e + 000 \\ 1.000000000000000e - 016 & 1.000000000000000e + 000 \end{pmatrix},$$

i.e., the impact of the rounding errors is practically absent.

8.1.3 A Numerical Example

Now we illustrate the performance of Gaussian elimination algorithms by solving the Dirichlet problem for Poisson's equation in two dimensions. Clearly, after a discretization of the problem using finite elements or finite differences we obtain a system of linear equations that can be solved by a number of different methods. In this section we present first the finite difference discretization of the problem and show how to construct a system of linear equations from this discretization. Next, our numerical example illustrates how Gaussian elimination algorithms can be used to solve this system.

The model problem is the following Dirichlet[3] problem for Poisson's[4] equation:

$$-\Delta u(x) = f(x) \text{ in } \Omega,$$
$$u = 0 \text{ on } \partial\Omega. \tag{8.11}$$

---

[3] Johann Peter Gustav Lejeune Dirichlet (1805–1859) was a German mathematician.
[4] Siméon Denis Poisson (1781–1840) was a French mathematician.

Here $f(x)$ is a given function, $u(x)$ is an unknown function, and the domain Ω is the unit square $\Omega = \{(x_1, x_2) \in (0, 1) \times (0, 1)\}$. To solve (8.11) numerically, we first discretize the domain Ω with $x_{1i} = ih_1$ and $x_{2j} = jh_2$, where $h_1 = 1/(n_i - 1)$ and $h_2 = 1/(n_j - 1)$ are the mesh sizes in the directions x_1, x_2, respectively, and n_i and n_j are the numbers of discretization points in the directions x_1, x_2, respectively. In computations we usually have the same mesh size $h = h_1 = h_2$. In this example, we choose $n_i = n_j = n$ with $n = N + 2$, where N is the number of inner mesh nodes in the directions x_1, x_2, respectively.

Indices (i, j) are such that $0 \leq i, j < n$, and they are associated with every global node n_{glob} of the finite difference mesh. Global nodes numbers n_{glob} in the two-dimensional case can be computed using the following formula:

$$n_{glob} = j + n_i(i - 1). \tag{8.12}$$

We use the standard finite difference discretization of the Laplace operator Δu in two dimensions and obtain the discrete Laplacian $\Delta u_{i,j}$:

$$\Delta u_{i,j} = \frac{u_{i+1,j} - 2u_{i,j} + u_{i-1,j}}{h^2} + \frac{u_{i,j+1} - 2u_{i,j} + u_{i,j-1}}{h^2}, \tag{8.13}$$

where $u_{i,j}$ is the solution at the discrete point (i, j). Using (8.13), we obtain the following scheme for solution of the problem (8.11):

$$-\left(\frac{u_{i+1,j} - 2u_{i,j} + u_{i-1,j}}{h^2} + \frac{u_{i,j+1} - 2u_{i,j} + u_{i,j-1}}{h^2}\right) = f_{i,j}, \tag{8.14}$$

where $f_{i,j}$ is the value of the function f at the discrete point (i, j). We observe that (8.14) can be rewritten as

$$-\left(u_{i+1,j} - 2u_{i,j} + u_{i-1,j} + u_{i,j+1} - 2u_{i,j} + u_{i,j-1}\right) = h^2 f_{i,j}, \tag{8.15}$$

or in a more convenient form as

$$-u_{i+1,j} + 4u_{i,j} - u_{i-1,j} - u_{i,j+1} - u_{i,j-1} = h^2 f_{i,j}. \tag{8.16}$$

System (8.16) can be written in the form $Au = b$. The vector b has the components $b_{i,j} = h^2 f_{i,j}$. The explicit elements of the matrix A are given by the following block matrix:

$$A = \begin{pmatrix} A_N & -I_N & & \\ -I_N & \ddots & \ddots & \\ & \ddots & \ddots & -I_N \\ & & -I_N & A_N \end{pmatrix},$$

with blocks A_N of order N given by

$$A_N = \begin{pmatrix} 4 & -1 & 0 & 0 & \cdots & 0 \\ -1 & 4 & -1 & 0 & \cdots & 0 \\ 0 & -1 & 4 & 0 & \cdots & 0 \\ \cdots & \cdots & \cdots & \cdots & \cdots & \cdots \\ 0 & \cdots & \cdots & 0 & -1 & 4 \end{pmatrix},$$

which are located on the main diagonal of the matrix A, and blocks with identity matrices $-I_N$ of order N on its off-diagonals. The matrix A is symmetric and positive definite (see Question 8.16, p. 289). Therefore, we can use the LU factorization algorithm without pivoting.

Suppose that we have discretized the two-dimensional domain Ω as described above, and the number of inner points in both directions is $N = 3$. We present the schematic discretization for the inner nodes of this domain and corresponding numbering for the global nodes using (8.12) in the following scheme:

$$\begin{pmatrix} a_{1,1} & a_{1,2} & a_{1,3} \\ a_{2,1} & a_{2,2} & a_{2,3} \\ a_{3,1} & a_{3,2} & a_{3,3} \end{pmatrix} \implies \begin{pmatrix} n_1 & n_2 & n_3 \\ n_4 & n_5 & n_6 \\ n_7 & n_8 & n_9 \end{pmatrix} \implies \begin{pmatrix} 1 & 2 & 3 \\ 4 & 5 & 6 \\ 7 & 8 & 9 \end{pmatrix}. \tag{8.17}$$

Then the explicit form of the block matrix A will be

$$A = \left(\begin{array}{ccc|ccc|ccc} 4 & -1 & 0 & -1 & 0 & 0 & 0 & 0 & 0 \\ -1 & 4 & -1 & 0 & -1 & 0 & 0 & 0 & 0 \\ 0 & -1 & 4 & 0 & 0 & -1 & 0 & 0 & 0 \\ \hline -1 & 0 & 0 & 4 & -1 & 0 & -1 & 0 & 0 \\ 0 & -1 & 0 & -1 & 4 & -1 & 0 & -1 & 0 \\ 0 & 0 & -1 & 0 & -1 & 4 & 0 & 0 & -1 \\ \hline 0 & 0 & 0 & -1 & 0 & 0 & 4 & -1 & 0 \\ 0 & 0 & 0 & 0 & -1 & 0 & -1 & 4 & -1 \\ 0 & 0 & 0 & 0 & 0 & -1 & 0 & -1 & 4 \end{array} \right).$$

Example 8.2 In this example we present the numerical solution of problem (8.11). We define the right-hand side $f(x)$ of (8.11) as

$$f(x_1, x_2) = A_f \exp\left(-\frac{(x_1 - c_1)^2}{2s_1^2} - \frac{(x_2 - c_2)^2}{2s_2^2} \right) \frac{1}{a(x_1, x_2)}. \tag{8.18}$$

The coefficient $a(x_1, x_2)$ in (8.18) is given by the following Gaussian function:

$$a(x_1, x_2) = 1 + A \exp\left(-\frac{(x_1 - c_1)^2}{2s_1^2} - \frac{(x_2 - c_2)^2}{2s_2^2} \right). \tag{8.19}$$

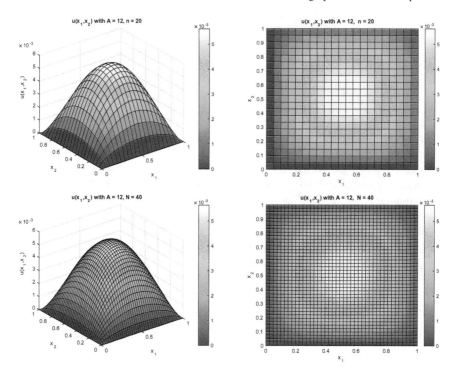

Fig. 8.1 Solution of problem (8.11) in the example of Section 8.1.3 on the unit square.

Here A, A_f are the amplitudes of these functions, c_1, c_2 are constants that show the location of the center of the Gaussian functions, and s_1, s_2 are constants that show the spreading of the functions in the x_1 and x_2 directions.

We produce a mesh with the points (x_{1i}, x_{2j}) such that $x_{1i} = ih$, $x_{2j} = jh$ with $h = 1/(N + 1)$, where N is the number of inner points in the x_1 and x_2 directions. We take the same number of points in the x_1 and x_2 directions: $n_i = n_j = N + 2$. The linear system of equations $Au = f$ is solved then via the LU factorization of the matrix A without pivoting. Figure 8.1 shows the results of numerical simulations for different discretizations of the unit square with the number of inner points $N = 20$, 40 and for $A = 12$, $A_f = 1$, $c_1 = c_2 = 0.5$, $s_1 = s_2 = 1$ in (8.81) and (8.18). The MATLAB® programs of Section 1.1 are available for running this test.[5]

8.2   Error Analysis

One of the main techniques for computing the error in the computed solution is checking its stability. This means that we need to check how much the computed solution

---

[5]The   MATLAB®   programs   can   be   found   on   the   online   version   of   Chapter   1 (doi:10.1007/978-3-319-57304-5_1).

is changed depending on changes in input data. We will begin with a derivation of perturbation theory in polynomial evaluation.

8.2.1 Rounding Analysis in Polynomial Evaluation

In this section we discuss the stability of polynomial evaluation by Horner's rule. Let there be given a polynomial

$$p(x) = \sum_{i=0}^{d} c_i x^i,$$

where c_i are the coefficients of the polynomial, and d is its degree. For polynomial evaluation we use Horner's rule, noting that the polynomial can be written in an alternative form as

$$\begin{aligned} p(x) &= c_0 + x(c_1 + c_2 x + ...c_d x^{d-1}) \\ &= c_0 + x(c_1 + c_2(x + ...c_d x^{d-2})) \\ &= c_0 + x(c_1 + x(....(c_{d-1} + c_d x)...)). \end{aligned} \quad (8.20)$$

Using (8.20), this rule can be programmed as the following iterative algorithm for every mesh point $x_j \in [x_{left}, x_{right}]$, $j \in 1, 2, ...N$, where N is the total number of discretization points.

Algorithm 8.4 Horner's rule for polynomial evaluation at the point $x_j \in [x_{left}, x_{right}]$, $j \in 1, 2, ...N$.

0. Initialize $p = c_d$. Set counter $i = d - 1$.
1. Compute $p = x_j p + c_i$.
2. Set $i = i - 1$ and go to step 1. Stop if $i = 0$.

To compute error bounds in the polynomial evaluation we insert a rounding term $1 + (\delta_{1,2})_i$ for every floating-point operation in Algorithm 8.4 to obtain the following algorithm.

Algorithm 8.5 Error bounds in polynomial evaluation at the point $x_j \in [x_{left}, x_{right}]$, $j \in 1, 2, ...N$.

0. Set counter $i = d - 1$ and initialize $p = c_d$.
1. Compute $p = \left(x_j p(1 + (\delta_1)_i) + c_i \right) (1 + (\delta_2)_i)$, where $|(\delta_1)_i|, |(\delta_2)_i| \le \varepsilon$.
2. Set $i := i - 1$ and go to step 1. Stop if $i = 0$.

In Algorithm 8.5 the number ε is the machine epsilon, and we define it as the maximum relative representation error $0.5b^{1-p}$, which is measured in floating-point arithmetic with base b and precision $p > 0$. Table 8.1 presents the values of the machine epsilon in standard floating-point formats.

Table 8.1 The values of machine epsilon in standard floating-point formats. The notation * means that one bit is implicit in precision p

IEEE 754 - 2008	Description	Base, b	Precision, p	$\varepsilon_1 = 0.5b^{-(p-1)}$	$\varepsilon_2 = b^{-(p-1)}$
binary16	Half precision	2	11*	$2^{-11} = 4.88e-04$	$2^{-10} = 9.77e-04$
binary32	Single precision	2	24*	$2^{-24} = 5.96e-08$	$2^{-23} = 1.19e-07$
binary64	Double precision	2	53*	$2^{-53} = 1.11e-16$	$2^{-52} = 2.22e-16$
binary80	Extended precision	2	64	$2^{-64} = 5.42e-20$	$2^{-63} = 1.08e-19$
binary128	Quad(ruple) precision	2	113*	$2^{-113} = 9.63e-35$	$2^{-112} = 1.93e-34$
decimal32	Single-precision decimal	10	7	5×10^{-7}	10^{-6}
decimal64	Double-precision decimal	10	16	5×10^{-16}	10^{-15}
decimal128	Quad(ruple)-precision decimal	10	34	5×10^{-34}	10^{-33}

Expanding the expression for p_i in Algorithm 8.5, we get the final value of p_0:

$$p_0 = \sum_{i=0}^{d-1} \left((1+(\delta_2)_i) \prod_{k=0}^{i-1} (1+(\delta_1)_k)(1+(\delta_2)_k) \right) c_i x^i$$

$$+ \left(\prod_{k=0}^{d-1} (1+(\delta_1)_k)(1+(\delta_2)_k) \right) c_d x^d. \quad (8.21)$$

Next, we write the upper and lower bounds for the products of $\delta = \delta_{1,2}$, provided that $k\varepsilon < 1$:

$$(1+\delta_1)\cdots(1+\delta_k) \leq (1+\varepsilon)^k \leq 1 + k\varepsilon + O(\varepsilon^2),$$
$$(1+\delta_1)\cdots(1+\delta_k) \geq (1-\varepsilon)^k \geq 1 - k\varepsilon. \quad (8.22)$$

Applying the above estimates, we get the following inequality:

$$1 - k\varepsilon \leq (1+\delta_1)\cdot\ldots\cdot(1+\delta_k) \leq 1 + k\varepsilon. \quad (8.23)$$

Using estimate (8.23), we can rewrite (8.21) as

$$p_0 \approx \sum_{i=0}^{d} (1+\tilde{\delta}_i)c_i x^i = \sum_{i=0}^{d} \tilde{c}_i x^i \quad (8.24)$$

with approximate coefficients $\tilde{c}_i = (1+\tilde{\delta}_i)c_i$ such that $|\tilde{\delta}_i| \leq 2k\varepsilon \leq 2d\varepsilon$. Now we can write the formula for the computing the error e_p in the polynomial:

$$e_p := |p_0 - p(x)| = \left| \sum_{i=0}^{d} (1+\tilde{\delta}_i)c_i x^i - \sum_{i=0}^{d} c_i x^i \right| = \left| \sum_{i=0}^{d} \tilde{\delta}_i c_i x^i \right|$$

$$\leq 2\sum_{i=0}^{d} d\varepsilon |c_i x^i| \leq 2d\varepsilon \sum_{i=0}^{d} |c_i x^i| = \Delta(x), \quad (8.25)$$

so the true value of the polynomial is in the interval $(p - \Delta, p + \Delta)$.

If we choose $\tilde{\delta}_i = \varepsilon\,\text{sign}(c_i x^i)$, then the error bound above can be attained within the factor $2d$. In this case, we can take

$$\text{cond}(p) := \frac{\sum_{i=0}^{d} |c_i x^i|}{\left| \sum_{i=0}^{d} c_i x^i \right|} \quad (8.26)$$

as the relative condition number for the case of polynomial evaluation.

In the following algorithm, we use (8.25) to compute the lower bound in polynomial evaluation.

Algorithm 8.6 Computation of the error $\Delta(x_j)$ in the polynomial evaluation at the point $x_j \in [x_{left}, x_{right}]$, $j \in 1, 2, ...N$.

0. Set counter $i = d - 1$ and initialize $p = c_d$, $\Delta = |c_d|$.
1. Compute $p = x_j p + c_i$, $\Delta = |x_j|\Delta + |c_i|$.
2. Set $i = i - 1$ and go to step 1. Stop if $i = 0$.
3. Set $\Delta(x_j) = 2d\varepsilon\Delta$ as the error bound at the point x_j.

Example 8.3 Figure 8.2-(a) shows the behavior of the computed solution using Horner's rule (Algorithm 8.4) for the evaluation of the polynomial

$$p(x) = (x - 9)^9 = x^9 - 81x^8 + 2916x^7 - 61236x^6 + 826686x^5 - 7440174x^4$$
$$+ 44641044x^3 - 172186884x^2 + 387420489x^1 - 387420489.$$
$$(8.27)$$

Figure 8.2-(b) shows the upper and lower bounds computed for the polynomial $p(x) = (x - 9)^9$ using Algorithm 8.6. We have performed all our computations taking $\varepsilon = 0.5e - 16$ in Algorithm 8.6. Using these figures, we observe that changing the argument x slightly can change computed values drastically. Figure 8.3 presents comparison of the theoretically estimated relative error $e_{ln} = -ln\left|\frac{\Delta}{p}\right|$ with the computed relative error $e_{comp} = -ln\left|\frac{p(x)-\tilde{p}(x)}{p(x)}\right|$ for two polynomials: for the polynomial $p(x) = (x - 1)^2(x - 2)(x - 3)(x - 4)(x - 5)$ (see Fig. 8.3a) and for the polynomial $p(x) = (x - 1)^2(x - 2)(x - 3)(x - 4)(x - 5)(x - 7)(x - 9)$ $(x - 11)(x - 15)(x - 17)$ (See Fig. 8.3b). We note that $\tilde{p}(x)$ is computed polynomial using the Horner's rule (Algorithm 8.4) and $p(x)$ is the exact polynomial.

Indeed, we encounter difficulties when we want to compute $p(x)$ with a high relative accuracy if $p(x)$ is close to zero. This is because small changes in ε result in an infinite relative error given by $\varepsilon/p(x) = \varepsilon/0$, which means that our relative condition number (8.26) is infinite; see also an illustration of this statement in Fig. 8.2-(c), (d). There is a simple geometric interpretation of this condition number: it tells us how far $p(x)$ is from a polynomial whose condition number at x is infinite. Now we introduce a necessary concept and prove the corresponding theorem.

Let $p(x) = \sum_{i=0}^{d} a_i x^i$ and $q(x) = \sum_{i=0}^{d} b_i x^i$ be two polynomials. Then the relative distance $\text{dist}(p, q)$ from $p(x)$ to $q(x)$ is defined as the smallest value such that

$$|a_i - b_i| \leq \text{dist}(p, q)|a_i|, \ i \leq 1 \leq d.$$

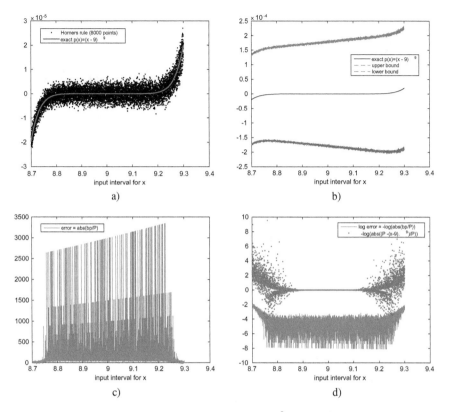

Fig. 8.2 *(a) Evaluation of the polynomial* $p(x) = (x - 9)^9$ *by Horner's rule (Algorithm 8.4) compared with the exact polynomial. (b) Computed upper and lower bounds for the polynomial* $p(x) = (x - 9)^9$ *using Algorithm 8.6. (c) Plot of the graph of the estimated relative error* $e = \left|\frac{\Delta}{p}\right|$. *(d) Plot of the graph of the estimated relative error* $e_{ln} = -ln\left|\frac{\Delta}{p}\right|$ *(presented in blue color) compared with the computed relative error* $e_{comp} = -ln\left|\frac{p(x)-(x-9)^9}{p(x)}\right|$ *(presented in red color). Here,* $p(x)$ *is computed by Horner's rule (Algorithm 8.4) and the exact polynomial* $(x - 9)^9$ *is computed in MATLAB®. Input interval for x in this example is* $x \in [8.7, 9.3]$.

If $a_i \neq 0$, $i \leq 1 \leq d$, the condition above can be rewritten as

$$\max_{0 \leq i \leq d} \frac{|a_i - b_i|}{|a_i|} = \text{dist}(p, q), \ i \leq 1 \leq d.$$

Theorem 8.2 *Let a polynomial* $p(x) = \sum_{i=0}^{d} c_i x^i$ *be not identically zero, and let q be another polynomial whose condition number at x is infinite, i.e.,* $q(x) = 0$. *Then*

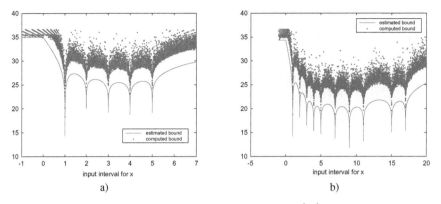

Fig. 8.3 *Plot of the graph of the estimated relative error $e_{ln} = -ln\left|\frac{\Delta}{p}\right|$ (shown in blue) compared with the computed relative error $e_{comp} = -ln\left|\frac{p(x)-\tilde{p}(x)}{p(x)}\right|$ (here $\tilde{p}(x)$ is computed by Horner's rule and $p(x)$ is the exact polynomial, which we compute in MATLAB®): (a) for the polynomial $p(x) = (x-1)^2(x-2)(x-3)(x-4)(x-5)$ and (b) for the polynomial $p(x) = (x-1)^2 (x-2)(x-3)(x-4)(x-5)(x-7)(x-9)(x-11)(x-15)(x-17)$.*

$$\min\{\text{dist}(p,q) : q(x) = 0\} = \frac{\left|\sum_{i=0}^{d} c_i x^i\right|}{\sum_{i=0}^{d} |c_i x^i|}. \tag{8.28}$$

Proof To prove this theorem, let us write $q(x) = \sum_{i=0}^{d} b_i x^i = \sum_{i=0}^{d}(1 + \varepsilon_i)c_i x^i$ such that $\text{dist}(p,q) = \max_{0 \le i \le d} |\varepsilon_i|$. Then $q(x) = 0$ implies that

$$|p(x)| = |q(x) - p(x)| = \left|\sum_{i=0}^{d} \varepsilon_i c_i x^i\right| \le \sum_{i=0}^{d} |\varepsilon_i c_i x^i| \le \max_{0 \le i \le d} |\varepsilon_i| \sum_{i=0}^{d} |c_i x^i|.$$

Thus,

$$\text{dist}(p,q) = \max_{0 \le i \le d} |\varepsilon_i| \ge \frac{|p(x)|}{\sum_{i=0}^{d} |c_i x^i|}.$$

There is a q that is close to p, for example the polynomial q with

$$\varepsilon_i = \frac{-p(x)}{\sum_{i=0}^{d} |c_i x^i|} \text{sign}(c_i x^i).$$

□

8.2.2 Error Analysis in Gaussian Elimination

In this section we derive an error analysis for LU decomposition and Gaussian elimination that is similar to the error analysis of polynomial evaluation of Section 8.2.1.

We assume that the matrix A has already been pivoted. We will simplify the error analysis only for two equations of Algorithm 8.1, one for a_{jk}, $j \leq k$, and one for $j > k$. Let us first analyze what this algorithm is doing with element a_{jk} when $j \leq k$. We observe that this element is repeatedly updated as

$$u_{jk} = a_{jk} - \sum_{i=1}^{j-1} l_{ji} u_{ik}.$$

If $j > k$, then we have

$$l_{jk} = \frac{a_{jk} - \sum_{i=1}^{k-1} l_{ji} u_{ik}}{u_{kk}}.$$

To do rounding error analysis of these two formulas, we use the following expression for floating-point approximations:

$$\mathrm{fl}\left(\sum_{i=1}^{d} x_i y_i\right) = \sum_{i=1}^{d} x_i y_i (1 + \delta_i), \quad |\delta_i| \leq d\varepsilon, \tag{8.29}$$

where ε is the machine epsilon or the relative representation error. The maximum of the relative representation error in a floating-point arithmetic with p digits and base b is $0.5b^{1-p}$; see also Table 8.1 for ε in standard floating-point formats.

We apply (8.29) to the formula for u_{jk}:

$$u_{jk} = \left(a_{jk} - \sum_{i=1}^{j-1} l_{ji} u_{ik}(1 + \delta_i)\right)(1 + \delta')$$

with $|\delta_i| \leq (j - 1)\varepsilon$ and $|\delta'| \leq \varepsilon$. Expressing a_{jk}, we get

$$a_{jk} = \frac{1}{1 + \delta'} u_{jk} l_{jj} + \sum_{i=1}^{j-1} l_{ji} u_{ik}(1 + \delta_i)$$

$$\leq \sum_{i=1}^{j} l_{ji} u_{ik} + \sum_{i=1}^{j} l_{ji} u_{ik}\delta_i = \sum_{i=1}^{j} l_{ji} u_{ik} + E_{jk}, \tag{8.30}$$

where we have used the fact that $l_{jj} = 1$ and the assumptions

$$|\delta_i| \le (j-1)\varepsilon, \quad 1+\delta_j := \frac{1}{1+\delta'}.$$

In the expression above we can bound E_{jk} by

$$|E_{jk}| = \left| \sum_{i=1}^{j} l_{ji} u_{ik} \delta_i \right| \le \sum_{i=1}^{j} |l_{ji}||u_{ik}| n\varepsilon = n\varepsilon (|L||U|)_{jk}.$$

Thus, we get the following estimate for a_{jk}:

$$a_{jk} \le \sum_{i=1}^{j} l_{ji} u_{ik} + E_{jk}.$$

We perform the same analysis for the formula for l_{jk} to get

$$l_{jk} = (1+\delta'') \left(\frac{(1+\delta')(a_{jk} - \sum_{i=1}^{k-1} l_{ji} u_{ik}(1+\delta_i))}{u_{kk}} \right),$$

where $|\delta_i| \le (k-1)\varepsilon$, $|\delta'| \le \varepsilon$, and $|\delta''| \le \varepsilon$. Expressing a_{jk}, we get

$$a_{jk} = \frac{1}{(1+\delta')(1+\delta'')} u_{kk} l_{jk} + \sum_{i=1}^{k-1} l_{ji} u_{ik}(1+\delta_i).$$

Defining $1+\delta_k := \frac{1}{(1+\delta')(1+\delta'')}$, we can rewrite the previous expression as

$$a_{jk} \le \sum_{i=1}^{k} l_{ji} u_{ik} + \sum_{i=1}^{k} l_{ji} u_{ik} \delta_i \equiv \sum_{i=1}^{k} l_{ji} u_{ik} + E_{jk}$$

with $|\delta_i| \le n\varepsilon$ and $|E_{jk}| \le n\varepsilon (|L||U|)_{jk}$ as before. We summarize this error analysis with the simple formula

$$A = LU + E,$$

where

$$|E| \le n\varepsilon |L||U|.$$

Taking norms, we get

$$\|E\| \le n\varepsilon \| |L| \| \| |U| \|.$$

If the norm does not depend on the sign of the entries of the matrix (this is valid for Frobenius, infinity, one-norms, but not for two-norms), we can simplify the expression above as

$$\|E\| \leq n\varepsilon \|L\|\|U\|. \tag{8.31}$$

Thus, in formula (8.31) we have obtained an error estimate for LU decomposition. The next step is to obtain errors in forward and backward substitutions. We solve $LUx = b$ via $Ly = b$ and $Ux = y$. Solving $Ly = b$ gives as a computed solution $\hat{y}$ such that $(L + \delta L)\hat{y} = b$, where $|\delta L| \leq n\varepsilon |L|$. The same is true for $(U + \delta U)\hat{x} = \hat{y}$ with $|\delta U| \leq n\varepsilon |U|$. Combining both estimates, we get

$$\begin{aligned}
b &= (L + \delta L)\hat{y} = (L + \delta L)(U + \delta U)\hat{x} \\
&= (LU + L\delta U + \delta LU + \delta L\delta U)\hat{x} \\
&= (A - E + L\delta U + \delta LU + \delta L\delta U)\hat{x} \\
&= (A + \delta A)\hat{x},
\end{aligned} \tag{8.32}$$

where $\delta A = -E + L\delta U + \delta LU + \delta L\delta U$. Now we combine all bounds for $E, \delta U, \delta L$ and use the triangle inequality to get

$$\begin{aligned}
|\delta A| &= |-E + L\delta U + \delta LU + \delta L\delta U| \\
&\leq |E| + |L||\delta U| + |\delta L||U| + |\delta L||\delta U| \\
&\leq n\varepsilon |L||U| + n\varepsilon |L||U| + n\varepsilon |L||U| + n^2\varepsilon^2 |L||U| \\
&\approx 3n\varepsilon |L||U|.
\end{aligned} \tag{8.33}$$

Assuming that $\| |X| \| = \|X\|$ is true (as before, this is valid for Frobenius, infinity and one-norms but not for two-norms), we obtain

$$\|\delta A\| \leq 3n\varepsilon \|L\|\|U\|. \tag{8.34}$$

Thus, Gaussian elimination is backward stable if (recall that in this analysis we have used $\delta b = 0$) the following condition holds:

$$3n\varepsilon \|L\|\|U\| = O(\varepsilon)\|A\|.$$

We note that GEPP allows us to estimate every entry of the matrix L by an entry in absolute value, so we need consider only $\|U\|$. The *pivot growth factor* for GEPP is the number

$$g = \frac{\|U\|_{max}}{\|A\|_{max}}, \tag{8.35}$$

where $\|A\|_{max} = \max\limits_{1 \leq i,j \leq n} |a_{ij}|$. In other words, stability is equivalent to g being small or growing slowly as a function of n.

Let us prove that the bound of the pivot growth factor for GEPP is

$$g \le 2^{n-1}. \tag{8.36}$$

Indeed, at the first step in GEPP, we perform an update for elements

$$\tilde{a}_{jk} = a_{jk} - l_{ji}u_{ik} \tag{8.37}$$

with $|l_{ji}| \le 1$, $|u_{ik}| = |a_{ik}| \le \max_{r,s} |a_{rs}|$. Substituting these estimates in (8.37), we obtain that

$$|\tilde{a}_{jk}| \le 2 \max_{r,s} |a_{rs}|. \tag{8.38}$$

Using estimate (8.38), we conclude that at every $(n-1)$ step of the Algorithm 8.1 we can double the size of the remaining matrix entries. Estimate (8.36) follows from this observation.

There are practical examples showing that the bound in (8.36) is attainable [23]. Now using the facts that $\|L\|_\infty \le n$ and $\|U\|_\infty \le ng\|A\|_\infty$ and substituting these estimates together with estimate (8.36) for g in (8.34), we obtain

$$\|\delta A\|_\infty \le 3n\varepsilon \|L\|_\infty \|U\|_\infty \le 3\, gn^3\varepsilon \|A\|_\infty. \tag{8.39}$$

For example, if $\varepsilon = 10^{-7}$ and the order of the matrix A is $n > 120$. Then using equality in (8.36), we can compute that $3gn^3\varepsilon > 1$. We observe that with such an estimate we lose all precision.

8.2.3 Estimating the Condition Number

Let us recall that in Theorem 7.12, p. 243, we established a connection between the relative perturbations of a matrix and the right-hand side of the system $Ax = b$ with the relative perturbations of its solution. An important role in the estimates obtained is played by the condition number $\mathrm{cond}(A) = \|A\|\|A^{-1}\|$ of A. To compute $\mathrm{cond}(A)$ we need to estimate $\|A^{-1}\|$, since $\|A\|$ is easy to compute.

If we compute A^{-1} explicitly and then compute its norm, that would cost $2n^3$ operations, but the number of operations in Gaussian elimination is approximately $2n^3/3$ (see p. 46). Therefore, we will seek a cheaper algorithm for an estimate of $\|A^{-1}\|$. Below we present this algorithm and will call it a *condition estimator* or Hager's algorithm. The algorithm was developed in [45, 50, 51] and has the following properties:

1. This estimator is guaranteed to produce only a lower bound of $\|A^{-1}\|$, not an upper bound.
2. The cost is $O(n^2)$ operations. This is negligible compared to the $2n^3/3$ cost of Gaussian elimination if the order n of the matrix A is large.
3. It provides an estimate that is almost always within a factor of 10 of $\|A^{-1}\|$ and usually within of factor 2 to 3.

The algorithm estimates the one-norm $\|B\|_1$ of a matrix B, under the condition that we can compute the products Bx and $B^T y$ for arbitrary vectors x and y. Recall that (see (6.26), p. 220) the norm $\|B\|_1$ is defined by

$$\|B\|_1 = \max_{x \in \mathbb{R}^n, \|x\|_1 = 1} \|Bx\|_1 = \max_{1 \le j \le n} \sum_{i=1}^{n} |b_{ij}|, \tag{8.40}$$

or as the maximum absolute column sum. As we saw on p. 220, the maximum over x is attained at the vector $x = i_{j_0} = (0, \ldots, 0, 1, 0, \ldots, 0)$ with component j_0 as the single nonzero entry, where $\max_j \sum_i |b_{ij}|$ occurs at $j = j_0$. However, direct searching over all such vectors $i_j, j = 1, \ldots, n$, is too expensive, because this means computing all columns of $B = A^{-1}$. Since $\|Bx\|_1 = \max_{\|x\|_1 \le 1} \|Bx\|_1$, we can use the gradient method to find the maximum of $f(x) = \|Bx\|_1$ inside the set $\|x\|_1 \le 1$. Here $\|x\|_1 \le 1$ is a convex set of vectors, and $f(x)$ is a convex function. Indeed, if $0 \le \alpha \le 1$, then

$$\begin{aligned} f(\alpha x + (1-\alpha)y) &= \|\alpha Bx + (1-\alpha)By\|_1 \\ &\le \alpha \|Bx\|_1 + (1-\alpha)\|By\|_1 = \alpha f(x) + (1-\alpha)f(y). \end{aligned} \tag{8.41}$$

If the gradient $\nabla f(x)$ of the function $f(x)$ exists then the convexity property of $f(x)$ means that $f(y) \ge f(x) + \nabla f(x)(y - x)$. To compute gradient $\nabla f(x)$ we assume that all $\sum_j^n b_{ij}x_j \ne 0$ (since $B = A^{-1}$) in $f(x) = \|Bx\|_1 = \sum_i^n \left|\sum_j^n b_{ij}x_j\right|$. Let $\zeta_i = \text{sign}(\sum_j^n b_{ij}x_j)$ such that $\zeta_i = \pm 1$ and $f(x) = \sum_i^n \sum_j^n \zeta_i b_{ij}x_j$. Then $\partial f/\partial x_k = \sum_i^n \zeta_i b_{ik}$ and $\nabla f = \zeta^T B = (B^T \zeta)^T$.

Algorithm 8.7 Hager's condition estimator returns a lower bound $\|w\|_1$ for $\|B\|_1$.

choose any x such that $\|x\|_1 = 1$ /* e.g., $x_i = 1/n$ */
repeat
    $w = Bx, \zeta = \text{sign}(w), z = B^T \zeta$,   /* $z^T = \nabla f$ */
    if $\|z\|_\infty \le z^T x$, then
        return $\|w\|_1$
    else
        $x = i_j$ where $|z_j| = \|z\|_\infty$ /* here 1 in i_j should be placed at index j where
                                  $|z_j| = \|z\|_\infty$ */
    end if
end repeat

Theorem 8.3 *The next two statements are valid for Algorithm 8.7.*

1. *If $\|z\|_\infty \le z^T x$ and $\|w\|_1$ is returned, then $\|w\|_1 = \|Bx\|_1$ is a local maximum of $\|Bx\|_1$.*
2. *Otherwise, the algorithm has made progress in maximizing $f(x)$.*

Proof 1. We have that $\|z\|_\infty \le z^T x$. Close to x,

$$f(x) = \|Bx\|_1 = \sum_i^n \sum_j^n \zeta_i b_{ij} x_j$$

is linear in x, and we can use a Taylor series to get

$$f(y) \approx f(x) + \nabla f(x)(y - x) = f(x) + z^T(y - x),$$

where $z^T = \nabla f(x)$. To show that x is a local maximum, we need to prove the inequality $f(y) \le f(x)$ or to show that $z^T(y - x) \le 0$ with $\|y\|_1 = 1$. We get

$$z^T(y - x) = z^T y - z^T x = \sum_i^n z_i y_i - z^T x \le \sum_i^n |z_i||y_i| - z^T x$$

$$\le \|z\|_\infty \|y\|_1 - z^T x = \|z\|_\infty - z^T x \le 0.$$
$$(8.42)$$

2. In this case, $\|z\|_\infty > z^T x$. We choose $\widetilde{x} = i_j \text{sign}(z_j)$, where the number j is such that $|z_j| = \|z\|_\infty$. Then

$$f(\widetilde{x}) \ge f(x) + \nabla f \times (\widetilde{x} - x) = f(x) + z^T(\widetilde{x} - x)$$
$$= f(x) + z^T \widetilde{x} - z^T x = f(x) + |z_j| - z^T x > f(x).$$
$$(8.43)$$

The last inequality holds, since $\|z\|_\infty > z^T x$. $\square$

In the example presented below, we will test the Algorithm 8.7 using the MATLAB® program of Section 1.3.[6]

Example 8.4 We test computation of Hager's condition estimator for the matrix

$$A = \begin{pmatrix} 16.5488 & 14.6149 & 4.3738 & 7.0853 & 2.3420 \\ 14.6149 & 3.4266 & 29.5580 & 23.7673 & 6.8848 \\ 4.3738 & 29.5580 & 0.1620 & 3.9291 & 6.7942 \\ 7.0853 & 23.7673 & 3.9291 & 6.5877 & 25.1377 \\ 2.3420 & 6.8848 & 6.7942 & 25.1377 & 1.4003 \end{pmatrix} \quad (8.44)$$

---

[6]The MATLAB® programs can be found on the online version of Chapter 1 (doi:10.1007/978-3-319-57304-5_1).

using Algorithm 8.7. We run the MATLAB® program of Section 1.3 and get the following results: the computed one-norm of the matrix (8.44) by Algorithm 8.7 is 78.2517 which is the same as the computed one-norm using the MATLAB® command norm(A,1).[7]

Remark 8.1 In [51, 52], a slightly improved version of this algorithm was tested on many random matrices A of orders 10, 25, and 50. These matrices had condition numbers $\text{cond}(A) = 10$, 10^3, 10^6, 10^9. Then using Hager's algorithm, in the worst case the computed $\text{cond}(A)$ underestimated the true $\text{cond}(A)$ by a factor of 0.44, which says something about the efficiency of Hager's algorithm for computation of $\text{cond}(A)$.

8.2.4 Estimating the Relative Condition Number

We can also apply Hager's algorithm to estimate the relative (or Bauer–Skeel) condition number $\kappa_{BS}(A) = \||A^{-1}||A|\|_\infty$ presented in Theorem 7.14, p. 245. To do so, we will estimate $\||A^{-1}|g\|_\infty$, where g is a vector with nonnegative entries. We explain this now. Let $e = (1, ..., 1)$ be the vector of all ones. Using properties of the infinity norm, we see that $\|X\|_\infty = \|Xe\|_\infty$ if the matrix X has nonnegative entries. Then

$$\kappa_{BS}(A) = \||A^{-1}||A|\|_\infty = \||A^{-1}||A|e\|_\infty = \||A^{-1}|g\|_\infty,$$

where $g = |A|e$.

Now we can estimate $\||A^{-1}|g\|_\infty$. Let $G = \text{diag}(g_1, \ldots, g_n)$ be the diagonal matrix with entries g_i on its diagonal. Then $g = Ge$ and thus

$$\||A^{-1}|g\|_\infty = \||A^{-1}|Ge\|_\infty = \||A^{-1}|G\|_\infty = \||A^{-1}G|\|_\infty = \|A^{-1}G\|_\infty. \quad (8.45)$$

The last equality holds because $\|Y\|_\infty = \||Y|\|_\infty$ for every matrix Y.

Now we will show how to estimate the infinity norm of the matrix $A^{-1}G$ by Hager's algorithm. Applying this algorithm to the matrix $(A^{-1}G)^T$, we can estimate $\|(A^{-1}G)^T\|_1 = \|A^{-1}G\|_\infty$. Thus, we will apply Hager's algorithm to $\|(A^{-1}G)^T\|_1$ and in this way compute $\|A^{-1}G\|_\infty$.

8.2.5 Practical Error Bounds

Below, we present two practical error bounds for an approximate solution $\tilde{x}$ of the equation $Ax = b$. In the first bound, we use inequality (7.53), p. 244, to get

$$\text{error} = \frac{\|\tilde{x} - x\|_\infty}{\|\tilde{x}\|_\infty} \leq \|A^{-1}\|_\infty \frac{\|r\|_\infty}{\|\tilde{x}\|_\infty}, \quad (8.46)$$

---

[7]The MATLAB® programs can be found on the online version of Chapter 1 (doi:10.1007/978-3-319-57304-5_1).

where $r = A\tilde{x} - b$ is the residual. To compute $\|A^{-1}\|_\infty$, we can apply Hager's algorithm for $B = (A^{-1})^T$, noting that $\|B\|_1 = \|(A^{-1})^T\|_1 = \|A^{-1}\|_\infty$. The second error bound follows from the inequality

$$\|\tilde{x} - x\|_\infty = \|A^{-1}r\|_\infty \le \|A^{-1}\|\|r\|_\infty,$$

where we have used the triangle inequality. This practical error bound has the form

$$\text{error} = \frac{\|\tilde{x} - x\|_\infty}{\|\tilde{x}\|_\infty} \le \frac{\|A^{-1}\|\|r\|_\infty}{\|\tilde{x}\|_\infty}. \tag{8.47}$$

The estimate of $\|A^{-1}\|\|r\|\|_\infty$ can be obtained using Hager's algorithm, taking into account (8.45).

Remark 8.2 1. Error bounds (8.46) and (8.47) cannot be guaranteed to provide bounds in all cases.
2. The estimate of $\|A^{-1}\|_1$ from Algorithm 8.7 provides only a lower bound.
3. There is a small probability that the rounding in the evaluation of the residual $r = A\tilde{x} - b$ can make $\|r\|$ artificially small. To take this into account, we add a small number to $|r|$ and replace $|r|$ with $|r| + (n+1)\varepsilon(|A||\tilde{x}| + |b|)$ in the bound (8.47) or $\|r\|$ with $\|r\| + (n+1)\varepsilon(\|A\|\|\tilde{x}\| + \|b\|)$ in the bound (8.46). This is done by noting that the rounding error in evaluating r is bounded by

$$|(A\tilde{x} - b) - \text{fl}(A\tilde{x} - b)| \le (n+1)\varepsilon(|A||\tilde{x}| + |b|). \tag{8.48}$$

4. The rounding in performing Gaussian elimination on very ill conditioned matrices A can get such inaccurate L and U that the bound (8.47) can be too low.

8.3 Algorithms for Improving the Accuracy of the Solution

If the error in the computed solution $\tilde{x} = x + \delta x$ is as large as $\text{cond}(A)\varepsilon$, we can try to use Newton's method to improve the solution. This means that to solve an equation $f(x) = 0$, we can construct an iterative procedure $x_{i+1} = x_i - f(x_i)/f'(x_i)$ and obtain the improved computed solution x_{i+1}. Applying the idea of this method to $f(x) = Ax - b$ yields the following algorithm:

Algorithm 8.8 Newton's algorithm.

```
repeat
    r = Ax_i - b
    solve Ad = r to compute d
    x_{i+1} = x_i - d
end repeat
```

If we could compute the residual $r = Ax_i - b$ in this algorithm exactly and solve the equation $Ad = r$ exactly, then we could finish the algorithm in one step. We expect such a solution from Newton's method applied to a linear problem. However, rounding errors prevent this immediate convergence. Algorithm 8.8 is useful when A is so ill conditioned that solving $Ad = r$ (and $Ax_0 = b$) is rather inaccurate.

Theorem 8.4 *Suppose that $r = Ax_i - b$ in Algorithm 8.8 is computed in double precision and $\text{cond}(A)\varepsilon < c = 1/(3n^3 g + 1) < 1$, where n is the dimension of A and g is the pivot growth factor given by (8.35). Then Algorithm 8.8 converges as*

$$\frac{\|x_i - A^{-1}b\|_\infty}{\|A^{-1}b\|_\infty} = O(\varepsilon)$$

and has the following relaxation property:

$$\|x_{i+1} - x\|_\infty \leq \frac{\text{cond}(A)\varepsilon}{c} \|x_i - x\|_\infty = \zeta \|x_i - x\|_\infty, \qquad (8.49)$$

with the relaxation parameter $\zeta = \text{cond}(A)\varepsilon/c < 1$.

Proof Let us denote here $\|\cdot\|_\infty$ by $\|\cdot\|$. Our goal is to show that (8.49) holds. By assumption, $\zeta < 1$, so this inequality implies that the error $\|x_{i+1} - x\|$ decreases monotonically to zero. We begin the proof by estimating the error in the computed residual r. Using estimate (8.29), we can write

$$r = \text{fl}(Ax_i - b) = Ax_i - b + f, \qquad (8.50)$$

where

$$|f| \leq n\varepsilon^2(|A||x_i| + |b|) + \varepsilon|Ax_i - b| \approx \varepsilon|Ax_i - b|. \qquad (8.51)$$

Here, as usual, ε is the relative representation error. The ε^2 term comes from the double-precision computation of r, and the ε term comes from rounding the double-precision result back to single precision. Since $\varepsilon^2 \ll \varepsilon$, we will neglect the ε^2 term in the bound on $|f|$. Next, from Newton's method we have

$$(A + \delta A)d = r. \qquad (8.52)$$

From the bound (8.39) we know that $\|\delta A\| \leq \gamma \varepsilon \|A\|$, where $\gamma = 3n^3 g$, although this is usually too large in reality. We assume that computations of $x_{i+1} = x_i - d$ in Algorithm 8.8 are performed exactly. Using (8.52) and substituting here (8.50) and ignoring all ε^2 terms, we get

$$d = (A + \delta A)^{-1} r = (I + A^{-1} \delta A)^{-1} A^{-1} r$$
$$= (I + A^{-1} \delta A)^{-1} A^{-1} (Ax_i - b + f) = (I + A^{-1} \delta A)^{-1} (x_i - x + A^{-1} f)$$
$$\approx (I - A^{-1} \delta A)(x_i - x + A^{-1} f) \approx x_i - x - A^{-1} \delta A (x_i - x) + A^{-1} f.$$

$$(8.53)$$

Next, subtracting x from both sides of $x_{i+1} = x_i - d$ and using then (8.53), we get

$$x_{i+1} - x = x_i - d - x = A^{-1} \delta A (x_i - x) - A^{-1} f. \qquad (8.54)$$

Taking then norms from (8.54), using (8.51) for estimation of f and the bound (8.39) for estimation of $\|\delta A\|$, we see that

$$\|x_{i+1} - x\| \leq \|A^{-1} \delta A (x_i - x)\| + \|A^{-1} f\|$$
$$\leq \|A^{-1}\| \|\delta A\| \|x_i - x\| + \|A^{-1}\| \|\varepsilon\| Ax_i - b\|$$
$$\leq \|A^{-1}\| \|\delta A\| \|x_i - x\| + \|A^{-1}\| \|\varepsilon\| A(x_i - x)\|$$
$$\leq \|A^{-1}\| \|\gamma \varepsilon\| A\| \|x_i - x\| + \|A^{-1}\| \|A\| \|\varepsilon\| \|x_i - x\|$$
$$= \|A^{-1}\| \|A\| \varepsilon (\gamma + 1) \|x_i - x\|.$$

$$(8.55)$$

Let

$$\zeta = \|A^{-1}\| \|A\| \varepsilon (\gamma + 1) = \text{cond}(A) \varepsilon / c < 1.$$

Then (8.49) is satisfied, and Newton's Algorithm 8.8 converges.   □

Note that the condition number in Theorem 8.4 does not appear in the final error bound. This means that we compute the answer accurately independent of the condition number, provided that $\text{cond}(A)\varepsilon$ is sufficiently less than 1. Usually, c is too conservative an upper bound, and the algorithm often succeeds even when $\text{cond}(A)\varepsilon > c$.

Sometimes we cannot run Algorithm 8.8 with double precision, and only computations with single precision are available. In this case, we still can run Algorithm 8.8 and compute the residual r in single precision. However, Theorem 8.4 is not valid for this case, and we have the following theorem.

Theorem 8.5 *Suppose that the residual $r = Ax_i - b$ in Algorithm 8.8 is computed in single precision and*

$$\|A^{-1}\|_\infty \|A\|_\infty \frac{\max_i (|A||x|)_i}{\min_i (|A||x|)_i} \varepsilon < 1.$$

Then one step of iterative refinement yields x_1 such that $(A + \delta A)x_1 = b + \delta b$ with $|\delta a_{ij}| = O(\varepsilon)|a_{ij}|$ and $|\delta b_i| = O(\varepsilon)|b_i|$.

For a proof, see [53] as well as [2], [104–106] for details. Theorem 8.5 says that the componentwise relative backward error is as small as possible. For example, this means that if A and b are sparse, then δA and δb have the same sparsity structures as A and b, respectively.

Now we present one more common technique for improving the error in solving a linear system: *equilibration*, which yields the following algorithm.

Algorithm 8.9 Choose an appropriate diagonal matrix D to solve $DAx = Db$ instead of $Ax = b$. The matrix D is chosen to try to make the condition number of DA smaller than that of A.

For example, if we choose d_{ii} to be the reciprocal of the two-norm of row i of A, that would make DA nearly equal to the identity matrix for very ill conditioned matrices. In [112] it was shown that choosing D in this way reduces the condition number of DA to within a factor of $\sqrt{n}$ of its smallest possible value for any diagonal D.

In computations we may also choose two diagonal matrices D_{row} and D_{col} and solve $(D_{row} A D_{col})\bar{x} = D_{row}b$, $x = D_{col}\bar{x}$, and thus $D_{row} Ax = D_{row}b$.

8.4 Special Linear Systems

It is important to exploit any special structure of the matrix to increase the speed of algorithms for linear systems $Ax = b$ and decrease storage of intermediate matrices and vectors. In this section we will discuss only real matrices, since extension to complex matrices is straightforward. Matrices A with the following structures will be considered:

- Symmetric positive definite matrices (s.p.d. matrices),
- Symmetric indefinite matrices,
- Band matrices.

8.4.1 Real Symmetric Positive Definite Matrices

Recall that a real matrix A is called s.p.d. if $A = A^T$ and $x^T Ax > 0$ for all $x \neq 0$.[8] In this subsection we show how to solve $Ax = b$ in half the time and half the space of Gaussian elimination when A is s.p.d.

As we know (see p. 150), if $A = A^T$, then A is s.p.d. if and only if all its eigenvalues are positive. Below we prove some other useful properties of s.p.d. matrices.

Proposition 8.1 *If X is nonsingular, then A is s.p.d. if and only if $X^T AX$ is s.p.d.*

Proof If X is nonsingular, then $Xx \neq 0$ for all $x \neq 0$, and thus $x^T X^T AXx > 0$ for all $x \neq 0$. Since is A s.p.d., this implies that $X^T AX$ is s.p.d. Use X^{-1} to deduce that if $X^T AX$ is s.p.d., then A is s.p.d. ☐

---

[8] See (4.130), p. 150.

Proposition 8.2 *If A is s.p.d. and* $H = A(j : k, j : k)$ *is any principal submatrix of A, then H is s.p.d.*

Proof Suppose first that $H = A(1 : m, 1 : m)$ is a leading principal submatrix of A. Then for any given vector y of size m, the n-vector $x = (y, 0)$ satisfies the equality $y^T H y = x^T A x$. Since A is s.p.d., $x^T A x > 0$ for all vectors $x \neq 0$, then $y^T H y > 0$ for all vectors $y \neq 0$, and thus H is s.p.d. If H is not a leading principal submatrix of A, let P be a permutation matrix such that H lies in the upper left corner of $P^T A P$. Then apply Proposition 8.1 to $P^T A P$.   □

Proposition 8.3 *If A is s.p.d., then all* $a_{ii} > 0$, *and* $\max_{ij} |a_{ij}| = \max_i a_{ii} > 0$.

Proof The first assertion is easy to check (see Property 5, p. 141). Let us prove the second assertion. Let as usual i_k be the kth standard unit vector. If $|a_{kl}| = \max_{ij} |a_{ij}|$ but $k \neq l$ (this means that we assume that $\max_{ij} |a_{ij}| \neq \max_i a_{ii}$), we choose the vector $x = e_k - \text{sign}(a_{kl})e_l$. Then $x^T A x = a_{kk} + a_{ll} - 2|a_{kl}| \leq 0$. But this is a contradiction to the positive definiteness of the matrix A, and thus $\max_{ij} |a_{ij}| = \max_i a_{ii}$.

Proposition 8.4 *A matrix A is s.p.d. if and only if there is a unique lower triangular nonsingular matrix L, with positive diagonal entries, such that* $A = L L^T$. *A matrix* $A = L L^T$ *is called the* Cholesky[9] *factorization of A, and L is called the* Cholesky factor *of A.*

Proof Sufficiency. Assume that there exists a factorization $A = L L^T$ with L nonsingular. Then $x^T A x = (x^T L)(L^T x) = \|L^T x\|_2^2 > 0$ for all $x \neq 0$, so A is s.p.d.

Necessity. If A is s.p.d., we show that L exists by induction on the dimension n. If we choose each $l_{ii} > 0$, our construction will determine L uniquely. If $n = 1$, choose $l_{11} = \sqrt{a_{11}}$, which exists since $a_{11} > 0$. Let us write

$$
\begin{aligned}
A &= \begin{pmatrix} a_{11} & A_{12} \\ A_{12}^T & A_{22} \end{pmatrix} = \begin{pmatrix} \sqrt{a_{11}} & 0 \\ y & \tilde{L}_{22} \end{pmatrix} \begin{pmatrix} \sqrt{a_{11}} & y^T \\ 0 & \tilde{L}_{22}^T \end{pmatrix} \\
&= \begin{pmatrix} a_{11} & \sqrt{a_{11}} y^T \\ \sqrt{a_{11}} y & y y^T + \tilde{L}_{22} \tilde{L}_{22}^T \end{pmatrix} = \begin{pmatrix} a_{11} & A_{12} \\ A_{12}^T & \tilde{A}_{22} + \frac{A_{12}^T A_{12}}{a_{11}} \end{pmatrix} \qquad (8.56) \\
&= \begin{pmatrix} \sqrt{a_{11}} & 0 \\ \frac{A_{12}^T}{\sqrt{a_{11}}} & I \end{pmatrix} \begin{pmatrix} 1 & 0 \\ 0 & \tilde{A}_{22} \end{pmatrix} \begin{pmatrix} \sqrt{a_{11}} & \frac{A_{12}}{\sqrt{a_{11}}} \\ 0 & I \end{pmatrix}.
\end{aligned}
$$

We observe that the $(n - 1)$-by-$(n - 1)$ matrix $\tilde{A}_{22} = A_{22} + A_{12}^T A_{12}/a_{11}$ is a symmetric matrix. Using Proposition 8.1 and expression (8.56), we conclude that $\begin{pmatrix} 1 & 0 \\ 0 & \tilde{A}_{22} \end{pmatrix}$ is s.p.d. By Proposition 8.2, the matrix $\tilde{A}_{22}$ is also s.p.d. Thus, by induction, there is an $\tilde{L}$ such that $\tilde{A}_{22} = \tilde{L} \tilde{L}^T$ and

---

[9]André-Louis Cholesky (1875–1918) was a French military officer and mathematician.

$$A = \begin{pmatrix} \sqrt{a_{11}} & 0 \\ \frac{A_{12}^T}{\sqrt{a_{11}}} & I \end{pmatrix} \begin{pmatrix} 1 & 0 \\ 0 & \tilde{L}\tilde{L}^T \end{pmatrix} \begin{pmatrix} \sqrt{a_{11}} & \frac{A_{12}}{\sqrt{a_{11}}} \\ 0 & I \end{pmatrix}$$

$$= \begin{pmatrix} \sqrt{a_{11}} & 0 \\ \frac{A_{12}^T}{\sqrt{a_{11}}} & \tilde{L} \end{pmatrix} \begin{pmatrix} \sqrt{a_{11}} & \frac{A_{12}}{\sqrt{a_{11}}} \\ 0 & \tilde{L}^T \end{pmatrix} = LL^T. \qquad (8.57)$$

□

We rewrite Proposition 8.4 as the following algorithm.

Algorithm 8.10 Cholesky algorithm.

for $j = 1$ to n
  $l_{jj} = (a_{jj} - \sum_{k=1}^{j-1} l_{jk}^2)^{1/2}$
  for $i = j + 1$ to n
    $l_{ij} = (a_{ij} - \sum_{k=1}^{j-1} l_{ik}l_{jk})/l_{jj}$
  end for
end for

Using this algorithm, we observe that if A is not positive definite, then the algorithm fails by attempting to compute the square root of a negative number in line $l_{jj} = (a_{jj} - \sum_{k=1}^{j-1} l_{jk}^2)^{1/2}$ or by dividing by zero in line $l_{ij} = (a_{ij} - \sum_{k=1}^{j-1} l_{ik}l_{jk})/l_{jj}$. We conclude that running this algorithm is the cheapest way to test whether a symmetric matrix is positive definite.

In the Cholesky algorithm, L can overwrite the lower half of A. Only the lower half of A is referred to in the algorithm, so in fact, only $n(n + 1)/2$ storage is needed instead of n^2. The number of FLOPS in the Cholesky algorithm is (see Question 8.8, p. 286)

$$\sum_{j=1}^{n} \left(2j + \sum_{i=j+1}^{n} 2j \right) = \frac{1}{3}n^3 + O(n^2). \qquad (8.58)$$

We see that the Cholesky algorithm requires just half the FLOPS of Gaussian elimination (see p. 46).

Pivoting is not necessary for Cholesky to be numerically stable. We show this as follows. The same analysis as for Gaussian elimination in Section 8.2.2 reveals that we will have a similar formula for the error E in Cholesky decomposition as in LU decomposition:

$$A = LL^T + E,$$

where the error in Cholesky decomposition will be bounded as

$$|E| \leq n\varepsilon|L||L^T|.$$

Taking norms, we get

$$\|E\| \leq n\varepsilon\| |L| \|\| |L^T| \|.$$

We can rewrite the expression above as

$$\|E\| \leq n\varepsilon \|L\| \|L^T\|. \tag{8.59}$$

Thus in formula (8.59), we have obtained an error estimate in the decomposition $A = LL^T$. The next step is to obtain the error in the Cholesky algorithm. We again solve $LL^T x = b$ via $Ly = b$ and $L^T x = y$. Solving $Ly = b$ gives as a computed solution $\hat{y}$ such that $(L + \delta L)\hat{y} = b$, where $|\delta L| \leq n\varepsilon |L|$. The same is true for $(L^T + \delta L^T)\hat{x} = \hat{y}$ with $|\delta L^T| \leq n\varepsilon |L^T|$. Combining both estimates, we get

$$
\begin{aligned}
b &= (L + \delta L)\hat{y} = (L + \delta L)(L^T + \delta L^T)\hat{x} \\
&= (LL^T + L\delta L^T + \delta L L^T + \delta L \delta L^T)\hat{x} \\
&= (A - E + L\delta L^T + \delta L L^T + \delta L \delta L^T)\hat{x} \\
&= (A + \delta A)\hat{x},
\end{aligned}
\tag{8.60}
$$

where $\delta A = -E + L\delta L^T + \delta L L^T + \delta L \delta L^T$. Now we combine all bounds for E, δL^T, and δL and use the triangle inequality to get

$$
\begin{aligned}
|\delta A| &\leq |-E + L\delta L^T + \delta L L^T + \delta L \delta L^T| \\
&\leq |E| + |L||\delta L^T| + |\delta L||L^T| + |\delta L||\delta L^T| \\
&\leq n\varepsilon |L||L^T| + n\varepsilon |L||L^T| + n\varepsilon |L||L^T| + n^2\varepsilon^2 |L||L^T| \\
&\approx 3n\varepsilon |L||L^T|.
\end{aligned}
\tag{8.61}
$$

Assuming that $\||X|\| = \|X\|$ is true (as before, this is valid for Frobenius, infinity and one-norms but not for two-norms), we obtain

$$\|\delta A\| \leq 3n\varepsilon \|L\| \|L^T\|. \tag{8.62}$$

Thus, it follows from (8.62) that the computed solution $\tilde{x}$ satisfies $(A + \delta A)\tilde{x} = b$ with $|\delta A| \leq 3n\varepsilon |L||L^T|$. But by the Cauchy–Schwartz inequality and Proposition 8.3, we see that for every entry (i, j) of $|L||L^T|$, we can write the estimate

$$(|L||L^T|)_{ij} = \sum_k^n |l_{ik}||l_{jk}| \leq \sqrt{\sum_k^n l_{ik}^2} \sqrt{\sum_k^n l_{jk}^2} = \sqrt{a_{ii}}\sqrt{a_{jj}} \leq \max_{ij} |a_{ij}|.$$

Then applying this estimate to all n entries of $|L||L^T|$, we have

$$\||L||L^T|\|_\infty \leq n\|A\|_\infty. \tag{8.63}$$

Substituting (8.63) into (8.62), we get the following estimate:

$$\|\delta A\|_\infty \leq 3n^2 \varepsilon \|A\|_\infty, \tag{8.64}$$

which says that $\|\delta A\|_\infty$ has an upper bound depending on $\|A\|_\infty$, but not on $\|L\|_\infty$. This estimate is also valid for Frobenius and one-norms.

8.4.2 Symmetric Indefinite Matrices

Let us consider now indefinite matrices that are neither positive definite nor negative definite. The question is whether there exists an algorithm that can solve a symmetric indefinite linear system of equations and save half the time and half the space. It turns out that this is possible with a more complicated pivoting scheme and factorization.

If A is nonsingular, one can show that there exist a permutation matrix P, a unit lower triangular matrix L, and a block diagonal matrix D with 1×1 and 2×2 blocks such that $PAP^T = LDL^T$. This algorithm is described in [16].

8.4.3 Band Matrices

A matrix A is called a *band matrix* with *lower bandwidth* b_L and *upper bandwidth* b_U if $a_{ij} = 0$ whenever $i > j + b_L$ or $i < j - b_U$:

$$A = \begin{pmatrix} a_{11} & \cdots & a_{1,b_U+1} & & & 0 \\ \vdots & & & a_{2,b_U+2} & & \\ a_{b_L+1,1} & & & & \ddots & \\ & a_{b_L+2,2} & & & & a_{n-b_U,n} \\ & & \ddots & & & \vdots \\ 0 & & & a_{n,n-b_L} & \cdots & a_{n,n} \end{pmatrix}.$$

Band matrices often arise in practice and are useful to recognize because their L and U factors are also "essentially banded." This makes them cheaper to compute and store.

Let the matrix A be banded with lower bandwidth b_L and upper bandwidth b_U. Then after Gaussian elimination with partial pivoting, U is banded with upper bandwidth at most $b_L + b_U$, and L is "essentially banded" with lower bandwidth b_L. This means that L has at most $b_L + 1$ nonzero entries in each column and so can be stored in the same space as a band matrix with lower bandwidth b_L.

For the case when $b_L = b_U = 1$, the band matrix A becomes a tridiagonal matrix. There exists a special tridiagonal matrix algorithm known as Thomas's[10] algorithm,

---

[10]Llewellyn Hilleth Thomas (1903—1992) was a British physicist and applied mathematician.

which solves such linear system of equations. This method is a simplified form of Gaussian elimination and was proposed by Gelfand[11] and Lokucievsky[12] in 1952 and then modified by different researchers. In general, this algorithm is stable only for s.p.d. or for diagonally dominant matrices; see details in [53].

The method is derived as follows. The tridiagonal system of linear equations $Ax = b$ can be written in the form

$$a_{i-1,i}x_{i-1} + a_{i,i}x_i + a_{i+1,i}x_{i+1} = b_i, \quad i = 1, ..., n - 1. \tag{8.65}$$

The tridiagonal matrix algorithm is based on the assumption that the solution can be obtained as

$$x_i = \alpha_{i+1}x_{i+1} + \beta_{i+1}, \quad i = n - 1, ..., 1. \tag{8.66}$$

Writing (8.66) for $i - 1$, we get

$$\begin{aligned} x_{i-1} &= \alpha_i x_i + \beta_i \\ &= \alpha_i(\alpha_{i+1}x_{i+1} + \beta_{i+1}) + \beta_i = \alpha_i\alpha_{i+1}x_{i+1} + \alpha_i\beta_{i+1} + \beta_i. \end{aligned} \tag{8.67}$$

Substituting (8.66) and (8.67) into (8.65), we obtain

$$\begin{aligned} a_{i-1,i}(\alpha_i\alpha_{i+1}x_{i+1} + \alpha_i\beta_{i+1} + \beta_i) \\ + a_{i,i}(\alpha_{i+1}x_{i+1} + \beta_{i+1}) + a_{i+1,i}x_{i+1} = b_i, \quad i = 1, ..., n - 1. \end{aligned} \tag{8.68}$$

The equation above can be rewritten in the form

$$\begin{aligned} (a_{i-1,i}\alpha_i\alpha_{i+1} + a_{i,i}\alpha_{i+1} + a_{i+1,i})x_{i+1} \\ + a_{i-1,i}\alpha_i\beta_{i+1} + a_{i-1,i}\beta_i + a_{i,i}\beta_{i+1} - b_i = 0, \quad i = 1, ..., n - 1. \end{aligned} \tag{8.69}$$

The Eq. (8.69) will be satisfied if we require that

$$\begin{aligned} a_{i-1,i}\alpha_i\alpha_{i+1} + a_{i,i}\alpha_{i+1} + a_{i+1,i} = 0, \\ a_{i-1,i}\alpha_i\beta_{i+1} + a_{i-1,i}\beta_i + a_{i,i}\beta_{i+1} - b_i = 0. \end{aligned} \tag{8.70}$$

From (8.70), it follows that

$$\begin{aligned} \alpha_{i+1} &= -\frac{a_{i+1,i}}{a_{i-1,i}\alpha_i + a_{i,i}}, \\ \beta_{i+1} &= \frac{b_i - a_{i-1,i}\beta_i}{a_{i-1,i}\alpha_i + a_{i,i}}. \end{aligned} \tag{8.71}$$

---

[11]Israel Moiseevich Gelfand (1913–2009) was a Russian mathematician.

[12]Oleg Vyacheslavovich Lokucievsky (1922—1990) was a Russian mathematician.

Starting from $i = 1$, we can find

$$\alpha_2 = -\frac{a_{2,1}}{a_{1,1}},$$

$$\beta_2 = \frac{b_1}{a_{1,1}}. \tag{8.72}$$

Then from (8.71), we can obtain all other coefficients $\alpha_{i+1}, \beta_{i+1}, i = 1, ..., n - 1$, recursively. Knowing all coefficients $\alpha_{i+1}, \beta_{i+1}, i = 1, ..., n - 1$, we can obtain the solution of the tridiagonal system of linear equations via (8.66).

Band matrices often arise from discretizing different physical problems in which mathematical models are usually described by ordinary differential equation (ODE) or by partial differential equations (PDE).

Example 8.5 Consider the following boundary value problem for the ODE

$$y''(x) - p(x)y'(x) - q(x)y(x) = r(x), \quad x \in [a, b], \tag{8.73}$$

$$y(a) = \alpha, \quad y(b) = \beta. \tag{8.74}$$

We assume $q(x) \geq q_m > 0$. Equation (8.73) models the heat flow in a long pipe, for example. To solve it numerically, we discretize it by seeking its solution only at mesh points $x_i = a + ih, i = 0, \ldots, N + 1$, where $h = (b - a)/(N + 1)$ is the mesh size. Define $p_i = p(x_i)$, $r_i = r(x_i)$, and $q_i = q(x_i)$.

To derive discretized equations for $y_i \approx y(x_i)$ with boundary conditions $y_0 = \alpha$, $y_{N+1} = \beta$ we approximate the derivative $y'(x_i)$ by the following *finite difference scheme* called *central difference*:

$$y'(x_i) \approx \frac{y_{i+1} - y_{i-1}}{2h}. \tag{8.75}$$

As the mesh size h gets smaller, (8.75) approximates $y'(x_i)$ more and more accurately. We can similarly approximate the second derivative by

$$y''(x_i) \approx \frac{y_{i+1} - 2y_i + y_{i-1}}{h^2}. \tag{8.76}$$

Inserting approximations (8.75), (8.76) into (8.73) yields

$$\frac{y_{i+1} - 2y_i + y_{i-1}}{h^2} - p_i \frac{y_{i+1} - y_{i-1}}{2h} - q_i y_i = r_i, \quad 1 \leq i \leq N. \tag{8.77}$$

Multiplying by $h^2/2$, we can rewrite (8.77) in the form of a linear system $Ay = b$ with

$$y = \begin{pmatrix} y_1 \\ \vdots \\ y_N \end{pmatrix}, \quad b = \frac{-h^2}{2} \begin{pmatrix} r_1 \\ \vdots \\ r_N \end{pmatrix} + \begin{pmatrix} \left(\frac{1}{2} + \frac{h}{4}p_1\right)\alpha \\ 0 \\ \vdots \\ 0 \\ \left(\frac{1}{2} - \frac{h}{4}p_N\right)\beta \end{pmatrix}, \qquad (8.78)$$

and

$$A = \begin{pmatrix} a_1 & -c_1 & & \\ -b_2 & \ddots & \ddots & \\ & \ddots & \ddots & -c_{N-1} \\ & & -b_N & a_N \end{pmatrix}, \quad \begin{aligned} a_i &= 1 + \tfrac{h^2}{2}q_i, \\ b_i &= \tfrac{1}{2}\left(1 + \tfrac{h}{2}p_i\right), \\ c_i &= \tfrac{1}{2}\left(1 - \tfrac{h}{2}p_i\right). \end{aligned} \qquad (8.79)$$

Since by our assumption we have $q_i > 0$, it follows from (11.36) that $a_i > 0$. For sufficiently small $h < 1$, also $b_i > 0$ and $c_i > 0$.

System $Ay = b$ is an asymmetric *tridiagonal* system. We will show how to change it to a symmetric positive definite tridiagonal system so that we may use *Cholesky decomposition* to solve it. Choose the diagonal matrix

$$D = \mathrm{diag}\left(1, \sqrt{\frac{c_1}{b_2}}, \sqrt{\frac{c_1 c_2}{b_2 b_3}}, \ldots, \sqrt{\frac{c_1 c_2 \cdots c_{N-1}}{b_2 b_3 \cdots b_N}}\right).$$

Then we may change $Ay = b$ to $(DAD^{-1})(Dy) = Db$ or $\tilde{A}\tilde{y} = \tilde{b}$, where

$$\tilde{A} = \begin{pmatrix} a_1 & -\sqrt{c_1 b_2} & & & \\ -\sqrt{c_1 b_2} & a_2 & -\sqrt{c_2 b_3} & & \\ & -\sqrt{c_2 b_3} & \ddots & & \\ & & \ddots & \ddots & -\sqrt{c_{N-1} b_N} \\ & & & -\sqrt{c_{N-1} b_N} & a_N \end{pmatrix}.$$

We observe that $\tilde{A}$ is symmetric, and it has the same eigenvalues as A, because A and $\tilde{A} = DAD^{-1}$ are similar. Let us prove that $\tilde{A}$ is positive definite. By the Gershgorin disk theorem (see Theorem 7.6, p. 238), all the eigenvalues λ of an arbitrary matrix B are located in the union of the n disks

$$|\lambda - b_{kk}| \leq \sum_{j \neq k} |b_{kj}|. \qquad (8.80)$$

We can take the mesh size h so small that for all i, we have $|h/2 p_i| < 1$ in all nondiagonal elements of the matrix A. Then

$$|b_i| + |c_i| = \frac{1}{2}\left(1 + \frac{h}{2}p_i\right) + \frac{1}{2}\left(1 - \frac{h}{2}p_i\right) = 1 < 1 + \frac{h^2}{2}q_m \leq 1 + \frac{h^2}{2}q_i = a_i.$$

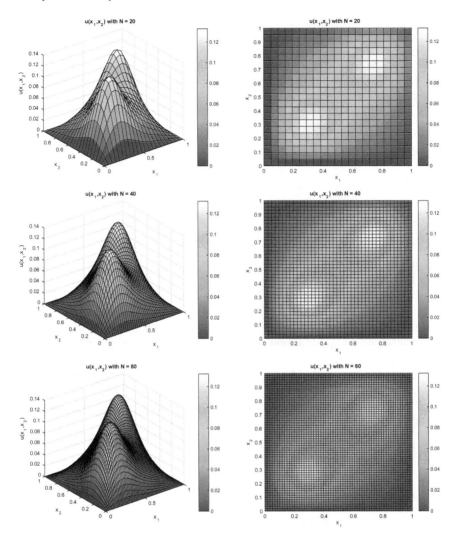

Fig. 8.4 Solution of Poisson's equation (8.11) using Cholesky decomposition in the example of Section 8.4.4.

Thus using (8.80), we see that all the eigenvalues of A lie inside the disks D_i centered at $a_i = 1 + h^2 q_i/2 \geq 1 + h^2 q_m/2$ with radius $R_i = 1$. They must all have positive real parts.

Since $\tilde{A}$ is symmetric, its eigenvalues are real and positive. Then the matrix $\tilde{A}$ is positive definite. Its smallest eigenvalue is bounded below by $q_m h^2/2$. Thus, it can be solved by Cholesky decomposition.

8.4.4   A Numerical Example

In this section we illustrate performance of the Cholesky algorithm on the solution of Poisson's equation (8.11) on the unit square $\{(x_1, x_2) : 0 \leq x_1, x_2 \leq 1\}$ with boundary conditions $u = 0$ on the boundary of this square. We take now in (8.11) the following function $f(x_1, x_2)$ given by two choices of Gaussians:

$$
\begin{aligned}
f(x_1, x_2) = 1 + A_1 \exp & \left(-\frac{(x_1 - 0.25)^2}{0.02} - \frac{(x_2 - 0.25)^2}{0.02} \right) \\
+ A_2 \exp & \left(-\frac{(x_1 - 0.75)^2}{0.02} - \frac{(x_2 - 0.75)^2}{0.02} \right),
\end{aligned}
$$
(8.81)

where A_1, A_2 are the amplitudes of these two Gaussians.

As in the example of Section 8.1.3, we construct a mesh with the points (x_{1i}, x_{2j}) such that $x_{1i} = ih$, $x_{2j} = jh$ with $h = 1/(N + 1)$, where N is the number of inner points in the x_1 and x_2 directions, and we take the same number of points in the x_1 and x_2 directions: $n_i = n_j = N + 2$. The linear system of equations $Au = f$ is solved then via Cholesky factorization (Algorithm 8.10). We can use Cholesky decomposition, since the matrix A is symmetric positive definite (see Question 8.16, p. 289).

Figure 8.4 shows results of numerical simulations for different discretizations of the unit square with number of inner points $N = 20, 40, 60$ and for $A_1 = A_2 = 10$ in (8.81). The MATLAB® program of Section 1.2 are available for running this test.[13]

Questions

8.1 (*Programming*)

Solve in MATLAB® Poisson's equation (8.11) on the unit square $\{(x_1, x_2) : 0 \leq x_1, x_2 \leq 1\}$ with boundary conditions $u = 0$ on the boundary of this square and with function $f(x_1, x_2)$ given by (8.81) using the lu function and programs of Section 1.1. Compare the obtained results with those of Section 8.4.4. Optional: extend these results to three-dimensional case.

8.2 (*Programming*)

Improve Algorithm 8.1, overwriting L and U on A. Write your own MATLAB® program and test it on your own examples.

8.3 (*Programming*)

Apply the bisection algorithm (see Algorithm 8.11 below) to find the roots of the polynomial $p(x) = (x - 2)^9$ and of some of your own polynomials, where $p(x)$ is

---

[13] All MATLAB® programs referenced on this page can be found on the online version of Chapter 1 (doi:10.1007/978-3-319-57304-5_1).

evaluated using Horner's rule (Algorithm 8.4). Write your own MATLAB® program. Confirm that changing the input interval for $x = [x_{left}, x_{right}]$ slightly changes the computed root drastically. Modify the algorithm to use the relative condition number for polynomial evaluation (8.26) to stop bisecting when the rounding error in the computed value of $p(x)$ gets so large that its sign cannot be determined. Present your results similarly to the results of Fig. 8.2.

Hint: use the MATLAB® function coeffs to compute the coefficients of a polynomial $p(x)$.

Here we present the bisection algorithm to find roots of a polynomial $p(x)$. Suppose that the input interval for x where we want to find roots of $p(x) = 0$ is $x \in [x_{left}, x_{right}]$. At every iteration, this algorithm divides the input interval in two by computing the midpoint $x_{middle} = (x_{left} + x_{right})/2$ of the input interval as well as the value of the polynomial $p(x_{middle})$ at that point. We will compute the value of the polynomial $p(x_{middle})$ using Horner's rule (Algorithm 8.4). Then if p_{left} and p_{mid} have opposite signs, the bisection algorithm sets x_{middle} as the new value for x_{right}, and if p_{right} and p_{mid} have opposite signs, then the method sets x_{middle} as the new x_{left}. If $p(x_{middle}) = 0$, then x_{middle} may be taken as the root of the polynomial and the algorithm stops.

Algorithm 8.11 Bisection algorithm to find zeros of the polynomial $p(x)$.

0. Initialization: set the left x_{left} and the right x_{right} bounds for the input interval for $x \in [x_{left}, x_{right}]$ where we will seek roots of the polynomial. Set the computational tolerance tol.

1. Evaluate the polynomial $p(x)$ at the points x_{left} and x_{right} to get $p_{left} = p(x_{left})$ and $p_{right} = p(x_{right})$ using Algorithm 8.4.
   Perform steps 2–3 while $x_{right} - x_{left} > 2tol$

2. Compute point $x_{mid} = (x_{left} + x_{right})/2$ and then $p_{mid} = p(x_{mid})$, using Algorithm 8.4.

3. Check:
   If $p_{left} p_{mid} < 0$, then we have a root in the interval $[x_{left}, x_{mid}]$. Assign $x_{right} = x_{mid}$ and $p_{right} = p_{mid}$.
   Else if $p_{right} p_{mid} < 0$, then we have a root in the interval $[x_{mid}, x_{right}]$. Assign $x_{left} = x_{mid}$ and $p_{left} = p_{mid}$.
   Else we have found a root at x_{mid} and assign $x_{left} = x_{mid}, x_{right} = x_{mid}$.

4. Compute the root as $(x_{left} + x_{right})/2$.

8.4 Write Algorithm 8.1 for the case $n = 3$. Using Algorithm 8.1, perform LU factorization of the matrix

$$A = \begin{pmatrix} 4 & 1 & 1 \\ 1 & 8 & 1 \\ 1 & 1 & 16 \end{pmatrix}.$$

8.5 Using Cholesky's algorithm (Algorithm 8.10), perform the factorization $A = LL^T$ of the matrix A in Question 8.4.

8.6 (*Programming*)

Implement Hager's algorithm, Algorithm 8.7, in MATLAB®. Test it on different matrices. Take, for example, $A = \text{hilb}(N)$ or $A = \text{rand}(N, N)$ for different N.

8.7 Let us consider the solution of the linear system $AX = B$, where A is an $n \times n$ matrix, B is an $n \times m$ matrix, and X is an unknown $n \times m$ matrix. We have two methods to solve it:

1. Factorization of $A = PLU$ and then using the algorithms of forward and backward substitution, Algorithms 8.2 and 8.3 to find every column of X.
2. Computation of A^{-1} by Gaussian elimination and then finding of $X = A^{-1}B$.

Count the number of FLOPS required for every algorithm. Show that the first algorithm requires fewer FLOPS than the second one.
Hint: use material of Section 8.2.2.

8.8 Derive formula (8.58) for the operation count in Cholesky decomposition. Hint: use the formula

$$\sum_{i=1}^{n} i^2 = \frac{(n+1)n(2n+1)}{6}$$

and the formula for the sum of an arithmetic progression,

$$\sum_{k=1}^{n} a_i = \frac{n(a_1 + a_n)}{2}.$$

8.9 Let A be an s.p.d. matrix. Show that $|a_{ij}| < \sqrt{a_{ii}a_{jj}}$.

8.10 Suppose A is an invertible nonsingular square matrix of order n and that u, v are vectors. Suppose furthermore that $1 + v^T A^{-1} u \neq 0$. Prove the Sherman–Morrison formula

$$(A + uv^T)^{-1} = A^{-1} - \frac{A^{-1}uv^T A^{-1}}{1 + v^T A^{-1} u}.$$

Here, uv^T is the outer product of two vectors u and v.

8.11 Suppose A is an invertible square matrix of order n and that U, V are $n \times k$ rectangular matrices with $k \leq n$. Prove the Sherman–Morrison–Woodburg formula, which states that $T = I + V^T A^{-1} U$ is nonsingular if and only if $A + UV^T$ is nonsingular and

$$(A + UV^T)^{-1} = A^{-1} - A^{-1}UT^{-1}V^T A^{-1}.$$

8.12 (*Programming*)

Similarly, with MATLAB® programs of Sections 1.1, 1.2, solve the three-dimensional problem[14]

---

[14]The MATLAB® programs can be found on the online version of Chapter 1 (doi:10.1007/978-3-319-57304-5_1).

$$-\Delta u(x) = f(x) \text{ in } \Omega,$$
$$u = 0 \text{ on } \partial\Omega \tag{8.82}$$

on the unit cube $\Omega = [0, 1] \times [0, 1] \times [0, 1]$. Choose an appropriate function $f(x)$.
Hint. We discretize the unit cube Ω with $x_{1i} = ih_1, x_{2j} = jh_2, x_{3k} = kh_3$, where

$$h_1 = \frac{1}{n_i - 1}, \quad h_2 = \frac{1}{n_j - 1}, \quad h_3 = \frac{1}{n_k - 1}$$

are the steps of the discrete finite difference mesh and n_i, n_j, n_k are the numbers of discretization points in the directions x_1, x_2, x_3, respectively. The indices (i, j, k) are such that $0 \leq i < n_i, 0 \leq j < n_j, 0 \leq j < n_k$. The global node numbers n_{glob} in the three-dimensional case can be computed as

$$n_{glob} = j + n_j \left((i - 1) + n_i (k - 1) \right). \tag{8.83}$$

We take $n_i = n_j = n_k = n = N + 2, h_1 = h_2 = h_3 = 1/(n - 1) = 1/(N + 1)$ and obtain the following scheme for the solution of Poisson's equation (8.11) in three dimensions:

$$-\frac{u_{i+1,j,k} - 2u_{i,j,k} + u_{i-1,j,k}}{h_1^2} - \frac{u_{i,j+1,k} - 2u_{i,j,k} + u_{i,j-1,k}}{h_2^2}$$
$$-\frac{u_{i,j,k+1} - 2u_{i,j,k} + u_{i,j,k-1}}{h_3^2} = \frac{f_{i,j,k}}{a_{i,j,k}}, \tag{8.84}$$

where $u_{i,j,k}, f_{i,j,k}, a_{i,j,k}$ are values of u, f, a, respectively, at the discrete point n_{glob} with indices (i, j, k). We rewrite Eq. (8.84) with $h = h_1 = h_2 = h_3$ as

$$6u_{i,j,k} - u_{i+1,j,k} - u_{i-1,j,k} - u_{i,j+1,k} - u_{i,j-1,k} - u_{i,j,k+1} - u_{i,j,k-1} = h^2 \frac{f_{i,j,k}}{a_{i,j,k}}. \tag{8.85}$$

Again, we recognize that the scheme (8.85) is a system of linear equations $Au = b$. The matrix A is of size $(n_i - 2)(n_j - 2)(n_k - 2) = N^3$, and on the unit cube it is given by the block matrix

$$A = \begin{pmatrix} A_N & -I_N & O_N & -I_N & \ddots \\ -I_N & A_N & -I_N & \ddots & \ddots \\ \ddots & \ddots & \ddots & \ddots & \ddots \\ -I_N & \ddots & -I_N & A_N & -I_N \\ \ddots & -I_N & O_N & -I_N & A_N \end{pmatrix}$$

with zero blocks O_N of order N. The blocks A_N of size $N \times N$ on the main diagonal of this matrix are given by

$$
A_N =
\begin{pmatrix}
6 & -1 & 0 & \cdots & \cdots & 0 \\
-1 & 6 & -1 & 0 & \cdots & 0 \\
0 & -1 & 6 & 0 & \cdots & 0 \\
\cdots & \cdots & \cdots & \cdots & \cdots & \cdots \\
0 & \cdots & \cdots & 0 & -1 & 6
\end{pmatrix}.
$$

8.13 (*Programming*)

Use the MATLAB® programs of Sections 1.1, 1.2 and solve the problem in Example 8.2 on an L-shaped 2D domain.[15]

8.14 (*Programming*)

Use formula (8.80) to estimate the eigenvalues of the matrix

$$
A =
\begin{pmatrix}
10 & -1 & 0 & 1 \\
0.2 & 8 & 0.2 & 0.2 \\
1 & 1 & 2 & 1 \\
-1 & -1 & -1 & -11
\end{pmatrix}. \tag{8.86}
$$

Write your own MATLAB® program to present results similar to those in Fig. 8.5, which shows Gershgorin disks together with their centers and the computed eigenvalues λ_i. Hint: use the MATLAB® function eigs(A) to compute eigenvalues.

8.15 (*Programming*)

Use formula (8.80) to estimate the eigenvalues of the matrix

$$
A =
\begin{pmatrix}
7 & 5 & 2 & 1 \\
2 & 8 & 3 & 2 \\
1 & 1 & 5 & 1 \\
1 & 1 & 1 & 6
\end{pmatrix}. \tag{8.87}
$$

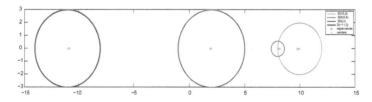

Fig. 8.5 Eigenvalues in Question 8.14 computed and estimated by the Gershgorin disk theorem. The computed eigenvalues of A are $\lambda_1 = 9.8218$, $\lambda_2 = 8.1478$, $\lambda_3 = 1.8995$, $\lambda_4 = -10.86$.

---

[15] The MATLAB® programs can be found on the online version of Chapter 1 (doi:10.1007/978-3-319-57304-5_1).

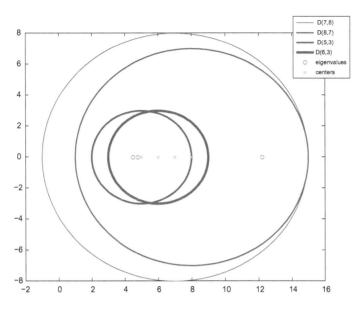

Fig. 8.6 Eigenvalues in Question 8.15 computed and estimated by the Gershgorin disk theorem. The computed eigenvalues are $\lambda_1 = 12.2249 + 0.0000i$, $\lambda_2 = 4.4977 + 0.6132i$, $\lambda_3 = 4.4977 - 0.6132i$, $\lambda_4 = 4.7797 + 0.0000i$.

Write your own MATLAB® program to present results similar to those in Fig. 8.6, which shows Gershgorin disks together with their centers and the computed eigenvalues λ_i. Hint: use the MATLAB® function `eigs(A)` to compute eigenvalues.

8.16 Prove that the matrix A in the numerical examples of Sections 8.1.3 and 8.4.4 is s.p.d. Hint: prove that for all u, $v \in \mathbb{R}^{(N+2)^2}$ that equal zero on the boundary of the mesh domain, we can write

$$(Au, v) = \sum_{i,j=1}^{N-1} (u_{i+1,j} - u_{i,j})(v_{i+1,j} - v_{i,j}) + \sum_{i,j=1}^{N-1} (u_{i,j+1} - u_{i,j})(v_{i,j+1} - v_{i,j}).$$

8.17 (*Programming*)
Write your own MATLAB® program and solve numerically the problem of Example 8.5 via the solution of the system of linear equations $Ay = b$ with b, A given in (8.78), (8.79), respectively. Use the recurrence formulas (8.66), (8.71) to obtain the solution of this system.

Chapter 9
Numerical Solution of Linear Least Squares Problems

In this chapter we present methods for numerical solution of linear least squares problems. These problems arise in many real-life applications such that curve fitting, statistical modelling and different inverse problems, when some model function should be fitted to the measured data.

Various matrix factorizations are usually applied to solve linear least squares problems (LLSP). In this chapter we will present several methods for the solution of LLSP:

1. Method of normal equations,
2. QR decomposition,
3. SVD decomposition.

The method of normal equations is widely used, since it is the fastest compared with all other methods for the solution of LLSP. However, this method is the least accurate and can be used only when the condition number of the matrix A is small. The method of factorization of a matrix A into two matrices Q and R such that $A = QR$, where Q is orthogonal and R is upper triangular, is called QR decomposition. This method is more accurate than the method of normal equations and is standard for the solution of LLSP. A drawback of this method is that it costs twice as much as the method of normal equations. When a matrix A is very ill conditioned, for example when A does not have full rank, then matrix factorization called SVD decomposition is commonly used for the solution of LLSP. However, like QR decomposition, this method is several times more expensive than the method of normal equations. Another method for the solution of very ill conditioned LLSP is iterative refinement, which iteratively improves the solution of a linear system of equations. This method can be adapted to deal efficiently with sparse matrices; see [14] for details.

In Section 9.1 we present the topic of LLSP and some typical examples of their application. In Section 9.2 we briefly present the main methods that also solve nonlin-

The original version of this chapter was revised. An Erratum to this book can be found at https://doi.org/10.1007/978-3-319-57304-5_9

ear least squares problem. Different methods for the solution of LLSP are described in the following sections: the method of normal equations is presented in Section 9.3, QR decomposition is outlined in Section 9.4, orthogonalization methods to perform QR decomposition are described in Section 9.5, SVD decomposition is presented in Section 9.6. Rank-deficient least squares problems and how to solve them are discussed in Sections 9.6.1 and 9.6.2, respectively. Available software for the solution of LLSP is outlined in Section 9.7.

9.1 Linear Least Squares Problems

Suppose that we have a matrix A of size $m \times n$ and a vector b of size $m \times 1$. The linear least square problem is to find a vector x of size $n \times 1$ that will minimize $\|Ax - b\|_2$. In the case that $m = n$ and the matrix A is nonsingular, we can obtain a solution to this problem as $x = A^{-1}b$. However, when $m > n$ (more equations than unknowns), the problem is called overdetermined. Conversely, when $m < n$ (more unknowns than equations), the problem is called underdetermined.

In real-life applications, engineers more commonly deal with overdetermined least squares problems in which the number of equations is much larger than the number of unknowns. This occurs because engineers usually take many more measurements than necessary to smooth out measurement error and remove noise from data. We will restrict our considerations to linear least squares problems. We refer to [14] and to the next section for the solution of nonlinear least squares problems.

Further, we assume that we are dealing with overdetermined problems, in which we have more equations than unknowns. This means that we will be interested in the solution of a linear system of equations

$$Ax = b, \tag{9.1}$$

where A is of size $m \times n$ with $m > n$, b is a vector of length m, and x is a vector of length n.

In a general case we are not able to get vector b of the size m as a linear combination of the n columns of the matrix A and n components of the vector x, from what follows that there is no solution to (9.1). In this chapter we will consider methods that can minimize the residual $r = b - Ax$ as a function of x in principle in any norm, but we will use only the 2-norm for theoretical and computational convenience (relationships between the 2-norm and the inner product and orthogonality, smoothness, and strict convexity properties). The 2-norm method is also called the method of least squares. We can write a least squares problem as a problem of minimizing the squared residuals

$$\|r\|_2^2 = \sum_{i=1}^{m} r_i^2 = \sum_{i=1}^{m} (Ax_i - b)^2. \tag{9.2}$$

In other words, our goal is to find the minimum of this residual using least squares:

$$\min_x \|r\|_2^2 = \min_x \sum_{i=1}^{m} r_i^2 = \min_x \sum_{i=1}^{m} (Ax_i - b)^2. \tag{9.3}$$

Example 9.1 Data fitting.

In this example we present a typical application of least squares called data fitting or curve fitting. This problem appear in statistical modeling and experimental engineering when data are generated by laboratory or other measurements.

Suppose that we have data points (x_i, y_i), $i = 1, ..., m$, and our goal is to find the vector of parameters c of size n that will fit best the data y_i of the model function $f(x_i, c)$, where $f : R^{n+1} \rightarrow R$, in the least squares sense:

$$\min_c \sum_{i=1}^{m} (y_i - f(x_i, c))^2. \tag{9.4}$$

If the function $f(x, c)$ is linear, then we can solve the problem (9.4) using the least squares method. The function $f(x, c)$ is linear if we can write it as a linear combination of the functions $\phi_j(x)$, $j = 1, ..., n$ as

$$f(x, c) = c_1\phi_1(x) + c_2\phi_2(x) + ... + c_n\phi_n(x). \tag{9.5}$$

The functions $\phi_j(x)$, $j = 1, ..., n$ are called basis functions.

Let now the matrix A have entries $a_{ij} = \phi_j(x_i)$, $i = 1, ..., m$; $j = 1, ..., n$, and the vector b will be such that $b_i = y_i$, $i = 1, ..., m$. Then a linear data-fitting problem takes the form of (9.1) with $x = c$:

$$Ac \approx b. \tag{9.6}$$

The elements of the matrix A are created by the basis functions $\phi_j(x)$, $j = 1, ..., n$. We will now consider different examples of choosing the basis functions $\phi_j(x)$, $j = 1, ..., n$.

Example 9.2 Problem of the fitting to a polynomial.

In the problem of fitting a polynomial

$$f(x, c) = \sum_{i=1}^{d} c_i x^{i-1} \tag{9.7}$$

of degree $d - 1$ to data points (x_i, y_i), $i = 1, ..., m$, the basis functions $\phi_j(x)$, $j = 1, ..., n$ can be chosen as $\phi_j(x) = x^{j-1}$, $j = 1, ..., n$. The matrix A constructed by these basis functions in a polynomial-fitting problem is a Vandermonde matrix:

$$A = \begin{pmatrix} 1 & x_1 & x_1^2 & \dots & x_1^{d-1} \\ 1 & x_2 & x_2^2 & \dots & x_2^{d-1} \\ 1 & x_3 & x_3^2 & \dots & x_3^{d-1} \\ \vdots & \vdots & \ddots & \ddots & \vdots \\ 1 & x_m & x_m^2 & \dots & x_m^{d-1} \end{pmatrix}. \tag{9.8}$$

Here $x_i, i = 1, ..., m$, are discrete points on the interval $x = [x_{left}, x_{right}]$. Suppose, that we choose $d = 4$ in (9.7). Then we can write the polynomial as $f(x, c) = \sum_{i=1}^{4} c_i x^{i-1} = c_1 + c_2 x + c_3 x^2 + c_4 x^3$, and our data-fitting problem (9.6) for this polynomial takes the form

$$\begin{pmatrix} 1 & x_1 & x_1^2 & x_1^3 \\ 1 & x_2 & x_2^2 & x_2^3 \\ 1 & x_3 & x_3^2 & x_3^3 \\ \vdots & \vdots & \ddots & \vdots \\ 1 & x_m & x_m^2 & x_m^3 \end{pmatrix} \cdot \begin{pmatrix} c_1 \\ c_2 \\ c_3 \\ c_4 \end{pmatrix} = \begin{pmatrix} b_1 \\ b_2 \\ b_3 \\ \dots \\ b_m \end{pmatrix}. \tag{9.9}$$

The right-hand side of the above system represents measurements of a function that we want to fit. Our goal is to find coefficients $c = \{c_1, c_2, c_3, c_4\}$ that will minimize the residual $r_i = f(x_i, c) - b_i, i = 1..., m$. Since we want to minimize the squared 2-norm of the residual, or $\|r\|_2^2 = \sum_{i=1}^{m} r_i^2$, we will solve a linear least squares problem.

Let us consider an example in which the right-hand side $b_i, i = 1, ...m$ is taken as a smooth function $b = sin(\pi x/5) + x/5$. Figure 9.1 shows fitting a polynomial to the function $b = sin(\pi x/5) + x/5$ for different values of d in (9.7) in the interval $x \in [-10, 10]$. We observe in this figure that with increasing degree $d - 1$ of the polynomial, we have a better fit to the exact function $b = sin(\pi x/5) + x/5$. However, for a degree of the polynomial greater than 18, we get an erratic fit to the function. Check this using the MATLAB® programs of Section 1.4.[1] This happens because the matrix A becomes more and more ill-conditioned with increasing degree d of the polynomial. And this, in turn, is because of the linear dependence of the columns in the Vandermonde matrix A.

Example 9.3 Approximation using linear splines.

When we want to solve the problem (9.4) of approximation to the data vector $y_i, i = 1, ..., m$, with linear splines, we use the basis functions $\phi_j(x), j = 1, ..., n$, in (9.5), which are also called hat functions:

$$\phi_j(x) = \begin{cases} \frac{x - T_{j-1}}{T_j - T_{j-1}}, & T_{j-1} \le x \le T_j, \\ \frac{T_{j+1} - x}{T_{j+1} - T_j}, & T_j \le x \le T_{j+1}. \end{cases} \tag{9.10}$$

Here, the column j in the matrix A is constructed by the given values of $\phi_j(x)$ at points $T_j, j = 1, .., n$, which are called conjunction points and are chosen by the

---

[1] The MATLAB® programs can be found on the online version of Chapter 1 (doi:10.1007/978-3-319-57304-5_1).

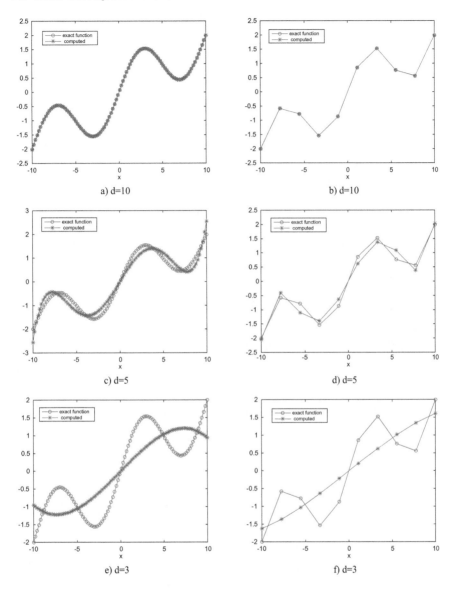

Fig. 9.1 Polynomial fitting for different degrees d in (9.7) to the function $b = sin(\pi x/5) + x/5$ on the interval $x \in [-10, 10]$ using the method of normal equations. Left-hand figures: fit to 100 points x_i, $i = 1, ..., 100$; Right-hand figures: fit to 10 points x_i, $i = 1, ..., 10$. Lines with *blue stars* represent the computed function, those with *red circles*, the exact function.

user. Using (9.10), we may conclude that the first basis function is $\phi_1(x) = \frac{T_2-x}{T_2-T_1}$ and the last one is $\phi_n(x) = \frac{x-T_{n-1}}{T_n-T_{n-1}}$.

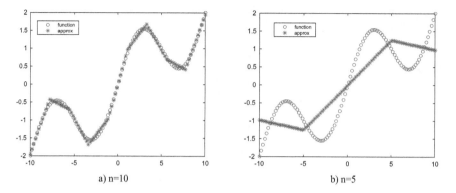

a) n=10   b) n=5

Fig. 9.2 Example 9.3. Polynomial fitting to the function $b = sin(\pi x/5) + x/5$ on the interval $x \in [-10, 10]$ using linear splines with different numbers n of conjunction points T_j, $j = 1, ..., n$, in (9.10). *Blue stars* represent the computed function, and *red circles*, the exact function.

Figure 9.2 shows an approximation of the function $b = sin(\pi x/5) + x/5$ on the interval $x \in [-10, 10]$ using linear splines with different numbers n of conjunction points T_j, $j = 1, ..., n$. A MATLAB® program is available for running this test in Section 1.7.[2]

Example 9.4 Approximation using bellsplines.

To solve the problem (9.4) using bellsplines, the number of bellsplines that can be constructed is $n + 2$, and the function $f(x, c)$ in (9.4) is written as

$$f(x, c) = c_1\phi_1(x) + c_2\phi_2(x) + ... + c_{n+2}\phi_{n+2}(x). \tag{9.11}$$

We define

$$\phi_j^0(x) = \begin{cases} 1, & T_j \leq x \leq T_{j+1}, \\ 0, & \text{otherwise.} \end{cases} \tag{9.12}$$

Then all other basis functions, or bellsplines, $\phi_j^k(x)$, $j = 1, ..., n + 2$; $k = 1, 2, 3$, are defined as follows:

$$\phi_j^k(x) = (x - T_k)\frac{\phi_j^{k-1}(x)}{T_{j+k} - T_j} + (T_{j+k+1} - x)\frac{\phi_{j+1}^{k-1}(x)}{T_{j+k+1} - T_{j+1}}. \tag{9.13}$$

Here, the column j in the matrix A is constructed by the given values of $\phi_j(x)$ at conjunction points T_j, $j = 1, .., n$, which are chosen by the user. If in (9.13) we obtain the ratio 0/0, then we assign $\phi_j^k(x) = 0$. We define three additional points T_{-2}, T_{-1}, T_0 at the left side of the input interval as $T_{-2} = T_{-1} = T_0 = T_1$, and correspondingly three points $T_{n+1}, T_{n+2}, T_{n+3}$ on the right side of the interval as $T_n = T_{n+1} = T_{n+2} = T_{n+3}$. Altogether, we have $n + 6$ conjunction points T_j, $j = 1, ..., n + 6$. The number of bellsplines that can be constructed is $n + 2$.

---

[2]The MATLAB® programs can be found on the online version of Chapter 1 (doi:10.1007/978-3-319-57304-5_1).

If the conjunction points T_j are distributed uniformly, then we can introduce the mesh size $h = T_{k+1} - T_k$, and the bellsplines can be written explicitly as

$$\phi_j(x) = \begin{cases} \frac{1}{6}t^3 & \text{if } T_{j-2} \leq x \leq T_{j-1}, \ t = \frac{1}{h}(x - T_{j-2}), \\ \frac{1}{6} + \frac{1}{2}(t + t^2 - t^3) & \text{if } T_{j-1} \leq x \leq T_j, \ t = \frac{1}{h}(x - T_{j-1}), \\ \frac{1}{6} + \frac{1}{2}(t + t^2 - t^3) & \text{if } T_j \leq x \leq T_{j+1}, \ t = \frac{1}{h}(T_{j+1} - x), \\ \frac{1}{6}t^3 & \text{if } T_{j+1} \leq x \leq T_{j+2}, t = \frac{1}{h}(T_{j+2} - x). \end{cases} \tag{9.14}$$

In the case of uniformly distributed bellsplines, we place additional points at the left side of the input interval as $T_0 = T_1 - h, T_{-1} = T_1 - 2h, T_{-2}T_1 - 3h$, and correspondingly on the right side of the interval as $T_{n+1} = T_n + h, T_{n+2} = T_n + 2h, T_{n+3} = T_n + 3h$. Then the function $f(x, c)$ in (9.4) will be the following linear combination of $n + 2$ functions $\phi_j(x)$ for indices $j = 0, 1, ..., n + 1$:

$$f(x, c) = c_1 \phi_0(x) + c_2 \phi_1(x) + ... + c_{n+2} \phi_{n+1}(x). \tag{9.15}$$

Figure 9.3 shows an approximation of the function $b = sin(\pi x/5) + x/5$ on the interval $x \in [-10, 10]$ using bellsplines. A MATLAB® program is available in Section 1.8 for running this test.[3]

9.2 Nonlinear Least Squares Problems

Suppose that for our data points $(x_i, y_i), i = 1, ..., m$, we want to find the vector of parameters $c = (c_1, ..., c_n)$ that will best fit the data $y_i, i = 1, ..., m$, of the model function $f(x_i, c), i = 1, ..., m$. We now consider the case that the model function $f : R^{n+1} \to R$ is nonlinear. Our goal is to find the minimum of the residual $r = y - f(x, c)$ in the least squares sense:

$$\min_c \sum_{i=1}^m (y_i - f(x_i, c))^2. \tag{9.16}$$

To solve problem (9.16) we can still use the linear least squares method if we can transform the nonlinear function $f(x, c)$ to a linear one. This can be done if the function $f(x, c)$ can be represented in the form $f(x, c) = Ae^{cx}, A = const$. Then taking the logarithm of $f(x, c)$, we get $\ln f = \ln A + cx$, which is already a linear function. Then the linear least squares problem after this transformation can be written as

$$\min_c \sum_{i=1}^m (\ln y_i - \ln f(x_i, c))^2. \tag{9.17}$$

---

[3]The MATLAB® programs can be found on the online version of Chapter 1 (doi:10.1007/978-3-319-57304-5_1).

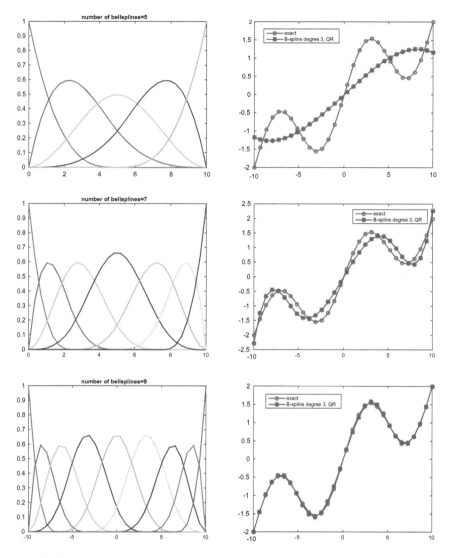

Fig. 9.3 Example 9.4. Polynomial fitting to the function $b = sin(\pi x/5) + x/5$ on the interval $x \in [-10, 10]$ with different numbers of bellsplines. *Blue stars* represent the computed function, and *red circles*, the exact function.

Another possibility for dealing with nonlinearity is to consider the least squares problem as an optimization problem. Let us define the residual $r : R^n \rightarrow R^m$ as

$$r_i(c) = y_i - f(x_i, c), \quad i = 1, ..., m. \tag{9.18}$$

Our goal is now to minimize the function

$$F(c) = \frac{1}{2} r(c)^T r(c) = \frac{1}{2} \|r(c)\|_2^2. \tag{9.19}$$

To find the minimum of (9.19), we should have

$$\nabla F(c) = \frac{\partial F(c)}{\partial c_i} = 0, \quad i = 1, ..., m. \tag{9.20}$$

Direct computations show that the gradient vector $\nabla F(c)$ is

$$\nabla F(c) = \frac{dF}{dc} = J^T(c)r(c), \tag{9.21}$$

where J^T is the transposed Jacobian matrix of the residual $r(c)$. For a sufficiently smooth function $F(c)$, we can write its Taylor expansion as

$$F(c) = F(c_0) + \nabla F(c_0)(c - c_0) + O(h^2), \tag{9.22}$$

with $|h| = \|c - c_0\|$. Since our goal is to find the minimum of $F(c)$, at a minimum point c^* we should have $\nabla F(c^*) = 0$. Taking the derivative with respect to c from (9.22), we obtain

$$H(F(c_0))(c - c_0) + \nabla F(c_0) = 0, \tag{9.23}$$

where H denotes the Hessian matrix of the function $F(c_0)$. Using (9.21) in (9.23), we obtain

$$H(F(c_0))(c - c_0) + J^T(c_0)r(c_0) = 0, \tag{9.24}$$

and from this expression we observe that we have obtained a system of linear equations

$$H(F(c_0))(c - c_0) = -J^T(c_0)r(c_0), \tag{9.25}$$

which can be solved again using the linear least squares method. The Hessian matrix $H(F(c_0))$ can be obtained from (9.21) as

$$H(F(c_0)) = J^T(c_0)J(c_0) + \sum_{i=1}^{m} r_i(c_0)H(r_i), \tag{9.26}$$

where $H(r_i)$ denotes the Hessian matrix of the residual function $r_i(c)$. These m matrices $H(r_i)$ are inconvenient to compute, but since they are multiplied by the small residuals $r_i(c_0)$, the second term in (9.26) is often very small in the solution c_0 and can be omitted. Then the system (9.25) is transformed to the linear system

$$J^T(c_0)J(c_0)(c - c_0) \approx -J^T(c_0)r(c_0), \tag{9.27}$$

which actually is a system of normal equations for the $m \times n$ linear least squares problem

$$J(c_0)(c - c_0) \approx -r(c_0). \qquad (9.28)$$

Using (9.27) we can write the iterative formula for the solution of the least squares problem

$$c^{k+1} = c^k - [J^T(c^k)J(c^k)]^{-1}J^T(c^k)r(c^k), \qquad (9.29)$$

where k is the number of the iteration, $c^0 = c_0$. Method which uses (9.29) in an iterative update is called the Gauss-Newton method.

An alternative to the Gauss–Newton method is the Levenberg[4]–Marquardt[5] method, which is used for ill-conditioned and rank-deficient problems. This method is similar to the problem of finding the minimum of the regularized function

$$F(c) = \frac{1}{2}r(c)^T r(c) + \frac{1}{2}\gamma(c - c_0)^T(c - c_0) = \frac{1}{2}\|r(c)\|_2^2 + \frac{1}{2}\gamma\|c - c_0\|_2^2, \quad (9.30)$$

where c_0 is a good initial guess for c and γ is a small regularization parameter. To employ the Levenberg–Marquardt method, we repeat all the steps that we performed for the Gauss–Newton method, see (9.21)–(9.26), but for the functional (9.30).

Finally, in the Levenberg–Marquardt method, the linear system that should be solved at every iteration k is

$$(J^T(c^k)J(c^k) + \gamma_k I)(c^{k+1} - c^k) \approx -J^T(c^k)r(c^k), \qquad (9.31)$$

and the corresponding linear least squares problem is

$$\begin{pmatrix} J(c^k) \\ \sqrt{\gamma_k}I \end{pmatrix} \cdot (c^{k+1} - c^k) \approx \begin{pmatrix} -r(c^k) \\ 0 \end{pmatrix}. \qquad (9.32)$$

In (9.31), (9.32), the number γ_k is an iteratively chosen regularization parameter, which can be computed as in [7] or using a trust region approach [13]. We refer to [31, 110] for various techniques for choosing the regularization parameter γ in (9.30), and to [74] for implementation and convergence analysis of this method.

Example 9.5 Let us consider the nonlinear model equation

$$Ae^{E/T - T_0} = y. \qquad (9.33)$$

Our goal is to determine parameters A, E, and T_0 in this equation from knowledge of y and T. We rewrite (9.33) as a nonlinear least squares problem in the form

---

[4]Kenneth Levenberg (1919–1973) was an American statistician.

[5]Donald W. Marquardt (1929–1997) was an American statistician.

$$\min_{A,E,T_0} \sum_{i=1}^{m}(y_i - Ae^{E/T_i-T_0})^2. \tag{9.34}$$

We will show how to obtain a linear problem from the nonlinear problem (9.34). We take the logarithm of (9.33) to get

$$\ln A + \frac{E}{T - T_0} = \ln y. \tag{9.35}$$

Now multiply both sides of (9.35) by $T - T_0$ to obtain

$$\ln A(T - T_0) + E = \ln y(T - T_0). \tag{9.36}$$

Rewrite the above equation as

$$T \ln A - T_0 \ln A + E + T_0 \ln y = T \ln y. \tag{9.37}$$

Let us now define the vector of parameters $c = (c_1, c_2, c_3)$ with $c_1 = T_0$, $c_2 = \ln A$, $c_3 = E - T_0 \ln A$. Now the problem (9.37) can be written as

$$c_1 \ln y + c_2 T + c_3 = T \ln y, \tag{9.38}$$

which is already a linear problem. We can rewrite (9.38), writing $f(c, y, T) = c_1 \ln y + c_2 T + c_3$ as a linear least squares problem in the form

$$\min_{c} \sum_{i=1}^{m}(T_i \ln y_i - f(c, y_i, T_i))^2. \tag{9.39}$$

The system of linear equations that we need to solve is

$$\begin{pmatrix} \ln y_1 & T_1 & 1 \\ \ln y_2 & T_2 & 1 \\ \vdots & \vdots & \vdots \\ \ln y_m & T_m & 1 \end{pmatrix} \cdot \begin{pmatrix} c_1 \\ c_2 \\ c_3 \end{pmatrix} = \begin{pmatrix} T_1 \ln y_1 \\ T_2 \ln y_2 \\ \vdots \\ T_m \ln y_m \end{pmatrix}. \tag{9.40}$$

Example 9.6 Suppose that a nonlinear model function is given as

$$f(x, c) = Ae^{c_1 x} + Be^{c_2 x}, A, B = \text{const} > 0, \tag{9.41}$$

and our goal is to fit this function using the Gauss–Newton method. In other words, we will use the iterative formula (9.28) for the iterative update of $c = (c_1, c_2)$. The residual function will be

$$r(c) = y - f(x, c), \tag{9.42}$$

where $y = y_i$, $i = 1, \ldots, m$, are data points. First, we compute the Jacobian matrix $J(c)$, where two columns in this matrix will be given by

$$
\begin{aligned}
J(c)_{i,1} &= \frac{\partial r_i}{\partial c_1} = -x_i A e^{c_1 x_i}, \, i = 1, \ldots, m, \\
J(c)_{i,2} &= \frac{\partial r_i}{\partial c_2} = -x_i B e^{c_2 x_i}, \, i = 1, \ldots, m.
\end{aligned}
\tag{9.43}
$$

If we make an initial guess for the parameters $c^0 = (c_1^0, c_2^0) = (1, 0)$, then we have to solve the following problem at iteration $k = 1$:

$$J(c^0)(c^1 - c^0) = -r(c^0). \tag{9.44}$$

The next update for the parameters $c^1 = (c_1^1, c_2^1)$ in the Gauss–Newton method can be computed as

$$c^1 = c^0 - [J^T(c^0)J(c^0)]^{-1} J^T(c^0)r(c^0). \tag{9.45}$$

Here $r(c^0)$ and $J(c^0)$ can be computed explicitly as follows:

$$r(c^0) = y_i - f(x_i, c^0) = y_i - (Ae^{1 \cdot x_i} + Be^{0 \cdot x_i}) = y_i - Ae^{x_i} - B, \, i = 1, \ldots, m, \tag{9.46}$$

where we note that $c^0 = (c_1^0, c_2^0) = (1, 0)$ and two columns in the Jacobian matrix $J(c^0)$ will be

$$
\begin{aligned}
J(c^0)_{i,1} &= -x_i A e^{1 \cdot x_i} = -x_i A e^{x_i}, \, i = 1, \ldots, m, \\
J(c^0)_{i,2} &= -x_i B e^{0 \cdot x_i} = -x_i B, \, i = 1, \ldots, m.
\end{aligned}
\tag{9.47}
$$

Substituting (9.46), (9.47) into (9.44) yields the following linear system of equations:

$$
\begin{pmatrix}
-x_1 A e^{x_1} & -x_1 B \\
-x_2 A e^{x_2} & -x_2 B \\
\vdots & \vdots \\
-x_m A e^{x_m} & -x_m B
\end{pmatrix}
\cdot
\begin{pmatrix}
c_1^1 - c_1^0 \\
c_2^1 - c_2^0
\end{pmatrix}
=
\begin{pmatrix}
-y_1 - A e^{x_1} - B \\
-y_2 - A e^{x_2} - B \\
\vdots \\
-y_m - A e^{x_m} - B
\end{pmatrix},
\tag{9.48}
$$

which is solved for $c^1 - c^0$ using the method of normal equations as

$$\begin{pmatrix} -x_1 Ae^{x_1} & -x_1 B \\ -x_2 Ae^{x_2} & -x_2 B \\ \vdots & \vdots \\ -x_m Ae^{x_m} & -x_m B \end{pmatrix}^T \begin{pmatrix} -x_1 Ae^{x_1} & -x_1 B \\ -x_2 Ae^{x_2} & -x_2 B \\ \vdots & \vdots \\ -x_m Ae^{x_m} & -x_m B \end{pmatrix} \cdot \begin{pmatrix} c_1^1 - c_1^0 \\ c_2^1 - c_2^0 \end{pmatrix}$$

$$= \begin{pmatrix} -x_1 Ae^{x_1} & -x_1 B \\ -x_2 Ae^{x_2} & -x_2 B \\ \vdots & \vdots \\ -x_m Ae^{x_m} & -x_m B \end{pmatrix}^T \begin{pmatrix} -y_1 - Ae^{x_1} - B \\ -y_2 - Ae^{x_2} - B \\ \vdots \\ -y_m - Ae^{x_m} - B \end{pmatrix}. \tag{9.49}$$

This system can be solved for $c^1 - c^0$, and the next values for c^1 are obtained using (9.45) as

$$\begin{pmatrix} c_1^1 \\ c_2^1 \end{pmatrix} = \begin{pmatrix} c_1^0 \\ c_2^0 \end{pmatrix} + \left(\begin{pmatrix} -x_1 Ae^{x_1} & -x_1 B \\ -x_2 Ae^{x_2} & -x_2 B \\ \vdots & \vdots \\ -x_m Ae^{x_m} & -x_m B \end{pmatrix}^T \begin{pmatrix} -x_1 Ae^{x_1} & -x_1 B \\ -x_2 Ae^{x_2} & -x_2 B \\ \vdots & \vdots \\ -x_m Ae^{x_m} & -x_m B \end{pmatrix} \right)^{-1} \cdot$$

$$\begin{pmatrix} -x_1 Ae^{x_1} & -x_1 B \\ -x_2 Ae^{x_2} & -x_2 B \\ \vdots & \vdots \\ -x_m Ae^{x_m} & -x_m B \end{pmatrix}^T \begin{pmatrix} -y_1 - Ae^{x_1} - B \\ -y_2 - Ae^{x_2} - B \\ \vdots \\ -y_m - Ae^{x_m} - B \end{pmatrix}. \tag{9.50}$$

9.3 Method of Normal Equations

The first method that we will consider for the solution of (9.2) will be the method of normal equations. This method is the fastest but is not as accurate as the QR or SVD decompositions. We can apply this method for the solution of linear least squares problems if the condition number of the matrix A is small.

Our goal is to minimize the function $F(x) = \|r(x)\|_2^2 = \|Ax - b\|_2^2$. To find minimum of this function and derive the *normal equations*, we look for the values of x for which the gradient of

$$F(x) = \|r(x)\|_2^2 = \|Ax - b\|_2^2 = (Ax - b)^T (Ax - b) \tag{9.51}$$

vanishes, or where $\nabla F(x) = 0$. Considering the standard definition of the Fréchet derivative we get

$$
\begin{aligned}
0 &= \lim_{h \to 0} \frac{(r(x+h))^T r(x+h) - (r(x))^T r(x)}{\|h\|_2} \\
&= \lim_{h \to 0} \frac{(A(x+h) - b)^T (A(x+h) - b) - (Ax - b)^T (Ax - b)}{\|h\|_2} \\
&= \lim_{h \to 0} \frac{2h^T (A^T Ax - A^T b) + h^T A^T A h}{\|h\|_2}.
\end{aligned}
\tag{9.52}
$$

We observe that the second term $\frac{|h^T A^T A h|}{\|h\|_2} \leq \frac{\|A\|_2^2 \|h\|_2^2}{\|h\|_2} = \|A\|_2^2 \|h\|_2$ goes to zero as $h \to 0$. This means that the term $A^T Ax - A^T b$ must also be zero, and thus we must have

$$
0 = \nabla F = 2A^T Ax - 2A^T b.
\tag{9.53}
$$

We conclude that

$$
A^T Ax = A^T b,
\tag{9.54}
$$

which is a symmetric linear system of $n \times n$ equations, commonly called a system of normal equations.

Using (9.53), we can compute the Hessian matrix $H = 2A^T A$. If the Hessian matrix $H = 2A^T A$ is positive definite, then x is indeed a minimum. This is a sufficient condition for x to be a minimum of (9.51). We can show that the matrix $A^T A$ is positive definite if and only if the columns of A are linearly independent, or when $r(A) = n$.

If the matrix A has a full rank ($r(A) = n$), the system (9.54) is of size $n \times n$ and is a symmetric positive definite system of normal equations. It has the same solution x as the least squares problem $\min_x \|Ax - b\|_2^2$ of size $m \times n$. To solve system (9.54) one can use the Cholesky decomposition

$$
A^T A = LL^T
\tag{9.55}
$$

with L a lower triangular matrix. Then the solution of (9.54) will be given by the solution of the triangular system

$$
\begin{aligned}
Ly &= A^T b, \\
L^T x &= y.
\end{aligned}
\tag{9.56}
$$

However, in practice, the method of normal equations can be inaccurate for two reasons.

- The condition number of $A^T A$ is the square of the condition number of the original matrix A:

$$
cond(A^T A) = cond(A)^2.
\tag{9.57}
$$

Thus, the method of normal equations can give a squared condition number even when the fit to data is good and the residual is small. This makes the computed solution more sensitive. In this sense, the method of normal equations is unstable.

- Information can be lost during computation of the product $A^T A$. For example, take

$$A = \begin{pmatrix} 1 & 1 \\ \delta & 0 \\ 0 & \delta \end{pmatrix} \qquad (9.58)$$

with $0 < \delta < \sqrt{\varepsilon}$ in a given floating-point system. In floating-point arithmetic we can compute $A^T A$:

$$A^T A = \begin{pmatrix} 1 & \delta & 0 \\ 1 & 0 & \delta \end{pmatrix} \cdot \begin{pmatrix} 1 & 1 \\ \delta & 0 \\ 0 & \delta \end{pmatrix} = \begin{pmatrix} 1+\delta^2 & 1 \\ 1 & 1+\delta^2 \end{pmatrix} = \begin{pmatrix} 1 & 1 \\ 1 & 1 \end{pmatrix}, \qquad (9.59)$$

which is a singular matrix in the working precision.

These inconveniences do not make the method of normal equations useless, but they provide motivation to seek more robust methods for linear least squares problems.

9.4   QR Decomposition

In this section we consider *QR decomposition* of a matrix A. QR decomposition of a matrix A can be computed, for example, using the Gram–Schmidt orthogonalization process [47]; see Section 9.5.3. *QR decomposition* of a matrix A means that a matrix A of size $m \times n$ with $m \geq n$ can be factorized as the product of a unitary matrix Q of size $m \times m$ and an upper triangular matrix R of size $m \times n$:

$$A = QR = Q \begin{pmatrix} R_1 \\ 0 \end{pmatrix} = (Q_1, Q_2) \begin{pmatrix} R_1 \\ 0 \end{pmatrix} = Q_1 R_1, \qquad (9.60)$$

where R_1 is an upper triangular matrix of size $n \times n$, 0 is the zero matrix of size $(m - n) \times n$, Q_1 is a matrix of size $m \times n$ with orthogonal columns, Q_2 is of size $m \times (m - n)$ with orthogonal columns. We note that $(m - n)$ rows of the upper triangular matrix R consist of zeros.

We can consider also an alternative definition of QR decomposition:

Theorem 9.1   *QR decomposition.* Let A be an $m \times n$ matrix with $m \geq n$. Suppose that A has full column rank. Then there exist a unique $m \times n$ orthogonal matrix $Q (Q^T Q = I_n)$ and a unique $n \times n$ upper triangular matrix R with positive diagonals $r_{ii} > 0$ such that $A = QR$.

Proof   The proof follows from the Gram–Schmidt orthogonalization process [47]; see Section 9.5.3. Another proof follows from the QR factorization using Householder reflection.   □

We will now show how to obtain a formula for the value of x that minimizes $\|Ax - b\|_2$ using three different decompositions the matrix A into matrices Q and R.

- In the first method, we choose $m - n$ more orthonormal vectors $\tilde{Q}$ such that $(Q, \tilde{Q})$ is a square orthogonal matrix, and therefore, $\tilde{Q}^T Q = 0$. One way in which to do this is to choose any $m - n$ independent vectors $\tilde{X}$ and then apply the Gram–Schmidt orthogonalization Algorithm 9.4 to the $n \times n$ nonsingular matrix $(Q, \tilde{X})$). Using the property

$$\|QAZ\|_2 = \|A\|_2 \tag{9.61}$$

of norms and matrices for any orthogonal or unitary matrices Q, Z and applying it to $\|Ax - b\|_2^2$, we can write

$$
\begin{aligned}
\|r(x)\|_2^2 = \|Ax - b\|_2^2 &= \|(Q, \tilde{Q})^T (Ax - b)\|_2^2 \\
&= \left\| \begin{pmatrix} Q^T \\ \tilde{Q}^T \end{pmatrix} (QRx - b) \right\|_2^2 \\
&= \left\| \begin{pmatrix} I^{n \times n} \\ O^{(m-n) \times n} \end{pmatrix} Rx - \begin{pmatrix} Q^T b \\ \tilde{Q}^T b \end{pmatrix} \right\|_2^2 \\
&= \left\| \begin{pmatrix} Rx - Q^T b \\ -\tilde{Q}^T b \end{pmatrix} \right\|_2^2 \\
&= \left\| Rx - Q^T b \right\|_2^2 + \|\tilde{Q}^T b\|_2^2 \\
&\geq \|\tilde{Q}^T b\|_2^2.
\end{aligned}
$$

We can solve the triangular linear system $Rx - Q^T b = 0$ for x, since A and R have the same rank, n, and so R is nonsingular and $Q^T b$ is a vector of length n. Then $x = R^{-1} Q^T b$, and the minimum residual norm of $\|Ax - b\|_2$ is given by $\|\tilde{Q}^T b\|_2$.

- The second method is slightly different and does not use the matrix $\tilde{Q}$. This method uses adding and subtracting the same term QQ^T to the expression for the residual $r(x) = Ax - b$:

$$
\begin{aligned}
r(x) = Ax - b = QRx - b = QRx - (QQ^T + I - QQ^T)b \\
= Q(Rx - Q^T b) - (I - QQ^T)b.
\end{aligned}
$$

Note that the vectors $Q(Rx - Q^T b)$ and $(I - QQ^T)b$ are orthogonal, because

$$
\begin{aligned}
(Q(Rx - Q^T b))^T ((I - QQ^T)b) &= (Rx - Q^T b)^T (Q^T (I - QQ^T))b \\
&= (Rx - Q^T b)^T (0)b = 0.
\end{aligned} \tag{9.62}
$$

Thus, we can use the Pythagorean theorem,

$$
\begin{aligned}
\|Ax - b\|_2^2 &= \|Q(Rx - Q^T b)\|_2^2 + \|(I - QQ^T)b\|_2^2 \\
&= \|Rx - Q^T b\|_2^2 + \|(I - QQ^T)b\|_2^2,
\end{aligned}
$$

where we have used the property of the norm $\|Qy\|_2^2 = \|y\|_2^2$. This sum of squares is minimized when the first term is zero, i.e., $x = R^{-1} Q^T b$.

- A third derivation uses the normal equations solution and then QR decomposition inside this solution:

$$\begin{aligned} x &= (A^T A)^{-1} A^T b \\ &= (R^T Q^T Q R)^{-1} R^T Q^T b = (R^T R)^{-1} R^T Q^T b \\ &= R^{-1} R^{-T} R^T Q^T b = R^{-1} Q^T b. \end{aligned}$$

9.5 Orthogonalization Methods

In this section we will present the main orthogonalization methods for computing the QR factorization of a matrix A, which include:

- Householder transformation (called also reflection),
- Givens transformation (called rotation),
- Gram–Schmidt orthogonalization.

9.5.1 Householder Transformations

A Householder transformation, which is also called reflection, is a matrix of the form

$$P = I - 2uu^T,$$

where $\|u\|_2 = 1$. We can see that $P = P^T$ and

$$P \cdot P^T = (I - 2uu^T)(I - 2uu^T) = I - 4uu^T + 4uu^T uu^T = I.$$

Using the above equations, we conclude that P is a symmetric orthogonal matrix. This matrix is called a reflection, because Px is the reflection of the vector x in the plane that passes through 0 and is perpendicular to the vector u.

For a given vector x, we can find a Householder reflection $P = I - 2uu^T$ that will zero out all except the first component of x and leave first entry of x as the nonzero value

$$Px = (c, 0, \ldots, 0)^T = c \cdot e_1$$

with $e_1 = (1, 0, \ldots, 0)$. We do this using the following procedure. First we apply P to x to get

$$Px = (I - 2uu^T)x = x - 2u(u^T x) = c \cdot e_1.$$

From the equation above, we get

$$u = \frac{1}{2(u^T x)}(x - c e_1), \tag{9.63}$$

i.e., u is a linear combination of x and e_1.

Since P is an orthogonal matrix, we can use the following property of the 2-norm: $\|x\|_2 = \|Px\|_2 = |c|$. Then the vector u in (9.63) must be parallel to the vector

$$\tilde{u} = x \pm \|x\|_2 e_1, \tag{9.64}$$

and thus the vector u can be computed as

$$u = \tilde{u}/\|\tilde{u}\|_2.$$

One can verify that as long as $\tilde{u} \neq 0$, the choice of sign in (9.64) yields a vector u satisfying $Px = c e_1$. We will determine the vector $\tilde{u}$ as

$$\tilde{u} = x + sign(x_1)\|x\|_2 e_1$$

and this means that there is no cancellation in computing the first component of u. Here, x_1 is the first coordinate in the vector x, after which all other entries of x in the matrix A are 0. Finally, the vector $\tilde{u}$ will have the following form:

$$\tilde{u} = \begin{pmatrix} x_1 + sign(x_1) \cdot \|x\|_2 \\ x_2 \\ \vdots \\ x_n \end{pmatrix} \quad where \ u = \frac{\tilde{u}}{\|\tilde{u}\|_2}.$$

We denote the procedure for obtaining the vector u by $u = House(x)$ and use it in the Algorithm 9.2. In computations, it is more efficient to store $\tilde{u}$ instead of u to save the work of computing u, and use the formula $P = I - (2/\|\tilde{u}\|_2^2)\tilde{u}\tilde{u}^T$ instead of $P = I - 2uu^T$.

Example 9.7 In this example, we present a general procedure for carrying out the QR decomposition of a matrix A of size 5×4 using Householder transformations. In all the matrices below, P_i denotes an orthogonal matrix, x denotes a generic nonzero entry, and 0 denotes a zero entry. Thus, for the decomposition $A = QR$, we need to perform the following steps:

- Choose the matrix P_1 such that

$$A_1 \equiv P_1 A = \begin{pmatrix} x & x & x & x \\ 0 & x & x & x \\ 0 & x & x & x \\ 0 & x & x & x \\ 0 & x & x & x \end{pmatrix}.$$

• Choose the matrix P_2 such that

$$P_2 = \begin{pmatrix} 1 & 0 & 0 & 0 \\ 0 & & & \\ 0 & & P_2' & \\ 0 & & & \\ 0 & & & \end{pmatrix}$$

and

$$A_2 \equiv P_2 A_1 = \begin{pmatrix} x & x & x & x \\ 0 & x & x & x \\ 0 & 0 & x & x \\ 0 & 0 & x & x \\ 0 & 0 & x & x \end{pmatrix}.$$

• Choose

$$P_3 = \begin{pmatrix} 1 & & & \\ & 1 & & 0 \\ \hline & 0 & & P_3' \end{pmatrix}$$

such that

$$A_3 \equiv P_3 A_2 = \begin{pmatrix} x & x & x & x \\ 0 & x & x & x \\ 0 & 0 & x & x \\ 0 & 0 & 0 & x \\ 0 & 0 & 0 & x \end{pmatrix}.$$

• Choose

$$P_4 = \begin{pmatrix} 1 & & & \\ & 1 & & 0 \\ & & 1 & \\ \hline & 0 & & P_4' \end{pmatrix}$$

such that

$$\tilde{R} := A_4 \equiv P_4 A_3 = \begin{pmatrix} x & x & x & x \\ 0 & x & x & x \\ 0 & 0 & x & x \\ 0 & 0 & 0 & x \\ 0 & 0 & 0 & 0 \end{pmatrix}.$$

In this example, we have chosen a Householder matrix P_i', $i = 2, 3, 4$, to zero out the subdiagonal entries in column i. We note that this does not disturb the zeros introduced in previous columns.

We observe that we have obtained the decomposition

$$A_4 = P_4 P_3 P_2 P_1 A. \tag{9.65}$$

Let us denote the final triangular matrix A_4 by $\tilde{R} \equiv A_4$. Then using (9.65), we observe that the matrix A is obtained via the decomposition

$$A = P_1^T P_2^T P_3^T P_4^T \tilde{R} = QR, \tag{9.66}$$

which is our desired QR decomposition. Here, the matrix Q comprises the first four columns of $P_1^T P_2^T P_3^T P_4^T = P_1 P_2 P_3 P_4$ (since all P_i are symmetric), and R comprises the first four rows of $\tilde{R}$.

QR factorization for a matrix A of order $m \times n$ is summarized in Algorithm 9.1. In this algorithm, a_k denotes the kth column of the matrix A. For simplicity, a rescaling procedure is omitted in this algorithm.

Algorithm 9.1   QR factorization using Householder reflections.

> for $k = 1$ to $\min(m - 1, n)$ /* loop over all columns */
> $\alpha_k = -sign(a_{kk})\sqrt{a_{kk}^2 + ... + a_{mk}^2}$
> $u_k = (0, ...0 \, a_{kk}....a_{mk})^T - \alpha_k e_k$
> $\beta_k = u_k^T u_k$
> if $\beta_k = 0$ then /* skip column k since it is already 0 */
> go to the next k
> for $j = k$ to n
> $\gamma_j = u_k^T a_j$ /* a_j are elements of column j of A */
> $a_j = a_j - \frac{2\gamma_j}{\beta_k} u_k$
> end
> end

Below we present an algorithm for obtaining the QR decomposition using Householder transformations in a more general form.

Algorithm 9.2   QR factorization using Householder reflections.

> for $i = 1$ to $\min(m - 1, n)$
> $\quad u_i = House(A(i : m, i))$
> $\quad P_i' = I - 2u_i u_i^T$
> $\quad A(i : m, i : n) = P_i' A(i : m, i : n)$
> end for

We can discuss some implementation issues of this algorithm. We note that we never form the matrix P_i explicitly but instead use the efficient multiplication

$$(I - 2u_i u_i^T)A(i : m, i : n) = A(i : m, i : n) - 2u_i(u_i^T A(i : m, i : n)).$$

To store P_i, we need only u_i, or $\tilde{u}_i$ and $\|\tilde{u}_i\|$. These values can be stored in column i of A, which means that the QR decomposition can be "overwritten" on A, where Q is stored in factored form $P_1, \ldots, P_{n-1}$, and P_i is stored as $\tilde{u}_i$ below the diagonal in column i of A.

Householder reflections can be applied for the solution of the following least squares problem:

$$\text{Find } x \text{ subject to } \min_x \|Ax - b\|_2^2. \tag{9.67}$$

To solve (9.67) using QR decomposition of the matrix A, we need to compute the vector $Q^T b$; see details in Section 9.4. We can do this computation in a following way: compute $Q^T b$ using Householder matrices $P_i, i = 1, \ldots, n$, as $Q^T b = P_n P_{n-1} \cdots P_1 b$, so we need only keep multiplying b by $P_1, P_2, \ldots, P_n$. We summarize this discussion in the following algorithm:

Algorithm 9.3 Computation of $Q^T b = P_n P_{n-1} \cdots P_1 b$

> for $i = 1$ to n
> $\quad \gamma = -2 \cdot u_i^T b(i : m)$
> $\quad b(i : m) = b(i : m) + \gamma u_i$
> end for

In MATLAB®, the command $A \setminus b$ solves the least squares problem if the matrix A is of size $m \times n$ with $m > n$. It is also possible to use the command $(Q, R) = qr(A)$ in MATLAB® to perform a QR decomposition of the matrix A.

Let us explain now in greater detail how to perform the Householder transformation $u = House(x)$ in Algorithm 9.2. First we introduce some notions:

- Let x be an arbitrary real m-dimensional column vector of A such that $\|x\| = |\alpha|$ for a scalar α.
- If the Householder algorithm is implemented using floating-point arithmetic, then α should get the opposite sign to that of the kth coordinate of x, where x_k is the pivot element after which all entries in the final upper triangular form of the matrix A are 0.

Then to compute the Householder matrix P, set

$$\begin{aligned} v &= x + \alpha e_1, \\ \alpha &= -sign(x_1)\|x\|, \\ u &= \frac{v}{\|v\|}, \\ P &= I - 2uu^T, \end{aligned} \tag{9.68}$$

where e_1 is the vector $(1, 0, \ldots, 0)^T$, $\| \cdot \|$ is the Euclidean norm, I is the $m \times m$ identity matrix, and x_1 is the first component of the vector x. The matrix P thus obtained is an $m \times m$ Householder matrix such that

$$Px = (\alpha, 0, \cdots, 0)^T .$$

To transform an $m \times n$ matrix A gradually to upper triangular form, we first left multiply A by the first Householder matrix P_1. This results in a matrix $P_1 A$ with zeros in the left column (except for the first row):

$$P_1 A = \begin{pmatrix} \alpha_1 & \star & \dots & \star \\ 0 & & & \\ \vdots & & A' & \\ 0 & & & \end{pmatrix} .$$

This can be repeated for the matrix A' that is obtained from $P_1 A$ by deleting the first row and first column, resulting in a Householder matrix P_2'. Note that P_2' is smaller than P_1. Since we want P_2' to operate on $P_1 A$ instead of A', we need to expand it on the upper left, or in general,

$$P_k = \begin{pmatrix} I_{k-1} & 0 \\ 0 & P_k' \end{pmatrix} .$$

After k iterations of this process, $k = \min(m - 1, n)$, we obtain an upper triangular matrix

$$R = P_k \cdots P_2 P_1 A .$$

Now choosing

$$Q = P_1^T P_2^T \cdots P_k^T,$$

we obtain the QR decomposition of A.

Example 9.8 Let us calculate the QR decomposition of the matrix

$$A = \begin{pmatrix} 12 & -51 & 4 \\ 6 & 167 & -68 \\ -4 & 24 & -41 \end{pmatrix}$$

using Householder reflection. First, we need to find a reflection that transforms the first column of the matrix A, the vector $x = a_1 = (12, 6, -4)^T$, to $\|x\| e_1 = \|a_1\| e_1 = (14, 0, 0)^T$.

Now using (9.68), we construct the vector

$$v = x + \alpha e_1,$$

where

$$\alpha = -sign(x_1) \|x\|$$

and

$$u = \frac{v}{\|v\|} .$$

We observe that in our example, $\|x\| = \|x\|_2 = \sqrt{12^2 + 6^2 + (-4)^2} = 14$,

$\alpha = -sign(12)\|x\| = -14$, and the vector αe_1 will be $\alpha e_1 = (14, 0, 0)^T$.

Therefore,

$$v = x + \alpha e_1 = (12, 6, -4)^T + (-14, 0, 0)^T = (-2, 6, -4)^T = 2(-1, 3, -2)^T,$$

and thus $u = \frac{v}{\|v\|} = \frac{1}{\sqrt{14}}(-1, 3, -2)^T$. Then the first Householder matrix will be

$$P_1 = I - \frac{2}{\sqrt{14}\sqrt{14}} \begin{pmatrix} -1 \\ 3 \\ -2 \end{pmatrix} (-1\ 3\ -2)$$

$$= I - \frac{1}{7} \begin{pmatrix} 1 & -3 & 2 \\ -3 & 9 & -6 \\ 2 & -6 & 4 \end{pmatrix}$$

$$= \begin{pmatrix} 6/7 & 3/7 & -2/7 \\ 3/7 & -2/7 & 6/7 \\ -2/7 & 6/7 & 3/7 \end{pmatrix}.$$

We computer the product $P_1 A$ to get the matrix

$$A_1 = P_1 A = \begin{pmatrix} 14 & 21 & -14 \\ 0 & -49 & -14 \\ 0 & 168 & -77 \end{pmatrix}, \tag{9.69}$$

which is almost a triangular matrix. We have only to zero the $(3, 2)$ entry.

Take the $(1, 1)$ minor of (9.69) and then apply the same process again to the matrix

$$A' = M_{11} = \begin{pmatrix} -49 & -14 \\ 168 & -77 \end{pmatrix}.$$

By the same method as above, we first need to find a reflection that transforms the first column of the matrix A', namely the vector $x = (-49, 168)^T$, to $\|x\|\ e_1 = (175, 0)^T$. Here $\|x\| = \sqrt{(-49)^2 + 168^2} = 175$,

$$\alpha = -sign(-49)\|x\| = 175 \text{ and } \alpha e_1 = (175, 0)^T.$$

Therefore,

$$v = x + \alpha e_1 = (-49, 168)^T + (175, 0)^T = (126, 168)^T,$$

$$\|v\| = \sqrt{126^2 + 168^2} = \sqrt{44100} = 210,$$

$$u = \frac{v}{\|v\|} = (126/210, 168/210)^T = (3/5, 4/5)^T.$$

Then

$$P_2' = I - 2 \begin{pmatrix} 3/5 \\ 4/5 \end{pmatrix} (3/5 \ 4/5)$$

or

$$P_2' = I - 2 \begin{pmatrix} 9/25 & 12/25 \\ 12/25 & 16/25 \end{pmatrix} = \begin{pmatrix} 7/25 & -24/25 \\ -24/25 & -7/25 \end{pmatrix}.$$

Finally, we obtain the matrix of the Householder transformation P_2 such that

$$P_2 = \begin{pmatrix} 1 & 0 \\ 0 & P_2' \end{pmatrix}$$

with P_2' given above and thus,

$$P_2 = \begin{pmatrix} 1 & 0 & 0 \\ 0 & 7/25 & -24/25 \\ 0 & -24/25 & -7/25 \end{pmatrix}.$$

Now we obtain

$$Q = P = P_1^T P_2^T = \begin{pmatrix} 6/7 & 69/175 & -58/175 \\ 3/7 & -158/175 & 6/175 \\ -2/7 & -6/35 & -33/35 \end{pmatrix}.$$

Thus, we have performed the QR decomposition of the matrix A with matrices Q and R given by

$$Q = P_1^T P_2^T = \begin{pmatrix} 0.8571 & 0.3943 & -0.3314 \\ 0.4286 & -0.9029 & 0.0343 \\ -0.2857 & -0.1714 & -0.9429 \end{pmatrix},$$

$$R = P_2 A_1 = P_2 P_1 A = Q^T A = \begin{pmatrix} 14 & 21 & -14 \\ 0 & -175 & 70 \\ 0 & 0 & 35 \end{pmatrix}.$$

We observe that the matrix Q is orthogonal and R is upper triangular, so $A = QR$ is the required QR decomposition. To obtain the matrices Q and R above, we have used the facts that

$$P_2 A_1 = P_2 P_1 A = R,$$
$$P_1^T P_2^T P_2 P_1 A = P_1^T P_2^T R,$$
$$A = P_1^T P_2^T R = QR,$$
$$\text{with } Q = P_1^T P_2^T.$$

We can also perform tridiagonalization of the matrix A using Householder reflection matrices. We follow [12] in the description of this procedure. In the first step of the tridiagonalization procedure, to form a Householder matrix at every step we need to determine constants α and r, which are given by the formulas

$$\alpha = -\text{sgn}(a_{21})\sqrt{\sum_{j=2}^{n} a_{j1}^2},$$

$$r = \sqrt{\frac{1}{2}(\alpha^2 - a_{21}\alpha)}.$$

(9.70)

Knowing α and r, we can construct the vector v such that

$$v^{(1)} = \begin{pmatrix} v_1 \\ v_2 \\ \dots \\ v_n \end{pmatrix},$$

(9.71)

where $v_1 = 0$, $v_2 = \frac{a_{21}-\alpha}{2r}$ and

$$v_k = \frac{a_{k1}}{2r} \quad \text{for each} \quad k = 3, 4, \dots, n.$$

Then we can compute the first Householder reflection matrix as

$$P^{(1)} = I - 2v^{(1)}(v^{(1)})^T$$

and obtain the matrix $A^{(1)}$ as

$$A^{(1)} = P^{(1)}AP^{(1)}.$$

Using $P^{(1)}$ and $A^{(1)}$ thus obtained, the process of tridiagonalization is repeated for $k = 2, 3, \dots, n$ as follows:

$$\alpha = -\text{sgn}(a_{k+1,k})\sqrt{\sum_{j=k+1}^{n} a_{jk}^2},$$

$$r = \sqrt{\frac{1}{2}(\alpha^2 - a_{k+1,k}\alpha)},$$

$$v_1^{(k)} = v_2^{(k)} = \dots = v_k^{(k)} = 0,$$

$$v_{k+1}^{(k)} = \frac{a_{k+1,k} - \alpha}{2r},$$

$$v_j^{(k)} = \frac{a_{jk}}{2r} \quad \text{for} \quad j = k+2; k+3, \dots, n,$$

$$P^{(k)} = I - 2v^{(k)}(v^{(k)})^T,$$

$$A^{(k+1)} = P^{(k)}A^{(k)}P^{(k)}.$$

(9.72)

In (9.72), the elements $a_{k+1,k}, a_{jk}$ are entries of the matrix $A^{(k)}$.

Example 9.9 In this example, the given matrix

$$A = \begin{pmatrix} 5 & 1 & 0 \\ 1 & 6 & 3 \\ 0 & 3 & 7 \end{pmatrix}$$

is transformed into a similar tridiagonal matrix A_1 using Householder's method.
We perform tridiagonalization by the following steps:

- First compute α via (9.70) as

$$\alpha = -\text{sgn}(a_{21}) \sqrt{\sum_{j=2}^{n} a_{j1}^2} = -\sqrt{(a_{21}^2 + a_{31}^2)} = -\sqrt{(1^2 + 0^2)} = -1.$$

- Using α, we obtain r via (9.70) as

$$r = \sqrt{\frac{1}{2}(\alpha^2 - a_{21}\alpha)} = \sqrt{\frac{1}{2}((-1)^2 - 1 \cdot (-1))} = 1.$$

- Using the known values of α and r, we construct the vector $v^{(1)}$ as in (9.71). Using (9.71), we compute

$$v_1 = 0,$$
$$v_2 = \frac{a_{21} - \alpha}{2r} = \frac{1 - (-1)}{2 \cdot 1} = 1,$$
$$v_3 = \frac{a_{31}}{2r} = 0.$$

Now we have obtained the vector

$$v^{(1)} = \begin{pmatrix} 0 \\ 1 \\ 0 \end{pmatrix}.$$

We compute the first Householder matrix P^1 as

$$P^{(1)} = I - 2v^{(1)}(v^{(1)})^T$$

and get

$$P^{(1)} = \begin{pmatrix} 1 & 0 & 0 \\ 0 & -1 & 0 \\ 0 & 0 & 1 \end{pmatrix}.$$

The tridiagonal matrix $A^{(1)}$ is obtained as

$$A^{(1)} = P^{(1)} A P^{(1)} = \begin{pmatrix} 5 & -1 & 0 \\ -1 & 6 & -3 \\ 0 & -3 & 7 \end{pmatrix}.$$

Example 9.10 In this example, the given 4×4 matrix

$$A = \begin{pmatrix} 4 & 1 & -2 & 2 \\ 1 & 2 & 0 & 1 \\ -2 & 0 & 3 & -2 \\ 2 & 1 & -2 & -1 \end{pmatrix}$$

is transformed into a similar tridiagonal matrix A_2 using Householder reflections. Similarly with the example above we perform the following steps:

• First compute α via (9.70) as

$$\alpha = -\text{sgn}(a_{21}) \sqrt{\sum_{j=2}^{n} a_{j1}^2} = (-1) \cdot \sqrt{(a_{21}^2 + a_{31}^2 + a_{41}^2)}$$
$$= -1 \cdot (1^2 + (-2)^2 + 2^2) = (-1) \cdot \sqrt{1 + 4 + 4} = -\sqrt{9} = -3.$$

• Using α, we obtain r as

$$r = \sqrt{\frac{1}{2}(\alpha^2 - a_{21}\alpha)} = \sqrt{\frac{1}{2}((-3)^2 - 1 \cdot (-3))} = \sqrt{6}.$$

• From α and r, we construct the vector $v^{(1)}$. Using (9.71), we compute

$$v_1 = 0,$$
$$v_2 = \frac{a_{21} - \alpha}{2r} = \frac{1 - (-3)}{2 \cdot \sqrt{6}} = \frac{2}{\sqrt{6}},$$
$$v_3 = \frac{a_{31}}{2r} = \frac{-2}{2 \cdot \sqrt{6}} = \frac{-1}{\sqrt{6}},$$
$$v_4 = \frac{a_{41}}{2r} = \frac{2}{2 \cdot \sqrt{6}} = \frac{1}{\sqrt{6}}.$$

Thus, we have obtained

$$v^{(1)} = \begin{pmatrix} 0 \\ \frac{2}{\sqrt{6}} \\ \frac{-1}{\sqrt{6}} \\ \frac{1}{\sqrt{6}} \end{pmatrix}.$$

Now we can compute the first Householder matrix $P^{(1)}$,

$$P^{(1)} = I - 2v^{(1)}(v^{(1)})^T = I - 2 \cdot \begin{pmatrix} 0 \\ \frac{2}{\sqrt{6}} \\ \frac{-1}{\sqrt{6}} \\ \frac{1}{\sqrt{6}} \end{pmatrix} \cdot \begin{pmatrix} 0 & \frac{2}{\sqrt{6}} & \frac{-1}{\sqrt{6}} & \frac{1}{\sqrt{6}} \end{pmatrix},$$

and obtain

$$P^{(1)} = \begin{pmatrix} 1 & 0 & 0 & 0 \\ 0 & -1/3 & 2/3 & -2/3 \\ 0 & 2/3 & 2/3 & 1/3 \\ 0 & -2/3 & 1/3 & 2/3 \end{pmatrix}.$$

After that, we compute the matrix $A^{(1)}$ as

$$A^{(1)} = P^{(1)} A P^{(1)}$$

to get

$$A^{(1)} = P^{(1)} A P^{(1)} = \begin{pmatrix} 4 & -3 & 0 & 0 \\ -3 & 10/3 & 1 & 4/3 \\ 0 & 1 & 5/3 & -4/3 \\ 0 & 4/3 & -4/3 & -1 \end{pmatrix}.$$

Next, having found $A^{(1)}$, we need to construct $A^{(2)}$ and $P^{(2)}$. Using formulas (9.72) for $k = 2$, we get

$$\alpha = -\text{sgn}(a_{3,2}) \sqrt{\sum_{j=3}^{4} a_{j,2}^2} = -\text{sgn}(1)\sqrt{a_{3,2}^2 + a_{4,2}^2} = -\sqrt{1 + \frac{16}{9}} = -\frac{5}{3};$$

$$r = \sqrt{\frac{1}{2}(\alpha^2 - a_{3,2} \cdot \alpha)} = \sqrt{\frac{20}{9}};$$

$$v_1^{(2)} = v_2^{(2)} = 0,$$

$$v_3^{(2)} = \frac{a_{3,2} - \alpha}{2r} = \frac{2}{\sqrt{5}},$$

$$v_4^{(2)} = \frac{a_{4,2}}{2r} = \frac{1}{\sqrt{5}},$$

and thus new vector v will be $v^{(2)} = (0, 0, \frac{2}{\sqrt{5}}, \frac{1}{\sqrt{5}})^T$, and the new Householder matrix $P^{(2)}$ will be

$$P^{(2)} = I - 2v^{(2)}(v^{(2)})^T = I - 2 \begin{pmatrix} 0 & 0 & 0 & 0 \\ 0 & 0 & 0 & 0 \\ 0 & 0 & 4/5 & 2/5 \\ 0 & 0 & 2/5 & 1/5 \end{pmatrix} = \begin{pmatrix} 1 & 0 & 0 & 0 \\ 0 & 1 & 0 & 0 \\ 0 & 0 & -3/5 & -4/5 \\ 0 & 0 & -4/5 & 3/5 \end{pmatrix}.$$

Finally, we obtain the tridiagonal matrix $A^{(2)}$ as

$$A^{(2)} = P^{(2)} A^{(1)} P^{(2)} = \begin{pmatrix} 4 & -3 & 0 & 0 \\ -3 & 10/3 & -5/3 & 0 \\ 0 & -5/3 & -33/25 & 68/75 \\ 0 & 0 & 68/75 & 149/75 \end{pmatrix}.$$

We observe that we have performed the process of tridiagonalization in two steps. The final result is a tridiagonal symmetric matrix $A^{(2)}$ that is similar to the original one A.

9.5.2 Givens Rotation

In the previous section we described the Householder transformation, which introduces many zeros in a column of matrix at once. But in some situations we need to introduce zeros one at a time, and in such cases, we should use Givens rotations.

A Givens rotation is represented by a matrix of the form

$$G(i, j, \theta) = \begin{pmatrix} 1 & \cdots & 0 & \cdots & 0 & \cdots & 0 \\ \vdots & \ddots & \vdots & & \vdots & & \vdots \\ 0 & \cdots & c & \cdots & -s & \cdots & 0 \\ \vdots & & \vdots & \ddots & \vdots & & \vdots \\ 0 & \cdots & s & \cdots & c & \cdots & 0 \\ \vdots & & \vdots & & \vdots & \ddots & \vdots \\ 0 & \cdots & 0 & \cdots & 0 & \cdots & 1 \end{pmatrix},$$

where $c = \cos \Theta$ and $s = \sin \Theta$ appear at the intersections of the ith and jth rows and columns. Here Θ is the angle of rotation. The nonzero elements of the Givens matrix are given by

$$g_{kk} = 1 \quad \text{for } k \neq i, j,$$
$$g_{ii} = c,$$
$$g_{jj} = c,$$
$$g_{ji} = -s,$$
$$g_{ij} = s \quad \text{for } i > j.$$

We note that the sign of the sine function switches for $j > i$. The orthogonality of matrix G implies that $c^2 + s^2 = 1$. This is true for the cosine and sine of an arbitrary angle. The product $G(i, j, \theta)x$ represents a counterclockwise rotation of the vector x in the (i, j)-plane through θ radians.

When a Givens rotation matrix G is multiplied by another matrix A from the right to form the product GA, only rows i and j of A are affected. Thus we restrict ourself to the solution of the following problem: Given a and b, find $c = \cos\theta$ and $s = \sin\theta$ such that

$$\begin{pmatrix} c & -s \\ s & c \end{pmatrix} \begin{pmatrix} a \\ b \end{pmatrix} = \begin{pmatrix} r \\ 0 \end{pmatrix}.$$

Explicit calculation of θ is rarely necessary or desirable. Instead, we directly seek c, s, and r. An obvious solution is

$$\begin{aligned} r &= \sqrt{a^2 + b^2}, \\ c &= a/r, \\ s &= -b/r. \end{aligned} \tag{9.73}$$

If $|a| > |b|$, then we work with the tangent of the rotation angle

$$t = s/c = b/a, \tag{9.74}$$

so that we have the following alternative formulas for computating c, s:

$$\begin{aligned} c &= 1/\sqrt{1 + t^2}, \\ s &= -ct. \end{aligned} \tag{9.75}$$

If $|b| > |a|$, then we can use the cotangent τ of the rotation angle,

$$\tau = s/c = a/b, \tag{9.76}$$

and obtain

$$\begin{aligned} c &= 1/\sqrt{1 + \tau^2}, \\ s &= -c\tau. \end{aligned} \tag{9.77}$$

Example 9.11   Given the 3×3 matrix

$$A = \begin{pmatrix} 6 & 5 & 0 \\ 5 & 1 & 4 \\ 0 & 4 & 3 \end{pmatrix},$$

we perform two iterations of the Givens rotation to bring the matrix into upper triangular form.

We must zero entries $(2, 1)$ and $(3, 2)$ of the matrix A. We first make element $(2, 1)$ equal to zero and construct a rotation matrix G_1:

$$G_1 = \begin{pmatrix} c & -s & 0 \\ s & c & 0 \\ 0 & 0 & 1 \end{pmatrix}.$$

We have to perform the matrix multiplication

$$A_1 = G_1 \cdot A = \begin{pmatrix} c & -s & 0 \\ s & c & 0 \\ 0 & 0 & 1 \end{pmatrix} \begin{pmatrix} 6 & 5 & 0 \\ 5 & 1 & 4 \\ 0 & 4 & 3 \end{pmatrix} \tag{9.78}$$

such that

$$\begin{pmatrix} c & -s \\ s & c \end{pmatrix} \begin{pmatrix} 6 \\ 5 \end{pmatrix} = \begin{pmatrix} r \\ 0 \end{pmatrix}. \tag{9.79}$$

Now we compute parameters c, s, and r in (9.79) using explicit formulas (9.73):

$$r = \sqrt{6^2 + 5^2} = 7.8102,$$
$$c = 6/r = 0.7682,$$
$$s = -5/r = -0.6402.$$

Plugging these values for c, s in (9.78) and performing matrix multiplication yields a new matrix A_1:

$$A_1 = \begin{pmatrix} 7.8102 & 4.4813 & 2.5607 \\ 0 & -2.4327 & 3.0729 \\ 0 & 4 & 3 \end{pmatrix}.$$

The next step will be to zero out the element $(3, 2)$. Using the same idea as before, we construct a rotation matrix G_2:

$$G_2 = \begin{pmatrix} 1 & 0 & 0 \\ 0 & c & -s \\ 0 & s & c \end{pmatrix}.$$

We have to perform the matrix multiplication

$$A_2 = G_2 \cdot A_1 = \begin{pmatrix} 1 & 0 & 0 \\ 0 & c & -s \\ 0 & s & c \end{pmatrix} \begin{pmatrix} 7.8102 & 4.4813 & 2.5607 \\ 0 & -2.4327 & 3.0729 \\ 0 & 4 & 3 \end{pmatrix}$$

such that

$$\begin{pmatrix} c & -s \\ s & c \end{pmatrix} \begin{pmatrix} -2.4327 \\ 4 \end{pmatrix} = \begin{pmatrix} r \\ 0 \end{pmatrix}. \tag{9.80}$$

The parameters c, s, and r in (9.80) are computed using the explicit formulas (9.73):

$$r = \sqrt{(-2.4327)^2 + 4^2} = 4.6817,$$
$$c = -2.4327/r = -0.5196,$$
$$s = -4/r = -0.8544.$$

Plugging in these values for c and s and performing matrix multiplication gives us a new matrix A_2 that is also the upper triangular matrix R:

$$R = A_2 = \begin{pmatrix} 7.8102 & 4.4813 & 2.5607 \\ 0 & 4.6817 & 0.9664 \\ 0 & 0 & -4.1843 \end{pmatrix}.$$

This new matrix R is the upper triangular matrix that is needed to perform an iteration of the QR decomposition. The matrix Q is now formed using the transpose of the rotation matrices as follows:

$$Q = G_1^T G_2^T.$$

We note that we have performed the following computations:

$$G_2 G_1 A = R,$$
$$G_1^T G_2^T G_2 G_1 A = G_1^T G_2^T R,$$

and thus

$$A = G_1^T G_2^T R = QR$$

with

$$Q = G_1^T G_2^T.$$

Performing this matrix multiplication yields the following matrix Q in the QR decomposition:

$$Q = \begin{pmatrix} 0.7682 & 0.3327 & 0.5470 \\ 0.6402 & -0.3992 & -0.6564 \\ 0 & 0.8544 & -0.5196 \end{pmatrix}.$$

The Givens rotation can also be applied to zero out any desired component of an m-vector. We illustrate how to zero out the element $(4, 4)$ of a 5×5 matrix A. We construct the Givens matrix

$$G(2, 4, \theta) = \begin{pmatrix} 1 & 0 & 0 & 0 & 0 \\ 0 & c & 0 & -s & 0 \\ 0 & 0 & 1 & 0 & 0 \\ 0 & s & 0 & c & 0 \\ 0 & 0 & 0 & 0 & 1 \end{pmatrix}$$

and compute the parameters r, c, s from

$$\begin{pmatrix} 1 & 0 & 0 & 0 & 0 \\ 0 & c & 0 & -s & 0 \\ 0 & 0 & 1 & 0 & 0 \\ 0 & s & 0 & c & 0 \\ 0 & 0 & 0 & 0 & 1 \end{pmatrix} \cdot \begin{pmatrix} a_1 \\ a_2 \\ a_3 \\ a_4 \\ a_5 \end{pmatrix} = \begin{pmatrix} a_1 \\ r \\ a_3 \\ 0 \\ a_5 \end{pmatrix}.$$

Using a sequence of a such Givens rotations, we can zero out individual entries of the matrix A and reduce it to upper triangualar form. In so doing, we should avoid reintroducing nonzero entries into the matrix entries that have already been zeroed out. This can be done by a number of different reorderings. The product of all rotations will be an orthogonal matrix Q in the QR factorization of the matrix A.

Implementation of a Givens rotation for solving linear least square problems is about 50 percent more expensive than doing Householder transformations. Givens rotations also require more disk space to store c, s, r. Therefore, Givens rotations are used in cases in which the matrix A is sparse.

9.5.3 Gram–Schmidt Orthogonalization

Gram–Schmidt orthogonalization is one more method for computing QR factorizations. If we apply Gram–Schmidt to the columns a_i of $A = (a_1, a_2, \ldots, a_n)$ from left to right, we get a sequence of orthonormal vectors q_1 through q_n spanning the same space. These orthogonal vectors are the columns of the matrix Q. The Gram–Schmidt orthogonalization process also computes the coefficients $r_{ji} = q_j^T a_i$ expressing each column a_i as a linear combination of q_1 through q_i: $a_i = \sum_{j=1}^{i} r_{ji} q_j$. The r_{ji} are just the entries of the upper triangular matrix R.

More precisely, in the Gram–Schmidt orthogonalization process, for a given pair of linearly independent vectors a_1 and a_2 of length m, we want to determine two orthonormal vectors q_1 and q_2 of length m that span the same subspace as the vectors a_1 and a_2. To do so, we first normalize a_1 and obtain $q_1 = a_1 / \|a_1\|_2$. Then we subtract from a_2 the values $(q_1^T a_2) q_1$. This is the same as the following $m \times 1$ least squares problem:

$$q_1 \gamma \approx a_2. \tag{9.81}$$

The solution of this problem is given by the method of normal equations as

$$\gamma \approx (q_1^T q_1)^{-1} q_1^T a_2 = q_1^T a_2. \tag{9.82}$$

Then the desired vector q_2 is obtained by normalizing the residual vector

$$r = a_2 - (q_1^T a_2)q_1.$$

This process, called the classical Gram–Schmidt (CGS) orthogonalization procedure, can be extended to any number of vectors $a_1, ..., a_k, 1 \leq k \leq m$; see Algorithm 9.4 for its implementation.

The classical Gram–Schmidt (CGS) procedure is unsatisfactory when implemented in finite-precision arithmetic. This is because the orthogonality among the computed vectors q_k will be lost due to rounding errors. CGS also requires separate storage for A, Q_1, and R, since the element a_k is used in the inner loop, and thus q_k cannot overwrite it (because q_k is used in the inner loop). The modified Gram–Schmidt (MGS) procedure overcomes these difficulties; see Algorithm 9.5.

The classical Gram–Schmidt (CGS) and modified Gram–Schmidt (MGS) algorithms for factoring $A = QR$ are the following:

Algorithm 9.4 The classical Gram–Schmidt (CGS) orthogonalization algorithm.

for $k = 1$ to n /* loop over columns 1...n */
$q_k = a_k$
for $j = 1$ to $k - 1$
$r_{jk} = q_j^T a_k$
$q_k = q_k - r_{jk} q_j$
end
$r_{kk} = \|q_k\|_2$
if $r_{kk} = 0$ then stop /* stop if linearly dependent */
$q_k = q_k / r_{kk}$
end

Algorithm 9.5 Modified Gram–Schmidt (MGS) orthogonalization algorithm.

for $k = 1$ to n /* loop over columns 1...n */
$r_{kk} = \|a_k\|_2$
if $r_{kk} = 0$ then stop /* stop if linearly dependent */
$q_k = a_k / r_{kk}$ /* normalize current column */
for $j = k + 1$ to n
$r_{kj} = q_k^T a_j$
$a_j = a_j - r_{kj} q_k$
end
end

If A has full column rank, then r_{kk} will not be zero. Although MGS is more stable than CGS, we still can have a matrix Q that is far from orthogonal. This is because $\|Q^T Q - I\|$ can be larger than ε when A is ill conditioned, though the loss is much less than with CGS. To avoid this difficulty, in solving the linear system of equations $Ax \approx b$ with MGS we should not compute the right-hand side c_1 as $c_1 = Q_1^T b$. Much better is to treat the vector b as a column of length $n + 1$ and use MGS to compute the reduced QR factorization for the following augmented matrix of order $m \times n + 1$:

$$(A b) = (Q_1 q_{n+1}) \begin{pmatrix} R & c_1 \\ 0 & \rho \end{pmatrix}. \tag{9.83}$$

Then the solution of the least squares problem can be found as the solution to the $n \times n$ triangular linear system $Rx = c_1$.

The orthogonality of the resulting matrix Q_1 can also be enhanced by a reorthogonalization process. This means that we need to repeat the orthogonalization procedure for Q_1 which can be considered as a form of iterative refinement. We refer to [14, 53] for further reading on this subject.

Example 9.12 Classical Gram–Schmidt (CGS) orthogonalization algorithm for the solution of least squares problems

We illustrate the CGS Algorithm 9.4 on the solution of the following least squares problem: find $x = (x_1, x_2, x_3)$ subject to $\min_x \|Ax - y\|_2^2$ when the matrix A is given by

$$A = \begin{pmatrix} 1 & 0 & 0 \\ 0 & 1 & 0 \\ 0 & 0 & 1 \\ -1 & 1 & 0 \\ -1 & 0 & 1 \\ 0 & -1 & 1 \end{pmatrix}$$

and the elements of the vector y are $y = (1237, 1941, 2417, 711, 1177, 475)^T$. We have implemented Algorithm 9.4 and applied it to the solution of this linear least square problem. Our QR decomposition of the matrix A is

$$Q = \begin{pmatrix} 0.577350269189626 & 0.204124145231932 & 0.353553390593274 \\ 0 & 0.612372435695794 & 0.353553390593274 \\ 0 & 0 & 0.707106781186548 \\ -0.577350269189626 & 0.408248290463863 & -0.000000000000000 \\ -0.577350269189626 & -0.204124145231932 & 0.353553390593274 \\ 0 & -0.612372435695794 & 0.353553390593274 \end{pmatrix},$$

$$R = \begin{pmatrix} 1.732050807568877 & -0.577350269189626 & -0.577350269189626 \\ 0 & 1.632993161855452 & -0.816496580927726 \\ 0 & 0 & 1.414213562373095 \end{pmatrix}.$$

After performing the QR decomposition of A, we solved the linear least squares problem transformed to the solution of the equation $Rx = Q^T y$ with upper triangular matrix R, by backward substitution. We have obtained the following solution of the least squares problem: $x = (1236, 1943, 2416)^T$.

The MATLAB® program of Section 1.4 is available for running this test.[6]

9.6 Singular Value Decomposition

In this section we will show how the singular value decomposition (SVD) of a matrix A allows us to reduce a linear least squares problem to a diagonal linear least squares problem that is easier to solve.

Let us recall (see Section 5.1.1) that the *singular value decomposition* of a matrix A of order $m \times n$ has the form

$$A = U \Sigma V^T, \tag{9.84}$$

where U is an $m \times m$ orthogonal matrix $U^T U = I$, V is an $n \times n$ orthogonal matrix such that $V^T V = I$, and Σ is an $m \times n$ diagonal matrix with elements σ_{ij} on its diagonal such that

$$\sigma_{ij} = \begin{cases} 0 & \text{for } i \neq j, \\ \sigma_i \geq 0 & \text{for } i = j. \end{cases} \tag{9.85}$$

The elements σ_i are called singular values of A. They are ordered such that $\sigma_1 \geq \cdots \geq \sigma_n \geq 0$. The columns $u_1, \ldots, u_m$ of U are called left singular vectors. The columns $v_1, \ldots, v_n$ of V are called right singular vectors. We note that if $m < n$, then the SVD is defined for A^T.

An alternative definition of SVD decomposition is formulated in the following theorem.

Theorem 9.2 *Let A be an arbitrary $m \times n$ matrix with $m \geq n$. Then the SVD decomposition has the form $A = U \Sigma V^T$, where U is of order $m \times n$ and satisfies $U^T U = I$, V is of order $n \times n$ and satisfies $V^T V = I$, and $\Sigma = diag(\sigma_1, \ldots, \sigma_n)$, where $\sigma_1 \geq \cdots \geq \sigma_n \geq 0$. The columns $u_1, \ldots, u_n$ of U are called left singular vectors. The columns $v_1, \ldots, v_n$ of V are called right singular vectors. The numbers σ_i are called singular values.*

Proof The proof of this theorem is done by induction on m and n. This means that we assume that the SVD decomposition exists for matrices of order $(m - 1) \times (n - 1)$, and our goal is to prove that the SVD decomposition exists also for matrices of order $m \times n$. In this proof we assume $A \neq 0$. If $A = 0$, we can take $\Sigma = 0$ and let U and V be arbitrary orthogonal matrices.

Since $m \geq n$, let us consider the case $n = 1$ and write the SVD decomposition as $A = U \Sigma V^T$ with $U = A/\|A\|_2$, $\Sigma = \|A\|_2$, and $V = 1$.

To apply the induction step, we choose the vector v such that $\|v\|_2 = 1$ and $\|A\|_2 = \|Av\|_2 > 0$. Such a vector v exists by the definition of the two-norm of the

---

[6]The MATLAB® programs can be found on the online version of Chapter 1 (doi:10.1007/978-3-319-57304-5_1).

matrix A: $\|A\|_2 = \max_{\|v\|_2=1} \|Av\|_2$. Let us define $u = \frac{Av}{\|Av\|_2}$, which is a unit vector. We choose now two matrices $\tilde{U}$ and $\tilde{V}$ such that $U = (u, \tilde{U})$ is an $m \times m$ orthogonal matrix, and $V = (v, \tilde{V})$ is an $n \times n$ orthogonal matrix. We now multiply the matrix A from the left by the matrix U^T and from the right by the matrix V to get

$$U^T A V = \begin{pmatrix} u^T \\ \tilde{U}^T \end{pmatrix} \cdot A \cdot (v \; \tilde{V}) = \begin{pmatrix} u^T A v & u^T A \tilde{V} \\ \tilde{U}^T A v & \tilde{U}^T A \tilde{V} \end{pmatrix}.$$

Since the vector u is chosen as $u = \frac{Av}{\|Av\|_2}$, we observe that

$$u^T A v = \frac{(Av)^T (Av)}{\|Av\|_2} = \frac{\|Av\|_2^2}{\|Av\|_2} = \|Av\|_2 = \|A\|_2 \equiv \sigma.$$

Next, we also observe that the following block is zero: $\tilde{U}^T A v = \tilde{U}^T u \|Av\|_2 = 0$. We want to prove that the block $u^T A \tilde{V}$ will be zero also: $u^T A \tilde{V} = 0$. To do so, we consider $\|(\sigma | u^T A \tilde{V})\|_2$. We observe that $\|(\sigma | u^T A \tilde{V})\|_2 > \sigma$ and $\|(1, 0, \ldots, 0) = U^T A V\|_2 = \|(\sigma | u^T A \tilde{V})\|_2$. Then using the properties of the two-norm, we can write $\sigma = \|A\|_2 = \|U^T A V\|_2 \geq \|(1, 0, \ldots, 0) U^T A V\|_2 = \|(\sigma | u^T A \tilde{V})\|_2 > \sigma$, a contradiction.

Collecting our observations above for blocks $u^T A v$, $u^T A \tilde{V}$ and $\tilde{U}^T A v$, we can rewrite the expression for $U^T A V$ as

$$U^T A V = \begin{pmatrix} \sigma & 0 \\ 0 & \tilde{U}^T A \tilde{V} \end{pmatrix} = \begin{pmatrix} \sigma & 0 \\ 0 & \tilde{A} \end{pmatrix}. \tag{9.86}$$

Now we use the induction hypothesis for $\tilde{A}$ to obtain the matrix $\tilde{A} = U_1 \Sigma_1 V_1^T$, where U_1 is a matrix of order $(m-1) \times (n-1)$, Σ_1 is a matrix of order $(n-1) \times (n-1)$, and V_1 is a matrix of order $(n-1) \times (n-1)$. Thus we can rewrite (9.86) as

$$U^T A V = \begin{pmatrix} \sigma & 0 \\ 0 & U_1 \Sigma_1 V_1^T \end{pmatrix} = \begin{pmatrix} 1 & 0 \\ 0 & U_1 \end{pmatrix} \begin{pmatrix} \sigma & 0 \\ 0 & \Sigma_1 \end{pmatrix} \begin{pmatrix} 1 & 0 \\ 0 & V_1 \end{pmatrix}^T.$$

Multiplying the above equation by the matrix U from the left and by the matrix V^T from the right, we obtain the desired SVD decomposition of the matrix A:

$$A = \left(U \begin{pmatrix} 1 & 0 \\ 0 & U_1 \end{pmatrix} \right) \cdot \begin{pmatrix} \sigma & 0 \\ 0 & \Sigma_1 \end{pmatrix} \cdot \left(V \begin{pmatrix} 1 & 0 \\ 0 & V_1 \end{pmatrix} \right)^T.$$

$\square$

The following theorems present some properties of the SVD decomposition that are frequently very important in computations. We note that analogous results also hold for the matrix A when $m < n$, only that we consider A^T instead of A.

Theorem 9.3 *Let $A = U \Sigma V^T$ be the SVD of the $m \times n$ matrix A, where $m \geq n$. Suppose that A is symmetric, with eigenvalues λ_i and orthonormal eigenvectors u_i. This means that $A = U \Lambda U^T$ is an eigendecomposition of A, with $\Lambda = diag(\lambda_1, \ldots, \lambda_n)$, $U = (u_1, \ldots, u_n)$, and $UU^T = I$. Then an SVD of A is $A = U \Sigma V^T$, where $\sigma_i = |\lambda_i|$ and $v_i = sign(\lambda_i) u_i$, where $sign(0) = 1$.*

Proof Since $A = U \Lambda U^T$ is an eigendecomposition of A, then by the definition an SVD of A is $A = U \Sigma V^T$, where $\sigma_i = |\lambda_i|$ and $v_i = sign(\lambda_i) u_i$ with $sign(0) = 1$.

$\square$

Theorem 9.4 *Let $A = U \Sigma V^T$ be an SVD of an $m \times n$ matrix A, where $m \geq n$. Then the eigenvalues of the symmetric matrix $A^T A$ are σ_i^2. The right singular vectors v_i are corresponding orthonormal eigenvectors.*

Proof Let us consider the SVD decomposition of $A = U \Sigma V^T$ and write it for $A^T A$:

$$A^T A = V \Sigma U^T U \Sigma V^T = V \Sigma^2 V^T. \tag{9.87}$$

We observe that by definition of the eigendecomposition, the above decomposition is an eigendecomposition of $A^T A$. In this decomposition, the columns of V are the eigenvectors, and the diagonal entries of Σ^2 are the eigenvalues. $\square$

Theorem 9.5 *Let $A = U \Sigma V^T$ be an SVD of an $m \times n$ matrix A, where $m \geq n$. Then the eigenvalues of the symmetric matrix AA^T are σ_i^2 and $m - n$ zeros. The left singular vectors u_i are corresponding orthonormal eigenvectors for the eigenvalues σ_i^2. We can take any $m - n$ other orthogonal vectors as eigenvectors for the eigenvalue 0.*

Proof Choose an $m \times (m - n)$ matrix $\tilde{U}$ such that $(U, \tilde{U})$ is square and orthogonal. Then we can write

$$AA^T = U \Sigma V^T V \Sigma U^T = U \Sigma^2 U^T = \left(U, \tilde{U}\right) \cdot \begin{pmatrix} \Sigma^2 & 0 \\ 0 & 0 \end{pmatrix} \cdot \left(U, \tilde{U}\right)^T.$$

We observe that the above decomposition is an eigendecomposition of AA^T. $\square$

Theorem 9.6 *Let $A = U \Sigma V^T$ be an SVD of an $m \times n$ matrix A, where $m \geq n$. Let the matrix H be constructed such that $H = \begin{pmatrix} 0 & A^T \\ A & 0 \end{pmatrix}$, where A is square and $A = U \Sigma V^T$ is the SVD of A. Let $\Sigma = diag(\sigma_1, \ldots, \sigma_n)$, $U = (u_1, \ldots, u_n)$, and $V = (v_1, \ldots, v_n)$. Then the $2n$ eigenvalues of H are $\pm \sigma_i$, with corresponding unit eigenvectors $\frac{1}{\sqrt{2}} \begin{pmatrix} v_i \\ \pm u_i \end{pmatrix}$.*

Proof The proof of this theorem is exercise given in Question 9.5. $\square$

Theorem 9.7 *Let $A = U \Sigma V^T$ be the SVD of the $m \times n$ matrix A, where $m \geq n$. If A has a full rank, then the solution of the linear least squares problem*

$$\min_x \|r(x)\|_2^2 = \min_x \|Ax - b\|_2^2$$

is $x = V \Sigma^{-1} U^T b$.

Proof Let us consider the two-norm of the residual $\|r(x)\|_2^2 = \|Ax - b\|_2^2 = \|U \Sigma V^T x - b\|_2^2$. Since A has full rank, Σ also has full rank, and thus Σ is invertible. Now let us construct the matrix $(U, \tilde{U})$, which will be square and orthogonal. We can write

$$\|U \Sigma V^T x - b\|_2^2 = \left\| \begin{pmatrix} U^T \\ \tilde{U}^T \end{pmatrix} (U \Sigma V^T x - b) \right\|_2^2$$

$$= \left\| \begin{pmatrix} \Sigma V^T x - U^T b \\ -\tilde{U}^T b \end{pmatrix} \right\|_2^2$$

$$= \|\Sigma V^T x - U^T b\|_2^2 + \|\tilde{U}^T b\|_2^2.$$

We observe that by making the first term zero, $\Sigma V^T x = U^T b$, we will find the minimum of the least squares problem given by $x = V \Sigma^{-1} U^T b$. $\qquad \square$

Theorem 9.8 *Let $A = U \Sigma V^T$ be the SVD of the $m \times n$ matrix A, where $m \geq n$, $\|A\|_2 = \sigma_1$. If A is square and nonsingular, then $\|A^{-1}\|_2^{-1} = \sigma_n$ and $\|A\|_2 \cdot \|A^{-1}\|_2 = \frac{\sigma_1}{\sigma_n}$.*

This assertion was proved in Section 6.4, p. 142.

Theorem 9.9 *Let $A = U \Sigma V^T$ be the SVD of the $m \times n$ matrix A, where $m \geq n$. Assume that $\sigma_1 \geq \cdots \geq \sigma_r > \sigma_{r+1} = \cdots = \sigma_n = 0$. Then the rank of A is r. The null space of A, i.e., the subspace of vectors v such that $Av = 0$, is the space spanned by columns $r + 1$ through r of $V : span(v_{r+1}, \ldots, v_n)$. The range space of A, the subspace of vectors of the form Aw for all w, is the space spanned by columns 1 through r of $U : span(u_1, \ldots, u_r)$.*

Proof Let us choose an $m \times (m - n)$ matrix $\tilde{U}$ such that the $m \times m$ matrix $\hat{U} = (U, \tilde{U})$ is orthogonal. Since $\hat{U}$ and V are nonsingular, A and

$$\hat{U}^T A V = \begin{pmatrix} \Sigma^{n \times n} \\ 0^{(m-n) \times n} \end{pmatrix} \equiv \hat{\Sigma} \tag{9.88}$$

have the same rank r. We may claim this using our assumption about entries of the matrix Σ.

Values of v are in the null space of A if and only if $V^T v$ is in the null space of $\hat{U}^T A V = \hat{\Sigma}$. This is true because $Av = 0$ if and only if $\hat{U}^T A V (V^T v) = 0$.

But the null space of $\hat{\Sigma}$ is spanned by columns $r + 1$ through n of the $n \times n$ identity matrix I_n. This means that the null space of A is spanned by V times these columns, i.e., v_{r+1} through v_n.

Similarly, we can show that the range space of A is the same as $\hat{U}$ times the range space of $\hat{U}^T A V = \hat{\Sigma}$, i.e., $\hat{U}$ times the first r columns of I_m, or u_1 through u_r.   □

Theorem 9.10 *Let $A = U \Sigma V^T$ be the SVD of the $m \times n$ matrix A, where $m \geq n$. Let B^{n-1} be the unit sphere in $\mathbb{R}^n$: $B^{n-1} = \{x \in \mathbb{R}^n : \|x\|_2 = 1\}$. Let $A \cdot B^{n-1}$ be the image of B^{n-1} under A: $A \cdot B^{n-1} = \{Ax : x \in \mathbb{R}^n \text{ and } \|x\|_2 = 1\}$. Then $A \cdot B^{n-1}$ is an ellipsoid centered at the origin of $\mathbb{R}^m$, with principal axes $\sigma_i u_i$.*

Proof We construct the set $A \cdot B^{n-1}$ by multiplying $A = U \Sigma V^T$ by B^{n-1} step by step. Let us assume for simplicity that A is square and nonsingular. Since V is orthogonal and thus maps unit vectors to other unit vectors, we can write

$$V^T \cdot B^{n-1} = B^{n-1}. \tag{9.89}$$

Now consider the product ΣB^{n-1}. Since $v \in B^{n-1}$ if and only if $\|v\|_2 = 1$, it follows that $w \in \Sigma B^{n-1}$ if and only if $\|\Sigma^{-1}w\|_2 = 1$ or

$$\sum_{i=1}^{n}(w_i/\sigma_i)^2 = 1. \tag{9.90}$$

The equation above defines an ellipsoid with principal axes $\sigma_i e_i$, where e_i is the ith column of the identity matrix. Finally, multiplying each $w = \Sigma v$ by U just rotates the ellipse so that each e_i becomes u_i, that is, the ith column of U.   □

Theorem 9.11 *Let $A = U \Sigma V^T$ be the SVD of the $m \times n$ matrix A, where $m \geq n$. Write $V = (v_1, v_2, \ldots, v_n)$ and $U = (u_1, u_2, \ldots, u_n)$, so $A = U \Sigma V^T = \sum_{i=1}^{n} \sigma_i u_i v_i^T$ (a sum of rank-1 matrices). Then a matrix of rank $k < n$ closest to A (measured with $\| \cdot \|_2$) is $A_k = \sum_{i=1}^{k} \sigma_i u_i v_i^T$ and $\|A - A_k\|_2 = \sigma_{k+1}$. We may also write $A_k = U \Sigma_k V^T$, where $\Sigma_k = diag(\sigma_1, \ldots, \sigma_k, 0, \ldots, 0)$.*

Proof The matrix A_k has rank k by construction. We can write the difference $\|A - A_k\|_2$ as

$$\|A - A_k\|_2 = \left\| \sum_{i=k+1}^{n} \sigma_i u_i v_i^T \right\| = \left\| U \begin{pmatrix} 0 & & & \\ & \sigma_{k+1} & & \\ & & \ddots & \\ & & & \sigma_n \end{pmatrix} V^T \right\|_2 = \sigma_{k+1}. \tag{9.91}$$

We need to show now that there is no other matrix closer to the matrix A than the matrix A_k. Let B be any matrix of rank k such that its null space has dimension $n - k$. The space spanned by $\{v_1, \ldots, v_{k+1}\}$ has dimension $k + 1$. Since the sum of their dimensions is $(n - k) + (k + 1) > n$, these two spaces must overlap. Let h be a unit vector in their intersection. Then

$$
\begin{aligned}
\|A - B\|_2^2 &\geq \|(A - B)h\|_2^2 = \|Ah\|_2^2 = \left\|U \Sigma V^T h\right\|_2^2 \\
&= \left\|\Sigma(V^T h)\right\|_2^2 \\
&\geq \sigma_{k+1}^2 \left\|V^T h\right\|_2^2 \\
&= \sigma_{k+1}^2.
\end{aligned}
$$
$\square$

Comparing the obtained expression $\|A - B\|_2^2 \geq \sigma_{k+1}^2$ with (9.91), we observe that $\|A - A_k\|_2^2 = \sigma_{k+1}^2$, and thus A_k is a best approximation to A. $\square$

Example 9.13 Image compression using SVD.

In this example we will demonstrate how to perform image compression using the standard demo-library in MATLAB® with pictures. For example, we can load the image of clown from this directory from the file `clown.mat`:

load `clown.mat`

The resulting image will be read into the array X. The size of this array will be (in pixels):

```
> Size(X) = m × n = 320 × 200
```

Now we can simply use the *svd* command to perform SVD decomposition of the matrix X:

```
> [U,S,V] = svd(X);
> colormap(map);
```

For example, to see rank $k = 20$ of this image, we write:

```
> k=20;
> image(U(:,1:k)*S(1:k,1:k)*V(:,1:k)');
```

Example 9.14 Image compression using SVD.

This is another example of how to compress an image in JPEG format that you can produce using any digital camera and then download in MATLAB®. To read any digital image from data file named `File.jpg`, we can write in the command line in MATLAB®:

```
> A = imread('File.jpg');
```

Then if you write

```
> size(A)
```

you will get the size of the obtained matrix A from your image. For the image presented in Fig. 9.4a we have obtained

```
> size(A)
> ans = 218 171 3
```

Thus, the matrix of the original image of Fig. 9.4a is of size $m \times n$ with $m = 218, n = 171$. We obtained a three-dimensional array A since the image presented in Fig. 9.4a is color, and the MATLAB®-command *imread* returned the value for A as an $m \times n \times 3$ array. We are not able simply to use the *svd* command in MATLAB® for such a matrix. If we tried to apply *svd*, we would get the following error message:

```
> [U3,S3,V3] = svd(A(:,:,3));
```

Undefined function 'svd' for input arguments of type 'uint8'.

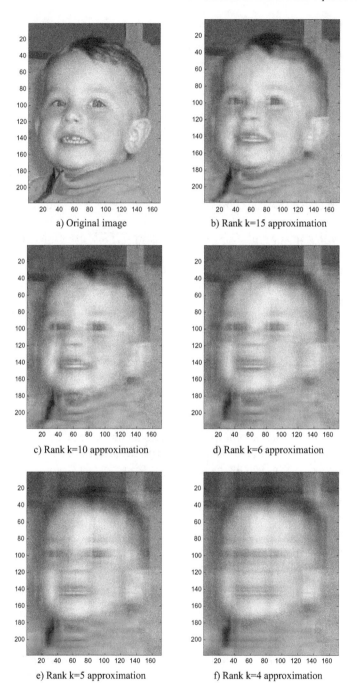

a) Original image

b) Rank k=15 approximation

c) Rank k=10 approximation

d) Rank k=6 approximation

e) Rank k=5 approximation

f) Rank k=4 approximation

Fig. 9.4 Example 9.14. Image compression using SVD decomposition for different rank-k approximations.

To avoid this error message, we need convert A from `'uint8'` format to the double format using the following command:

```
> DDA = im2double(A);
```

Then the size of the matrix DDA will be $m \times n \times 3$. At the next step, we perform SVD decomposition for every three entries of DDA:

```
> [U1,S1,V1] = svd(DDA(:,:,1));
> [U2,S2,V2] = svd(DDA(:,:,2));
> [U3,S3,V3] = svd(DDA(:,:,3));
```

Finally, we can perform image compression for different rank-k approximations. For example, let us choose rank $k = 15$. Then using the following commands, we can compute new approximation matrices svd1, svd2, svd3:

```
> svd1 = U1(:,1:k)* S1(1:k,1:k)* V1(:,1:k)';
> svd2 = U2(:,1:k)* S2(1:k,1:k)* V2(:,1:k)';
> svd3 = U3(:,1:k)* S3(1:k,1:k)* V3(:,1:k)';
```

To obtain different compressed images that are similar to the images of Fig. 9.4, we write:

```
> DDAnew = zeros(size(DDA));
> DDAnew(:,:,1) = svd1;
> DDAnew(:,:,2) = svd2;
> DDAnew(:,:,3) = svd3;
```

Then to see the approximated image we use the following command:

```
> image(DDAnew);
```

9.6.1 Rank-Deficient Least Squares Problems

In all our considerations above, we have assumed that our matrix A has full column rank, or $r(A) = n$. if A has linearly dependent columns, so that $r(A) < n$, then it is still possible to perform QR factorization, but the matrix R will be singular. This means that many vectors can give the minimal norm $\|Ax - b\|_2$, and the least squares solution is not unique. This can happen in the case of insufficient data collection, digital image restoration, computing the inverse Laplace transform, in other words, in ill-posed problems [110].

The next proposition says that in the case of a nearly rank-deficient matrix A, the least squares solution is not unique.

Proposition 9.1 *Let A be an $m \times n$ matrix with $m \geq n$ and rank $A = r < n$. Then there is an $(n - r)$-dimensional set of vectors that minimizes $\|Ax - b\|_2$.*

Proof Let $Az = 0$. Then if x minimizes $\|Ax - b\|_2$, $x + z$ also minimizes $\|A(x + z) - b\|_2$.

This means that the least squares solution is not unique. $\square$

Below, we define the Moore–Penrose[7] pseudoinverse A^+ for a full-rank matrix A. The pseudoinverse allows us to write the solution of the full-rank overdetermined least squares problem $\min_x \|Ax - b\|_2$ simply as $x = A^+ b$.

Suppose that A is $m \times n$ with $m > n$ and has full rank, with $A = QR = U\Sigma V^T$ representing QR and SVD decompositions of A, respectively. Then

$$A^+ \equiv (A^T A)^{-1} A^T = R^{-1} Q^T = V\Sigma^{-1} U^T$$

is called the *Moore–Penrose pseudoinverse* of A. If $m < n$, then $A^+ \equiv A^T (AA^T)^{-1}$.

If A is square and has full rank, then the solution of the full-rank overdetermined least squares problem $\min_x \|Ax - b\|_2$ reduces to $x = A^{-1} b$. The matrix A^+ is computed by the function *pinv(A)* in MATLAB®.

In the case of a rank-deficient matrix A, we have the following definition of the Moore–Penrose pseudoinverse A^+.

Suppose that A is $m \times n$ with $m > n$ and is rank-deficient with rank $r < n$. Let $A = U\Sigma V^T = U_1 \Sigma_1 V_1^T$ be SVD decompositions of A such that

$$A = (U_1, U_2) \begin{pmatrix} \Sigma_1 & 0 \\ 0 & 0 \end{pmatrix} (V_1, V_2)^T = U_1 \Sigma_1 V_1^T.$$

Here, (Σ_1) is of size $r \times r$ and is nonsingular; U_1 and V_1 have r columns. Then

$$A^+ \equiv V_1 \Sigma_1^{-1} U_1^T$$

is called the *Moore–Penrose pseudoinverse* for the rank-deficient matrix A.

The solution of the least-squares problem is always $x = A^+ b$. The next proposition states that if A is nearly rank-deficient, then the solution x of $Ax = b$ will be ill conditioned and very large.

Proposition 9.2 *Let $\sigma_{min} > 0$ be the smallest singular value of the nearly rank-deficient matrix A. Then*

- *If x minimizes $\|Ax - b\|_2$, then $\|x\|_2 \geq \frac{|u_n^T b|}{\sigma_{min}}$, where u_n is the last column of U in the SVD decomposition of $A = U\Sigma V^T$.*
- *Changing b to $b + \delta b$ can change x to $x + \delta x$, where $\|\delta x\|_2$ can be estimated as $\frac{\|\delta b\|_2}{\sigma_{min}}$, or the solution is very ill conditioned.*

Proof 1. By Theorem 9.7, we have that for the case of a full-rank matrix A, the solution of $Ax = b$ is given by $x = (U\Sigma V^T)^{-1} b = V\Sigma^{-1} U^T b$. The matrix $A^+ = V\Sigma^{-1} U^T$ is the Moore–Penrose pseudoinverse of A. Thus, we can write this solution also as

$$x = V\Sigma^{-1} U^T b = A^+ b.$$

---

[7]Eliakim Hastings Moore (1862–1932) was an American mathematician. Sir Roger Penrose (born 1931) is an English mathematical physicist, mathematician, and philosopher of science.

Then taking norms of both sides of the above expression yields

$$\|x\|_2 = \|\Sigma^{-1} U^T b\|_2 \geq |(\Sigma^{-1} U^T b)_n| = \frac{|u_n^T b|}{\sigma_{min}}, \tag{9.92}$$

where $|(\Sigma^{-1} U^T b)_n|$ is the nth column of this product.

2. We now apply (9.92) for $\|x + \delta x\|$ instead of $\|x\|$ to get

$$\|x + \delta x\|_2 = \|\Sigma^{-1} U^T (b + \delta b)\|_2 \geq |(\Sigma^{-1} U^T (b + \delta b))_n|$$
$$= \frac{|u_n^T (b + \delta b)|}{\sigma_{min}} = \frac{|u_n^T b + u_n^T \delta b|}{\sigma_{min}}. \tag{9.93}$$

We observe that

$$\frac{|u_n^T b|}{\sigma_{min}} + \frac{|u_n^T \delta b|}{\sigma_{min}} \leq \|x + \delta x\|_2 \leq \|x\|_2 + \|\delta x\|_2.$$

Choosing δb parallel to u_n and applying again (9.92) for estimating $\|x\|_2$, we have

$$\|\delta x\|_2 \geq \frac{\|\delta b\|_2}{\sigma_{min}}. \tag{9.94}$$

$\square$

In the next proposition we prove that the minimum-norm solution x is unique and may be well conditioned if the smallest nonzero singular value in Σ is not too small.

Proposition 9.3 *Let a matrix A be singular and suppose that x minimizes $\|Ax - b\|_2$. Let $A = U \Sigma V^T$ have rank $r < n$. Write the SVD decomposition of A as*

$$A = (U_1, U_2) \begin{pmatrix} \Sigma_1 & 0 \\ 0 & 0 \end{pmatrix} (V_1, V_2)^T = U_1 \Sigma_1 V_1^T.$$

Here, (Σ_1) is of size $r \times r$ and is nonsingular; U_1 and V_1 have r columns. Let $\sigma = \sigma_{min}(\Sigma_1)$. Then:

- *All solutions x can be written as $x = V_1 \Sigma_1^{-1} U_1^T + V_2 z$.*
- *The solution x has minimal norm $\|x\|_2$ when $z = 0$. Then $x = V_1 \Sigma_1^{-1} U_1^T$ and $\|x\|_2 \leq \frac{\|b\|_2}{\sigma}$.*
- *Changing b to $b + \delta b$ can change x to $\frac{\|\delta b\|_2}{\sigma}$.*

Proof We choose the matrix $\tilde{U}$ such that $(U, \tilde{U}) = (U_1, U_2, \tilde{U})$ is an $m \times m$ orthogonal matrix. Then using the property of the norm, we can write

$$\|Ax - b\|_2^2 = \|(U_1, U_2, \tilde{U})^T (Ax - b)\|_2^2$$

$$= \left\| \begin{pmatrix} U_1^T \\ U_2^T \\ \tilde{U}^T \end{pmatrix} (U_1 \Sigma_1 V_1^T x - b) \right\|_2^2$$

$$= \left\| \begin{pmatrix} I^{r \times r} \\ O^{m \times (n-r)} \\ O^{m \times m-n} \end{pmatrix} (\Sigma_1 V_1^T x - (U_1, U_2, \tilde{U})^T \cdot b) \right\|_2^2$$

$$= \left\| \begin{pmatrix} \Sigma_1 V_1^T x - U_1^T b \\ -U_2^T b \\ -\tilde{U}^T b \end{pmatrix} \right\|_2^2$$

$$= \|\Sigma_1 V_1^T x - U_1^T b\|_2^2 + \|U_2^T b\|_2^2 + \|\tilde{U}^T b\|_2^2.$$

To prove part 1, we observe that $\|Ax - b\|_2$ is minimized when $\Sigma_1 V_1^T x - U_1^T b = 0$. Using Proposition 9.1, we can also write that the vector $x = (\Sigma_1 V_1^T)^{-1} U_1^T b + V_2 z$ or $x = V_1 \Sigma_1^{-1} U_1^T b + V_2 z$ is also a solution of this minimization problem, where $V_1^T V_2 z = 0$, since the columns of V_1 and V_2 are orthogonal.

To prove part 2, we note that since the columns of V_1 and V_2 are orthogonal, it follows by the Pythagorean theorem that we have

$$\|x\|_2^2 = \|V_1 \Sigma_1^{-1} U_1^T b\|^2 + \|V_2 z\|^2, \tag{9.95}$$

which is minimized for $z = 0$.

For a proof of part 3, we change b to δb in (9.95) to get

$$\|V_1 \Sigma_1^{-1} U_1^T \delta b\|_2 \leq \|V_1 \Sigma_1^{-1} U_1^T\|_2 \cdot \|\delta b\|_2 = \|\Sigma_1^{-1}\|_2 \cdot \|\delta b\|_2 = \frac{\|\delta b\|_2}{\sigma}, \tag{9.96}$$

where σ is the smallest nonzero singular value of A. In this proof we used the properties of the norm that $\|QAZ\|_2 = \|A\|_2$ if Q, Z are orthogonal.   □

9.6.2 How to Solve Rank-Deficient Least Squares Problems

In this section we discuss how to solve rank-deficient least squares problems using QR decomposition with pivoting. QR decomposition with pivoting is cheaper but can be less accurate than the SVD technique for the solution of rank-deficient least squares problems. If A has rank $r < n$ with r independent columns, then the QR decomposition can look like this:

$$A = QR = Q \cdot \begin{pmatrix} R_{11} & R_{12} \\ 0 & 0 \\ 0 & 0 \end{pmatrix} \tag{9.97}$$

with nonsingular R_{11} of order $r \times r$ and R_{12} of order $r \times (n - r)$. We can try to obtain a matrix

$$R = \begin{pmatrix} R_{11} & R_{12} \\ 0 & R_{22} \\ 0 & 0 \end{pmatrix}, \tag{9.98}$$

in which the elements of R_{22} are very small and are of order $\varepsilon \|A\|_2$. If we set $R_{22} = 0$ and choose $(Q, \tilde{Q})$, which is square and orthogonal, then we will minimize

$$\begin{aligned} \|Ax - b\|_2^2 &= \left\| \begin{pmatrix} Q^T \\ \tilde{Q}^T \end{pmatrix} (Ax - b) \right\|_2^2 \\ &= \left\| \begin{pmatrix} Q^T \\ \tilde{Q}^T \end{pmatrix} (QRx - b) \right\|_2^2 \\ &= \left\| \begin{pmatrix} Rx - Q^T b \\ -\tilde{Q}^T b \end{pmatrix} \right\|_2^2 \\ &= \|Rx - Q^T b\|_2^2 + \|\tilde{Q}^T b\|_2^2. \end{aligned} \tag{9.99}$$

Here we again used the properties of the norm that $\|QAZ\|_2 = \|A\|_2$ if Q, Z are orthogonal.

Let us now decompose $Q = (Q_1, Q_2)$ with $x = (x_1, x_2)^T$ and

$$R = \begin{pmatrix} R_{11} & R_{12} \\ 0 & 0 \end{pmatrix} \tag{9.100}$$

such that Eq. (9.99) becomes

$$\begin{aligned} \|Ax - b\|_2^2 &= \left\| \begin{pmatrix} R_{11} & R_{12} \\ 0 & 0 \end{pmatrix} \cdot \begin{pmatrix} x_1 \\ x_2 \end{pmatrix} - \begin{pmatrix} Q_1^T b \\ Q_2^T b \end{pmatrix} \right\|_2^2 + \|\tilde{Q}^T b\|_2^2 \\ &= \|R_{11}x_1 + R_{12}x_2 - Q_1^T b\|_2^2 + \|Q_2^T b\|_2^2 + \|\tilde{Q}^T b\|_2^2. \end{aligned} \tag{9.101}$$

We now take the derivative with respect to x to get $(\|Ax - b\|_2^2)'_x = 0$. We see that the minimum is achieved when

$$x = \begin{pmatrix} R_{11}^{-1}(Q_1^T b - R_{12}x_2) \\ x_2 \end{pmatrix} \tag{9.102}$$

for every vector x_2. If R_{11} is well conditioned and $R_{11}^{-1} R_{12}$ is small, then the choice $x_2 = 0$ will be a good one.

The method just described is not reliable for all rank-deficient least squares problems. This is because R can be nearly rank-deficient in case when we cannot construct R_{22} with all small elements. In this case, QR decomposition with column pivoting can help: we factorize $AP = QR$ with a permutation matrix P. To compute this permutation, we proceed as follows:

- In all columns from 1 to n, at step i we select from the unfinished decomposition of part A in columns i to n and rows i to m the column with largest norm, and exchange it with the ith column.
- Then we compute the usual Householder transformation to zero out column i in entries $i + 1$ to m.

Much research has been devoted to more advanced algorithms called rank-revealing QR algorithms, which detect the rank faster and more efficiently; see [15, 18] for details.

9.7   Software for the Solution of Linear Least Squares Problems

We list available packages and routines that solve linear least squares problems in Table 9.1. Usually, we use MATLAB® for implementation of the solution of least squares problems. Here, the backslash \ is used for the solution of square and rectangular linear systems of the form $Ax = b$. The solution is given as $x = A\backslash b$. The QR decomposition of a matrix A in MATLAB® is computed as the function $[Q, R] = qr(A)$, and the SVD decomposition is given as the function $[U, S, V] = svd(A)$.

Many statistical packages such as BMDP, Minitab, Omnitab, S, S-plus, SAS, SPSS, as well as the statistical toolbox in MATLAB®, have extensive software for solving least squares problems.

The programs of Sections 1.4–1.8 solve linear least squares problem of polynomial fitting.[8] see Questions 1 through 4 for details.

Table 9.1 Software for linear least squares problems (LLSP)

Package	Factorization	Solution of LLSP	Rank-deficient LLSP
MATLAB®	qr	\	svd
FMM [58]	svd		svd
IMSL	lqrrr	lqrsl	lsqrr
KMN [61]	sqrls	sqrls	ssvdc
LAPACK [3]	sqeqrf	sormqr/strtrs	sgeqpf/stzrqf
Lawson and Hanson [68]	hft	hs1	hfti
LINPACK [69]	sqrdc	sqrsl	sqrst
NAPACK [88]	qr	over	sing/rsolve
NUMAL [89]	lsqortdec	lsqsol	solovr
SOL [114]	hredl	qrvslv	mnlnls

---

[8]The MATLAB® programs can be found on the online version of Chapter 1 (doi:10.1007/978-3-319-57304-5_1).

Questions

9.1 (*Programming*)

Solve the least squares problem $\min_c \|Ac - y\|_2^2$ of Example 9.2 by the method of normal equations and QR decomposition (either Algorithm 9.4 or Algorithm 9.5). The matrix A in this least squares problem is a Vandermonde matrix (9.8). This means that the columns of the matrix A are powers of the vector x such that $y(x) = \sum_{i=0}^{d} c_i x^{i-1}$, d is the degree of the polynomial, and (x_i, y_i), $i = 1, ..., m$.

Use your own code or the programs of Section 1.4.[9] Show that we get an erratic fit to the function for polynomials of degree greater than 18. Compare both methods for different values of d by computing the relative error

$$e = \frac{\|y - y^*\|_2}{\|y^*\|_2}. \tag{9.103}$$

Here, the y_i^*, $i = 1, \ldots, m$ are the exact values of the function y^*, and the y_i, $i = 1, \ldots, m$ are the computed values of generated data y. Report your results in a table for different discretizations of the interval for x and different values of d.

9.2 (*Programming*)

Solve the least squares problem

$$\min_c \sum_{i=1}^{m} (y_i - f(x_i, c))^2 \tag{9.104}$$

of Example 9.2 by approximating the function $f(x_i, c)$ by linear splines; see Example 9.3.

9.3 (*Programming*)

Solve the least squares problem

$$\min_c \sum_{i=1}^{m} (y_i - f(x_i, c))^2 \tag{9.105}$$

of Example 9.2 by approximating the function $f(x_i, c)$ by bellsplines; see Example 9.4.

9.4 (*Programming*)

Solve the problem of fitting a polynomial $p(x) = \sum_{i=0}^{d} c_i x^{i-1}$ of degree d to data points (x_i, y_i), $i = 1, ..., m$, in the plane by the method of normal equations and QR decomposition (either Algorithm 9.4, 9.5). Choose the degree of the polynomial to be $d = 5$ and then $d = 14$, choose the interval $x \in [0, 1]$, discretize it using N points,

---

[9] The MATLAB® programs can be found on the online version of Chapter 1 (doi:10.1007/978-3-319-57304-5_1).

and compute discrete values of $y(x)$ as $y_i = y(x_i) = p(x_i)$. Our goal is to recover the coefficients c_i of the polynomial $p(x) = \sum_{i=0}^{d} c_i x^{i-1}$ by solving the system

$$Ac = y \tag{9.106}$$

using the method of normal equations and QR decomposition (Algorithm 9.4 or Algorithm 9.5). Here, the columns of the matrix A are powers of the vector x that create the Vandermonde matrix (9.8). Compare both methods for $d = 5$ and then for $d = 14$ by computing the relative error

$$e = \frac{\|c - c^*\|_2}{\|c^*\|_2}. \tag{9.107}$$

Here, the c_i^* are the exact values of the computed coefficients c_i.
Hints:

- Compute first the values of the right-hand side of (9.106), the vector y_i, at the points x_i, $i = 1, \ldots, m$, with known values of the coefficients c_i. Take exact values $c_i = 1$.
- The matrix A is a Vandermonde matrix:

$$A = \begin{pmatrix} 1 & x_1 & x_1^2 & \cdots & x_1^d \\ 1 & x_2 & x_2^2 & \cdots & x_2^d \\ 1 & x_3 & x_3^2 & \cdots & x_3^d \\ \vdots & \vdots & \vdots & \ddots & \vdots \\ 1 & x_m & x_m^2 & \cdots & x_m^d \end{pmatrix}. \tag{9.108}$$

Here, x_i, $i = 1, \ldots, m$, are points in the interval $x \in [0, 1]$, and d is the degree of the polynomial.
- Use the method of normal equations and QR decomposition to solve the resulting system $Ax = y$. Compare your results in a table by computing the relative error (12.24) for both methods for different discretizations of the interval $x \in [0, 1]$.

9.5 Prove Theorem 9.6:
Let $H = \begin{pmatrix} 0 & A^T \\ A & 0 \end{pmatrix}$, where A is square and $A = U \Sigma V^T$ is the SVD of A. Let $\Sigma = diag(\sigma_1, \ldots, \sigma_n)$, $U = (u_1, \ldots, u_n)$, and $V = (v_1, \ldots, v_n)$. Then the $2n$ eigenvalues of H are $\pm \sigma_i$, with corresponding unit eigenvectors $\frac{1}{\sqrt{2}} \begin{pmatrix} v_i \\ \pm u_i \end{pmatrix}$.

9.6 (*Programming*)
We define the covariance matrix of the $m \times n$ least squares problem $\min_x \|Ax - b\|_2^2$ by

$$\delta^2 (A^T A)^{-1}, \tag{9.109}$$

where

$$\delta^2 = \frac{\|b - Ax\|_2^2}{m - n} \tag{9.110}$$

at the least squares solution x.

The inverse of the covariance matrix cannot be computed explicitly. Instead, for computing $A^T A$ we use

$$(A^T A)^{-1} = (R^T R)^{-1}, \qquad (9.111)$$

where R is the upper triangular matrix in the QR decomposition of A.

Implement the computation of the covariance matrix (9.109) using only the computed matrix R and then (9.111). Test your implementation on some examples to show that it gives the same result as computing $(A^T A)^{-1}$.

9.7 Let the matrix A be defined as

$$\mathbf{A} = \begin{pmatrix} 5 & 4 & 3 \\ 4 & 6 & 1 \\ 3 & 1 & 7 \end{pmatrix}. \qquad (9.112)$$

1. Transform the matrix A to tridiagonal form using Householder reflection. Describe all steps of this transformation.
2. Transform the matrix A to tridiagonal form using Givens rotation. Describe this procedure step by step.

9.8 Let us consider a weighted least squares problem. When some entries of $Ax - b$ are more important than other components, we can use scale factors d_i to weight them. Then instead of the solution $\min_x \|Ax - b\|_2$, we are interested in the solution $\min_x \|D(Ax - b)\|_2$. Here D is a diagonal matrix with entries d_i on its diagonal.

Derive the method of normal equations for this problem.

9.9 Let A be of order $m \times n$ with SVD $A = U \Sigma V^T$. Compute the SVDs of the following matrices in terms of U, Σ, and V:

1. $(A^T A)^{-1}$,
2. $(A^T A)^{-1} A^T$,
3. $A(A^T A)^{-1}$,
4. $A(A^T A)^{-1} A^T$.

9.10 Assume that we have three data points $(x_i, y_i) = (0, 1), (1, 2), (3, 3)$ and we want fit them by a polynomial $y = c_0 + c_1 x$.

1. Formulate an overdetermined linear system for the least squares problem.
2. Write corresponding normal equations.
3. Compute the least squares solution using Cholesky factorization.

9.11 Let A be of order $n \times n$. Prove that:

1. $A^T = A$,
2. $A^T A = I$,
3. $A^2 = I$.

What is a nontrivial class of matrices that have all these properties called? Give an example of a 3×3 matrix that has all three properties (other than I or a permutation of it).

9.12 Show that if a vector u is nonzero, then the matrix

$$P = I - 2\frac{uu^T}{u^T u}$$

is orthogonal and symmetric.

9.13 Let a be any nonzero vector such that $u = a - \alpha e_1$ with $\alpha = \pm \|a\|_2$ and

$$P = I - 2\frac{uu^T}{u^T u}.$$

Show that $Pa = \alpha e_1$.

9.14 Prove that the pseudoinverse A^+ of a matrix A of order $m \times n$ satisfies the following Moore–Penrose conditions:

1. $AA^+A = A$,
2. $A^+AA^+ = A^+$,
3. $(AA^+)^T = AA^+$,
4. $(A^+A)^T = A^+A$.

9.15 Let A^+ be the pseudoinverse of a matrix A of order $m \times n$. Prove:

1. If $m = n$ and A is nonsingular, then $A^+ = A^{-1}$.
2. If $m > n$ and A has rank n, then $A^+ = (A^T A)^{-1} A^T$.
3. If $m < n$ and A has rank m, then $A^+ = A^T (AA^T)^{-1}$.

9.16 (*Programming*)
Consider the nonlinear model problem:

$$y(T) = A \cdot \exp^{-\frac{E}{T-T_0}}.$$

Determine parameters A, E, T_0, which are positive constants by knowing T and output data $y(T)$.

Hint: Transform first the nonlinear function $y(T)$ to a linear one and then solve a linear least squares problem. Discretize T by N points and compute discrete values of $y(T)$ as $y_i = y(T_i)$ for the known values of the parameters A, E, T_0. Then forget about these parameters (we will call them exact parameters A^*, E^*, T_0^*) and solve the linear least squares problem using the method of normal equations (optionally QR decomposition) in order to recover these exact parameters.

Try adding random noise δ to data $y(T)$ using the formula $y_\sigma(T) = y(T)(1 + \delta\alpha)$, where $\alpha \in (-1, 1)$ is a randomly distributed number and δ is the noise level (if noise in the data is 5%, then $\delta = 0.05$).

Analyze your results by computing the relative errors e_A, e_E, e_{T_0} in the computed parameters as

$$e_A = \frac{|A - A^*|}{|A^*|},$$

$$e_E = \frac{|E - E^*|}{|E^*|},\quad\quad\quad (9.113)$$

$$e_{T_0} = \frac{|T_0 - T_0^*|}{|T_0^*|}.$$

Here, A^*, E^*, T_0^* are exact values, and A, E, T_0 are computed values.

9.17 (*Programming*)

Suppose that a nonlinear model function is given as

$$f(x, c) = Ae^{c_1 x} + Be^{c_2 x}, \text{A, B} = \text{const} > 0, \quad\quad (9.114)$$

and our goal is to fit this function using the Gauss–Newton method. In other words, we want to use following formula for iteratively updating of $c = (c_1, c_2)$:

$$c^{k+1} = c^k - [J^T(c_k)J(c_k)]^{-1}J^T(c_k)r(c_k), \quad\quad (9.115)$$

where k is the iteration number and $J(c_k)$ is the Jacobian matrix of the residual $r(c_k)$. We define the residual function

$$r(c) = y - f(x, c), \quad\quad\quad (9.116)$$

where $y = y_i$, $i = 1, ..., m$, are known data points (use information in Question 9.16 to generate data $y = y_i$, $i = 1, ..., m$).

Add random noise δ to data $y = y_i$, $i = 1, ..., m$, using the formula $y_\sigma(x, c) = f(x, c)(1 + \delta\alpha)$, where $\alpha \in (-1, 1)$ is a randomly distributed number and δ is the noise level (if the noise in the data is 5%, then $\delta = 0.05$).

Analyze the results obtained by computing the relative errors e_c in the computed parameters $c = (c_1, c_2)$ as

$$e_c = \frac{||c - c^*||_2}{||c^*||_2}. \quad\quad\quad (9.117)$$

Here $c^* = (c_1^*, c_2^*)$ are exact values.

Chapter 10
Algorithms for the Nonsymmetric Eigenvalue Problem

In this chapter, we will present the main algorithms for solving the nonsymmetric eigenvalue problem using direct methods. Direct methods are usually applied to dense matrices, and iterative methods such as the Rayleigh–Ritz method and Lanczos's algorithm are applied to sparse matrices. Iterative methods usually can compute not all of the eigenvalues and eigenvectors, but only some subset, and their convergence depends on the structure of the matrix. We will begin with an analysis of the simplest direct method, called the power method, and then continue to consider more complicated methods such as inverse iteration, orthogonal iteration, QR iteration, QR iteration with shifts, and Hessenberg reduction. To simplify our presentation in this chapter, we will assume that the matrix A is real. We will illustrate the performance of every method by MATLAB® programs which are available in the online version of Chapter 1 (doi:10.1007/978-3-319-57304-5_1). We will test these programs by computing eigenvalues and eigenvectors for different kinds of matrices, such as matrices with real and complex eigenvalues, and matrices with different multiplicities of eigenvalues.

10.1 Power Method

This method can find only the largest absolute eigenvalue and corresponding eigenvector for a matrix A.

Algorithm 10.1 The power method.

0. Set $i = 0$ and initialize x_0.
1. Compute $y_{i+1} = Ax_i$.

The original version of this chapter was revised. An Erratum to this book can be found at https://doi.org/10.1007/978-3-319-57304-5_10

© Springer International Publishing AG 2017
L. Beilina et al., *Numerical Linear Algebra: Theory and Applications*,
DOI 10.1007/978-3-319-57304-5_10

2. Compute the approximate normalized eigenvector as $x_{i+1} = y_{i+1}/\|y_{i+1}\|$.[1]
3. Compute the approximate eigenvalue $\tilde{\lambda}_{i+1} = x_{i+1}^T A x_{i+1}$.
4. Stop updating the approximate eigenvalue and set $\tilde{\lambda}_M = \tilde{\lambda}_{i+1}, M = i + 1$, if either $|\tilde{\lambda}_{i+1} - \tilde{\lambda}_i| \le \theta$ or the absolute values of the differences $|\tilde{\lambda}_{i+1} - \tilde{\lambda}_i|$ stabilize. Here θ is a tolerance number. Otherwise, set $i = i + 1$ and go to step 1.

Theorem 10.1 *Let A be a diagonalizable matrix,*[2] *i.e., $A = S\Lambda S^{-1}$, where the matrix Λ is $\mathrm{diag}(\lambda_1, \lambda_2, \ldots, \lambda_n)$ and the eigenvalues satisfy the inequalities*

$$|\lambda_1| > |\lambda_2| \ge \cdots \ge |\lambda_n|.$$

Write the matrix $S = (s_1, s_2, \ldots, s_n)$, where the columns s_i are the corresponding eigenvectors that also satisfy $\|s_i\| = 1$. Then the approximate eigenvector computed in step 2 of Algorithm 10.1 converges to the eigenvector s_1, which corresponds to λ_1, and the approximate eigenvalue converges to λ_1.

Proof First we will prove the theorem for the case that A is a diagonal matrix. Let $A = \mathrm{diag}(\lambda_1, \ldots, \lambda_n)$, with $|\lambda_1| > |\lambda_2| \ge \cdots \ge |\lambda_n|$. In this case, the eigenvectors x_i of A are columns $i_1, i_2, \ldots, i_n$ of the identity matrix I. We note that the factors $1/\|y_{i+1}\|$ in step 2 of Algorithm 10.1 scale x_{i+1} to be a unit vector and do not change its direction. Then x_i can also be written as $x_i = A^i x_0 / \|A^i x_0\|$. Let us write the vector x_0 in the form $x_0 = S(S^{-1}x_0) = S((x_1^{(0)}, \ldots, x_n^{(0)})^T)$, where the matrix S is the identity and $x_0 = (x_1^{(0)}, \ldots, x_n^{(0)})^T$. Assuming that $x_1^{(0)} \ne 0$, we get

$$A^i x_0 \equiv A^i \begin{pmatrix} x_1^{(0)} \\ x_2^{(0)} \\ \vdots \\ x_n^{(0)} \end{pmatrix} = \begin{pmatrix} x_1^{(0)} \lambda_1^i \\ x_2^{(0)} \lambda_2^i \\ \vdots \\ x_n^{(0)} \lambda_n^i \end{pmatrix} = x_1^{(0)} \lambda_1^i \begin{pmatrix} 1 \\ \left(x_2^{(0)}/x_1^{(0)}\right)(\lambda_2/\lambda_1)^i \\ \vdots \\ \left(x_n^{(0)}/x_1^{(0)}\right)(\lambda_n/\lambda_1)^i \end{pmatrix}. \tag{10.1}$$

Using (10.1), we observe that all the fractions $|\lambda_j/\lambda_1|, j = 2, \ldots, n$, are less than 1. At every ith iteration, $A^i x_0$ becomes more nearly parallel to i_1, so that $x_i = A^i x_0 / \|A^i x_0\|$ will be closer to $\pm i_1$, which is an eigenvector corresponding to the largest eigenvalue λ_1. Since x_i converges to $\pm i_1$, the computed eigenvalue $\tilde{\lambda}_i = x_i^T A x_i$ converges to the largest eigenvalue λ_1.

Consider now the general case in which the matrix $A = S\Lambda S^{-1}$ is diagonalizable. We write again the vector x_0 as $x_0 = S(S^{-1}x_0) = S((x_1^{(0)}, \ldots, x_n^{(0)})^T)$ to get

$$A^i = \underbrace{(S\Lambda S^{-1}) \cdots (S\Lambda S^{-1})}_{i \text{ times}} = S\Lambda^i S^{-1}.$$

Here we have used the fact that all $S^{-1}S$ pairs cancel. Because of that, we can write

---

[1] In this chapter, the norm of a vector is $\|\cdot\|_2$, as defined on p. 211.
[2] See Section 4.2.3, p. 120.

$$A^i x_0 = (S\Lambda^i S^{-1})S \begin{pmatrix} x_1^{(0)} \\ x_2^{(0)} \\ \vdots \\ x_n^{(0)} \end{pmatrix} = S \begin{pmatrix} x_1^{(0)}\lambda_1^i \\ x_2^{(0)}\lambda_2^i \\ \vdots \\ x_n^{(0)}\lambda_n^i \end{pmatrix} = x_1^{(0)}\lambda_1^i S \begin{pmatrix} 1 \\ \left(x_2^{(0)}/x_1^{(0)}\right)(\lambda_2/\lambda_1)^i \\ \vdots \\ \left(x_n^{(0)}/x_1^{(0)}\right)(\lambda_n/\lambda_1)^i \end{pmatrix}.$$

Similarly to the first case, the vector in parentheses converges to $i_1 = (1, 0, ..., 0)$, so $A^i x_0$ gets closer and closer to a multiple of $Si_1 = s_1$, the eigenvector corresponding to λ_1. Therefore, $\tilde{\lambda}_i = x_i^T A x_i$ converges to $s_1^T A s_1 = s_1^T \lambda_1 s_1 = \lambda_1$. $\qquad\square$

Remark 10.1

1. One of the drawbacks of the power method is the assumption that the element $x_1^{(0)}$ is nonzero, i.e., that x_0 is not from the invariant subspace span$\{s_2, \ldots, s_n\}$.[3] If x_0 is chosen as a random vector, this is true with very high probability.
2. A major drawback is that the power method converges only to the eigenvalue of the largest absolute magnitude. Theorem 10.1 states that the power method converges under the assumption that all other eigenvalues of the matrix are smaller in absolute value.
3. The rate of convergence depends on the ratios $|\lambda_2/\lambda_1| \geq \cdots \geq |\lambda_n/\lambda_1|$. If the ratios $|\lambda_2/\lambda_1| \geq \cdots \geq |\lambda_n/\lambda_1|$ are much smaller than 1, then we will get faster convergence. Otherwise, convergence will be slower.

Below we present some examples that illustrate convergence of the power method for different kinds of matrices A. As an initial guess for the approximate eigenvector x_0, we take normalized randomly distributed numbers in the interval $(0, 1)$. The MATLAB® programs of Section 1.9 are available for running all tests of this section.[4]

Example 10.1 In this example, we test the matrix

$$A = \begin{pmatrix} 5 & 0 & 0 \\ 0 & 2 & 0 \\ 0 & 0 & -5 \end{pmatrix}$$

with the exact eigenvalues 5, 2, and -5. The MATLAB® program of Section 1.9 is available for running of this test. The power method can converge to the exact first eigenvalue 5, as well as to a completely erroneous eigenvalue. This is because two eigenvalues of this matrix, 5 and -5, have the same absolute values, $|5| = |-5|$, as well as because the initial guess $x_1^{(0)}$ in the MATLAB® program is chosen randomly. Thus, assumptions 1, 2 of Remark 10.1 about the convergence of the power method are not fulfilled.

---

[3] See Theorem 3.4, p. 83.
[4] All MATLAB® programs referenced on this page can be found in the online version of Chapter 1 (doi:10.1007/978-3-319-57304-5_1).

Example 10.2 In this example, the matrix A is given by

$$A = \begin{pmatrix} 3 & 7 & 8 & 9 \\ 5 & -7 & 4 & -7 \\ 1 & -1 & 1 & -1 \\ 9 & 3 & 2 & 5 \end{pmatrix}.$$

This matrix has four different real reference eigenvalues[5] $\lambda = (\lambda_1, ..., \lambda_4)$ given by $\lambda = (12.3246, -11.1644, -0.3246, 1.1644)$. Thus, all assumptions of Theorem 10.1 about the matrix A are fulfilled, and we run the MATLAB® program of Section 1.9 to get the reference eigenvalue 12.3246.[6]

Example 10.3 Now we take the matrix

$$A = \begin{pmatrix} 0 & -5 & 2 \\ 6 & 0 & -12 \\ 1 & 3 & 0 \end{pmatrix}$$

with the following one real and two complex eigenvalues (with the largest absolute value): $\lambda = (1.4522, -0.7261 + 8.0982i, -0.7261 - 8.0982i)$. We run the MATLAB® program of Section 1.9 and observe that the power method does not converge in this case. Clearly, starting from a real initial approximation $x_0 \in \mathbb{R}^3$ in Algorithm 10.1, we cannot compute an approximate complex eigenvalue.

Example 10.4 In this example, the matrix A has order 5. The elements of this matrix are uniformly distributed pseudorandom numbers in the open interval $(0, 1)$. We run the MATLAB® program of Section 1.9 and observe that sometimes we can obtain a good approximation to the eigenvalue 2.9. In a second round of computations, we can get a completely different erroneous eigenvalue. This is because we randomly generate elements of the matrix A as well as because the initial guess $x_1^{(0)}$ in the MATLAB® program is chosen randomly. Thus, assumptions 1, 2 of Remark 10.1 about the convergence of the power method cannot be fulfilled. This example is similar to Example 10.1, where convergence was not achieved.

10.2 Inverse Iteration

The method of inverse iteration can find all eigenvalues and eigenvectors obtained by applying the power method for $(A - \sigma I)^{-1}$ for some shift σ. This means that we will apply the power method to the matrix $(A - \sigma I)^{-1}$ instead of A, which will converge to the eigenvalue closest to σ, rather than just λ_1. This method is called the *method of inverse iteration* or the *inverse power method*.

---

[5]We get the reference eigenvalues in all examples using the command `eig(A)` in MATLAB®. These eigenvalues are computed to high precision.

[6]All MATLAB® programs referenced on this page can be found on the online version of Chapter 1 (doi:10.1007/978-3-319-57304-5_1).

Algorithm 10.2 The method of inverse iteration.

0. Set $i = 0$ and initialize x_0. Choose a shift σ.
1. Compute $y_{i+1} = (A - \sigma I)^{-1} x_i$.
2. Compute the approximate normalized eigenvector as $x_{i+1} = y_{i+1}/\|y_{i+1}\|$.
3. Compute the approximate eigenvalue $\tilde{\lambda}_{i+1} = x_{i+1}^T A x_{i+1}$.
4. Stop updating the approximate eigenvalue and set $\tilde{\lambda}_M = \tilde{\lambda}_{i+1}, M = i + 1$, if either $|\tilde{\lambda}_{i+1} - \tilde{\lambda}_i| \le \theta$ or the absolute values of the differences $|\tilde{\lambda}_{i+1} - \tilde{\lambda}_i|$ have stabilized. Here θ is a tolerance number. Otherwise, set $i = i + 1$ and go to step 1.

Theorem 10.2 *Let A be a diagonalizable matrix, i.e., $A = S \Lambda S^{-1}$, where the matrix Λ is* $\mathrm{diag}(\lambda_1, \lambda_2, \ldots, \lambda_n)$. *Assume that for a given shift σ, the following inequalities hold:*

$$|\lambda_k - \sigma| < |\lambda_i - \sigma| \quad \forall i \ne k. \tag{10.2}$$

Write $S = (s_1, s_2, \ldots, s_n)$, where the columns s_i are the corresponding eigenvectors and they also satisfy $\|s_i\| = 1$. Then the approximate eigenvector computed at step 2 of Algorithm 10.2 converges to the eigenvector s_k, which corresponds to λ_k, and the approximate eigenvalue converges to λ_k.

Proof We begin the proof by noting that the matrix $A = S \Lambda S^{-1}$ is diagonalizable. Then $A - \sigma I = S(\Lambda - \sigma I)S^{-1}$, and hence $(A - \sigma I)^{-1} = S(\Lambda - \sigma I)^{-1}S^{-1}$. Thus the matrix $(A - \sigma I)^{-1}$ has the same eigenvectors s_i as A with the corresponding eigenvalues $((\Lambda - \sigma I)^{-1})_{jj} = (\lambda_j - \sigma)^{-1}$. By assumption, $|\lambda_k - \sigma|$ is smaller than all the other $|\lambda_i - \sigma|$. This means that $(\lambda_k - \sigma)^{-1}$ is the largest eigenvalue in absolute value. As in the proof of Theorem 10.1, we write $x_0 = S(S^{-1}x_0) = S((x_1^{(0)}, \ldots, x_n^{(0)})^T)$ and assume $x_k^{(0)} \ne 0$. Then we get

$$(A - \sigma I)^i = \underbrace{(S(\Lambda - \sigma I)S^{-1}) \cdots (S(\Lambda - \sigma I)S^{-1})}_{i \text{ times}} = S(\Lambda - \sigma I)^i S^{-1},$$

where all $S^{-1}S$ pairs cancel. This means that

$$(A - \sigma I)^{-i} x_0 = (S(\Lambda - \sigma I)^{-i} S^{-1}) S \begin{pmatrix} x_1^{(0)} \\ x_2^{(0)} \\ \vdots \\ x_n^{(0)} \end{pmatrix}$$

$$= S \begin{pmatrix} x_1^{(0)}(\lambda_1 - \sigma)^{-i} \\ \vdots \\ x_n^{(0)}(\lambda_n - \sigma)^{-i} \end{pmatrix} = x_k^{(0)}(\lambda_k - \sigma)^{-i} S \begin{pmatrix} \left(x_1^{(0)}/x_k^{(0)}\right)(\lambda_k - \sigma)^i/(\lambda_1 - \sigma)^i \\ \vdots \\ 1 \\ \vdots \\ \left(x_n^{(0)}/x_k^{(0)}\right)(\lambda_k - \sigma)^i/(\lambda_n - \sigma)^i \end{pmatrix},$$

where we put 1 in the kth entry. Since by assumption (10.2), all the fractions $|(\lambda_k - \sigma)/(\lambda_i - \sigma)|$ are less than 1, and the vector in parentheses will approximate i_k such that $\|(A - \sigma I)^{-i} x_0\|$ will be closer to a multiple of $S i_k = s_k$, which is the eigenvector corresponding to λ_k. As in Theorem 10.1, we see that $\tilde{\lambda}_i = x_i^T A x_i$ also converges to λ_k.   □

Remark 10.2

1. The advantage of inverse iteration over the power method is the ability to converge to any desired eigenvalue (the one nearest the shift σ).
2. By choosing σ very close to a desired eigenvalue, we can achieve rapid convergence and are thus not as limited by the proximity of nearby eigenvalues as in the original power method.
3. The method is very effective when we have a good approximation to an eigenvalue and want only its corresponding eigenvector.

To test the performance of the inverse iteration method, we run the MATLAB® program of Section 1.10 with different shifts σ.[7] We tested the same matrices as in the power method of Section 10.1, except Example 10.1.

Example 10.5 In this example, we tested the matrix

$$A = \begin{pmatrix} 0 & 10 \\ 0 & 0 \end{pmatrix},$$

which has the exact eigenvalues $\lambda = (0, 0)$ with multiplicity $m = 2$. Note that in this example and in all other examples of this section, we made the additional transformation of the original matrix A as $\tilde{A} = Q^T A Q$, where Q was an orthogonal matrix that was generated in MATLAB® as $Q = \text{orth}(\text{rand}(n, n))$, where n is the size of the matrix A. Running the MATLAB® program of Section 1.10, we observe that the inverse iteration method could converge to the reference eigenvalues for both shifts $\sigma = 2$ and $\sigma = 10$. We also note that by applying the power method to this matrix, we could get only *NaN* as a result.

Example 10.6 We tested the matrix of Example 10.2. Let us recall that the reference eigenvalues in this example are $\lambda = (12.3246, -11.1644, -0.3246, 1.1644)$. Running the MATLAB® program of Section 1.10, we observe nice convergence. For $\sigma = 2$ we were able to obtain the eigenvalue 1.1644, which is the same as the last reference eigenvalue. This is because the shift $\sigma = 2$ is closer to this eigenvalue than to all others. For the shift $\sigma = 10$, the algorithm converged to the first reference eigenvalue 12.3246, as expected.

This test confirms that the inverse iteration method converges to the eigenvalue that is closest to the shift σ.

Example 10.7 We tested the matrix of Example 10.3. Running the MATLAB® program of Section 1.10 allowed us to obtain nice convergence in this case, too, for both shifts σ. Recall that the power method does not converge at all in Example 10.3.

---

[7]All MATLAB® programs referenced on this page can be found on the online version of Chapter 1 (doi:10.1007/978-3-319-57304-5_1).

Example 10.8 We tested the matrix of Example 10.4. Again, running the MATLAB®
program of Section 1.10, we observe nice convergence to the first eigenvalue of the
matrix A for both shifts $\sigma = 2, 10.$[8]

10.3 Orthogonal Iteration

In this section we will consider the method of *orthogonal iteration*, which converges
to a p-dimensional invariant subspace (with $p > 1$) rather than to one eigenvector as
in the two previous methods. The method of orthogonal iteration is sometimes called
the method of *subspace iteration* or *simultaneous iteration*.

Let Q_0 be an $n \times p$ orthogonal matrix with $p \le n$. Our goal is to compute eigenvalues and eigenvectors of the square matrix A of order n. To do so, we perform the
following iterative algorithm.

Algorithm 10.3 Orthogonal iteration.

0. Set $i = 0$ and initialize a matrix Q_0.
1. Compute $Y_{i+1} = AQ_i$.
2. Factorize Y_{i+1} using QR decomposition (see Section 9.4) to obtain the matrices
    Q_{i+1} and R_{i+1}. The matrix Q_{i+1} spans an approximate invariant subspace.
3. Compute $T_{i+1} = Q_{i+1}^T A \, Q_{i+1}$.
4. Compute the vector of approximate eigenvalues $\tilde{\lambda}_{i+1} = (\tilde{\lambda}_1^{(i+1)}, ..., \tilde{\lambda}_p^{(i+1)})$ from
    the real Schur block (see Theorem 4.27, p. 131) of the matrix T_{i+1}. The approxi-
    mate eigenvectors will be the columns of Q_{i+1}.
5. Stop updating the eigenvalues and set $\tilde{\lambda}_M = \tilde{\lambda}_{i+1}, M = i+1$, if either the norm
    $\|\tilde{\lambda}_{i+1} - \tilde{\lambda}_i\| \le \theta$ or the differences $\|\tilde{\lambda}_{i+1} - \tilde{\lambda}_i\|$ are stabilized or the subdiagonal
    entries of T_i are small enough (smaller than the rounding errors of size $O(\varepsilon\|T_i\|)$).
    Here θ is a tolerance number, and ε is machine epsilon.[9] Otherwise, set $i = i + 1$
    and go to step 1.

Theorem 10.3 *Assume that $A = S\Lambda S^{-1}$ is diagonalizable, $\Lambda = \text{diag}(\lambda_1, \lambda_2,$
$..., \lambda_n)$, the eigenvalues sorted so that $|\lambda_1| \ge |\lambda_2| \ge \cdots \ge |\lambda_n|$ and $|\lambda_n| > 0$ or
$|\lambda_p| > |\lambda_{p+1}|$, where $p \in [1, n-1]$ is an integer. Write $S = (s_1, s_2, ..., s_n)$, where
the columns s_i are the corresponding eigenvectors, and they also satisfy $\|s_i\| = 1$. Let
$Q_{i+1}R_{i+1}$ be the QR decomposition of the matrix $Y_{i+1} = AQ_i$ on step 2 and iteration
i in Algorithm 10.3. Then $\text{span}(Q_i)$ converges to $\text{span}(S_p)$, the invariant subspace
spanned by the first p eigenvectors, $1 \le p \le n$.*

Proof We assume that $|\lambda_p| > |\lambda_{p+1}|$. If we set $p = 1$, then the method of orthogonal
iteration and its analysis are identical to the power method.

---

[8]The MATLAB® programs can be found on the online version of Chapter 1
(doi:10.1007/978-3-319-57304-5_1).

[9]The machine epsilon ε represents the upper bound on the relative error due to rounding in floating-
point arithmetic.

Consider now the case $p > 1$. Using step 1 of Algorithm 10.3, we can write $\text{span}(Q_{i+1}) = \text{span}(Y_{i+1}) = \text{span}(AQ_i)$. Thus, we conclude that the following equalities hold: $\text{span}(Q_i) = \text{span}(A^i Q_0) = \text{span}(S\Lambda^i S^{-1} Q_0)$. We also note that

$$S\Lambda^i S^{-1} Q_0 = S \operatorname{diag}(\lambda_1^i, \ldots, \lambda_n^i) S^{-1} Q_0$$

$$= \lambda_p^i S \begin{pmatrix} (\lambda_1/\lambda_p)^i & & & & \\ & \ddots & & & \\ & & 1 & & \\ & & & \ddots & \\ & & & & (\lambda_n/\lambda_p)^i \end{pmatrix} S^{-1} Q_0. \quad (10.3)$$

By the assumption that $|\lambda_p| > |\lambda_{p+1}|$, we have $|\lambda_j/\lambda_p| \geq 1$ for $j \leq p$ and $|\lambda_j/\lambda_p| < 1$ if $j > p$. Then for the entries of the matrix Λ, we get

$$\begin{pmatrix} (\lambda_1/\lambda_p)^i & & \\ & \ddots & \\ & & (\lambda_n/\lambda_p)^i \end{pmatrix} S^{-1} Q_0 = \begin{pmatrix} V_i^{p \times p} \\ W_i^{(n-p) \times p} \end{pmatrix} = X_i,$$

where the elements of the submatrix W_i tend to zero like $(\lambda_{p+1}/\lambda_p)^i$, and the elements of the submatrix V_i do not converge to zero. This is true, since if V_0 has full rank (by assumption, we have $\lambda_p \neq 0$), then the V_i have full rank, too. Now we write the matrix of eigenvectors $S = (s_1, \ldots, s_n)$ as $\left(S_p^{n \times p}, \hat{S}_p^{n \times (n-p)} \right)$ or $S_p = (s_1, \ldots, s_p)$. Then we get

$$S\Lambda^i S^{-1} Q_0 = \lambda_p^i S \begin{pmatrix} V_i^{p \times p} \\ W_i^{(n-p) \times p} \end{pmatrix} = \lambda_p^i \left(S_p^{n \times p} V_i^{p \times p} + \hat{S}_p^{n \times (n-p)} W_i^{(n-p) \times p} \right).$$

Thus,

$$\text{span}(Q_i) = \text{span}\left(S\Lambda^i S^{-1} Q_0 \right)$$

$$= \text{span}\left(S_p^{n \times p} V_i^{p \times p} + \hat{S}_p^{n \times (n-p)} W_i^{(n-p) \times p} \right) = \text{span}(S_p X_i) \quad (10.4)$$

converges to $\text{span}\left(S_p V_i \right) = \text{span}\left(S_p \right)$, the invariant subspace spanned by the first p eigenvectors, as stated in the theorem. $\square$

The next theorem states that under certain assumptions, by the method of orthogonal iteration we can compute eigenvalues of A from the Schur form of A.

Theorem 10.4 *Let us consider Algorithm 10.3 applied to the matrix A with $p = n$ and $Q_0 = I$. If all the eigenvalues of A have distinct absolute values and if all the principal submatrices $S(1 : j, 1 : j)$ are nonsingular, then the sequence of the matrices $T_i = Q_i^T A Q_i$ converges to the Schur form of A, i.e., an upper triangular* *matrix with the eigenvalues on the diagonal. The eigenvalues will appear in decreasing order of absolute value.*

Proof Using the assumption of the nonsingularity of $S(1:j, 1:j)$ for all j, we have that X_0 in the proof of Theorem 10.3 is nonsingularity. This means that no vector in the invariant subspace span$\{s_1, \ldots, s_j\}$ is orthogonal to span$\{i_1, \ldots, i_j\}$, which is the space spanned by the first j columns of $Q_0 = I$. First note that Q_i is a square orthogonal matrix, so A and $T_i = Q_i^T A Q_i$ are similar. We can decompose the matrix Q_i into two submatrices as $Q_i = (Q_{1i}, Q_{2i})$, where Q_{1i} has j columns and Q_{2i} has $n - j$ columns such that

$$T_i = Q_i^T A Q_i = \begin{pmatrix} Q_{1i}^T A Q_{1i} & Q_{1i}^T A Q_{2i} \\ Q_{2i}^T A Q_{1i} & Q_{2i}^T A Q_{2i} \end{pmatrix}. \tag{10.5}$$

Since span(Q_{1i}) converges to an invariant subspace of A, span(AQ_{1i}) converges to the same subspace. Next, $Q_{2i}^T A Q_{1i}$ converges to $Q_{2i}^T Q_{1i} = 0$. This is because we have $Q_i^T = (Q_{1i}, Q_{2i})^T$ and $Q_i^T Q_i = I$, which means that

$$I = Q_i^T Q_i = (Q_{1i}, Q_{2i})^T (Q_{1i}, Q_{2i}) = \begin{pmatrix} Q_{1i}^T Q_{1i} & Q_{1i}^T Q_{2i} \\ Q_{2i}^T Q_{1i} & Q_{2i}^T Q_{2i} \end{pmatrix} = \begin{pmatrix} I & 0 \\ 0 & I \end{pmatrix}.$$

Since $Q_{2i}^T A Q_{1i}$ converges to zero in (10.5) for all $j < n$, every subdiagonal entry of T_i converges to zero, and thus T_i converges to upper triangular form, i.e., Schur form. We see that the submatrix $Q_{2i}^T A Q_{1i} = T_i(j+1:n, 1:j)$ should converge to zero like $|\lambda_{j+1}/\lambda_j|^i$. Thus, λ_j should appear as the (j,j) entry of T_i and converge like $\max(|\lambda_{j+1}/\lambda_j|^i, |\lambda_j/\lambda_{j-1}|^i)$.   □

Remark 10.3

1. The use of the QR decomposition keeps the vectors spanning span$(A^i Q_0)$ of full rank despite rounding.
2. The method of orthogonal iteration is effectively running the algorithm for all $\tilde{p} = 1, 2, \ldots, p$ at the same time. If all the eigenvalues have distinct absolute values, the same convergence analysis as in Theorem 10.3 implies that the first $\tilde{p} \leq p$ columns of Q_i converge to span$\{s_1, \ldots, s_{\tilde{p}}\}$ for all $\tilde{p} \leq p$.
3. If all assumptions of Theorem 10.4 hold, then we can set $p = n$ and $Q_0 = I$ in Algorithm 10.3 in order to obtain all eigenvalues and corresponding eigenvectors of the matrix A.

We test the performance of the method of orthogonal iteration using the MATLAB® program of Section 1.11.[10] In this program we compute the eigenvalues and corresponding eigenvectors in six different cases which are described below.

Example 10.9 In this example, we tested the Hilbert matrix (3.46), p. 90, of order 10. Let us recall that the elements of this matrix are given by $1/(i + j - 1)$, where $i, j = 1, 2, \ldots, n$. From Fig. 10.1, Example 1, we observe that we have obtained all

---

[10]The MATLAB® programs can be found on the online version of Chapter 1 (doi:10.1007/978-3-319-57304-5_1).

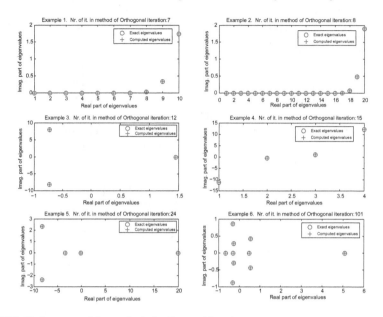

Fig. 10.1 Performance of the method of orthogonal iteration

computed eigenvalues of this matrix, which coincide with the reference eigenvalues from the seventh iteration.

Example 10.10 Here we tested the Hilbert matrix of order 20. Again we computed almost exact eigenvalues of this matrix, this time at the eighth iteration; see Fig. 10.1, Example 2.

Example 10.11 This is the same as Example 10.3 with the power method. Figure 10.1, Example 3 shows the nice convergence to the one real and two complex eigenvalues of the matrix A at the twelfth iteration.

Example 10.12 This is the same as Example 10.2 with the power method. Figure 10.1, Example 4 shows convergence to the four real eigenvalues of the matrix A at the fifteenth iteration.

Example 10.13 Here we tested the matrix

$$A = \begin{pmatrix} 3 & 7 & 8 & 9 & 12 \\ 5 & -7 & 4 & -7 & 8 \\ 1 & 1 & -1 & 1 & -1 \\ 4 & 3 & 2 & 1 & 7 \\ 9 & 3 & 2 & 5 & 4 \end{pmatrix},$$

which has three real and two complex reference eigenvalues:

$$\lambda = (19.9655, -8.2137 + 2.3623i, -8.2137 - 2.3623i, -3.4043, -0.1337).$$

From Fig. 10.1, Example 5 we observe the convergence of all the computed eigenvalues to the reference eigenvalues at the 24th iteration.

Example 10.14 Here we choose a square matrix of order 10 whose elements are uniformly distributed pseudorandom numbers in the open interval $(0, 1)$. Using Fig. 10.1, Example 6 we observe the convergence of the computed eigenvalues to the reference ones at the 101th iteration.

10.4 QR Iteration

Now we will consider an improvement of the method of orthogonal iteration, namely the method of QR iteration. This method reorganizes the method of orthogonal iteration and is more efficient, since for the variant with shifts (see the next section) it does not requires the assumption of distinct absolute eigenvalues of A, in contrast to Theorem 10.4.

Let A_0 be an $n \times n$ matrix, and our goal is to compute eigenvalues and eigenvectors of this matrix. To do so, we perform the following iterative algorithm.

Algorithm 10.4 The method of QR iteration.

0. Set $i = 0$ and initialize a matrix A_0.
1. Compute the QR decomposition of A_i such that $A_i = Q_i R_i$.
2. Compute $A_{i+1} = R_i Q_i$.
3. Compute the vector of the approximate eigenvalues $\tilde{\lambda}_{i+1} = (\tilde{\lambda}_1^{(i+1)}, ..., \tilde{\lambda}_p^{(i+1)})$ from the real Schur block of the matrix A_{i+1}. The approximate eigenvectors will be the columns of Q_i.
4. Stop updating the eigenvalues and set $\tilde{\lambda}_M = \tilde{\lambda}_{i+1}$, $M = i + 1$, if either the norm $\|\tilde{\lambda}_{i+1} - \tilde{\lambda}_i\| \leq \theta$ or the differences $\|\tilde{\lambda}_{i+1} - \tilde{\lambda}_i\|$ are stabilized or the subdiagonal elements of A_{i+1} are small enough (smaller than the rounding errors of size $O(\varepsilon \|A_{i+1}\|)$. Here θ is a tolerance number, and ε is the machine epsilon. Otherwise, set $i = i + 1$ and go to step 1.

By step 2 of Algorithm 10.4, we have $A_{i+1} = R_i Q_i$. Using step 1, we can also write that $R_i Q_i = Q_i^T (Q_i R_i) Q_i = Q_i^T A_i Q_i$. From both equalities, we see that the matrices A_{i+1} and A_i are orthogonally similar. The next theorem states that the matrix A_i computed by QR iteration is identical to the matrix $Q_i^T A Q_i$ implicitly computed by the method of orthogonal iteration.

Theorem 10.5 *Let A_i be the matrix computed by Algorithm 10.4. Then $A_i = Q_i^T A Q_i$, where Q_i is the matrix computed from the method of orthogonal iteration (Algorithm 10.3) starting with $Q_0 = I$. Thus, A_i converges to the Schur form if all the eigenvalues have different absolute values.*

Proof We use induction. Assume $A_i = Q_i^T A Q_i$. Using step 2 of Algorithm 10.3, we can write $A Q_i = Q_{i+1} R_{i+1}$, where Q_{i+1} is orthogonal and R_{i+1} is upper triangular. Hence,

$$A_i = Q_i^T A Q_i = Q_i^T (Q_{i+1} R_{i+1}) = (Q_i^T Q_{i+1}) R_{i+1} = QR.$$

This is the product of an orthogonal matrix $Q = Q_i^T Q_{i+1}$ and an upper triangular matrix $R = R_{i+1} = Q_{i+1}^T A Q_i$ (this is because $A Q_i = Q_{i+1} R_{i+1}$, and thus multiplying both sides of this equality by Q_{i+1}^T, we get $R = R_{i+1} = Q_{i+1}^T A Q_i$). Since the QR decomposition is unique (except for possibly multiplying each column of Q and row of R by -1), this is the QR decomposition $A_i = QR$. Then

$$Q_{i+1}^T A Q_{i+1} = Q_{i+1}^T (A Q_i Q_i^T) Q_{i+1} = (Q_{i+1}^T A Q_i)(Q_i^T Q_{i+1}) = R_{i+1}(Q_i^T Q_{i+1}) = RQ.$$

This is precisely how the QR iteration maps A_i to A_{i+1}, and therefore, $A_{i+1} = Q_{i+1}^T A Q_{i+1}$, as desired. Thus, the convergence of the method of QR iteration follows from the convergence of the method of orthogonal iteration. If all the eigenvalues of A have different absolute values, the proof is similar to the proof of Theorem 10.4 applied to the matrix $A_{i+1} = Q_{i+1}^T A Q_{i+1}$. $\square$

Example 10.15 We test the performance of the method of QR iteration for six different matrices that are chosen the same as for the method of orthogonal iteration. Running the MATLAB® program of Section 1.12, we observe that in the method of QR iteration we obtain the same rate of convergence as in the method of orthogonal iteration.[11]

10.5 QR Iteration with Shifts

From previous sections we know that the convergence rate depends on the ratios of eigenvalues. In order to speed convergence of the method of QR iteration, we can use shifts. Let A_0 be an $n \times n$ matrix, and our goal is to compute eigenvalues and eigenvectors of this matrix. To do so, we perform the following iterative algorithm.

Algorithm 10.5 The method of QR iteration with shifts.

0. Set $i = 0$ and initialize a matrix A_0. Choose an initial shift σ_0.
1. Compute the QR decomposition of $A_i - \sigma_i I$ such that $A_i - \sigma_i I = Q_i R_i$.
2. Compute $A_{i+1} = R_i Q_i + \sigma_i I$.
3. Compute the vector of the approximate eigenvalues $\tilde{\lambda}_{i+1} = (\tilde{\lambda}_1^{(i+1)}, ..., \tilde{\lambda}_p^{(i+1)})$ from the real Schur block of the matrix A_{i+1}. The approximate eigenvectors will be the columns of Q_i.
4. Stop updating the eigenvalues and set $\tilde{\lambda}_M = \tilde{\lambda}_{i+1}$, $M = i + 1$, if either the norm $\|\tilde{\lambda}_{i+1} - \tilde{\lambda}_i\| \le \theta$ or the differences $\|\tilde{\lambda}_{i+1} - \tilde{\lambda}_i\|$ have stabilized or the subdiagonal elements of A_{i+1} are small enough (smaller than the rounding errors of size $O(\varepsilon \|A_{i+1}\|)$). Here θ is a tolerance number, and ε is the machine epsilon. Otherwise, set $i = i + 1$, choose a shift σ_i, and go to step 1.

---

[11] The MATLAB® programs can be found on the online version of Chapter 1 (doi:10.1007/978-3-319-57304-5_1).

Lemma 10.1 *The matrices A_i and A_{i+1} in Algorithm 10.5 are orthogonally similar.*

Proof From step 2 of Algorithm 10.5 we see that

$$A_{i+1} = R_i Q_i + \sigma_i I = Q_i^T Q_i R_i Q_i + \sigma_i Q_i^T Q_i = Q_i^T (Q_i R_i + \sigma_i I) Q_i.$$

Using step 1 of Algorithm 10.5, we observe that $Q_i^T (Q_i R_i + \sigma_i I) Q_i = Q_i^T A_i Q_i$, and thus $A_{i+1} = Q_i^T A_i Q_i$. □

If R_i is nonsingular, we can also get

$$A_{i+1} = R_i Q_i + \sigma_i I = R_i Q_i R_i R_i^{-1} + \sigma_i R_i R_i^{-1} = R_i (Q_i R_i + \sigma_i I) R_i^{-1} = R_i A_i R_i^{-1}.$$

Remark 10.4

1. If σ_i is an exact eigenvalue of A_i, then the method of QR iteration with the shift σ_i converges in one step. This is because if σ_i is an eigenvalue, then $A_i - \sigma_i I$ is singular and R_i is singular, which means that some diagonal entry of R_i must be zero. Assume that the (n, n)th element of the matrix R_i is zero, $R_i(n, n) = 0$. Then the last row of $R_i Q_i$ is zero, and the last row of the matrix $A_{i+1} = R_i Q_i + \sigma_i I$ equals $\sigma_i i_n^T$, where i_n is the nth column of A_i.

   We can also say that the last row of A_{i+1} is zero except for the eigenvalue σ_i appearing in the (n, n)th entry. This means that Algorithm 10.5 has converged, because we have obtained that A_{i+1} is a block upper triangular matrix, with a 1×1 block σ_i: $A_{i+1} = \begin{pmatrix} A' & a \\ 0 & \sigma_i \end{pmatrix}$. In this matrix, the leading $(n - 1) \times (n - 1)$ block A' is a new matrix, where QR iteration can be used again without changing σ_i.

2. When σ_i is not an exact eigenvalue, we will have convergence to the matrix $A_{i+1}(n, n)$ when the lower left block $A_{i+1}(n, 1 : n - 1)$ is small enough. Recall that in the convergence of the method of inverse iteration (see Theorem 10.1), we expect that $A_{i+1}(n, 1 : n - 1)$ will shrink by a factor $|\lambda_k - \sigma_i| / \min_{j \neq k} |\lambda_j - \sigma_i|$, where $|\lambda_k - \sigma_i| = \min_j |\lambda_j - \sigma_i|$. This means that if σ_i is a very good approximation to the eigenvalue λ_k, then we will have fast convergence.

Now we will concentrate on how to choose shifts σ_i in Algorithm 10.5 in order to get accurate approximate eigenvalues. When we want to get good convergence to the n-eigenvalue of the matrix A, then the choice $\sigma_i = A_i(n, n)$ for a shift is a good one. Such a choice of shift means local quadratic convergence to a real eigenvalue in Algorithm 10.5. This means that the number of correct digits doubles at every step i of Algorithm 10.5. However, it is difficult to get global convergence with this shift, and there exist examples for which the algorithm of QR iteration with this shift does not converge [94].

Another choice of shift is the *Francis*[12] *shift*, in which double shifts $\sigma, \bar{\sigma}$ are chosen as eigenvalues of the 2×2 corner of the matrix A_i:

---

[12]John G.F. Francis (born 1934) is an English computer scientist.

$$\begin{pmatrix} a_{n-1,n-1} & a_{n-1,n} \\ a_{n,n-1} & a_{n,n} \end{pmatrix}.$$

Such a choice of shifts allows convergence to either two real eigenvalues in the bottom 2×2 corner of the matrix A_i or the single 2×2 block with complex conjugate eigenvalues. Such a choice leads to quadratic convergence asymptotically, which means that if the values of $a_{n-1,n-2}$ are small enough, its amplitude will rapidly decrease to zero. However, the method of QR iteration with Francis shift can often fail to converge; see [8, 22].

There is another option for choosing a shift that is called *Wilkinson's*[13] *shift*: the shift σ_i is chosen as an eigenvalue of the matrix

$$\begin{pmatrix} a_{n-1,n-1} & a_{n-1,n} \\ a_{n,n-1} & a_{n,n} \end{pmatrix}$$

that is closest to the value $a_{n,n}$ of the matrix A_i.

Theorem 10.6 *The method of QR iteration with Wilkinson's shift is globally and at least linearly convergent. It is asymptotically cubically convergent for almost all matrices.*

A proof of this theorem can be found in [94].

Example 10.16 We test the performance of the method of QR iteration with shift for the same matrices as in the method of orthogonal iteration. Running the MATLAB® program of Section 1.13, we observe good convergence to the reference eigenvalues for the shift chosen as $\sigma = A_{nn}$ as well as for the Wilkinson's shift.[14]

10.6 Hessenberg Reduction

All QR algorithms are computationally expensive: one iteration of the QR decomposition costs $O(n^3)$ FLOPS. Assume that we can do only one iteration to find one eigenvalue. Then in this case, the cost will be $O(n^4)$. The goal of this section is to present one more technique for reducing computations. It turns out that if we first reduce the original matrix A to upper Hessenberg form and then apply the method of QR iteration without computing Q, we dramatically reduce computations, and instead of $O(n^4)$ FLOPS, we perform our computations in $O(n^3)$ FLOPS.

A Hessenberg matrix is a special kind of square matrix, one that is "almost" triangular. More precisely, an upper Hessenberg matrix has zero entries below the first subdiagonal, and a lower Hessenberg matrix has zero entries above the first superdiagonal. They are named after Karl Hessenberg.[15] For example:

---

[13] James Hardy Wilkinson (1919–1986) was an English mathematician.

[14] The MATLAB® programs can be found on the online version of Chapter 1 (doi:10.1007/978-3-319-57304-5_1).

[15] Karl Adolf Hessenberg (1904–1959) was a German mathematician.

$$\begin{pmatrix} 5 & 7 & 2 & 3 \\ 2 & 5 & 1 & 7 \\ 0 & 2 & 3 & 4 \\ 0 & 0 & 1 & 3 \end{pmatrix}$$

is upper Hessenberg and

$$\begin{pmatrix} 1 & 2 & 0 & 0 \\ 5 & 2 & 3 & 0 \\ 3 & 4 & 3 & 7 \\ 5 & 6 & 1 & 1 \end{pmatrix}$$

is lower Hessenberg.

When the matrix A is upper Hessenberg, the setting to zero of the element $a_{p+1,p}$ of this matrix will bring A into a block upper triangular matrix of the form

$$A = \begin{pmatrix} A_{11}^{p \times p} & A_{12}^{p \times (n-p)} \\ 0^{(n-p) \times (p-1)} & A_{22}^{(n-p) \times (n-p+1)} \end{pmatrix}$$

with the upper Hessenberg matrices A_{11} and A_{22}. This decomposition of A means that we can independently find eigenvalues of A_{11} and A_{22}. If in the process of Hessenberg reduction any subdiagonal or superdiagonal entry of the matrix A_i is smaller than the rounding errors of size $O(\varepsilon \|A\|)$, then we set this value to zero. We stop our computations when all these diagonal blocks are of size 1×1 or 2×2, and our algorithm for finding the eigenvalues of A is finished.

Below we present an algorithm for the reduction of a matrix A of order n to an upper Hessenberg matrix. Given a real matrix A, we seek an orthogonal matrix Q such that the matrix QAQ^T is an upper Hessenberg matrix.

Algorithm 10.6 Reduction to an upper Hessenberg matrix.

0. Initialize the matrix $Q = I$ and perform steps 1–7 in a loop from $i = 1$ to $n - 2$.
1. Take the elements of the vector $u_i = A(i + 1 : n, i)$.
2. Obtain the first element of the vector u_i as $u_i(1) = u_i(1) + \text{sign}(u_i(1)) \|u_i\|$.
3. Compute the elements of the vector $u_i = u_i / \|u_i\|$.
4. Compute the elements of the matrix $P_i = I^{(n-i) \times (n-i)} - 2u_i u_i^T$.
5. Compute the elements of the matrix $A(i + 1 : n, i : n) = P_i A(i + 1 : n, i : n)$.
6. Compute the elements of the matrix $A(1 : n, i + 1 : n) = A(1 : n, i + 1 : n) P_i$.
7. Compute the elements of the matrix Q as $Q(i + 1 : n, i : n) = P_i Q(i + 1 : n, i : n)$.

Proposition 10.1 *Hessenberg form is preserved by QR iteration.*

Proof If A_i is upper Hessenberg, then $A_i - \sigma I$ is also upper Hessenberg. Let us consider Algorithm 10.5 and perform QR decomposition of the matrix $A_i - \sigma I$. Since the jth column of Q is a linear combination of the leading j columns of the matrix $A_i - \sigma I$, the QR decomposition yields an upper Hessenberg matrix Q. Then

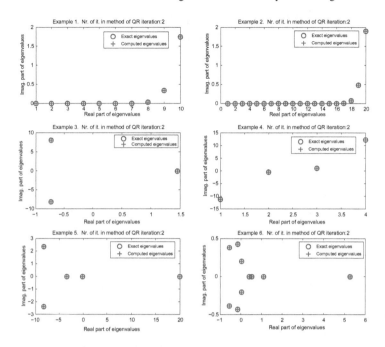

Fig. 10.2 Performance of the method of QR iteration: first we reduce the original matrix to upper Hessenberg form and then we apply the method of QR iteration

RQ is also upper Hessenberg, as well as $RQ + \sigma I$. This means that Hessenberg form is preserved by QR iteration.   □

Clearly, the convergence analysis of Hessenberg reduction follows from the convergence of the method of QR iteration.

Example 10.17 We test performance of the method of QR iteration via reducing first the original matrix to an upper Hessenberg matrix, using the MATLAB® program of Section 1.15.[16] We again test the same matrices as in the method of orthogonal iteration. In Fig. 10.2 we observe the rapid convergence for all examples.

Now we illustrate the general pattern of Hessenberg reduction with a matrix A of order 4. Every matrix Q_i below is a Householder reflection matrix[17] of order 4.

1. Choose Q_1 such that

$$Q_1 A = \begin{pmatrix} x\ x\ x\ x \\ x\ x\ x\ x \\ 0\ x\ x\ x \\ 0\ x\ x\ x \end{pmatrix} \text{ such that } A_1 = Q_1 A Q_1^T = \begin{pmatrix} x\ x\ x\ x \\ x\ x\ x\ x \\ 0\ x\ x\ x \\ 0\ x\ x\ x \end{pmatrix}.$$

---

[16]The MATLAB® programs can be found on the online version of Chapter 1 (doi:10.1007/978-3-319-57304-5_1).

[17]See p. 161.

The matrix Q_1 leaves the first row of $Q_1 A$ unchanged, and Q_1^T leaves the first column of $Q_1 A Q_1^T$ unchanged, including the zeros.

2. Choose Q_2 such that

$$Q_2 A_1 = \begin{pmatrix} x\ x\ x\ x \\ x\ x\ x\ x \\ 0\ x\ x\ x \\ 0\ 0\ x\ x \end{pmatrix} \text{ and obtain } A_2 = Q_2 A_1 Q_2^T = \begin{pmatrix} x\ x\ x\ x \\ x\ x\ x\ x \\ 0\ x\ x\ x \\ 0\ 0\ x\ x \end{pmatrix}.$$

The matrix Q_2 changes only the last two rows of A_1, and Q_2^T leaves the first two columns of $Q_2 A_1 Q_2^T$ unchanged, including the zeros. The matrix A_2 is upper Hessenberg. Combining steps 1 and 2, we get $A_2 = (Q_2 Q_1) A (Q_2 Q_1)^T = Q A Q^T$.

Let us consider an example of obtaining an upper Hessenberg matrix using Householder reflection.

Example 10.18 In this example we will use Householder reflection to get an upper Hessenberg matrix from the matrix

$$A = \begin{pmatrix} 12 & -51 & 4 \\ 6 & 167 & -68 \\ -4 & 24 & -41 \end{pmatrix}.$$

To do so, we need zero out the value of entry $(3, 1)$ of this matrix. First, we need to find the Householder reflection that transforms the first column of the matrix A, i.e., the vector $x = (6, -4)^T$, to the form $\|x\| i_1 = (\sqrt{6^2 + (-4)^2}, 0)^T = (2\sqrt{13}, 0)^T$. Now,

$$u = x + \alpha i_1,$$

and

$$v = \frac{u}{\|u\|}.$$

Here, $\alpha = -\text{sign}(x_1) \|x\| = -2\sqrt{13}$ and $x = (6, -4)^T$. Therefore,

$$u = (6 - 2\sqrt{13}, -4)^T \approx (-1.2111, -4)^T$$

and $v = u/\|u\| \approx (-0.2898, -0.9571)^T$, and then

$$Q_1 = I - 2 \begin{pmatrix} -0.2898 \\ -0.9571 \end{pmatrix} (-0.2898\ -0.9571) = \begin{pmatrix} 0.8321 & -0.5547 \\ -0.5547 & -0.8321 \end{pmatrix}.$$

Now observe that $Q_1 A$ preserves the first row of the matrix A,

$$Q_1 A = \begin{pmatrix} 1 & 0 & 0 \\ 0 & 0.8321 & -0.5547 \\ 0 & -0.5547 & -0.8321 \end{pmatrix} \begin{pmatrix} 12 & -51 & 4 \\ 6 & 167 & -68 \\ -4 & 24 & -41 \end{pmatrix}$$

$$= \begin{pmatrix} 12 & -51 & 4 \\ 7.2111 & 125.6396 & -33.8367 \\ 0.0 & -112.6041 & 71.8337 \end{pmatrix},$$

and the matrix $Q_1 A Q_1^T$ preserves the first column of the matrix $Q_1 A$,

$$A_1 = Q_1 A Q_1^T = \begin{pmatrix} 12 & -44.6534 & 24.9615 \\ 7.2111 & 123.3077 & -41.5385 \\ 0.0 & -133.5385 & 2.6923 \end{pmatrix},$$

which is an upper Hessenberg matrix.

10.7   Tridiagonal and Bidiagonal Reduction

Suppose now that the matrix A is symmetric. Then Hessenberg reduction leaves the matrix A symmetric at every step in such a way that zero elements will be created in symmetric positions. This will reduce the number of operations to $(4/3)n^3 + O(n^2)$ or $(8/3)n^3 + O(n^2)$ to form matrices $Q_{n-1}, \ldots, Q_1$ (see [23]). This procedure is called *tridiagonal reduction*. We recall that the eigenvalues of the symmetric matrix $A^T A$ are the squares of the singular values of A.[18] The SVD algorithms that we consider in Section 11.6 use this fact, and the goal of this section is to find a form for A that implies that $A^T A$ is tridiagonal. Our goal is to compute orthogonal matrices Q and V such that QAV is an upper bidiagonal matrix, or nonzero only on the diagonal and the first superdiagonal. This algorithm is called *bidiagonal reduction*.

Below we present the general procedure of bidiagonal reduction, which is illustrated on a matrix A of size 5×5.

1. Choose Q_1 and V_1 such that

$$Q_1 A = \begin{pmatrix} x & x & x & x & x \\ 0 & x & x & x & x \\ 0 & x & x & x & x \\ 0 & x & x & x & x \\ 0 & x & x & x & x \end{pmatrix} \text{ and } A_1 = Q_1 A V_1 = \begin{pmatrix} x & x & 0 & 0 & 0 \\ 0 & x & x & x & x \\ 0 & x & x & x & x \\ 0 & x & x & x & x \\ 0 & x & x & x & x \end{pmatrix}.$$

Here Q_1 is the matrix obtained after Householder reflection, and V_1 is another matrix obtained after Householder reflection that leaves the first column of $Q_1 A$ unchanged.

---

[18] See Section 5.1.1, p. 163.

2. Choose Q_2 and V_2 such that

$$Q_2A_1 = \begin{pmatrix} x\,x\,0\,0\,0 \\ 0\,x\,x\,x\,x \\ 0\,0\,x\,x\,x \\ 0\,0\,x\,x\,x \\ 0\,0\,x\,x\,x \end{pmatrix} \text{ and } A_2 = Q_2A_1V_2 = \begin{pmatrix} x\,x\,0\,0\,0 \\ 0\,x\,x\,0\,0 \\ 0\,0\,x\,x\,x \\ 0\,0\,x\,x\,x \\ 0\,0\,x\,x\,x \end{pmatrix}.$$

Here Q_2 is the matrix obtained after Householder reflection that leaves the first row of A_1 unchanged. The matrix V_2 is a Householder reflection that leaves the first two columns of Q_2A_1 unchanged.

3. Choose Q_3 and V_3 such that

$$Q_3A_2 = \begin{pmatrix} x\,x\,0\,0\,0 \\ 0\,x\,x\,0\,0 \\ 0\,0\,x\,x\,x \\ 0\,0\,0\,x\,x \\ 0\,0\,0\,x\,x \end{pmatrix} \text{ and } A_3 = Q_3A_2V_3 = \begin{pmatrix} x\,x\,0\,0\,0 \\ 0\,x\,x\,0\,0 \\ 0\,0\,x\,x\,0 \\ 0\,0\,0\,x\,x \\ 0\,0\,0\,x\,x \end{pmatrix}.$$

Here Q_3 is a Householder reflection that leaves the first two rows of A_2 unchanged. The matrix V_3 is a Householder reflection that leaves the first three columns of Q_3A_2 unchanged.

4. Choose Q_4 such that

$$Q_4A_3 = \begin{pmatrix} x\,x\,0\,0\,0 \\ 0\,x\,x\,0\,0 \\ 0\,0\,x\,x\,0 \\ 0\,0\,0\,x\,x \\ 0\,0\,0\,0\,x \end{pmatrix} \text{ and } V_4 = I \text{ so } A_4 = Q_4A_3.$$

Here Q_4 is a Householder reflection that leaves the first three rows of A_3 unchanged. Then we obtain the tridiagonal matrix as

$$A_4^TA_4 = \begin{pmatrix} x\,x\,0\,0\,0 \\ x\,x\,x\,0\,0 \\ 0\,x\,x\,x\,0 \\ 0\,0\,x\,x\,x \\ 0\,0\,0\,x\,x \end{pmatrix}.$$

In the general case, the matrix A has order n. Then applying the above procedure to this matrix, we get orthogonal matrices $Q = Q_{n-1}\cdots Q_1$ and $V = V_1\cdots V_{n-2}$ such that $QAV = A'$ is upper bidiagonal. Note that $A'^TA' = V^TA^TQ^TQAV = V^TA^TAV$, so A'^TA' has the same eigenvalues as A^TA; i.e., A' has the same singular values as A. The cost of the algorithm of bidiagonal reduction is $(8/3)n^3 + 4n^3 + O(n^2)$, where $4n^3 + O(n^2)$ counts for computations of Q and V.

Example 10.19 In this example we apply the above procedure to the bidiagonal reduction of the matrix

$$A = \begin{pmatrix} 4\ 4\ 3 \\ 3\ 6\ 1 \\ 0\ 1\ 7 \end{pmatrix}$$

using Householder transformation. We proceed in following steps.

First, we need to zero out the second entry in the first column of the matrix A, the vector $x = (4, 3, 0)^T$. We compute first $\alpha = -\text{sign}(x_1)\|x\| = -5$, and then the vectors $u = x + \alpha i_1 = (-1, 3, 0)^T$ and $v = u/\|u\| = (-1, 3, 0)^T/\sqrt{10}$. Compute the Householder matrix P_1 as

$$P_1 = I - 2vv^T = \begin{pmatrix} 0.8\ \ 0.6\ \ 0 \\ 0.6\ -0.8\ \ 0 \\ 0\ \ \ \ 0\ \ \ \ 1 \end{pmatrix}.$$

Compute P_1A to zero out the two entries below 5 in the first column:

$$P_1A = \begin{pmatrix} 5\ \ 6.8\ \ 3 \\ 0\ -2.4\ \ 1 \\ 0\ \ \ 1\ \ \ 7 \end{pmatrix}. \tag{10.6}$$

Now we want to zero out the $(1, 3)$ entry of the matrix (10.6). To do so, we take the minor

$$M = \begin{pmatrix} 6.8\ \ 3 \\ -2.4\ \ 1 \end{pmatrix}$$

and compute again for $x = (6.8, 3)^T$ the number $\alpha = -\text{sign}(x_1)\|x\| = -7.4324$ and then the vectors $u = x + \alpha i_1 = (-0.6324, 3)^T$, $v = u/\|u\| = (-0.2063, 0.9785)^T$. Compute the matrix

$$V_1' = I - 2vv^T = \begin{pmatrix} 0.9149\ \ \ 0.4037 \\ 0.4037\ -0.9149 \end{pmatrix}.$$

Construct V_1 such that

$$V_1 = \begin{pmatrix} 1\ 0\ 0 \\ 0\ V_1' \\ 0\ \ \ \ \ \end{pmatrix} = \begin{pmatrix} 1\ \ \ \ 0\ \ \ \ \ \ 0 \\ 0\ 0.9149\ \ \ 0.4037 \\ 0\ 0.4037\ -0.9149 \end{pmatrix}.$$

Compute P_1AV_1 to zero out the $(1, 3)$ entry:

$$P_1AV_1 = \begin{pmatrix} 5\ \ 7.4324\ \ \ 0.0005 \\ 0\ -1.7921\ -1.8838 \\ 0\ \ 3.7408\ \ -6.0006 \end{pmatrix}. \tag{10.7}$$

It remains only to zero out the $(3, 2)$ entry of the matrix in (10.7). We take the minor

$$M = \begin{pmatrix} -1.7921 & -1.8838 \\ 3.7408 & -6.0006 \end{pmatrix}$$

and compute for $x = (-1.7921, 3.7408)^T$ the number $\alpha = -\text{sign}(x_1)\|x\| = 4.1479$ and the vectors $u = x + \alpha i_1 = (2.3558, 3.7408)^T$, $v = u/\|u\| = (0.5329, 0.8462)^T$. Compute the 2×2 matrix P_2':

$$P_2' = I - 2vv^T = \begin{pmatrix} 0.4320 & -0.9019 \\ -0.9019 & -0.4321 \end{pmatrix}.$$

Construct P_2 such that the matrix P_2' is inserted into the 3×3 identity matrix:

$$P_2 = \begin{pmatrix} 1 & 0 & 0 \\ 0 & P_2' & \\ 0 & & \end{pmatrix} = \begin{pmatrix} 1.0000 & 0 & 0 \\ 0 & 0.4320 & -0.9019 \\ 0 & -0.9019 & -0.4321 \end{pmatrix}.$$

Finally, multiply the matrix P_2 by the matrix P_1AV_1 obtained in (10.7) to get the bidiagonal matrix

$$P_2P_1AV_1 = \begin{pmatrix} 5.0000 & 7.4324 & 0.0005 \\ -0.0000 & -4.1480 & 4.5981 \\ 0.0000 & -0.0001 & 4.2918 \end{pmatrix}. \tag{10.8}$$

10.7.1 Tridiagonal Reduction Using Householder Transformation

In this section we present an alternative procedure that can be used for tridiagonal reduction using Householder transformation. This procedure is taken from [12]. To form the Householder matrix in this procedure, in each step we need to determine α and r, which are given by

$$\alpha = -\text{sign}(a_{21}) \sqrt{\sum_{j=2}^{n} a_{j1}^2}, \quad r = \sqrt{\frac{1}{2}(\alpha^2 - a_{21}\alpha)}.$$

From α and r, we construct the vector v:

$$v^{(1)} = \begin{pmatrix} v_1 \\ v_2 \\ \vdots \\ v_n \end{pmatrix},$$

where $v_1 = 0$, $v_2 = (a_{21} - \alpha)/(2r)$, and

$$v_k = \frac{a_{k1}}{2r}$$

for $k = 3, 4, \ldots n$. Then we compute the matrix

$$P^1 = I - 2v^{(1)}(v^{(1)})^T$$

and obtain the matrix $A^{(1)}$ as

$$A^{(1)} = P^1 A P^1.$$

Having found P^1 and computed $A^{(1)}$, we repeat the process for $k = 2, 3, \ldots, n$ as follows:

$$\alpha = -\text{sign}(a_{k+1,k}) \sqrt{\sum_{j=k+1}^{n} a_{jk}^2}, \quad r = \sqrt{\frac{1}{2}(\alpha^2 - a_{k+1,k}\alpha)},$$

$$v_1^k = v_2^k = .. = v_k^k = 0, \quad v_{k+1}^k = \frac{a_{k+1,k}^k - \alpha}{2r}, \quad v_j^k = \frac{a_{jk}^k}{2r} \text{ for } j = k + 2, \ k + 3, \ \ldots, \ n,$$

$$P^k = I - 2v^{(k)}(v^{(k)})^T, \quad A^{(k+1)} = P^k A^{(k)} P^k.$$

Example 10.20 In this example we apply the above algorithm to perform tridiagonal reduction of the matrix

$$A = \begin{pmatrix} 5 & 4 & 3 \\ 4 & 6 & 1 \\ 3 & 1 & 7 \end{pmatrix}, \tag{10.9}$$

using Householder transformation. To do so, we proceed in the following steps.

First we compute α as

$$\alpha = -\text{sign}(a_{21}) \sqrt{\sum_{j=2}^{n} a_{j1}^2} = -\sqrt{(a_{21}^2 + a_{31}^2)} = -\sqrt{4^2 + 3^2} = -5.$$

Using α, we find r as

$$r = \sqrt{\frac{1}{2}(\alpha^2 - a_{21}\alpha)} = \sqrt{\frac{1}{2}((-5)^2 - 4 \cdot (-5))} = \frac{3\sqrt{5}}{\sqrt{2}}.$$

Then we compute the components of the vector v:

$$v_1 = 0,$$

$$v_2 = \frac{a_{21} - \alpha}{2r} = \frac{3\sqrt{2}}{2\sqrt{5}},$$

$$v_3 = \frac{a_{31}}{2r} = \frac{\sqrt{2}}{2\sqrt{5}},$$

and we get

$$v^{(1)} = \left(0, \frac{3\sqrt{2}}{2\sqrt{5}}, \frac{\sqrt{2}}{2\sqrt{5}}\right)^T.$$

Now we compute the first Householder matrix $P^1 = I - 2v^{(1)}(v^{(1)})^T$ to get

$$P^1 = \begin{pmatrix} 1 & 0 & 0 \\ 0 & -4/5 & -3/5 \\ 0 & -3/5 & 4/5 \end{pmatrix}.$$

Finally, we obtain the tridiagonal matrix $A^{(1)}$ as

$$A^{(1)} = P^1 A P^1 = \begin{pmatrix} 5 & -5 & 0 \\ -5 & 7.32 & -0.76 \\ 0 & -0.76 & 5.68 \end{pmatrix}.$$

10.7.2 Tridiagonal Reduction Using Givens Rotation

To make the tridiagonal matrix from the matrix A using Givens rotation,[19] we first recall that a Givens rotation is represented by a matrix of the form

$$G(i, j, \theta) = \begin{pmatrix} 1 & \cdots & 0 & \cdots & 0 & \cdots & 0 \\ \vdots & \ddots & \vdots & & \vdots & & \vdots \\ 0 & \cdots & c & \cdots & -s & \cdots & 0 \\ \vdots & & \vdots & \ddots & \vdots & & \vdots \\ 0 & \cdots & s & \cdots & c & \cdots & 0 \\ \vdots & & \vdots & & \vdots & \ddots & \vdots \\ 0 & \cdots & 0 & \cdots & 0 & \cdots & 1 \end{pmatrix},$$

---

[19] See p. 160.

where $c = cos(\Theta)$ and $s = sin(\Theta)$ appear at the intersections of the ith and jth rows and columns. The elements in the Givens matrix are

$$g_{kk} = 1 \quad \text{for } k \neq i, j,$$
$$g_{ii} = c,$$
$$g_{jj} = c,$$
$$g_{ji} = -s,$$
$$g_{ij} = s \quad \text{for } i > j.$$

We note that the sign of elements g_{ji}, g_{ji} switches for $j > i$. For given a and b, our goal is to find $c = \cos\theta$ and $s = \sin\theta$ such that

$$\begin{pmatrix} c & -s \\ s & c \end{pmatrix} \begin{pmatrix} a \\ b \end{pmatrix} = \begin{pmatrix} r \\ 0 \end{pmatrix},$$

where

$$r = \sqrt{a^2 + b^2},$$
$$c = a/r,$$
$$s = -b/r.$$

Example 10.21 To obtain a tridiagonal matrix from the matrix (10.9) using Givens rotation we have to zero out the $(3, 1)$ and $(1, 3)$ elements of the matrix A.

Thus, we use the above expressions to construct the Givens rotation matrix G_1 of the form

$$G_1 = \begin{pmatrix} 1 & 0 & 0 \\ 0 & c & -s \\ 0 & s & c \end{pmatrix}.$$

We compute then the product $G_1 A$ to get

$$G_1 A = \begin{pmatrix} 1 & 0 & 0 \\ 0 & c & -s \\ 0 & s & c \end{pmatrix} \begin{pmatrix} 5 & 4 & 3 \\ 4 & 6 & 1 \\ 3 & 1 & 7 \end{pmatrix} = \begin{pmatrix} 5 & 4 & 3 \\ 4c - 3s & 6c - sc & 3 - 7s \\ 4s + 3c & 6s + cs & 7 + 7c \end{pmatrix}.$$

The element $(3, 1)$ of the matrix will be zero if $4s + 3c = 0$. This is true when $c = 4/5$ and $s = -3/5$. To compute c, s we have used the formulas

$$r = \sqrt{a^2 + b^2} = \sqrt{4^2 + 3^2} = 5, \quad c = \frac{a}{r} = \frac{4}{5}, \quad s = -\frac{b}{r} = -\frac{3}{5}.$$

Next, to get a tridiagonal matrix, we have to compute $G_1 A G_1{}^T$:

$$A_1 = G_1 A G_1{}^T = \begin{pmatrix} 1 & 0 & 0 \\ 0 & 4/5 & 3/5 \\ 0 & -3/5 & 4/5 \end{pmatrix} \begin{pmatrix} 5 & 4 & 3 \\ 4 & 6 & 1 \\ 3 & 1 & 7 \end{pmatrix} \begin{pmatrix} 1 & 0 & 0 \\ 0 & 4/5 & -3/5 \\ 0 & 3/5 & 4/5 \end{pmatrix} = \begin{pmatrix} 5 & 5 & 0 \\ 5 & 7.32 & 0.76 \\ 0 & 0.76 & 5.68 \end{pmatrix}.$$

$$(10.10)$$

Example 10.22 As another example, let us now make an upper triangular matrix from the matrix (10.10), using Givens rotation. To do so, we need to zero out the elements (2, 1) and (3, 2) of (10.10). To zero out the element (2, 1), we compute the numbers c and s from the known $a = 5$ and $b = 5$ as

$$\begin{pmatrix} c & -s \\ s & c \end{pmatrix} \begin{pmatrix} a \\ b \end{pmatrix} = \begin{pmatrix} r \\ 0 \end{pmatrix}$$

to get

$$r = \sqrt{a^2 + b^2} = \sqrt{5^2 + 5^2} = 5\sqrt{2} \approx 7.0711,$$

$$c = \frac{a}{r} \approx 0.7071, \quad s = \frac{-b}{r} \approx -0.7071.$$

The Givens matrix will be

$$G_2 = \begin{pmatrix} c & -s & 0 \\ s & c & 0 \\ 0 & 0 & 1 \end{pmatrix} = \begin{pmatrix} 0.7071 & 0.7071 & 0 \\ -0.7071 & 0.7071 & 0 \\ 0 & 0 & 1 \end{pmatrix}.$$

Finally, we obtain the matrix

$$A_2 = G_2 A_1 = \begin{pmatrix} 0.7071 & 0.7071 & 0 \\ -0.7071 & 0.7071 & 0 \\ 0 & 0 & 1 \end{pmatrix} \begin{pmatrix} 5 & 5 & 0 \\ 5 & 7.32 & 0.76 \\ 0 & 0.76 & 5.68 \end{pmatrix}$$

$$= \begin{pmatrix} 7.7071 & 8.7116 & 0.5374 \\ 0 & 1.6405 & 0.5374 \\ 0 & 0.7600 & 5.6800 \end{pmatrix}. \quad (10.11)$$

Now to zero out the element (3, 2), we compute c, s from the known $a = 1.6405$ and $b = 0.76$ to get

$$r = \sqrt{a^2 + b^2} = \sqrt{1.6405^2 + 0.76^2} = 1.8080,$$

$$c = \frac{a}{r} \approx 0.9074, \quad s = \frac{-b}{r} \approx -0.4204.$$

The last Givens matrix will be

$$G_3 = \begin{pmatrix} 1 & 0 & 0 \\ 0 & c & -s \\ 0 & s & c \end{pmatrix} = \begin{pmatrix} 1 & 0 & 0 \\ 0 & 0.9074 & 0.4204 \\ 0 & -0.4204 & 0.9074 \end{pmatrix}.$$

Finally, we obtain the upper triangular matrix

$$A_3 = G_3 A_2 = \begin{pmatrix} 7.0711 & 8.7116 & 0.5374 \\ 0 & 1.8080 & 2.8752 \\ 0 & 0.0000 & 4.9279 \end{pmatrix}. \tag{10.12}$$

10.8   QR Iteration with Implicit Shifts

In this section we will first reduce the matrix A to an upper Hessenberg matrix and then compute its QR factorization implicitly. This means that QR factorization will be computed by construction of the matrix Q using the implicit Q theorem. This theorem improves the efficiency of Hessenberg's QR iteration algorithm. Next, we will present how to choose a single shift to accelerate convergence of the method of QR iteration.

We say that an upper Hessenberg matrix H is *unreduced* if all elements on its subdiagonal are nonzero.

Theorem 10.7 *Let H and G be unreduced upper Hessenberg matrices of order n such that $H = Q^T A Q$ and $G = V^T A V$. Here, Q and V are orthogonal matrices of order n, where the first columns are the same, or $Q i_1 = V i_1$ with $i_1 = (1, 0, ..., 0)$. Let $X(:, i)$ denote the ith column of the matrix X. Then the columns of Q and V are the same up to the sign, or $Q(:, i) = \pm V(:, i)$ for $i = 2, 3, ..., n$.*

Proof The assertion regarding the first columns of the matrices Q and V is obvious. Our goal is to prove that $Q(:, i) = \pm V(:, i)$ for $i > 1$. This is equivalent to proving that the matrix W satisfies $W = V^T Q = \mathrm{diag}(\pm 1, ..., \pm 1)$. Since by assumption, $W = V^T Q$, we can write $GW = GV^T Q = V^T A V V^T Q = V^T A Q = V^T Q Q^T A Q = V^T Q H = WH$. Since $GW = WH$, we have $GW(:, i) = (GW)(:, i) = (WH)(:, i) = \sum_{j=1}^{i+1} H_{ji} W(:, j)$, and thus, $H_{i+1,i} W(:, i+1) = GW(:, i) - \sum_{j=1}^{i} H_{ji} W(:, j)$. The first column of W is $W(:, 1) = (1, 0, ..., 0)^T$ (this is because Q and V are orthogonal and $Q(:, 1) = V(:, 1)$), and G is an upper Hessenberg matrix. We can use induction on the index of column i to show that W_i is nonzero only for entries from 1 to i. Thus, W is an upper triangular matrix. But because W is also an orthogonal matrix, it must be diagonal, or $W = \mathrm{diag}(\pm 1, ..., \pm 1)$.   □

Algorithm 10.7 The single shift QR algorithm.

Theorem 10.7 implies that to compute $A_{i+1} = Q_i^T A_i Q_i$ from A_i in the QR algorithm, we will need to perform only the following two steps.

1. Compute the first column of the matrix Q_i. This column is parallel to the first column of $A_i - \sigma_i I$ and thus can be obtained just by normalizing this column vector.
2. Choose other columns of Q_i such that Q_i is an orthogonal matrix and $A_{i+1} = Q_i^T A_i Q_i$ is an unreduced upper Hessenberg matrix.

Using the above theorem, we can conclude that the matrix A_{i+1} in the algorithm is computed correctly because the matrix Q_i is computed uniquely up to signs. The choice of sign does not matter. This is true because if we change signs in the columns of Q_i, then the signs in $A_i - \sigma_i I = Q_i R_i$ also will be changed: $A_i - \sigma_i I = Q_i R_i = Q_i D_i D_i R_i$, where $D_i = \text{diag}(\pm 1, ..., \pm 1)$. Then we can write

$$A_{i+1} = Q_i^T A_i Q_i = Q_i^T (Q_i R_i + \sigma_i I) Q_i = Q_i^T (Q_i D_i D_i R_i + \sigma_i I) Q_i$$
$$= D_i R_i Q_i D_i + \sigma_i I = D_i (R_i Q_i + \sigma_i I) D_i, \tag{10.13}$$

and this is an orthogonal similarity that changes only the signs in the columns and rows of A_{i+1}.

As an example, we will now illustrate how the above algorithm works for the computation of $A_{i+1} = Q_i^T A_i Q_i$ for $i = 4$. In all matrices Q_i^T below, the values c_i and s_i can be computed using the Givens rotation algorithm. The symbols $*$ in the matrices A_i should be removed during the iterations on index i in order to restore Hessenberg form.

1. Choose Q_1 such that

$$Q_1^T = \begin{pmatrix} c_1 & s_1 & 0 & 0 & 0 \\ -s_1 & c_1 & 0 & 0 & 0 \\ 0 & 0 & 1 & 0 & 0 \\ 0 & 0 & 0 & 1 & 0 \\ 0 & 0 & 0 & 0 & 1 \end{pmatrix} \quad \text{to get } A_1 = Q_1^T A_0 Q_1 = \begin{pmatrix} x & x & x & x & x \\ x & x & x & x & x \\ * & x & x & x & x \\ 0 & 0 & x & x & x \\ 0 & 0 & 0 & x & x \end{pmatrix}.$$

2. Choose Q_2 such that

$$Q_2^T = \begin{pmatrix} 1 & 0 & 0 & 0 & 0 \\ 0 & c_2 & s_2 & 0 & 0 \\ 0 & -s_2 & c_2 & 0 & 0 \\ 0 & 0 & 0 & 1 & 0 \\ 0 & 0 & 0 & 0 & 1 \end{pmatrix} \quad \text{to get } A_2 = Q_2^T A_1 Q_2 = \begin{pmatrix} x & x & x & x & x \\ x & x & x & x & x \\ 0 & x & x & x & x \\ 0 & * & x & x & x \\ 0 & 0 & 0 & x & x \end{pmatrix}.$$

We note that

$$Q_2^T A_1 = \begin{pmatrix} x & x & x & x & x \\ x & x & x & x & x \\ 0 & x & x & x & x \\ 0 & 0 & x & x & x \\ 0 & 0 & 0 & x & x \end{pmatrix}.$$

3. Choose Q_3 such that

$$Q_3^T = \begin{pmatrix} 1 & 0 & 0 & 0 & 0 \\ 0 & 1 & 0 & 0 & 0 \\ 0 & 0 & c_3 & s_3 & 0 \\ 0 & 0 & -s_3 & c_3 & 0 \\ 0 & 0 & 0 & 0 & 1 \end{pmatrix} \quad \text{to get } A_3 = Q_3^T A_2 Q_3 = \begin{pmatrix} x & x & x & x & x \\ x & x & x & x & x \\ 0 & x & x & x & x \\ 0 & 0 & x & x & x \\ 0 & 0 & * & x & x \end{pmatrix}.$$

We note that

$$Q_3^T A_2 = \begin{pmatrix} x & x & x & x & x \\ x & x & x & x & x \\ 0 & x & x & x & x \\ 0 & 0 & x & x & x \\ 0 & 0 & 0 & x & x \end{pmatrix}.$$

4. Choose Q_4 such that

$$Q_4^T = \begin{pmatrix} 1 & 0 & 0 & 0 & 0 \\ 0 & 1 & 0 & 0 & 0 \\ 0 & 0 & 1 & 0 & 0 \\ 0 & 0 & 0 & c_4 & s_4 \\ 0 & 0 & 0 & -s_4 & c_4 \end{pmatrix} \quad \text{to get } A_4 = Q_4^T A_3 Q_4 = \begin{pmatrix} x & x & x & x & x \\ x & x & x & x & x \\ 0 & x & x & x & x \\ 0 & 0 & x & x & x \\ 0 & 0 & 0 & x & x \end{pmatrix}.$$

We note that

$$Q_4^T A_3 = \begin{pmatrix} x & x & x & x & x \\ x & x & x & x & x \\ 0 & x & x & x & x \\ 0 & 0 & x & x & x \\ 0 & 0 & 0 & x & x \end{pmatrix}.$$

At step 4 we have obtained the upper Hessenberg matrix A_4. We observe that combining steps 1–4, we get $A_4 = Q_4^T Q_3^T Q_2^T Q_1^T A Q_1 Q_2 Q_3 Q_4 = Q^T A Q$, where the matrix $Q = Q_1 Q_2 Q_3 Q_4$ is such that

$$Q = \begin{pmatrix} c_1 & x & x & x & x \\ s_1 & x & x & x & x \\ 0 & s_2 & x & x & x \\ 0 & 0 & s_3 & x & x \\ 0 & 0 & 0 & s_4 & x \end{pmatrix}.$$

If we now choose the first column of Q, which is $Q(:, 1) = (c_1, s_1, 0, 0, 0)$, proportional to the first column of the matrix $A - \sigma I$, which is

$$(A - \sigma I)(:, 1) = (a_{11} - \sigma, a_{21}, 0, 0, 0)^T,$$

then this matrix Q will be the same as in the QR decomposition of $A - \sigma I$.

We can choose the single shift σ as $\sigma = a_{n,n}$ for the matrix A_i. This will result in asymptotic quadratic convergence to a real eigenvalue; see [23] for details on how to choose shifts.

Questions

Do the following exercises, explaining the results that you have obtained.

10.1 (*Programming*)
   Use the MATLAB® program `PowerM.m` of Section 1.9 to test the power method and to compute the largest eigenvalue of the matrix A.[20] Try the following examples when the matrix A and the tolerance Θ in Algorithm 10.1, p. 345, are defined as follows:

1. $A = \mathrm{randn}(5)$ and tolerance $\Theta = \{1e-5, 1e-4, 1e-3, 1e-2, 0.1\}$.
2. $A = \mathrm{diag}(\mathrm{ones}(2n, 1)) + \mathrm{diag}(\mathrm{ones}(2n-1, 1), 1) + \mathrm{diag}(\mathrm{ones}(2n-1, 1), -1)$
   for each number $n = 3, 4, 5$ and tolerance $\Theta = \{1e-12, 1e-10, 1e-8, 1e-7\}$.
3.

$$A = \begin{pmatrix} 1 & 1e6 & 0 & 0 \\ 0 & 2 & 1 & 0 \\ 1 & 2 & 3 & 10 \\ 0 & 0 & -1 & 4 \end{pmatrix} \tag{10.14}$$

   and tolerance $\Theta = \{1e-12, 1e-10, 1e-8\}$.
4.

$$A = \begin{pmatrix} 1 & 0 & 0 & 0 & 0 & 0 \\ 0 & 2 & 1 & 0 & 0 & 0 \\ 0 & 0 & 2 & 0 & 0 & 0 \\ 0 & 0 & 0 & 3 & 1e2 & 1e4 \\ 0 & 0 & 0 & 0 & 3 & 1e2 \\ 0 & 0 & 0 & 0 & 0 & 3 \end{pmatrix} \tag{10.15}$$

   and tolerance $\Theta = \{1e-10, 1e-8, 1e-6, 1e-4, 1e-3\}$.

10.2 (*Programming*)
   Use the MATLAB® program `InverseIteration.m` of Section 1.10 to test the inverse iteration method for the computation of the eigenvalue of the matrix A that is closest to the shift σ. Try all examples of matrix A and tolerance Θ as in Question 10.1. Choose also different shifts σ. For a stopping criterion using tolerance Θ, we refer to Algorithm 10.2, p. 348.

10.3 (*Programming*)
   Use the MATLAB® program `MethodOrtIter.m` of Section 1.11 to test the method of orthogonal iteration for the computation of the eigenvalues

---

[20] All MATLAB® programs referenced on this page can be found on the online version of Chapter 1 (doi:10.1007/978-3-319-57304-5_1).

of the matrix A.[21] Try all examples of matrix A and tolerance Θ defined in Question 10.1. For a stopping criterion using tolerance Θ, we refer to Algorithm 10.3, p. 351.

10.4 (*Programming*)

Use the MATLAB® program `MethodQR_iter.m` of Section 1.12 to test the method of QR iteration for the computation of the eigenvalues of the matrix A. Try all examples of matrix A and tolerance Θ defined in Question 10.1. For a stopping criterion using tolerance Θ, we refer to Algorithm 10.4, p. 355.

10.5 (*Programming*)

Use the MATLAB® program `MethodQR_shift.m` of Section 1.13 to test the method of QR iteration with the shift $\sigma = A(n, n)$ for the computation of the eigenvalues of the matrix A. Try all examples of matrix A and tolerance Θ defined in Question 10.1. For a stopping criterion using tolerance Θ, we refer to Algorithm 10.5, p. 356.

10.6 (*Programming*)

Use the MATLAB® program `MethodQR_Wshift.m` of Section 1.14 to test the method of QR iteration with Wilkinson's shift for the computation of the eigenvalues of the matrix A. Try all examples of matrix A and tolerance Θ defined in Question 10.1. For a stopping criterion using tolerance Θ, we refer to Algorithm 10.5, p. 356.

10.7 (*Programming*)

Use the MATLAB® program `HessenbergQR.m` of Section 1.15 to test the reduction of the matrix A to an upper Hessenberg matrix. Try the following examples when the matrix A and the tolerance Θ are defined thus:

1. $A = \text{randn}(5)$ and the tolerance is $\Theta = \{1e - 7, 1e - 5, 1e - 4, 1e - 3, 1e - 2, 0.1\}$.
2. $A = \text{diag}(\text{ones}(2n, 1)) + \text{diag}(\text{ones}(2n - 1, 1), 1) + \text{diag}(\text{ones}(2n - 1, 1), -1)$ for each number $n = 3, 4, 5$ and tolerance $\Theta = \{1e - 12, 1e - 10, 1e - 8, 1e - 7\}$.
3.
$$A = \begin{pmatrix} 1 & 1e6 & 0 & 0 \\ 0 & 2 & 1e - 3 & 0 \\ 0 & 0 & 3 & 10 \\ 0 & 0 & -1 & 4 \end{pmatrix} \tag{10.16}$$

and tolerance $\Theta = \{1e - 12, 1e - 10, 1e - 8\}$.
4. Each tolerance from the set $\Theta = \{1e - 10, 1e - 8, 1e - 6, 1e - 4, 1e - 3\}$ and the matrix $A = (1, 0, 0, 0, 0, 0; 0, 2, 1, 0, 0, 0; 0, 0, 2, 0, 0, 0; 0, 0, 0, 1e4, 1, 1; 0, 0, 0, 0, 1e2, 1; 0, 0, 0, 0, 0, 1e4) + \text{diag}(\text{ones}(5, 1), -1)$.

---

[21] All MATLAB® programs referenced on this page can be found on the online version of Chapter 1 (doi:10.1007/978-3-319-57304-5_1).

Chapter 11
Algorithms for Solution of Symmetric Eigenvalue Problems

In this chapter we will discuss algorithms which can solve only symmetric eigenvalue problems using direct noniterative methods. Recall that in the previous chapter, the algorithms that can find all eigenvalues and eigenvectors for an nonsymmetric eigenvalue problem were based only on the method of QR iteration. However, there exist many algorithms for the solution of symmetric eigenvalue problems that are more efficient than the algorithms for the solution of nonsymmetric eigenvalue problems. We list here the main algorithms and main advantages of every algorithm that we will consider in this chapter:

1. *Tridiagonal QR iteration*. This algorithm can be used to find all the eigenvalues, and if needed, all the eigenvectors, of a symmetric tridiagonal matrix. This method is the fastest numerical method that can compute all the eigenvalues of a symmetric tridiagonal matrix. If we want to apply this algorithm to find all the eigenvectors as well, then it will be efficient only for small matrices of dimension up to 25. We note that this algorithm is used in the Matlab® command eig.
2. *Rayleigh quotient iteration*. This algorithm is similar to the algorithm of QR iteration, and we present it here to analyze its extremely rapid cubic convergence.
3. *Divide and conquer*. This is the fastest method for finding all the eigenvalues and eigenvectors of symmetric tridiagonal matrices that have dimensions larger than 25.
4. *Bisection and inverse iteration*. Bisection may be used to find a subset of the eigenvalues of a symmetric tridiagonal matrix on some subinterval of the interval on which the eigenvalues are located. The algorithm of Inverse iteration described in the previous chapter can then be used to find the corresponding eigenvectors. In [29, 42, 90–92, 96, 97, 115], the inverse iteration method is developed further to find close eigenvalues and eigenvectors as fast as possible.

The original version of this chapter was revised. An Erratum to this book can be found at https://doi.org/10.1007/978-3-319-57304-5_11

© Springer International Publishing AG 2017
L. Beilina et al., *Numerical Linear Algebra: Theory and Applications*,
DOI 10.1007/978-3-319-57304-5_11

5. *Jacobi's method.* This is the first known method for the solution of eigenvalue problems. It was developed in 1846 by Jacobi. Compared with all previous methods, Jacobi's method is much slower. However, it is very accurate and can find tiny eigenvalues much more precisely than the previous methods; see details in [25].

11.1 Tridiagonal QR Iteration

The tridiagonal QR iteration algorithm is very similar to the usual algorithm of QR iteration for the solution of nonsymmetric eigenproblems that we considered in the previous chapter. That algorithm consists of two stages:

- First we constructed an orthogonal matrix Q via Algorithm 10.6 (reduction to an upper Hessenberg matrix) such that $QAQ^T = H$ is an upper Hessenberg matrix.
- Then we applied QR iteration to the resulting matrix H and obtained a sequence of upper Hessenberg matrices $H_0, H_1, H_2, \ldots$, that converged to real Schur form.

The algorithm of QR iteration for a symmetric tridiagonal matrix is very similar to the procedure described above and consists of the following steps:

- Use a modified Algorithm 10.6 (reduction to upper Hessenberg form) to find an orthogonal Q such that $QAQ^T = T$ is tridiagonal.
- Apply the algorithm of QR iteration on the resulting matrix T to obtain a sequence of tridiagonal matrices $T_0, T_1, T_2, \ldots$ that will converge to diagonal form.

We note that the algorithm of QR iteration keeps all matrices T_i tridiagonal. This is because the matrix $QAQ^T = T$ is symmetric and upper Hessenberg, and thus also lower Hessenberg, or tridiagonal.

We now describe how to choose the shifts at every QR iteration. Let us denote by T_i the tridiagonal matrix obtained at iteration i in the algorithm of QR iteration:

$$
T_i = \begin{pmatrix}
a_1 & b_1 & & \\
b_1 & \ddots & \ddots & \\
& \ddots & \ddots & b_{n-1} \\
& & b_{n-1} & a_n
\end{pmatrix}.
$$

We can choose as a shift the single shift $\sigma_i = a_n$ in the algorithm of QR iteration; see Section 10.8. Then the method is cubically convergent for almost all matrices. However, in some special cases, the method of QR iteration does not converge; see p. 76 in [94]. Thus, to get global convergence of the method, one needs to compute a shift in a more complicated manner. Let the shift σ_i, called *Wilkinson's shift*, be the eigenvalue of

$$
\begin{pmatrix}
a_{n-1} & b_{n-1} \\
b_{n-1} & a_n
\end{pmatrix}
$$

closest to a_n.

Theorem 11.1 (Wilkinson).

The algorithm of QR iteration with Wilkinson's shift is globally and at least linearly convergent. This algorithm is asymptotically cubically convergent for almost all matrices.

We refer to [94] for a proof of this theorem. Efficient implementation of this algorithm is studied in [43, 95].

11.2 Rayleigh Quotient Iteration

The *Rayleigh*[1] *quotient* of a symmetric matrix A and a nonzero real vector u is

$$\rho(u, A) \equiv \frac{u^T A u}{u^T u}.$$

Evident properties of the Rayleigh quotient $\rho(u, A)$ are (see Section 4.3.11 for variational properties of eigenvalues and self-adjoint operators):

- $\rho(\gamma u, A) = \rho(u, A)$ for every nonzero scalar γ.
- If $A q_i = \tilde{\lambda}_i q_i$, then $\rho(q_i, A) = \tilde{\lambda}_i$.

Algorithm 11.1 Rayleigh quotient iteration.

0. Initialization: set $i = 0$, stopping tolerance θ.
1. Compute $y_{i+1} = (A - \rho_i I)^{-1} x_i$.
2. Compute $x_{i+1} = y_{i+1} / \|y_{i+1}\|_2$.
3. Compute the approximate Rayleigh quotient $\rho_{i+1} = \rho(x_{i+1}, A)$.
4. Stop updating the Rayleigh quotient and set $\rho_M = \rho_{i+1}$, $M = i + 1$, if $\|A x_{i+1} - \rho_{i+1} x_{i+1}\|_2 < \theta$. Otherwise, set $i = i + 1$ and go to step 1.

From results of Section 7.1, p. 217, it follows that when the stopping criterion $\|A x_i - \rho_i x_i\|_2 < \theta$ in the above algorithm is satisfied, then the computed Rayleigh quotient ρ_i is within tolerance θ of an eigenvalue of A.

If we will take the shift $\sigma_i = a_{nn}$ in Algorithm 10.5 and then run Algorithm 11.1 with $x_0 = (0, ..., 0, 1)^T$, then $\sigma_i = \rho_i$.

Theorem 11.2 *Rayleigh quotient iteration is locally cubically convergent.*

Proof We will analyze only the case that A is diagonal. This is enough, since writing $Q^T A Q = \Lambda$, where Q is an orthogonal matrix whose columns are eigenvectors, and $\Lambda = \mathrm{diag}(\tilde{\lambda}_1, \ldots, \tilde{\lambda}_n)$ is a diagonal matrix of eigenvalues, we can write the Rayleigh quotient ρ_i computed at iteration i as

$$\rho_i = \rho(x_i, A) = \frac{x_i^T A x_i}{x_i^T x_i} = \frac{\hat{x}_i^T Q^T A Q \hat{x}_i}{\hat{x}_i^T Q^T Q \hat{x}_i} = \frac{\hat{x}_i^T \Lambda \hat{x}_i}{\hat{x}_i^T \hat{x}_i} = \rho(\hat{x}_i, \Lambda),$$

[1] John William Strutt, 3rd Baron Rayleigh (1842–1919), was an English physicist.

where $\hat{x}_i \equiv Q^T x_i$ and $\hat{y}_i \equiv Q^T y_i$. We observe also that $Q\hat{y}_{i+1} = (A - \rho_i I)^{-1} Q\hat{x}_i$, so

$$\hat{y}_{i+1} = Q^T (A - \rho_i I)^{-1} Q\hat{x}_i = (Q^T A Q - \rho_i I)^{-1} \hat{x}_i = (\Lambda - \rho_i I)^{-1} \hat{x}_i.$$

We see that running Algorithm 11.1 for Rayleigh quotient iteration with A and x_0 is the same as running Rayleigh quotient iteration with Λ and $\hat{x}_0$. Thus we will assume that $A = \Lambda$ is already diagonal and that the eigenvectors of A are e_i, or the columns of the identity matrix I.

Assume that x_i converges to e_1, so we can write $x_i = e_1 + d_i$, where $\|d_i\|_2 \equiv \varepsilon \ll 1$. To prove cubic convergence, we need to show that $x_{i+1} = e_1 + d_{i+1}$ with $\|d_{i+1}\|_2 = O(\varepsilon^3)$. We first note that

$$1 = x_i^T x_i = (e_1 + d_i)^T (e_1 + d_i) = e_1^T e_1 + 2e_1^T d_i + d_i^T d_i = 1 + 2d_{i1} + \varepsilon^2,$$

so that $d_{i1} = -\varepsilon^2/2$. Therefore,

$$\begin{aligned} \rho_i &= x_i^T \Lambda x_i = (e_1 + d_i)^T \Lambda (e_1 + d_i) = e_1^T \Lambda e_1 + 2e_1^T \Lambda d_i + d_i^T \Lambda d_i \\ &= \tilde{\lambda}_1 - (-2e_1^T \Lambda d_i - d_i^T \Lambda d_i) = \tilde{\lambda}_1 - \eta = \tilde{\lambda}_1 - \tilde{\lambda}_1 \varepsilon^2 - d_i^T \Lambda d_i, \end{aligned} \tag{11.1}$$

and since $-2e_1^T d_i = \varepsilon^2$, we have that $\eta \equiv -2e_1^T \Lambda d_i - d_i^T \Lambda d_i = \tilde{\lambda}_1 \varepsilon^2 - d_i^T \Lambda d_i$. We see that

$$|\eta| \le |\tilde{\lambda}_1| \varepsilon^2 + \|\Lambda\|_2 \|d_i\|_2^2 \le |\tilde{\lambda}_1| \varepsilon^2 + \|\Lambda\|_2 \varepsilon^2 \le 2\|\Lambda\|_2 \varepsilon^2,$$

so $\rho_i = \tilde{\lambda}_1 - \eta = \tilde{\lambda}_1 + O(\varepsilon^2)$ is a very good approximation to the eigenvalue $\tilde{\lambda}_1$.

Using Algorithm 11.1, we see that $y_{i+1} = (A - \rho_i I)^{-1} x_i$. Thus, using the remark above, we can write

$$\begin{aligned} y_{i+1} = (\Lambda - \rho_i I)^{-1} x_i &= \operatorname{diag} \frac{1}{\tilde{\lambda}_j - \rho_i} x_i \\ &= \left(\frac{x_{i1}}{\tilde{\lambda}_1 - \rho_i}, \frac{x_{i2}}{\tilde{\lambda}_2 - \rho_i}, \dots, \frac{x_{in}}{\tilde{\lambda}_n - \rho_i} \right)^T. \end{aligned}$$

Since $x_i = e_1 + d_i$ we get

$$y_{i+1} = \left(\frac{1 + d_{i1}}{\tilde{\lambda}_1 - \rho_i}, \frac{d_{i2}}{\tilde{\lambda}_2 - \rho_i}, \dots, \frac{d_{in}}{\tilde{\lambda}_n - \rho_i} \right)^T.$$

Next, because $\rho_i = \tilde{\lambda}_1 - \eta$ and $d_{i1} = -\varepsilon^2/2$ we obtain

$$\begin{aligned} y_{i+1} &= \left(\frac{1 - \varepsilon^2/2}{\eta}, \frac{d_{i2}}{\tilde{\lambda}_2 - \tilde{\lambda}_1 + \eta}, \dots, \frac{d_{in}}{\tilde{\lambda}_n - \tilde{\lambda}_1 + \eta} \right)^T \\ &= \frac{1 - \varepsilon^2/2}{\eta} \left(1, \frac{d_{i2}\eta}{(1 - \varepsilon^2/2)(\tilde{\lambda}_2 - \tilde{\lambda}_1 + \eta)}, \dots, \frac{d_{in}\eta}{(1 - \varepsilon^2/2)(\tilde{\lambda}_n - \tilde{\lambda}_1 + \eta)} \right)^T \\ &\equiv \frac{1 - \varepsilon^2/2}{\eta} (e_1 + \hat{d}_{i+1}). \end{aligned}$$

To bound $\|\hat{d}_{i+1}\|_2$, we will bound every denominator in the above expression using the inequality (see Section 7.1, p. 217, for the definition of gap)

$$|\tilde{\lambda}_j - \tilde{\lambda}_1 + \eta| \geq \text{gap}(1, \Lambda) - |\eta|,$$

as well as the following estimate for $|\eta|$ in every numerator,

$$|\eta| \leq |\tilde{\lambda}_1|\varepsilon^2 + \|\Lambda\|_2\|d_i\|_2^2 \leq 2\|\Lambda\|_2\varepsilon^2,$$

to get

$$
\begin{aligned}
\|\hat{d}_{i+1}\|_2 &\leq \frac{\|d_i\|_2|\eta|}{(1 - \varepsilon^2/2)(\text{gap}(1, \Lambda) - |\eta|)} \\
&\leq \frac{2\|\Lambda\|_2\varepsilon^3}{(1 - \varepsilon^2/2)(\text{gap}(1, \Lambda) - 2\|\Lambda\|_2\varepsilon^2)}.
\end{aligned}
\tag{11.2}
$$

In other words, inequality (11.2) means that $\|\hat{d}_{i+1}\|_2 = O(\varepsilon^3)$. Finally, by Algorithm 11.1, we have that $x_i = y_i/\|y_i\|_2$, and thus

$$x_{i+1} = e_1 + d_{i+1} = y_{i+1}/\|y_{i+1}\|_2, \tag{11.3}$$

or

$$x_{i+1} = \frac{y_{i+1}}{\|y_{i+1}\|_2} = \frac{\left(\dfrac{1 - \varepsilon^2/2}{\eta}(e_1 + \hat{d}_{i+1})\right)}{\left\|\dfrac{1 - \varepsilon^2/2}{\eta}(e_1 + \hat{d}_{i+1})\right\|_2} = (e_1 + \hat{d}_{i+1})/\|e_1 + \hat{d}_{i+1}\|_2.$$

Comparing the above expression with (11.3), we see that $\|d_{i+1}\|_2 = O(\varepsilon^3)$. ☐

Below we present an example of Algorithm 11.1 used for the computation of the Rayleigh quotient of some predefined symmetric tridiagonal matrix A.

Example 11.1 We compute the Rayleigh quotient of the matrix

$$A = \begin{pmatrix} 1 & 5 & 0 & 0 \\ 5 & 7 & 1 & 0 \\ 0 & 1 & 3 & 4 \\ 0 & 0 & 4 & 2 \end{pmatrix}$$

using Algorithm 11.1. This matrix has four different eigenvalues,

$$\lambda = (-2.0607, -1.3469, 6.4239, 9.9837),$$

which we obtained using the command `eig(A)` in MATLAB®. The computed Rayleigh quotient is $\rho = -1.3469$, which is one of the eigenvalues of A. The MATLAB® program `RayleighQuotient.m` of Section 1.16 is available for running this test.[2]

---

[2]The MATLAB® programs can be found on the online version of Chapter 1 (doi:10.1007/978-3-319-57304-5_1).

11.3 Divide and Conquer

The main advantage of the method of divide and conquer is that it is the fastest among existing methods to compute all eigenvalues and eigenvectors of a tridiagonal matrix of size n greater than about $n = 25$. The divide and conquer method is not easy to implement efficiently in a stable way. It was introduced first in [21], and the first efficient and stable implementation of it was presented in [30, 38].

The structure of the algorithm is the following. Let T be a tridiagonal symmetric matrix,

$$
T = \begin{pmatrix}
a_1 & b_1 & 0 & \cdots & \cdots & 0 \\
b_1 & a_2 & b_2 & 0 & \cdots & 0 \\
\cdots & \cdots & \cdots & \cdots & \cdots & \cdots \\
\cdots & a_{m-1} & b_{m-1} & \cdots & \cdots & 0 \\
0 & b_{m-1} & a_m & b_m & 0 & 0 \\
0 & 0 & b_m & a_{m+1} & b_{m+1} & 0 \\
0 & 0 & 0 & b_{m+1} & \cdots & \cdots \\
\cdots & \cdots & \cdots & \cdots & \cdots & \cdots \\
0 & 0 & 0 & \cdots & \cdots & b_{n-1} \\
0 & 0 & 0 & 0 & b_{n-1} & a_n
\end{pmatrix},
$$

such that we can decompose it in the following way:

$$
T = \begin{pmatrix}
a_1 & b_1 & 0 & & \cdots & & 0 \\
b_1 & a_2 & b_2 & & 0 & & 0 \\
\cdots & \cdots & \cdots & & \cdots & & \cdots \\
\cdots & a_{m-1} & b_{m-1} & & \cdots & & 0 \\
0 & b_{m-1} & a_m - b_m & 0 & & 0 & 0 \\
0 & 0 & 0 & a_{m+1} - b_m & b_{m+1} & 0 \\
0 & 0 & 0 & b_{m+1} & \cdots & \cdots \\
\cdots & \cdots & \cdots & & \cdots & \cdots \\
0 & 0 & 0 & & \cdots & b_{n-1} \\
0 & 0 & 0 & 0 & b_{n-1} & a_n
\end{pmatrix} + \begin{pmatrix}
\cdots & \cdots \\
\ldots b_m & b_m \ldots \\
\ldots b_m & b_m \ldots \\
\cdots & \cdots
\end{pmatrix}
$$

(11.4)

$$
= \left(\begin{array}{c|c} T_1 & 0 \\ \hline 0 & T_2 \end{array} \right) + b_m \begin{pmatrix} 0 \\ \vdots \\ 0 \\ 1 \\ 1 \\ 0 \\ \vdots \\ 0 \end{pmatrix} (0...0\ 1\ 1\ 0....0) = \left(\begin{array}{c|c} T_1 & 0 \\ \hline 0 & T_2 \end{array} \right) + b_m v v^T.
$$

Assume that we have an eigendecomposition of T_1, T_2 such that $T_1 = Q_1 \Lambda_1 Q_1^T$ and $T_2 = Q_2 \Lambda_2 Q_2^T$. Then we can write that

$$
T = \begin{pmatrix} T_1 & 0 \\ 0 & T_2 \end{pmatrix} + b_m v v^T = \begin{pmatrix} Q_1 \Lambda_1 Q_1^T & 0 \\ 0 & Q_2 \Lambda_2 Q_2^T \end{pmatrix} + b_m v v^T
$$
$$
= \begin{pmatrix} Q_1 & 0 \\ 0 & Q_2 \end{pmatrix} \left(\begin{pmatrix} \Lambda_1 & 0 \\ 0 & \Lambda_2 \end{pmatrix} + b_m u u^T \right) \begin{pmatrix} Q_1^T & 0 \\ 0 & Q_2^T \end{pmatrix}.
$$
(11.5)

Let us define the diagonal matrix

$$
D = \begin{pmatrix} \Lambda_1 & 0 \\ 0 & \Lambda_2 \end{pmatrix}
$$

and rewrite (11.5) as

$$
T = \begin{pmatrix} Q_1 & 0 \\ 0 & Q_2 \end{pmatrix} (D + b_m u u^T) \begin{pmatrix} Q_1^T & 0 \\ 0 & Q_2^T \end{pmatrix}.
$$
(11.6)

We observe that the eigenvalues of T are the same as those of

$$
D + b_m u u^T = D + \rho u u^T
$$
(11.7)

with scalar $\rho = b_m$. Thus, our goal now is to find the eigenvalues of (11.7). To do so, we proceed in the following steps:

Step 1. We assume that the diagonal elements of D are sorted such that $d_1 \geq \dots \geq d_n$ and $D + \lambda I$ is nonsingular.

Step 2. To find the eigenvalues of $D + \rho u u^T$, we compute the characteristic polynomial

$$
\det(D + \rho u u^T - \lambda I) = 0,
$$

noting that

$$
\det(D + \rho u u^T - \lambda I) = \det((D - \lambda I)(I + \rho(D - \lambda I)^{-1} u u^T)).
$$
(11.8)

Step 3. By assumption, in Step 1 we have

$$
\det(D - \lambda I) \neq 0,
$$

and thus in (11.8) we should have

$$
\det(I + \rho(D - \lambda I)^{-1} u u^T) = 0.
$$

Finally, the eigenvalues of $D + \rho u u^T$ are found from the above expression.

Lemma 11.1 *If x and y are vectors, then $\det(I + x y^T) = 1 + y^T x$.*

The proof of this lemma is left as Exercise 11.9. Thus, using this lemma we can get that

$$\det(I + \rho(D - \lambda I)^{-1}uu^T) = 1 + u^T\rho(D - \lambda I)^{-1}u = 1 + \rho\sum_{i=1}^{n}\frac{u_i^2}{d_i - \lambda} = f(\lambda).$$
(11.9)

We see that the eigenvalues of T are the roots of the so-called secular equation $f(\lambda) = 0$.

The secular equation can be solved using Newton's method

$$\lambda^{k+1} = \lambda^k - \frac{f(\lambda^k)}{f'(\lambda^k)},$$
(11.10)

where

$$f'(\lambda^k) = \rho\sum_{i=1}^{n}\frac{u_i^2}{(d_i - \lambda^k)^2},$$
(11.11)

and k is the number of the iteration in Newton's method. We observe that $f'(\lambda^k)$ is positive and increasing when $\rho = b_m > 0$ except at the point $\lambda = d_i$. Thus, the roots of $f(\lambda)$ are seperated by the d_i where $\lambda = d_i$ be a vertical asymptotes, and $f(\lambda)$ is monotonic and smooth on every interval (d_i, d_{i+1}). Then Newton's method (11.10) will converge for a starting point $\lambda^0 \in (d_i, d_{i+1})$.

To obtain the eigenvectors x for the eigenvalues λ that we have found via solution of the secular equation (11.9), we need the following lemma.

Lemma 11.2 *If $\tilde{\lambda}$ is an eigenvalue of $D + \rho uu^T$, then*

$$x = (D - \tilde{\lambda}I)^{-1}u$$
(11.12)

is its eigenvector.

Proof If $\tilde{\lambda}$ is an eigenvalue of $D + \rho uu^T$ and (11.12) is its eigenvector, then we must have

$$(D + \rho uu^T)((D - \tilde{\lambda}I)^{-1}u) = ((D - \tilde{\lambda}I) + \tilde{\lambda}I + \rho uu^T)((D - \tilde{\lambda}I)^{-1}u)$$
$$= u + \tilde{\lambda}(D - \tilde{\lambda}I)^{-1}u + u(\rho u^T(D - \tilde{\lambda}I)^{-1}u).$$
(11.13)

Now we use the expression for the secular equation, or

$$\rho u^T(D - \tilde{\lambda}I)^{-1}u + 1 = f(\tilde{\lambda}) = 0,$$

and thus

$$\rho u^T(D - \tilde{\lambda}I)^{-1}u = -1,$$

or

$$(D + \rho u u^T)((D - \tilde{\lambda}I)^{-1}u) = u + \tilde{\lambda}(D - \tilde{\lambda}I)^{-1}u - u$$
$$= \tilde{\lambda}(D - \tilde{\lambda}I)^{-1}u = \tilde{\lambda}x. \tag{11.14}$$

□

Algorithm 11.2 Finding the eigenvalues and eigenvectors of a symmetric tridiagonal matrix using divide and conquer.

1. If T is 1×1
   return $Q = 1, \Lambda = T$
   else
   form $T = \begin{pmatrix} T_1 & 0 \\ 0 & T_2 \end{pmatrix} + b_m v v^T$
2. Compute output matrices Q_1 and Λ_1 by eigendecomposition of T_1.
3. Compute output matrices Q_2 and Λ_2 by eigendecomposition of T_2.
4. Form $D + \rho u u^T$ from $\Lambda_1, \Lambda_2, Q_1, Q_2$.
5. Find eigenvalues Λ and eigenvectors Q' of $D + \rho u u^T$ as roots of the secular equation (11.9) using Newton's method (11.10).
6. Form $Q = \begin{pmatrix} Q_1 & 0 \\ 0 & Q_2 \end{pmatrix} Q' = $ eigenvectors of T.
6. Return Q, Λ and stop.

Remark 11.1

- Eigenvectors can be computed by formula (11.12).
- This formula is not stable when two eigenvalues $(\tilde{\lambda}_i, \tilde{\lambda}_{i+1})$ are located close to each other. This means that $(D - \tilde{\lambda}_i)^{-1}u$ and $(D - \tilde{\lambda}_{i+1})^{-1}u$ are inaccurate and far from orthogonal.
- Löwner's theorem is used to compute eigenvectors for two eigenvalues $(\tilde{\lambda}_i, \tilde{\lambda}_{i+1})$ that are close to each other.

Theorem 11.3 (Löwner)[3].
   Let $D = \text{diag}(d_1, \ldots, d_n)$ be diagonal with $d_n < \ldots < d_1$. Let $\tilde{\lambda}_n < \ldots < \tilde{\lambda}_1$ be given, satisfying the alternating property

$$d_n < \tilde{\lambda}_n < \cdots < d_{i+1} < \tilde{\lambda}_{i+1} < d_i < \tilde{\lambda}_i < \cdots < d_1 < \tilde{\lambda}_1.$$

   Then there is a vector $\hat{u}$ such that the $\tilde{\lambda}_i$ are the exact eigenvalues of $\hat{D} \equiv D + \hat{u}\hat{u}^T$. The entries of $\hat{u}$ are

$$|\hat{u}_i| = \left(\frac{\prod_{j=1}^n (\tilde{\lambda}_j - d_i)}{\prod_{j=1, j \neq i}^n (d_j - d_i)} \right)^{1/2}. \tag{11.15}$$

---

[3]Charles Löwner (1893–1968) was an American mathematician.

Proof The characteristic polynomial of $\hat{D}$ can be written in two ways, as

$$\det(\hat{D} - \lambda I) = \prod_{j=1}^{n}(\tilde{\lambda}_j - \lambda) \tag{11.16}$$

and using

$$\det(\hat{D} - \lambda I) = \det(D + \hat{u}\hat{u}^T - \lambda I) = \det(D(I + D^{-1}\hat{u}\hat{u}^T) - \lambda I)$$
$$= \det((D - \lambda I)(I + (D - \lambda I)^{-1}\hat{u}\hat{u}^T))$$

as

$$\det(\hat{D} - \lambda I) = \left(\prod_{j=1}^{n}(d_j - \lambda)\right)\left(1 + \sum_{j=1}^{n}\frac{\hat{u}_j^2}{d_j - \lambda}\right)$$

$$= \left(\prod_{j=1}^{n}(d_j - \lambda)\right)\left(1 + \sum_{\substack{j=1 \\ j \neq i}}^{n}\frac{\hat{u}_j^2}{d_j - \lambda} + \frac{\hat{u}_i^2}{d_i - \lambda}\right)$$

$$= \left(\prod_{j=1}^{n}(d_j - \lambda)\right)\left(1 + \sum_{\substack{j=1 \\ j \neq i}}^{n}\frac{\hat{u}_j^2}{d_j - \lambda}\right) + \left(\prod_{j=1}^{n}(d_j - \lambda)\right)\frac{\hat{u}_i^2}{d_i - \lambda}$$

$$= \left(\prod_{j=1}^{n}(d_j - \lambda)\right)\left(1 + \sum_{\substack{j=1 \\ j \neq i}}^{n}\frac{\hat{u}_j^2}{d_j - \lambda}\right) + \left(\prod_{\substack{j=1 \\ j \neq i}}^{n}(d_j - \lambda)\right)\hat{u}_i^2. \tag{11.17}$$

Now we choose $\lambda = d_i$ in (11.16) and in (11.17) for $\det(\hat{D} - \lambda I)$ to get

$$\prod_{j=1}^{n}(\tilde{\lambda}_j - d_i) = \hat{u}_i^2 \prod_{\substack{j=1 \\ j \neq i}}^{n}(d_j - d_i),$$

or

$$\hat{u}_i^2 = \frac{\prod_{j=1}^{n}(\tilde{\lambda}_j - d_i)}{\prod_{j=1, j \neq i}^{n}(d_j - d_i)}.$$

Using the alternating property, we can show that the right-hand side in the above expression is positive, and thus we get (11.15). □

Below we give a stable algorithm for computing the eigenvalues and eigenvectors where we have assumed that $\rho = 1$.

Algorithm 11.3 Compute the eigenvalues and eigenvectors of $D + uu^T$.

1. Solve the secular equation $1 + \sum\limits_{i=1}^{n} \frac{u_i^2}{d_i - \tilde{\lambda}} = 0$ to get the eigenvalues $\tilde{\lambda}_i$ of $D + uu^T$.

2. Use Löwner's theorem to compute $\hat{u}$ so that the $\tilde{\lambda}_i$ are "exact" eigenvalues of $D + \hat{u}\hat{u}^T$.

3. Use formula (11.12) in Lemma 11.2 to compute the eigenvectors of $\hat{D} = D + \hat{u}\hat{u}^T$.

Below we present an example using Algorithm 11.2 on the computation of eigenvalues and eigenvectors of a predefined symmetric tridiagonal matrix A. The MATLAB® program of Section 1.17 is available for running this test.[4]

Example 11.2 We compute the eigenvalues and eigenvectors of the matrix

$$
A = \begin{pmatrix}
10.8901 & 9.5557 & 0 & 0 & 0 \\
9.5557 & 10.6813 & 2.6985 & 0 & 0 \\
0 & 2.6985 & 2.2341 & 4.0888 & 0 \\
0 & 0 & 4.0888 & 13.5730 & 14.8553 \\
0 & 0 & 0 & 14.8553 & 3.7942
\end{pmatrix}
$$

using Algorithm 11.2. This matrix has five different eigenvalues,

$$
\lambda = (-7.5981, -0.1710, 3.5923, 20.5154, 24.8341),
$$

obtained using the command `eig(A)` in MATLAB®. We apply the MATLAB® program of Section 1.17 and compute the eigenvalues and eigenvectors of the matrix A above. It turns out that the computed eigenvalues and eigenvectors of the above matrix using the MATLAB® program of Section 1.17 are the same as those obtained by the command `eig(A)` in MATLAB®.

11.4 Bisection and Inverse Iteration

The bisection algorithm uses Sylvester's inertia theorem, Theorem 4.44, to find only the k eigenvalues that we want.

Recall that Inertia$(A) = (\nu, \zeta, \pi)$, where ν, ζ, and π are the numbers of negative, zero, and positive eigenvalues of A, respectively. Suppose that X is nonsingular. Using Sylvester's inertia theorem, we have that Inertia$(A) = $ Inertia$(X^T A X)$.

---

[4]All MATLAB® programs referenced on this page can be found on the online version of Chapter 1 (doi:10.1007/978-3-319-57304-5_1).

Now suppose that we use Gaussian elimination to factorize $A - zI = LDL^T$, where L is nonsingular and D is diagonal. Then Inertia$(A - zI) =$ Inertia(D). The inertia of D is very easily computable, since D is diagonal.

Further in our considerations of this section we use the notation

$$\#d_{ii} < 0,$$

which means "the number of values of d_{ii} less than zero." Then

$$
\begin{aligned}
\text{Inertia}(A - zI) &= (\#d_{ii} < 0, \#d_{ii} = 0, \#d_{ii} > 0) \\
&= (\# \text{ negative eigenvalues of } A - zI, \\
&\quad\ \# \text{ zero eigenvalues of } A - zI, \\
&\quad\ \# \text{ positive eigenvalues of } A - zI) \\
&= (\# \text{ eigenvalues of } A < z, \\
&\quad\ \# \text{ eigenvalues of } A = z, \\
&\quad\ \# \text{ eigenvalues of } A > z).
\end{aligned}
$$

Let us define
$$\text{NrofEig}(A, z) = \# \text{ eigenvalues of } A < z.$$

Suppose $z_1 < z_2$, and we compute Inertia$(A - z_1 I)$ and Inertia$(A - z_2 I)$. Then the number of eigenvalues $N_{[z_1, z_2)}$ in the interval $[z_1, z_2)$ equals (# eigenvalues of $A < z_2$) $-$ (# eigenvalues of $A < z_1$), or

$$N_{[z_1, z_2)} = \text{NrofEig}(A, z_2) - \text{NrofEig}(A, z_1).$$

Algorithm 11.4 Bisection: find all eigenvalues of A inside $[a, b]$ to a given error tolerance θ.

$n_a = \text{NrofEig}(A, a)$
$n_b = \text{NrofEig}(A, b)$
if $n_a = n_b$, quit (because there are no eigenvalues in $[a, b)$)
put $[a, n_a, b, n_b]$ onto WorkingArray
 /* WorkingArray contains all subintervals of $[a, b)$ containing
  eigenvalues from $n - n_a$ through $n - n_b + 1$, which the algorithm
  will update until they are less than tolerance θ. */
while WorkingArray is not empty
 remove $[low, n_{low}, up, n_{up}]$ from WorkingArray
 if $up - low < \theta$ then
  print "there are $n_{up} - n_{low}$ eigenvalues in [low, up)"
 else
  $mid = (low + up)/2$
  $n_{mid} = \text{NrofEig}(A, mid)$
  if $n_{mid} > n_{low}$ then print "there are eigenvalues in $[low, mid)$"
   put $[low, n_{low}, mid, n_{mid}]$ onto WorkingArray
  end if

```
      if n_up > n_mid then print "there are eigenvalues in [mid, up)"
         put [mid, n_mid, up, n_up] onto WorkingArray
      end if
   end if
end if
end while
```

if $n_{up} > n_{mid}$ then print "there are eigenvalues in $[mid, up)$"
    put $[mid, n_{mid}, up, n_{up}]$ onto WorkingArray
    end if
  end if
end if
end while

From NrofEig(A, z) it is easy to use Gaussian elimination, provided that

$$A - zI = \begin{pmatrix} a_1 - z & b_1 & \dots & \dots \\ b_1 & a_2 - z & \dots & \dots \\ \dots & \dots & \dots & \dots \\ \dots & b_{n-2} & a_{n-1} - z & b_{n-1} \\ \dots & \dots & b_{n-1} & a_n - z \end{pmatrix}$$

$$= LDL^T = \begin{pmatrix} 1 & \dots & \dots \\ l_1 & 1 & \dots \\ \dots & \dots & \dots \\ \dots & l_{n-1} & 1 \end{pmatrix} \begin{pmatrix} d_1 & \dots & \dots \\ \dots & d_2 & \dots \\ \dots & \dots & \dots \\ \dots & \dots & d_n \end{pmatrix} \begin{pmatrix} 1 & l_1 & \dots & \dots \\ \dots & 1 & \dots \\ \dots & \dots & l_{n-1} \\ \dots & \dots & 1 \end{pmatrix}. \tag{11.18}$$

Using (11.18), we observe that

$$\begin{aligned} a_1 - z &= d_1, \\ d_1 l_1 &= b_1, \\ l_{i-1}^2 d_{i-1} + d_i &= a_i - z, \\ d_i l_i &= b_i. \end{aligned} \tag{11.19}$$

Substitute $l_i = b_i / d_i$ into $l_{i-1}^2 d_{i-1} + d_i = a_i - z$ to obtain the recurrence formula

$$d_i = (a_i - z) - \frac{b_{i-1}^2}{d_{i-1}}, \tag{11.20}$$

from which is easy to compute the values d_i of the matrix D from previously known values d_{i-1} and known values a_i, b_i. In [26, 27], it was shown that since $A - zI$ is a tridiagonal matrix, the formula (11.20) is stable.

Below we present an example using Algorithm 11.4. The MATLAB® program of Section 1.18 is available for running this test.[5]

Example 11.3 We compute the eigenvalues of the matrix

$$A = \begin{pmatrix} 16.1984 & 2.8029 & 0 & 0 & 0 \\ 2.8029 & 9.0301 & 23.0317 & 0 & 0 \\ 0 & 23.0317 & 12.5310 & 24.2558 & 0 \\ 0 & 0 & 24.2558 & 10.5238 & 17.5216 \\ 0 & 0 & 0 & 17.5216 & 10.4891 \end{pmatrix}$$

---

[5]The MATLAB® programs can be found on the online version of Chapter 1 (doi:10.1007/978-3-319-57304-5_1).

using the bisection algorithm, Algorithm 11.4. This matrix has five different eigenvalues $\lambda = (-25.0154, -1.2034, 15.9244, 21.8223, 47.2444)$, which we obtained using the command $\texttt{eig(A)}$ in MATLAB®.

We compute now the eigenvalues of the matrix A using the MATLAB® program of Section 1.18.[6] Since in the MATLAB® program of Section 1.18 we require that the left- and right-hand sides of the input interval in Algorithm 11.4 have difference no more than the given error $\texttt{tol}$, it follows that this interval will contain one eigenvalue, and the left- or right-hand side of this interval can be taken as our desired eigenvalue. The output information obtained by the MATLAB® program of Section 1.18 for the matrix A defined above is the following:

```
There is 1 eigenvalue in the interval [-25.0154,-25.0154)
There is 1 eigenvalue in the interval [-1.2034,-1.2034)
There is 1 eigenvalue in the interval [15.9244,15.9244)
There is 1 eigenvalue in the interval [21.8223,21.8223)
There is 1 eigenvalue in the interval [47.2444,47.2444)
```

Comparing the above results with the exact ones, we observe that the computed eigenvalues using the MATLAB® program of Section 1.18 are the same as those produced by the command $\texttt{eig(A)}$.

11.5 Jacobi's Method

We will not reduce the original matrix A to a tridiagonal matrix T as in all previous methods, but will work on the original A. Jacobi's[7] method produces a sequence A_i, $i = 0, ..., m$, of orthogonally similar matrices for a given matrix $A = A_0$, which will converge to a diagonal matrix with the eigenvalues on the diagonal. The next matrix A_{i+1} is obtained from the previous one A_i by the recurrence formula

$$A_{i+1} = J_i^T A_i J_i,$$

where J_i is an orthogonal matrix called a *Jacobi rotation*. Thus

$$\begin{aligned} A_m &= J_{m-1}^T A_{m-1} J_{m-1} \\ &= J_{m-1}^T J_{m-2}^T A_{m-2} J_{m-2} J_{m-1} = \cdots \\ &= J_{m-1}^T \cdots J_0^T A_0 J_0 \cdots J_{m-1} \\ &= J^T A J. \end{aligned}$$

If we choose every J_i in some special way, then A_m will converge to a diagonal matrix Λ for large m. Thus we can write

$$\Lambda \approx J^T A J,$$

---

[6] All MATLAB® programs referenced on this page can be found on the online version of Chapter 1 (doi:10.1007/978-3-319-57304-5_1).

[7] Carl Gustav Jacob Jacobi (1804–1851) was a German mathematician.

or

$$J \Lambda J^T \approx A,$$

from which we see that the columns of J are approximate eigenvectors.

To make $J^T A J$ nearly diagonal, we will construct the J_i iteratively to make *one* pair of off-diagonal entries of $A_{i+1} = J_i^T A_i J_i$ zero at a time. We will do this by taking J_i to be a Givens rotation, or

$$J_i = R(j, k, \theta) \equiv \begin{pmatrix} 1 & & & & & & & & \\ & 1 & & & & & & & \\ & & \ddots & & & & & & \\ & & & \cos\theta & & -\sin\theta & & & \\ & & & & \ddots & & & & \\ & & & \sin\theta & & \cos\theta & & & \\ & & & & & & \ddots & & \\ & & & & & & & 1 & \\ & & & & & & & & 1 \end{pmatrix}, \qquad (11.21)$$

where θ is chosen such that the (j, k) and (k, j) entries of A_{i+1} will be zero. To determine θ (or actually $\cos\theta$ and $\sin\theta$), let us consider

$$\begin{aligned}
\begin{pmatrix} a_{jj}^{(i+1)} & a_{jk}^{(i+1)} \\ a_{kj}^{(i+1)} & a_{kk}^{(i+1)} \end{pmatrix} &= \begin{pmatrix} \cos\theta & -\sin\theta \\ \sin\theta & \cos\theta \end{pmatrix}^T \begin{pmatrix} a_{jj}^{(i)} & a_{jk}^{(i)} \\ a_{kj}^{(i)} & a_{kk}^{(i)} \end{pmatrix} \begin{pmatrix} \cos\theta & -\sin\theta \\ \sin\theta & \cos\theta \end{pmatrix} \\
&= \begin{pmatrix} c & -s \\ s & c \end{pmatrix}^T \begin{pmatrix} a_{jj}^{(i)} & a_{jk}^{(i)} \\ a_{kj}^{(i)} & a_{kk}^{(i)} \end{pmatrix} \begin{pmatrix} c & -s \\ s & c \end{pmatrix} = \begin{pmatrix} \lambda_1 & 0 \\ 0 & \lambda_2 \end{pmatrix},
\end{aligned} \qquad (11.22)$$

where λ_1 and λ_2 are the eigenvalues of

$$\begin{pmatrix} a_{jj}^{(i)} & a_{jk}^{(i)} \\ a_{kj}^{(i)} & a_{kk}^{(i)} \end{pmatrix}.$$

It is easy to compute $c = \cos\theta$ and $s = \sin\theta$ from (11.22) on every iteration i:

$$\begin{pmatrix} \lambda_1 & 0 \\ 0 & \lambda_2 \end{pmatrix} = \begin{pmatrix} a_{jj}c^2 + a_{kk}s^2 + 2sca_{jk} & sc(a_{kk} - a_{jj}) + a_{jk}(c^2 - s^2) \\ sc(a_{kk} - a_{jj}) + a_{jk}(c^2 - s^2) & a_{jj}s^2 + a_{kk}c^2 - 2sca_{jk} \end{pmatrix}.$$

Setting the off-diagonals to zero and solving for θ, we get

$$0 = sc(a_{kk} - a_{jj}) + a_{jk}(c^2 - s^2),$$

or

$$\frac{a_{jj} - a_{kk}}{2a_{jk}} = \frac{c^2 - s^2}{2sc} = \frac{\cos 2\theta}{\sin 2\theta} = \cot 2\theta \equiv \tau.$$

We now introduce the notation $t = \frac{s}{c} = \tan \theta$, noting that $t^2 + 2\tau t - 1 = 0$. Solving this quadratic equation, we get

$$t = \frac{\text{sign}(\tau)}{|\tau| + \sqrt{1 + \tau^2}},$$

$$c = \frac{1}{\sqrt{1 + \tau^2}}, \qquad (11.23)$$

$$s = tc.$$

Algorithm 11.5 Compute and apply a Jacobi rotation to A for indices (j, k).
    function Jacobi-Rotation(A, j, k)
    if $|a_{jk}|$ is not too small
        $\tau = (a_{jj} - a_{kk})/(2a_{jk})$
        $t = \text{sign}(\tau)/(|\tau| + \sqrt{1 + \tau^2})$
        $c = 1/\sqrt{1 + \tau^2}$
        $s = tc$
        $A = R^T(j, k, \theta) A R(j, k, \theta)$ /* here, $c = \cos \theta$ and $s = \sin \theta$ */
            $J = J \ R(j, k, \theta)$ /* if eigenvectors are desired */
    end if
    end if

The general Jacobi algorithm is given below.

Algorithm 11.6 Jacobi's method to find the eigenvalues of a symmetric matrix. Perform the following steps in a loop:

1. Choose (j, k).
2. Call the function Jacobi-Rotation(A, j, k) until A is sufficiently diagonal.

There are different ways to choose the pairs (j, k). To measure the progress of convergence, we define

$$\text{off}(A) \equiv \sqrt{\sum_{1 \le j < k \le n} a_{jk}^2}.$$

Thus, off(A) is the root-sum-of-squares of the (upper) off-diagonal entries of A, so A is diagonal if and only if off$(A) = 0$. We want to make off$(A) = 0$ as quickly as possible.

The next lemma shows that off(A) decreases monotonically with every iteration of the Jacobi rotation.

Lemma 11.3 *Let A' be the matrix obtained after calling the procedure Jacobi-Rotation (A, j, k) for $j \neq k$. Then*

$$\text{off}^2(A') = \text{off}^2(A) - a_{jk}^2.$$

The proof of this lemma can be found in [23].

The next algorithm is the original version of the Jacobi algorithm developed in 1846. However, in practical computations, this algorithm is too slow.

Algorithm 11.7 Classical Jacobi's algorithm.

0. Set $i = 0$ and tolerance θ.
1. Choose (j, k) such that a_{jk} is the largest off-diagonal entry in magnitude.
2. Call Jacobi-Rotation(A, j, k).
3. Compute $\text{off}_i(A)$.
4. Stop and set $\text{off}_M(A) = \text{off}_i(A)$, $M = i$, if $\text{off}_i(A) < \theta$. Otherwise, set $i = i + 1$ and go to step 1.

Theorem 11.4 *Let A' be the matrix obtained after calling Jacobi-Rotation(A, j, k) for $j \neq k$. After one step of calling the Jacobi-Rotation procedure in the classical Jacobi's algorithm, Algorithm 11.7, we have*

$$\text{off}(A') \leq \sqrt{1 - \frac{1}{N}} \, \text{off}(A),$$

where $N = \frac{n(n-1)}{2}$ is the number of superdiagonal entries of A. After k steps of calling the Jacobi-Rotation procedure, we have

$$\text{off}(A') \leq \left(1 - \frac{1}{N}\right)^{k/2} \text{off}(A).$$

Proof By Lemma 11.3, after one step of Jacobi rotation, we have

$$\text{off}^2(A') = \text{off}^2(A) - a_{jk}^2,$$

where a_{jk} is the largest off-diagonal entry. Thus,

$$\text{off}^2(A) \leq \frac{n(n-1)}{2} a_{jk}^2,$$

or

$$a_{jk}^2 \geq \frac{1}{n(n-1)/2} \text{off}^2(A),$$

so that

$$\text{off}^2(A) - a_{jk}^2 \leq \left(1 - \frac{1}{N}\right) \text{off}^2(A),$$

from which follows the statements of Theorem 11.4.   □

Summarizing, we have that the classical Jacobi's algorithm converges at least linearly with the error decreasing by a factor of at least $\sqrt{1 - \frac{1}{N}}$ at each step.

Theorem 11.5 *Jacobi's method is locally quadratically convergent after N steps. This means that for a large i,*

$$\text{off}(A_{i+N}) = O(\text{off}^2(A_i)).$$

The proof of this theorem is given in [113]. In practice, we do not use the classical Jacobi's algorithm, because searching for the largest entry is too slow. We use the following simple method to choose j and k.

Algorithm 11.8 Cyclic-by-row-Jacobi: run through the off-diagonals of A rowwise.
   Loop
      for $j = 1$ to $n - 1$
      for $k = j + 1$ to n
         call Jacobi-Rotation(A, j, k)
      end for
      end for
   until A is sufficiently diagonal.

The matrix A no longer changes when Jacobi-Rotation(A, j, k) chooses only $c = 1$ and $s = 0$ for an entire pass through the inner loop. As was shown in [113], the cyclic Jacobi's algorithm is also asymptotically quadratically convergent, like the classical Jacobi's algorithm.

The cost of one loop in Algorithm 11.8 is around half the cost of the tridiagonal reduction and the computation of eigenvalues and eigenvectors via QR iteration, as well as more than the cost using divide and conquer. For convergence of Jacobi's method, one needs to perform from five to ten loops in Algorithm 11.8, and thus this method is much slower than other methods.

Below we present an example using the classical Algorithm 11.7. The MATLAB® program of Section 1.19 is available for running this test.[8]

Example 11.4 We compute the eigenvalues of the matrix

$$A = \begin{pmatrix} 14.7776 & 4.9443 & 0 & 0 & 0 \\ 4.9443 & 18.2496 & 28.3358 & 0 & 0 \\ 0 & 28.3358 & 10.8790 & 2.5361 & 0 \\ 0 & 0 & 2.5361 & 11.0092 & 18.9852 \\ 0 & 0 & 0 & 18.9852 & 15.0048 \end{pmatrix} \tag{11.24}$$

---

[8]The MATLAB® programs can be found on the online version of Chapter 1 (doi:10.1007/978-3-319-57304-5_1).

using the classical Jacobi algorithm, Algorithm 11.7. This matrix has five different eigenvalues,

$$\lambda = (-14.6416, -5.8888, 14.6644, 32.0314, 43.7547),$$

which we obtained using the command eig(A) in MATLAB®. We run the MATLAB® program of Section 1.19 until the matrix A is sufficiently diagonal, i.e., until $off(A) < tol$ for $tol = 0.005$.[9] The computed final matrix A obtained after all Jacobi rotations is the following:

$$A = \begin{pmatrix} 14.6644 & 0.0000 & -0.0000 & 0.0029 & 0.0001 \\ 0.0000 & 43.7547 & -0.0000 & 0.0000 & -0.0008 \\ -0.0000 & -0.0000 & -14.6416 & 0.0000 & -0.0000 \\ 0.0029 & 0.0000 & 0.0000 & -5.8888 & 0.0000 \\ 0.0001 & -0.0008 & -0.0000 & 0.0000 & 32.0314 \end{pmatrix}.$$

We observe that the values lying on the diagonal of the above matrix A are the eigenvalues of the initial matrix A given in (11.24).

Comparing these values with the exact ones, we observe that the computed eigenvalues using the MATLAB® program of Section 1.19 are almost the same (depending on the input tolerance tol) as produced by the command $eig(A)$.

11.6 Algorithms for the Singular Value Decomposition

Algorithms for the solution of the symmetric eigenvalue problem can be transformed to the algorithms for the SVD of a symmetric matrix A. Eigendecomposition of a symmetric matrix A, except for Jacobi's method, can be performed in the following steps:

1. Reduce A to tridiagonal form T with an orthogonal matrix Q_1:

$$A = Q_1 T Q_1^T.$$

2. Find the eigendecomposition of T:

$$T = Q_2 \Lambda Q_2^T,$$

where Λ is the diagonal matrix of eigenvalues and Q_2 is the orthogonal matrix whose columns will be eigenvectors.

3. Combine these decompositions to get

$$A = (Q_1 Q_2)\Lambda(Q_1 Q_2)^T.$$

---

[9]The MATLAB® programs referenced on this page can be found on the online version of Chapter 1 (doi:10.1007/978-3-319-57304-5_1).

The columns of $Q = Q_1 Q_2$ will be the eigenvectors of A.

All the algorithms for the SVD of a general matrix G, except Jacobi's method, have an analogous structure, which is the following:

1. Reduce G to bidiagonal form B, which has nonzero elements only on the main diagonal and first superdiagonal, with orthogonal matrices U_1 and V_1 such that

$$G = U_1 B V_1^T.$$

2. Find the SVD of B:

$$B = U_2 \Sigma V_2^T,$$

where Σ is the diagonal matrix of singular values, and U_2 and V_2 are orthogonal matrices whose columns are left and right singular vectors, respectively.
3. Combine these decompositions to get

$$G = (U_1 U_2) \Sigma (V_1 V_2)^T.$$

The columns of $U = U_1 U_2$ and $V = V_1 V_2$ are left and right singular vectors of G, respectively.

Lemma 11.4 *Let B be an $n \times n$ bidiagonal matrix such that*

$$B = \begin{pmatrix} a_1 & b_1 & \ldots & \ldots \\ \ldots & a_2 & \ldots & \ldots \\ \ldots & \ldots & \ldots & \ldots \\ \ldots & \ldots & a_{n-1} & b_{n-1} \\ \ldots & \ldots & \ldots & a_n \end{pmatrix}. \tag{11.25}$$

There are the following possibilities for converting the problem of finding the SVD of B to that of finding the eigenvalues and eigenvectors of a symmetric tridiagonal matrix.

1. Let the matrix A be such that

$$A = \begin{pmatrix} 0 & B^T \\ B & 0 \end{pmatrix}.$$

Let P be the permutation matrix

$$P = (e_1, e_{n+1}, e_2, e_{n+2}, \ldots, e_n, e_{2n}).$$

Here e_i denote the i-th column of the $2n \times 2n$ identity matrix. Then the matrix

$$T_{P^T A P} \equiv P^T A P$$

is symmetric tridiagonal such that

$$T_{P^T AP} = \begin{pmatrix} 0 & a_1 & \dots & \dots \\ a_1 & 0 & \dots & \dots \\ \dots & \dots & \dots & \dots \\ \dots & b_{n-1} & 0 & a_n \\ \dots & \dots & a_n & 0 \end{pmatrix}.$$

The matrix $T_{P^T AP}$ has all zeros on its main diagonal, and its superdiagonal and subdiagonal are $a_1, b_1, a_2, b_2, \dots, b_{n-1}, a_n$.
If $(\tilde{\lambda}_i, x_i)$ is an eigenpair for $T_{P^T AP}$, with x_i a unit vector such that

$$T_{P^T AP} x_i = \tilde{\lambda}_i x_i,$$

then $\tilde{\lambda}_i = \pm \sigma_i$, where σ_i is a singular value of B, and $P x_i = \frac{1}{\sqrt{2}} \begin{pmatrix} v_i \\ \pm u_i \end{pmatrix}$. Here, u_i and v_i are left and right singular vectors of B, respectively.
2. *Let $T_{BB^T} \equiv B B^T$. Then T_{BB^T} is symmetric tridiagonal,*

$$T_{BB^T} = \begin{pmatrix} a_1^2 + b_1^2 & a_2 b_1 & \dots & \dots \\ a_2 b_1 & a_2^2 + b_2^2 & \dots & \dots \\ \dots & \dots & \dots & \dots \\ \dots & \dots & a_{n-1}^2 + b_{n-1}^2 & a_n b_{n-1} \\ \dots & \dots & a_n b_{n-1} & a_n^2 \end{pmatrix},$$

with diagonal $a_1^2 + b_1^2, a_2^2 + b_2^2, \dots, a_{n-1}^2 + b_{n-1}^2, a_n^2$, and superdiagonal and subdiagonal $a_2 b_1, a_3 b_2, \dots, a_n b_{n-1}$. The singular values of B are the square roots of the eigenvalues of T_{BB^T}, and the left singular vectors of B are the eigenvectors of T_{BB^T}.
3. *Let $T_{B^T B} \equiv B^T B$. Then $T_{B^T B}$ is symmetric tridiagonal,*

$$T_{B^T B} = \begin{pmatrix} a_1^2 & a_1 b_1 & \dots & \dots \\ a_1 b_1 & a_2^2 + b_1^2 & \dots & \dots \\ \dots & \dots & \dots & \dots \\ \dots & \dots & a_{n-1}^2 + b_{n-2}^2 & a_{n-1} b_{n-1} \\ \dots & \dots & a_{n-1} b_{n-1} & a_n^2 + b_{n-1}^2 \end{pmatrix},$$

with diagonal $a_1^2, a_2^2 + b_1^2, a_3^2 + b_2^2, \dots, a_n^2 + b_{n-1}^2$ and superdiagonal and subdiagonal $a_1 b_1, a_2 b_2, \dots, a_{n-1} b_{n-1}$. The singular values of B are the square roots of the eigenvalues of $T_{B^T B}$, and the right singular vectors of B are the eigenvectors of $T_{B^T B}$.

The proof of the first statement of this lemma follows from Theorem 9.6, and the proof of the second and third statements follows from Theorem 9.4.

However, direct application of Lemma 11.4 for computing the SVD of a symmetric tridiagonal matrix using the algorithms of QR iteration, divide and conquer, or bisection is inefficient, since by Lemma 11.4, in the case of a matrix $T_{P^T AP}$, we need to compute not all eigenvalues, but only positive ones, and in addition, there are difficulties in computing singular vectors for tiny singular values: numerical computing of entries for T_{BB^T}, $T_{B^T B}$ is unstable because of rounding in floating-point arithmetic.

There exist, however, the following stable algorithms for computing the SVD numerically:

1. Different versions of QR iteration. This is the fastest algorithm for small matrices up to size 25 to find all the singular values of a bidiagonal matrix.
2. Divide and conquer. This is the fastest method for finding all singular values and singular vectors for matrices larger than 25×25.
3. Bisection and inverse iteration. In this algorithm, the first part of Lemma 11.4 is applied for $T_{P^T AP} = P^T AP$ to find only the singular values in a desired interval. Singular values are computed with high accuracy, but singular vectors can lose orthogonality.
4. Jacobi's method. The SVD of a dense matrix G is computed implicitly by applying Jacobi's method, Algorithm 11.5, to GG^T or $G^T G$.

11.7   Different Versions of QR Iteration for the Bidiagonal SVD

In this section we will present the so-called algorithm dqds (differential quotient–difference algorithm with shifts [99]), which was originally derived in [28] and later updated in [32] for the case of computing only singular values. We refer to [95] for a survey of different versions of QR Iteration for the SVD.

To derive the dqds algorithm, we will begin with the algorithm of LR iteration, which can be applied to symmetric positive definite (s.p.d.) matrices. Let T_0 be any symmetric positive definite matrix. The following algorithm produces a sequence of similar symmetric positive definite matrices T_i:

Algorithm 11.9 LR iteration.

0. Set $i = 0$ and initialize s.p.d. T_0. Perform steps 1–4 in a loop:
1. Compute a shift τ_i^2 such that it is smaller than the smallest eigenvalue of T_i.
2. Compute the Cholesky factorization of $T_i - \tau_i^2 I = B_i^T B_i$, where B_i is an upper triangular matrix with all positive elements on the main diagonal.
3. Update $T_{i+1} = B_i B_i^T + \tau_i^2 I$.
4. Stop updating T_i and set $T_M = T_{i+1}$, $M = i + 1$, if $\|T_{i+1} - T_i\|_2 \leq \theta$. Here, θ is a tolerance number. Otherwise, set $i = i + 1$ and go to step 1.

The algorithm of LR iteration is very similar to that of QR iteration: we compute a factorization and multiply the factors in reverse order to get the next iterate T_{i+1}. It is easy to see that T_{i+1} and T_i are similar:

$$
\begin{aligned}
T_{i+1} &= B_i B_i^T + \tau_i^2 I = B_i^{-T} B_i^T B_i B_i^T + \tau_i^2 B_i^{-T} B_i^T \\
&= B_i^{-T} (B_i^T B_i + \tau_i^2) B_i^T = B_i^{-T} T_i B_i^T .
\end{aligned}
\tag{11.26}
$$

The following lemma states that when we choose the shift $\tau_i^2 = 0$, then two steps of LR iteration produce the same T_2 as one step of QR iteration.

Lemma 11.5 *Let T_2 be the matrix produced by two steps of Algorithm 11.9 with $\tau_i^2 = 0$. Let T' be the matrix produced by one step of QR iteration such that $QR = T_0$, $T' = RQ$. Then $T_2 = T'$.*

Proof Using the property that T_0 is symmetric, we factorize T_0^2 in two ways:

1. The first factorization is

$$
T_0^2 = T_0^T T_0 = (QR)^T QR = R^T R,
$$

where R^T is a lower triangular matrix. We assume that $R_{ii} > 0$. By uniqueness of the Cholesky factorization, it is unique.

2. The second factorization is

$$
T_0^2 = B_0^T B_0 B_0^T B_0.
$$

Using Algorithm 11.9, we have

$$
T_1 = B_0 B_0^T = B_1^T B_1.
$$

Using the second factorization and then the above expression, we can rewrite

$$
T_0^2 = B_0^T B_0 B_0^T B_0 = B_0^T (B_1^T B_1) B_0 = (B_1 B_0)^T B_1 B_0,
$$

where $(B_1 B_0)^T$ is a lower triangular matrix. This must be the Cholesky factorization, since T_0 is s.p.d. By uniqueness of the Cholesky factorization, we conclude that $R = B_1 B_0$ and that two steps of LR iteration equal one step of QR iteration. We can prove this also in the following way: since $T_0 = QR$ we have

$$
T' = RQ = RQ(RR^{-1}) = R(QR)R^{-1} = RT_0 R^{-1}.
$$

Substituting $R = B_1 B_0$ and $T_0 = B_0^T B_0$ into the right-hand side of the above equation, we get

$$
T' = (B_1 B_0)(B_0^T B_0)(B_1 B_0)^{-1} = B_1 B_0 B_0^T B_0 B_0^{-1} B_1^{-1} = B_1 (B_0 B_0^T) B_1^{-1}.
$$

Using the fact $B_0 B_0^T = T_1 = B_1^T B_1$, we finally obtain

$$T' = B_1(B_1^T B_1)B_1^{-1} = B_1 B_1^T = T_2.$$

$\square$

Remark 11.2

- We observe that Algorithm 11.9 and Lemma 11.5 depend on the s.p.d. T_0, which should not be tridiagonal.
- Because of the similarity of LR iteration and QR iteration stated in Lemma 11.5, the analysis of LR iteration follows from the analysis of QR iteration.
- We observe that the matrices $T_{i+1} = B_i B_i^T + \tau_i^2 I$ in Algorithm 11.9 are constructed explicitly, and that can be an unstable procedure because of rounding errors in floating-point arithmetic.

The next dqds algorithm is mathematically the same as the algorithm of LR iteration. However, in the dqds algorithm, the matrices B_{i+1} are computed directly from B_i without constructing $T_{i+1} = B_i B_i^T + \tau_i^2 I$.

Let B_i have diagonal $a_1, \ldots, a_n$ and superdiagonal $b_1, \ldots, b_{n-1}$, and let B_{i+1} have diagonal $\hat{a}_1, \ldots, \hat{a}_n$ and superdiagonal $\hat{b}_1, \ldots, \hat{b}_{n-1}$. We assume that $b_0 = \hat{b}_0 = b_n = \hat{b}_n = 0$. Using Algorithm 11.9, we have that

$$B_{i+1}^T B_{i+1} + \tau_{i+1}^2 I = T_{i+1} = B_i B_i^T + \tau_i^2 I. \tag{11.27}$$

Writing (11.27) for the (j, j) entries for $j < n$, we get

$$\hat{a}_j^2 + \hat{b}_{j-1}^2 + \tau_{i+1}^2 = a_j^2 + b_j^2 + \tau_i^2$$

and expressing $\hat{a}_j^2$ from this, we have

$$\hat{a}_j^2 = a_j^2 + b_j^2 - \hat{b}_{j-1}^2 - \delta, \tag{11.28}$$

where $\delta = \tau_{i+1}^2 - \tau_i^2$. The shift τ_i^2 should be chosen in a special way; see step 1 in Algorithm 11.9. Writing (11.27) for the squares of $(j, j + 1)$, we have

$$\hat{a}_j^2 \hat{b}_j^2 = a_{j+1}^2 b_j^2,$$

and expressing $\hat{b}_j^2$ from this, we obtain

$$\hat{b}_j^2 = a_{j+1}^2 b_j^2 / \hat{a}_j^2. \tag{11.29}$$

Combining the two Eqs. (11.28) and (11.29), we get the intermediate algorithm.

Algorithm 11.10 Intermediate algorithm.

for $j = 1$ to $n - 1$
$$\hat{a}_j^2 = a_j^2 + b_j^2 - \hat{b}_{j-1}^2 - \delta$$
$$\hat{b}_j^2 = b_j^2(a_{j+1}^2/\hat{a}_j^2)$$
end for
$$\hat{a}_n^2 = a_n^2 - \hat{b}_{n-1}^2 - \delta$$

We observe that Algorithm 11.10 maps the *squares* of the entries of B_i directly to the *squares* of the entries of B_{i+1}. Thus, the square roots are taken only at the end of the algorithm.

In the next algorithm, we rewrite Algorithm 11.10 in the classical notation of [99] using the change of variables

$$q_j = a_j^2, e_j = b_j^2.$$

Algorithm 11.11 One step of the qds algorithm.

for $j = 1$ to $n - 1$
$$\hat{q}_j = q_j + e_j - \hat{e}_{j-1} - \delta$$
$$\hat{e}_j = e_j(q_{j+1}/\hat{q}_j)$$
end for
$$\hat{q}_n = q_n - \hat{e}_{n-1} - \delta$$

The final dqds algorithm is the same as qds except that it will be more accurate. To derive the dqds algorithm, we take the part $q_j - \hat{e}_{j-1} - \delta$ from $\hat{q}_j$ of Algorithm 11.11 and use first (11.29) to express $\hat{e}_{j-1}$ and then (11.28) to express $\hat{q}_{j-1}$ to obtain

$$d_j \equiv q_j - \hat{e}_{j-1} - \delta = q_j - \frac{q_j e_{j-1}}{\hat{q}_{j-1}} - \delta$$

$$= q_j \left(\frac{\hat{q}_{j-1} - e_{j-1}}{\hat{q}_{j-1}} \right) - \delta = q_j \left(\frac{q_{j-1} - \hat{e}_{j-2} - \delta}{\hat{q}_{j-1}} \right) - \delta \qquad (11.30)$$

$$= \frac{q_j}{\hat{q}_{j-1}} d_{j-1} - \delta.$$

Using (11.30), we can rewrite the inner loop of Algorithm 11.11 as

$$\hat{q}_j = d_j + e_j,$$
$$\hat{e}_j = e_j(q_{j+1}/\hat{q}_j), \qquad (11.31)$$
$$d_{j+1} = d_j(q_{j+1}/\hat{q}_j) - \delta.$$

To get the final algorithm, we note that d_{j+1} can overwrite d_j:

Algorithm 11.12 One step of the dqds algorithm.
$$d = q_1 - \delta$$
$$\text{for } j = 1 \text{ to } n - 1$$
$$\hat{q}_j = d + e_j$$
$$t = (q_{j+1}/\hat{q}_j)$$
$$\hat{e}_j = e_j t$$
$$d = d t - \delta$$
$$\text{end for}$$
$$\hat{q}_n = d$$

The dqds algorithm, Algorithm 11.12, has the same number of floating-point operations in its inner loop as the qds algorithm, Algorithm 11.11. How to choose a shift τ_i in $\delta = \tau_{i+1}^2 - \tau_i^2$ and an analysis of the convergence of these algorithms are presented in [32].

11.8   Jacobi's Method for the SVD

In this section, we will present algorithms that can determine the SVD of a dense matrix. These algorithms will use Jacobi's algorithm, Algorithm 11.8, for a symmetric matrix
$$A = G^T G.$$

Like Algorithm 11.8, the algorithms of this section are very slow in computing the SVD compared with other methods that we have considered. However, Jacobi's method can compute the singular values and singular vectors much more accurately than those other algorithms.

The first Jacobi algorithm computes a Jacobi rotation matrix J at every iteration step and updates $G^T G$ to $J^T G^T G J$. Since we compute only GJ instead of $G^T G$ or $J^T G^T G J$, this algorithm is called one-sided Jacobi rotation.

Algorithm 11.13 One-sided Jacobi rotation of G.
function One-Sided-Jacobi-Rotation (G, j, k)
    Compute $a_{jj} = (G^T G)_{jj}$, $a_{jk} = (G^T G)_{jk}$, and $a_{kk} = (G^T G)_{kk}$
    if $|a_{jk}| > \varepsilon \sqrt{a_{jj} a_{kk}}$
        $\tau = (a_{jj} - a_{kk})/(2a_{jk})$
        $t = sign(\tau)/(|\tau| + \sqrt{1 + \tau^2})$
        $c = 1/\sqrt{1 + t^2}$
        $s = c t$
        $G = G \ R(j, k, \theta)$ /* here $c = \cos\theta$ and $s = \sin\theta$ */
    /* if right singular vectors are desired */
        $J = J \ R(j, k, \theta)$
    end if
end if

We note that the entries a_{jj}, a_{jk}, and a_{kk} of $A = G^T G$ are computed by Algorithm 11.13, where the Jacobi rotation $R(j, k, \theta)$ is computed using Algorithm 11.5.

In the next algorithm we assume that G is of order $n \times n$. We compute the singular values σ_i, the left singular vector matrix U, and the right singular vector matrix V such that $G = U \Sigma V^T$, where $\Sigma = \text{diag}(\sigma_i)$.

Algorithm 11.14 One-sided Jacobi.

> Loop
> for $j = 1$ to $n - 1$
> for $k = j + 1$ to n
> call One-Sided-Jacobi-Rotation (G, j, k)
> end for
> end for
> until $G^T G$ is diagonal enough
> Set $\sigma_i = \|G(:, i)\|_2$ (the 2-norm of column i of G)
> Set $U = [u_1, \ldots, u_n]$, where $u_i = G(:, i)/\sigma_i$
> Set $V = J$ (product of Jacobi rotations)

The following theorem shows that the one-sided Jacobi algorithm, Algorithm 11.14, can compute the SVD with high accuracy.

Theorem 11.6 Let $G = DX$ be an $n \times n$ matrix, where D is diagonal and nonsingular, and X is nonsingular. Let $\hat{G}$ be the matrix after calling One-Sided-Jacobi-Rotation (G, j, k) m times in floating-point arithmetic. Let $\sigma_1 \geq \ldots \geq \sigma_n$ be the singular values of G, and let $\hat{\sigma}_1 \geq \ldots \geq \hat{\sigma}_n$ be the singular values of $\hat{G}$. Then

$$\frac{|\sigma_i - \hat{\sigma}_i|}{\sigma_i} \leq O(m\varepsilon)\kappa(X),$$

where $\kappa(X) = \|X\| \|X^{-1}\|$ is the condition number of X.

A proof can be found in [23].

In the example presented below we will illustrate the performance of the one-sided Jacobi algorithm, Algorithm 11.14, using the MATLAB® program of Section 1.20.[10]

Example 11.5 We compute the SVD decomposition of the matrix

$$A = \begin{pmatrix} 3.8373 & 16.5466 & 0 & 0 & 0 \\ 16.5466 & 17.7476 & 5.5205 & 00 & \\ 0 & 5.5205 & 11.4120 & 7.1830 & 0 \\ 0 & 0 & 7.1830 & 11.4657 & 8.7969 \\ 0 & 0 & 0 & 8.7969 & 18.5031 \end{pmatrix} \tag{11.32}$$

using the one-sided Jacobi algorithm, Algorithm 11.14. We run the MATLAB® program of Section 1.20 until the matrix $A^T A$ is sufficiently diagonal, i.e., until $off(A^T A) < tol$ for $tol = 0.005$. The computed SVD decomposition of the matrix

[10] All MATLAB® programs referenced on this page can be found on the online version of Chapter 1 (doi:10.1007/978-3-319-57304-5_1).

$A = U \Sigma V^T$ obtained by applying the MATLAB® program of Section 1.20 is the following:[11]

$$U = \begin{pmatrix} -0.8000 & 0.4934 & -0.2008 & -0.2157 & -0.1723 \\ 0.5608 & 0.7867 & 0.0260 & -0.1254 & -0.2241 \\ -0.1934 & 0.2983 & 0.5262 & 0.7440 & 0.2077 \\ 0.0853 & 0.1764 & -0.7317 & 0.3083 & 0.5754 \\ -0.0286 & 0.1324 & 0.3830 & -0.5377 & 0.7388 \end{pmatrix},$$

$$\Sigma = \begin{pmatrix} 7.7615 & 0 & 0 & 0 & 0 \\ 0 & 30.2188 & 0 & 0 & 0 \\ 0 & 0 & 1.6960 & 0 & 0 \\ 0 & 0 & 0 & 13.4582 & 0 \\ 0 & 0 & 0 & 0 & 25.3541 \end{pmatrix},$$

$$V = \begin{pmatrix} 0.8000 & 0.4934 & -0.2008 & -0.2157 & -0.1723 \\ -0.5608 & 0.7867 & 0.0260 & -0.1254 & -0.2241 \\ 0.1934 & 0.2983 & 0.5262 & 0.7440 & 0.2077 \\ -0.0853 & 0.1764 & -0.7317 & 0.3083 & 0.5754 \\ 0.0286 & 0.1324 & 0.3830 & -0.5377 & 0.7388 \end{pmatrix}.$$

For comparison, the computed SVD decomposition of the matrix $A = U \Sigma V^T$ given in (11.32) using the svd command in MATLAB® gives the following result:

$$U = \begin{pmatrix} -0.4934 & -0.1723 & -0.2157 & -0.8000 & -0.2008 \\ -0.7867 & -0.2241 & -0.1254 & 0.5608 & 0.0260 \\ -0.2983 & 0.2077 & 0.7440 & -0.1934 & 0.5262 \\ -0.1764 & 0.5754 & 0.3083 & 0.0853 & -0.7317 \\ -0.1324 & 0.7388 & -0.5377 & -0.0286 & 0.3830 \end{pmatrix},$$

$$\Sigma = \begin{pmatrix} 30.2188 & 0 & 0 & 0 & 0 \\ 0 & 25.3541 & 0 & 0 & 0 \\ 0 & 0 & 13.4582 & 0 & 0 \\ 0 & 0 & 0 & 7.7615 & 0 \\ 0 & 0 & 0 & 0 & 1.6960 \end{pmatrix},$$

$$V = \begin{pmatrix} -0.4934 & -0.1723 & -0.2157 & 0.8000 & -0.2008 \\ -0.7867 & -0.2241 & -0.1254 & -0.5608 & 0.0260 \\ -0.2983 & 0.2077 & 0.7440 & 0.1934 & 0.5262 \\ -0.1764 & 0.5754 & 0.3083 & -0.0853 & -0.7317 \\ -0.1324 & 0.7388 & -0.5377 & 0.0286 & 0.3830 \end{pmatrix}.$$

---

[11] The MATLAB® programs can be found on the online version of Chapter 1 (doi:10.1007/978-3-319-57304-5_1).

Questions

11.1 Prove that 5 is an eigenvalue of the matrix

$$A = \begin{pmatrix} 6\ 3\ 3\ 1 \\ 0\ 7\ 4\ 5 \\ 0\ 0\ 5\ 4 \\ 0\ 0\ 0\ 8 \end{pmatrix}. \tag{11.33}$$

Compute an eigenvector of A that corresponds to the eigenvalue 5.

11.2 Compute the eigenvalues and corresponding eigenvectors of the matrix

$$A = \begin{pmatrix} 1\ 2\ -4 \\ 0\ 2\ 1 \\ 0\ 0\ 3 \end{pmatrix}. \tag{11.34}$$

11.3 Compute the eigenvalues and corresponding eigenvectors of the matrix

$$A = \begin{pmatrix} 1\ 4 \\ 1\ 1 \end{pmatrix}. \tag{11.35}$$

(a) Compute an estimate to an eigenvalue of A by the Rayleigh quotient with vector $x = (1, 1)^T$.

(b) If we apply the method of inverse iteration to A, to which one eigenvector of A will this method converge?

(c) If we apply the method of inverse iteration with a shift $\sigma = 2$, what eigenvalue of A will be obtained?

(d) If we apply the method of QR iteration to A, to what form will this matrix converge: diagonal or triangular? Why?

11.4 Assume that $G^T G$ converges to a diagonal matrix. Prove that Algorithm 11.14 implements the SVD decomposition of the matrix G.

11.5 Let x be a unit vector and y a vector orthogonal to x. Prove that $\|(x+y)x^T - I\|_2 = \|x+y\|_2$.

11.6 Let $A = D + \rho uu^T$, where D is the diagonal matrix $D = \text{diag}(d_1, ..., d_n)$ and u is the vector $u = (u_1, ..., u_n)^T$.

(a) Prove that d_i is an eigenvalue of A if $d_i = d_{i+1}$ or $u_i = 0$.

(b) Prove that an eigenvector corresponding to d_i is e_i (the ith column of I) if $u_i = 0$.

11.7 Show how to compute scalars c and $\tilde{c}$ in the function $f(\lambda) = \tilde{c} + \frac{c}{d-\lambda}$ if we know that at $\lambda = \xi$ we have $f(\xi) = \psi$ and $f'(\xi) = \psi'$. Here ψ, ψ' are known scalars.

11.8 Let $A = G^T G$ in Algorithm 11.14. Here A and G are of order $n \times n$. Assume that $|a_{jk}| \leq \varepsilon \sqrt{a_{jj}a_{kk}}$ for all $j \neq k$. Let $\sigma_n \leq \sigma_{n-1} \leq \dots \leq \sigma_1$ be the singular values of G, and $\lambda_n^2 \leq \dots \leq \lambda_1^2$ the sorted diagonal entries of A. Prove that $|\sigma_i - \lambda_i| \leq n\varepsilon|\lambda_i|$, where the λ_i are the singular values computed with high relative accuracy.

11.9 Prove Lemma 11.1.

11.10 Let A be a symmetric matrix and consider Algorithm 10.5 with a Rayleigh quotient shift $\sigma_i = a_{nn}$. Consider also the algorithm of Rayleigh quotient iteration, Algorithm 11.1, starting with $x_0 = (0, \dots, 0, 1)^T$, which computes Rayleigh quotients ρ_i. Show that the sequences $\sigma_i = \rho_i$ are the same for all i. Hint: to prove this statement we can use the same arguments as those used to prove the connection between the algorithm of QR iteration and the algorithm of inverse iteration.

11.11 Prove part 1 of Lemma 11.4.

11.12 Prove parts 2 and 3 of Lemma 11.4.

11.13 Let the matrix A be defined as

$$A = \begin{pmatrix} I & B \\ \bar{B}^T & I \end{pmatrix}, \tag{11.36}$$

where B is a Hermitian matrix with $\|B\|_2 < 1$. Prove that

$$\kappa(A) = \|A^{-1}\|_2 \|A\|_2 = \frac{1 + \|B\|_2}{1 - \|B\|_2}.$$

11.14 (*Programming*)
Use the MATLAB® program `RayleighQuotient.m` of Section 1.16 to test the Rayleigh quotient iteration algorithm, Algorithm 11.1. Try your own examples of a symmetric matrix A and different tolerances *tol*.

11.15 (*Programming*)
Use the MATLAB® program `DivideandConq.m` of Section 1.17 to test the divide and conquer algorithm, Algorithm 11.3.[12] Try your own examples of a symmetric matrix A and different tolerances in Newton's method for the solution of the secular equation.

11.16 (*Programming*)
Use the MATLAB® programs of Section 1.18 to test the inverse iteration algorithm, Algorithm 11.4. Try your own examples of a symmetric matrix A and different tolerances *tol*.

---

[12] All MATLAB® programs referenced on this page can be found on the online version of Chapter 1 (doi:10.1007/978-3-319-57304-5_1).

11.17 (*Programming*)

Use the MATLAB® programs of Section 1.19 to test the classical Jacobi algorithm, Algorithm 11.4.[13] Try your own examples of a symmetric matrix A and different tolerances *tol*.

11.18 (*Programming*)

Use the MATLAB® programs of Section 1.20 to test the SVD decomposition of a symmetric matrix A using the one-sided Jacobi algorithm, Algorithm 11.14. Try your own examples of a matrix A and different tolerances *tol*.

---

[13] All MATLAB® programs referenced on this page can be found on the online version of Chapter 1 (doi:10.1007/978-3-319-57304-5_1).

Chapter 12
Introduction to Iterative Methods for the Solution of Linear Systems

In this chapter we will discuss iterative algorithms for the solution of linear systems of equations (LSE) $Ax = b$. These algorithms are used when direct methods take a lot of time and computer space to solve this system, or in other words, when they are not efficient.

Most of the methods presented in this chapter are described in greater detail in [9, 23]. We also refer to the books on the iterative methods [4, 49, 100]. Parallel implementation of many of the iterative methods discussed here is presented in [98]. The goal of this chapter is to introduce the reader to the topic of iterative algorithms.

In Sections 12.1–12.6, we will discuss basic iterative methods such as Jacobi, Gauss–Seidel, and successive overrelaxation, and in Section 12.7, we introduce Krylov subspace methods. Further, the conjugate gradient method (CG) and preconditioned conjugate gradient method (PCG) are presented in Sections 12.8, 12.9, respectively. We refer to [4, 44, 49, 100] for a survey of Krylov subspace methods and different preconditioning techniques.

12.1 Basic Iterative Methods

The basic iterative methods for the solution of a system of linear equations $Ax = b$ are:

1. *Jacobi.*
2. *Gauss–Seidel.*
3. *Successive overrelaxation (SOR).*

These methods produce a sequence of iterative solutions x_m of a linear system $Ax = b$ that converge to the solution $x = A^{-1}b$, provided that there exists an initial guess x_0. To use iterative methods we will introduce a splitting:

The original version of this chapter was revised. An Erratum to this book can be found at https://doi.org/10.1007/978-3-319-57304-5_12

L. Beilina et al., *Numerical Linear Algebra: Theory and Applications*,
DOI 10.1007/978-3-319-57304-5_12

$$A = M - K,$$

where $\det M \neq 0$. Applying this splitting to $Ax = b$, we get

$$Ax = Mx - Kx = b.$$

From the equation above, we have

$$Mx = b + Kx,$$

and thus

$$x = M^{-1}(b + Kx) = M^{-1}b + M^{-1}Kx.$$

Let us define

$$Rx = M^{-1}Kx, \ c = M^{-1}b.$$

The iterative update for x_m can be written as

$$x_{m+1} = Rx_m + c, \tag{12.1}$$

where m is the number of the iteration.

Lemma 12.1 *Let* $\|R\| = \max_{x \neq 0} \frac{\|Rx\|}{\|x\|}$. *If* $\|R\| < 1$ *then the iterations (12.1) will converge for all initial guesses* x_0.

Proof For exact x, we have
$$x = Rx + c. \tag{12.2}$$

Subtracting (12.2) from (12.1), we get

$$x_{m+1} - x = R(x_m - x). \tag{12.3}$$

Taking norms, we have

$$\|x_{m+1} - x\| = \|R(x_m - x)\| \leq \|R\| \, \|x_m - x\| \leq \|R^{m+1}\| \, \|x_0 - x\|. \tag{12.4}$$

This inequality will converge to zero, since $\|R\| < 1$. $\square$

Another convergence criterion requires the introduction of the definition of spectral radius for R. Let $(\lambda_1, ..., \lambda_n)$ be the (real or complex) eigenvalues of a matrix R. The spectral radius of R is $\rho(R) \equiv \max_{\lambda_i, i=1,...,n} |\lambda_i|$.

Lemma 12.2 *For all operator norms,* $\rho(R) \leq \|R\|$. *Then* $\forall R$ *and* $\forall \varepsilon > 0$, *there exists an operator norm* $\| \cdot \|_{(R,\varepsilon)}$ *such that* $\|R\|_{(R,\varepsilon)} \leq \rho(R) + \varepsilon$.

A proof of this lemma can be found in [23].

Theorem 12.1 *Let* $\|R\| = \max_{x \neq 0} \frac{\|Rx\|}{\|x\|}$. *If* $\rho(R) < 1$ *then the iterations (12.1) will converge for all initial guesses* x_0.

Proof Using (12.4), we have

$$\|x_{m+1} - x\| = \|R(x_m - x)\| \leq \|R\| \|x_m - x\| \leq \|R^{m+1}\| \|x_0 - x\|.$$
(12.5)

Since $\rho(R) < 1$, using Lemma 12.2 we may choose an operator norm such that $\|R\|_{(R,\varepsilon)} < 1$. Then by Lemma 12.1, the iterations (12.1) will converge for all initial guesses x_0. $\qquad\square$

The rate of convergence $r(R)$ of the iterative procedure $x_{m+1} = Rx_m + c$ is defined as

$$r(R) = -\log_{10} \rho(R).$$

In the iterative methods considered below, we want to have a splitting $A = M - K$ as efficient as possible. Let us introduce the following notation. If a matrix A has no zeros on its diagonal, we will write the splitting as

$$A = D - \tilde{L} - \tilde{U} = D(I - L - U),$$
(12.6)

where D is a diagonal matrix, $-\tilde{L}$ is the strictly lower triangular part of A such that

$$DL = \tilde{L},$$

and $-\tilde{U}$ is the strictly upper triangular part of A such that

$$DU = \tilde{U}.$$

12.2 Jacobi Method

The splitting for the Jacobi method is

$$A = D - (\tilde{L} + \tilde{U}).$$
(12.7)

Applying it to the solution of $Ax = b$, we have

$$Ax = Dx - (\tilde{L}x + \tilde{U}x) = b.$$

From the equation above, we obtain

$$Dx = b + \tilde{L}x + \tilde{U}x,$$

and thus

$$x = D^{-1}(b + \tilde{L}x + \tilde{U}x) = D^{-1}b + D^{-1}\tilde{L}x + D^{-1}\tilde{U}x.$$

Let us define

$$\begin{aligned} R_J &\equiv D^{-1}(\tilde{L} + \tilde{U}) = L + U, \\ c_J &\equiv D^{-1}b. \end{aligned} \tag{12.8}$$

Then the iterative update in the Jacobi method can be written as

$$x_{m+1} = R_J x_m + c_R. \tag{12.9}$$

Formula (12.9) can be also written as

$$Dx_{m+1} = b + \tilde{L}x_m + \tilde{U}x_m, \tag{12.10}$$

or using the definition of the matrix D at the element level, the same formula can be represented as

$$a_{j,j} x_{m+1,j} = b_j - \sum_{\substack{k=1 \\ k \neq j}}^{n} a_{j,k} x_{m,k}. \tag{12.11}$$

Algorithm 12.1 One step in the Jacobi method.

    for $j = 1$ to n

$$x_{m+1,j} = \frac{b_j - \sum_{\substack{k=1 \\ k \neq j}}^{n} a_{j,k} x_{m,k}}{a_{j,j}}$$

    end

In the case of the model problem for the Poisson's equation of Section 8.1.3 implemented on a square, we will have the following Jacobi method:

Algorithm 12.2 One step in the Jacobi method for the two-dimensional Poisson's equation.

    for $i = 1$ to N
    for $j = 1$ to N

$$u_{m+1,i,j} = \frac{u_{m,i-1,j} + u_{m,i+1,j} + u_{m,i,j-1} + u_{m,i,j+1} + h^2 f_{i,j}}{4}$$

    end
    end

Example 12.1 In this example we present the numerical solution of the Dirichlet problem for the Poisson's equation (8.11) in two dimensions using the iterative

Jacobi method. We define the right-hand side $f(x)$ and the coefficient $a(x_1, x_2)$ in (8.11) to be the same as in Example 8.2 of Chapter 8. We produce the same mesh as in this example and then solve the linear system of equations $Au = f$. The MATLAB® program of Section 1.21 is available for running this test.[1] We have implemented three different versions of the Jacobi method in this program: the first version uses the formula (12.9), the second version employs Algorithm 12.1, and the third version employs Algorithm 12.2. For all three algorithms we have used the stopping criterion $\|u^{m+1} - u^m\|_2 < tol$, where the chosen tolerance was $tol = 10^{-9}$.

The results of our numerical simulations are the same as those presented in Fig. 8.1 for the number of inner points $N = 20$ and for the tolerance $tol = 10^{-9}$ in the iterative update (check it by running the MATLAB® program of Section 1.21).

12.3 Gauss–Seidel Method

To get the Gauss–Seidel[2] method, we use the same splitting (12.7) as for the Jacobi method. Applying it to the solution of $Ax = b$, we have

$$Ax = Dx - (\tilde{L}x + \tilde{U}x) = b.$$

Next, we rearrange terms on the right-hand side of the above equation to get

$$Dx - \tilde{L}x = b + \tilde{U}x, \tag{12.12}$$

and thus the solution of (12.12) is computed as

$$x = (D - \tilde{L})^{-1}(b + \tilde{U}x) = (D - \tilde{L})^{-1}b + (D - \tilde{L})^{-1}\tilde{U}x.$$

We can rewrite the above equation using the notation $DL = \tilde{L}$ and $DU = \tilde{U}$ to get

$$
\begin{aligned}
x &= (D - \tilde{L})^{-1}b + (D - \tilde{L})^{-1}\tilde{U}x \\
  &= (D - DL)^{-1}b + (D - DL)^{-1}\tilde{U}x \\
  &= (I - L)^{-1}D^{-1}b + (I - L)^{-1}D^{-1}\tilde{U}x \\
  &= (I - L)^{-1}D^{-1}b + (I - L)^{-1}Ux.
\end{aligned}
\tag{12.13}
$$

Let us define

$$
\begin{aligned}
R_{GS} &\equiv (I - L)^{-1}U, \\
c_{GS} &\equiv (I - L)^{-1}D^{-1}b.
\end{aligned}
\tag{12.14}
$$

Then iterative update in the Gauss–Seidel method can be written as

$$x_{m+1} = R_{GS}x_m + c_{GS}. \tag{12.15}$$

---

[1] All MATLAB® programs referenced on this page can be found on the online version of Chapter 1 (doi:10.1007/978-3-319-57304-5_1).

[2] Philipp Ludwig von Seidel (1821–1896) was a German mathematician.

We can also write the formula (12.13) via an iterative update as

$$(I - L)Dx_{m+1} = b + DUx_m, \tag{12.16}$$

or using the definition of matrices D, L, U at the element level as

$$a_{j,j}x_{m+1,j} = b_j - \sum_{k=1}^{j-1} a_{j,k}x_{m+1,k} - \sum_{k=j+1}^{n} a_{j,k}x_{m,k}. \tag{12.17}$$

Here $\sum_{k=1}^{j-1} a_{j,k}x_{m+1,k}$ represents already updated terms with values of x_{m+1}, and terms $\sum_{k=j+1}^{n} a_{j,k}x_{m,k}$ are those with older values of x_m that we have updated on iteration m.

Algorithm 12.3 One step in the Gauss–Seidel method.

   for $j = 1$ to n

$$x_{m+1,j} = \frac{b_j - \sum_{k=1}^{j-1} a_{j,k}x_{m+1,k} - \sum_{k=j+1}^{n} a_{j,k}x_{m,k}}{a_{j,j}}$$

   end

If we want apply the Gauss–Seidel method to the solution of the model problem for the Poisson's equation of Section 8.13, we need to organize the ordering for the new $m + 1$ variables and old already computed values m. We will use a so-called red–black ordering based on a chessboard-like coloring. Let **B** nodes correspond to the black squares on a chessboard, and **R** nodes correspond to the weight squares. The Gauss–Seidel method for the solution of the two-dimensional Poisson's equation on a square becomes the following.

Algorithm 12.4 One step in the Gauss–Seidel method for the two-dimensional Poisson's equation.

   for all **R** red nodes i, j

$$u_{m+1,i,j} = \frac{u_{m,i-1,j} + u_{m,i+1,j} + u_{m,i,j-1} + u_{m,i,j+1} + h^2 f_{i,j}}{4}$$

   end
   for all **B** black nodes i, j

$$u_{m+1,i,j} = \frac{u_{m+1,i-1,j} + u_{m+1,i+1,j} + u_{m+1,i,j-1} + u_{m+1,i,j+1} + h^2 f_{i,j}}{4}$$

   end

Example 12.2 Here we present the numerical solution of the Dirichlet problem for the Poisson's equation (8.11) in two dimensions using the iterative Gauss–Seidel method. The setup for our numerical experiments is the same as in Example 8.2 of Chapter 8. The MATLAB® programs of Sections 1.22 and 1.23 are available for running this test.[3] The MATLAB® program of Section 1.22 implements Algorithm 12.3, while the MATLAB® program of Section 1.23 implements Algorithm 12.4, the Gauss–Seidel method with red–black ordering. In both cases, we have used the computation of the residual in the stopping criterion $\|Au^{m+1} - b\|_2 < tol$, where the chosen tolerance was $tol = 10^{-9}$.

The results of our numerical simulations are the same as in Fig. 8.1 for the number of inner points $N = 20$ and tolerance $tol = 10^{-9}$ in the iterative update (check it by running the MATLAB® programs of Sections 1.22, 1.23). However, the convergence of the Gauss–Seidel method (665 iterations in the usual Gauss–Seidel method and 634 iterations in Gauss–Seidel with red–black ordering) is much faster than in the usual Jacobi method, which converged after 1204 iterations.

12.4 Successive Overrelaxation SOR(ω) Method

The method of successive overrelaxation improves the Gauss–Seidel method in the following way: it takes the weighted average of values x_{m+1} and x_m such that

$$x_{m+1,j} = (1 - \omega)x_{m,j} + \omega x_{m+1,j}, \tag{12.18}$$

where ω is a weight, also called a relaxation parameter. When $\omega = 1$, then we get the usual Gauss–Seidel method; when $\omega < 1$, we get the underrelaxation method; and when $\omega > 1$, we have the overrelaxation method. We will investigate all three cases in Section 12.6.

To get the SOR(ω) method in matrix form, we again apply the splitting (12.7) and obtain an equation similar to (12.12), but only in the iterative form

$$(D - \tilde{L})x_{m+1} = b + \tilde{U}x_m. \tag{12.19}$$

Applying now the weighted average (12.18) to this equation, we have

$$(D - \omega\tilde{L})x_{m+1} = \omega b + ((1 - \omega)D + \omega\tilde{U})x_m. \tag{12.20}$$

Using the notation $DL = \tilde{L}$ and $DU = \tilde{U}$, Eq. (12.20) can be rewritten as

$$\begin{aligned} x_{m+1} &= (D - \omega\tilde{L})^{-1}\omega b + (D - \omega\tilde{L})^{-1}((1 - \omega)D + \omega\tilde{U})x_m \\ &= (I - \omega L)^{-1}D^{-1}\omega b + (I - \omega L)^{-1}((1 - \omega)I + \omega U)x_m. \end{aligned} \tag{12.21}$$

Now defining

---

[3] All MATLAB® programs referenced on this page can be found on the online version of Chapter 1 (doi:10.1007/978-3-319-57304-5_1).

$$R_{SOR} = (I - \omega L)^{-1}((1 - \omega)I + \omega U),$$
$$c_{SOR} = (I - \omega L)^{-1}D^{-1}\omega b,$$
(12.22)

we can rewrite (12.21) in the form

$$x_{m+1} = R_{SOR}x_m + c_{SOR}.$$
(12.23)

To get SOR(ω) for implementation, we take $x_{m+1,j}$ on the right-hand side of (12.18) from the Gauss–Seidel algorithm, Algorithm 12.3, and obtain the following algorithm:

Algorithm 12.5 One step in the SOR(ω) method.

    for $j = 1$ to n

$$x_{m+1,j} = (1 - \omega)x_{m,j} + \omega \left[\frac{b_j - \sum_{k=1}^{j-1} a_{j,k}x_{m+1,k} - \sum_{k=j+1}^{n} a_{j,k}x_{m,k}}{a_{j,j}} \right]$$

    end

To apply the SOR(ω) method for the solution of the model problem for the Poisson's equation of Section 8.1.3, we will use the red–black ordering as in the Gauss–Seidel method. The SOR(ω) method will be the following.

Algorithm 12.6 One step in the SOR(ω) method for the two-dimensional Poisson's equation.

    for all **R** red nodes i, j

$$u_{m+1,i,j} = (1 - \omega)u_{m,i,j} + \frac{\omega(u_{m,i-1,j} + u_{m,i+1,j} + u_{m,i,j-1} + u_{m,i,j+1} + h^2 f_{i,j})}{4}$$

    end

    for all **B** black nodes i, j

$$u_{m+1,i,j} = (1 - \omega)u_{m,i,j} + \frac{\omega(u_{m+1,i-1,j} + u_{m+1,i+1,j} + u_{m+1,i,j-1} + u_{m+1,i,j+1} + h^2 f_{i,j})}{4}$$

    end

12.5   Symmetric Successive Overrelaxation SSOR(ω) Method

The main scheme of all iterative methods that we studied before was to construct an iterative procedure $x_{i+1} = Rx_i + c$ such that all x_i will converge to the exact solution x of the system of linear equations $Ax = b$ under the condition that $\rho(R) < 1$.

The method of symmetric successive overrelaxation constructs such a sequence of solutions of $Ax = b$ that improves the approximations x_i already obtained. In other words, we are interested in the answer to the following question: for already computed approximations x_i, can we construct a linear combination $\hat{x}_n = \sum_{i=0}^{n} \alpha_{i,n} x_i$ whose coefficients $\alpha_{i,n}$ satisfy $\sum_{i=0}^{n} \alpha_{i,n} = 1$ and that will be a better approximation of the exact solution x?

Suppose that we have a positive answer to this question. Then the error $e = x - \hat{x}_n$ in the new computed $\hat{x}_n$ can be computed as

$$e = \hat{x}_n - x = \sum_{i=0}^{n} \alpha_{i,n} x_i - x = \sum_{i=0}^{n} \alpha_{i,n}(x_i - x) = \sum_{i=0}^{n} \alpha_{i,n} R^i (x_0 - x) = P_n(R)(x_0 - x), \quad (12.24)$$

where $P_n(R) = \sum_{i=0}^{n} \alpha_{i,n} R^i$ is a polynomial of degree n such that $P_n(1) = \sum_{i=0}^{n} \alpha_{i,n} = 1$. The classical Chebyshev polynomials usually satisfy this condition. The Chebyshev polynomials of the first kind are defined by the recurrence formula

$$\begin{aligned} T_0(x) &= 1, \\ T_1(x) &= x, \\ T_{n+1}(x) &= 2x T_n(x) - T_{n-1}(x). \end{aligned} \quad (12.25)$$

Properties of the Chebyshev polynomials are listed in [109]. We are going to construct a polynomial of the form

$$P_n(x) = \frac{T_n(x/\rho)}{T_n(1/\rho)} \quad (12.26)$$

such that the spectral radius of $P_n(R)$ is as small as possible. The polynomial $P_n(x)$ in (12.26) has the following properties:

$$\begin{aligned} P_n(1) &= 1, \\ \max_{-\rho(R) < x < \rho(R)} |P_n(x)| &< \varepsilon \end{aligned} \quad (12.27)$$

for small tolerances $\varepsilon > 0$. By the spectral mapping theorem (since $P_n(R) = \sum_{i=0}^{n} \alpha_{i,n} R^i$ is a polynomial of degree n), we have that the eigenvalues of $P_n(R)$ are $P_n(\lambda(R))$, where λ are the eigenvalues of R. Combining this observation with (12.27), we can conclude that the spectral radius of R will be small, and thus polynomials of the form (12.26) are suitable for our purposes.

Let us define now

$$\mu_n := \frac{1}{T_n(1/\rho)}. \quad (12.28)$$

Then (12.26) can be written for $x = R$ in terms of μ_n as

$$P_n(R) = \mu_n T_n(R/\rho). \quad (12.29)$$

Writing (12.25) for $x = 1/\rho$ and using (12.28), we get

$$\frac{1}{\mu_n} = T_n(1/\rho) = 2/\rho T_{n-1}(1/\rho) - T_{n-2}(1/\rho). \tag{12.30}$$

Writing (12.25) for $x = R/\rho$, we have

$$T_n(R/\rho) = \frac{2R}{\rho} T_{n-1}(R/\rho) - T_{n-2}(R/\rho). \tag{12.31}$$

We now substitute Eqs. (12.29), (12.30) into the error Eq. (12.24) to obtain

$$\begin{aligned}
e = \hat{x}_n - x &= P_n(R)(x_0 - x) = \mu_n T_n\left(\frac{R}{\rho}\right)(x_0 - x) \\
&= \mu_n\left(\frac{2R}{\rho}T_{n-1}(R/\rho)(x_0 - x) - T_{n-2}(R/\rho)(x_0 - x)\right).
\end{aligned} \tag{12.32}$$

Writing (12.28) for $x = R$ and different indices n, we get

$$\begin{aligned}
P_{n-1}(R) &= \mu_{n-1}T_{n-1}(R/\rho), \\
P_{n-2}(R) &= \mu_{n-2}T_{n-2}(R/\rho).
\end{aligned} \tag{12.33}$$

We use Eq. (12.33) in the last row of (12.32) to obtain

$$e = \hat{x}_n - x = \mu_n\left(\frac{2R}{\rho}\frac{P_{n-1}(R/\rho)(x_0 - x)}{\mu_{n-1}} - \frac{P_{n-2}(R/\rho)(x_0 - x)}{\mu_{n-2}}\right). \tag{12.34}$$

Using (12.24), we can write the error for different indices n as

$$\begin{aligned}
e = \hat{x}_{n-1} - x &= P_{n-1}(R)(x_0 - x), \\
e = \hat{x}_{n-2} - x &= P_{n-2}(R)(x_0 - x).
\end{aligned} \tag{12.35}$$

Substituting (12.35) in (12.34), we have

$$\begin{aligned}
e = \hat{x}_n - x &= \mu_n\left(\frac{2R}{\rho}\frac{\hat{x}_{n-1} - x}{\mu_{n-1}} - \frac{\hat{x}_{n-2} - x}{\mu_{n-2}}\right) \\
&= \mu_n\frac{2R}{\rho}\frac{\hat{x}_{n-1}}{\mu_{n-1}} - \mu_n\frac{2R}{\rho}\frac{x}{\mu_{n-1}} \\
&\quad - \mu_n\frac{\hat{x}_{n-2}}{\mu_{n-2}} + \mu_n\frac{x}{\mu_{n-2}} \\
&= \mu_n\frac{2R}{\rho}\frac{\hat{x}_{n-1}}{\mu_{n-1}} - \mu_n\frac{\hat{x}_{n-2}}{\mu_{n-2}} + C_1(x),
\end{aligned} \tag{12.36}$$

where

$$C_1(x) = \mu_n \frac{x}{\mu_{n-2}} - \mu_n \frac{2 R}{\rho} \frac{x}{\mu_{n-1}}. \tag{12.37}$$

Adding x to both parts of (12.36), we get

$$\hat{x}_n = \mu_n \frac{2 R}{\rho} \frac{\hat{x}_{n-1}}{\mu_{n-1}} - \mu_n \frac{\hat{x}_{n-2}}{\mu_{n-2}} + C_2(x), \tag{12.38}$$

where

$$C_2(x) = x + \mu_n \frac{x}{\mu_{n-2}} - \mu_n \frac{2 R}{\rho} \frac{x}{\mu_{n-1}}. \tag{12.39}$$

Further, since for the exact value of x we have $x = Rx + c$ and thus $R = (x - c)/x$, the function $C_2(x)$ can be written as

$$\begin{aligned} C_2(x) &= x + \mu_n \frac{x}{\mu_{n-2}} - \frac{2 (x - c)}{\rho} \frac{\mu_n}{\mu_{n-1}} \\ &= x \mu_n \left(\frac{1}{\mu_n} + \frac{1}{\mu_{n-2}} - \frac{2}{\rho \mu_{n-1}} \right) + \frac{2\mu_n c}{\rho \mu_{n-1}}. \end{aligned} \tag{12.40}$$

Since by (12.28) and (12.25) we have

$$\frac{1}{\mu_n} = \frac{2}{\rho \mu_{n-1}} - \frac{1}{\mu_{n-2}}, \tag{12.41}$$

it follows that (12.40) can be simplified to

$$C_2(x) = \frac{2\mu_n c}{\rho \mu_{n-1}}. \tag{12.42}$$

Combining (12.38), (12.41), and (12.42), we can formulate the following accelerating algorithm for iterations $x_{i+1} = Rx_i + c$.

Algorithm 12.7 Chebyshev acceleration algorithm.

    Step 0: Initialization:
    set $N, \varepsilon, \mu_0 = 1, \mu_1 = \rho(R), \hat{x}_0 = x_0; \hat{x}_1 = Rx_1 + c$.
    Step 1: for $n = 1$ to N
    $\mu_n = \dfrac{1}{\frac{2}{\rho(R)\mu_{n-1}} - \frac{1}{\mu_{n-2}}}$
    $\hat{x}_n = \mu_n \dfrac{2 R}{\rho(R)\mu_{n-1}} \hat{x}_{n-1} - \dfrac{\mu_n}{\mu_{n-2}} \hat{x}_{n-2} + \dfrac{2\mu_n c}{\rho(R)\mu_{n-1}}$
    if $\|\hat{x}_n - \hat{x}_{n-1}\| < \varepsilon$ quit
    else set $n := n + 1$ and go to Step 1.
    end

Algorithm 12.7 requires that the matrix R has only real eigenvalues. Thus, this algorithm cannot be applied to $SOR(\omega)$, since the matrix R_{SOR} defined in (12.22) can have complex eigenvalues. However, if we write the iterations in $SOR(\omega)$ as $x_{i+1} = \hat{R}x_i + c$ with a matrix $\hat{R}$ that has real eigenvalues, then Algorithm 12.7 can be used.

Assume that we have a symmetric matrix A such that $A = D(I - L - U)$ with $U = L^T$. Recall now the iterations in (12.21) in $SOR(\omega)$ and let us write them in two steps:

- Step 1:

$$
\begin{aligned}
x_i^{new} &= (I - \omega L)^{-1}D^{-1}\omega b + (I - \omega L)^{-1}((1 - \omega)I + \omega U)x_i \\
&= (I - \omega L)^{-1}((1 - \omega)I + \omega U)x_i + const. := \hat{L}x_i + const.
\end{aligned}
\tag{12.43}
$$

- Step 2:

$$
x_{i+1} = (I - \omega U)^{-1}((1 - \omega)I + \omega L)x_i^{new} + const. := \hat{U}x_i^{new} + const.
\tag{12.44}
$$

Substituting (12.43) into (12.44), we get iterations $x_{i+1} = \hat{R}x_i + c$, where the matrix $\hat{R} := \hat{L}\hat{U}$ has only real eigenvalues, since the similar symmetric matrix $(I - \omega U)\hat{R}(I - \omega U)^{-1}$ has real eigenvalues:

$$
\begin{aligned}
(I - \omega U)\hat{R}(I - \omega U)^{-1} &= I + (2 - \omega)^2(I - \omega L)^{-1}(I - \omega U)^{-1} \\
&\quad + (\omega - 2)(I - \omega U)^{-1} + (\omega - 2)(I - \omega L)^{-1} \\
&= I + (2 - \omega)^2(I - \omega L)^{-1}(I - \omega L^T)^{-1} \\
&\quad + (\omega - 2)(I - \omega L^T)^{-1} + (\omega - 2)(I - \omega L)^{-1}.
\end{aligned}
\tag{12.45}
$$

Thus, we can formulate the following symmetric successive overrelaxation $SSOR(\omega)$ algorithm for iterations $x_{i+1} = \tilde{R}x_i + c$.

Algorithm 12.8 $SSOR(\omega)$ algorithm for $x_{i+1} = \tilde{R}x_i + c$

Step 1.   Apply one step of the $SOR(\omega)$ algorithm, Algorithm 12.5, for $j = 1$ to n to compute $x_{i,1}, ..., x_{i,n}$.

Step 2.   Apply one step of Algorithm 12.5 backward for $j = n$ to 1 to compute $x_{i,n}, ..., x_{i,1}$.

12.6   Convergence of Main Iterative Methods

Theorem 12.2 *If the matrix A is strictly row diagonally dominant (i.e., such that $|a_{ii}| > \sum_{i \neq j}|a_{ij}|$), then the Jacobi and Gauss–Seidel methods converge such that*

$$
\|R_{GS}\|_\infty < \|R_J\|_\infty < 1,
\tag{12.46}
$$

where R_{GS} and R_J are defined in (12.14), (12.8), respectively.

Proof We can rewrite (12.46) with $e = (1, ..., 1)^T$ as

$$\|R_{GS}\|_\infty = \| |R_{GS}| e\|_\infty < \|R_J\|_\infty = \| |R_J| e\|_\infty < 1. \qquad (12.47)$$

Using definitions (12.14) and (12.8), we can get from the above inequality,

$$\| |R_{GS}| e\|_\infty = \| |(I-L)^{-1}U| e\|_\infty < \| |R_J| e\|_\infty = \| |L+U| e\|_\infty < 1. \qquad (12.48)$$

Further, the triangle inequality along with the facts that $L^n = 0$ and $(I - |L|)^{-1} \approx \sum_{i=0}^{n-1} |L|^i$ imply that

$$|(I-L)^{-1}U| e \le |(I-L)^{-1}| |U| e \approx | \sum_{i=0}^{n-1} L^i | |U| e$$

$$\le \sum_{i=0}^{n-1} |L|^i |U| e \approx (I - |L|)^{-1} |U| e. \qquad (12.49)$$

Using the assumption

$$\|R_J\|_\infty = \rho < 1 \qquad (12.50)$$

together with the fact that all entries of $(I - |L|)^{-1} \approx \sum_{i=0}^{n-1} |L|^i$ are positive, we have that

$$0 \le (I - |L| - |U|) e. \qquad (12.51)$$

By assumption (12.50), we have

$$|R_J| e = (|L| + |U|) e \le e.$$

Next, multiplying (12.51) by $|L|$, we obtain

$$0 \le |L|(I - |L| - |U|) e = (|L| - |L|^2 - |L||U|) e.$$

Then adding $|U| e$ to both sides of the above inequality, we get

$$|U| e \le (I - |L|)(|L| + |U|) e = (|L| - |L|^2 - |L||U| + |U|) e,$$

from which follows (12.47) and thus (12.46). □

Theorem 12.3 *Let the spectral radius of R_{SOR} be such that $\rho(R_{SOR}) \ge |\omega - 1|$. Then $0 < \omega < 2$ is required for convergence of $SOR(\omega)$.*

Proof We write the characteristic polynomial for R_{SOR} as

$$\varphi(\lambda) = \det(\lambda I - R_{SOR}) = \det((I - \omega L)(\lambda I - R_{SOR}))$$
$$= \det((\lambda + \omega - 1)I - \omega \lambda L - \omega U). \tag{12.52}$$

From the equation above, we have

$$\varphi(0) = \pm \prod \lambda_i(R_{SOR}) = \pm \det((\omega - 1)I) = \pm(\omega - 1)^n,$$

and thus

$$\max_i |\lambda_i(R_{SOR})| \geq |\omega - 1|,$$

from which it follows that $\rho(R_{SOR}) \geq |\omega - 1|$. □

Theorem 12.4 *If A is an s.p.d. matrix, then $\rho(R_{SOR}) < 1$ for all $0 < \omega < 2$, and thus SOR converges for all $0 < \omega < 2$. If we choose $\omega = 1$, then we obtain the usual Gauss–Seidel method, which also converges.*

A proof of this theorem can be found in [23].
Assume that for every matrix M that can be written as

$$M = D - \tilde{L} - \tilde{U},$$

$R_J(\alpha)$ is the matrix that is defined as

$$R_J(\alpha) = \alpha D^{-1} \tilde{L} + \frac{1}{\alpha} D^{-1} \tilde{U}$$

for every scalar $\alpha > 0$. The matrix M is said to be *consistently ordered* if the eigenvalues of $R_J(\alpha)$ are independent of α.

Theorem 12.5 *Assume that A is consistently ordered and $\omega \neq 0$. Then the following statements are true:*

1. *The eigenvalues of R_J appear in pairs with positive and negative signs.*
2. *Assume that $\tilde{\lambda}$ is an eigenvalue of R_j and the following equation is true:*

$$(\lambda + \omega - 1)^2 = \lambda \omega^2 \tilde{\lambda}^2. \tag{12.53}$$

   Then λ is an eigenvalue of R_{SOR}.
3. *If $\lambda \neq 0$ is an eigenvalue of R_{SOR}, then $\tilde{\lambda}$ in (12.53) is an eigenvalue of R_j.*

Proof 1. Since A is consistently ordered, $R_J(1) = R_J$ and $R_J(-1) = -R_J$ have the same eigenvalues and thus appear in $\pm$ pairs.
2. Assume that $\lambda = 0$. Then from (12.53), we have

$$(\omega - 1)^2 = 0. \tag{12.54}$$

We see that $\omega = 1$ and $R_{SOR(1)} = R_{GS} = (I - L)^{-1}U$. Thus, $\lambda = 0$ is an eigenvalue of R_{SOR}. Otherwise, we can write

$$
\begin{aligned}
0 = \det(\lambda I - R_{SOR}) &= \det((I - \omega L)(\lambda I - R_{SOR})) \\
&= \det((\lambda + \omega - 1)I - \omega\lambda L - \omega U) \\
&= \det\left(\sqrt{\lambda}\omega\left(\left(\frac{\lambda + \omega - 1}{\sqrt{\lambda}\omega}\right)I - \sqrt{\lambda}L - \frac{1}{\sqrt{\lambda}}U\right)\right) \\
&= \det\left(\left(\frac{\lambda + \omega - 1}{\sqrt{\lambda}\omega}\right)I - L - U\right)(\sqrt{\lambda}\omega)^n.
\end{aligned}
\tag{12.55}
$$

Defining

$$\tilde{\lambda} = \frac{\lambda + \omega - 1}{\sqrt{\lambda}\omega}, \tag{12.56}$$

we see that it is an eigenvalue of $L + U = R_J$, and thus equation (12.53) is valid.
3. If $\lambda \neq 0$, then we use the previous proof in the opposite direction. □

From Theorem 12.5 it follows that if A is consistently ordered, then $\rho(R_{GS}) = \rho(R_J)^2$. We can see this from (12.53), since for $\omega = 1$, we have the Gauss–Seidel method, and thus (12.53) can be written as

$$\lambda^2 = \lambda\tilde{\lambda}^2, \tag{12.57}$$

or

$$\lambda = \tilde{\lambda}^2. \tag{12.58}$$

Thus, $\rho(R_{GS}) = \rho(R_J)^2$, and the Gauss–Seidel method is twice as fast as the Jacobi method.

Theorem 12.6 *Assume that A is consistently ordered and that R_J has real eigenvalues such that $\tilde{\lambda} = \rho(R_J) < 1$. Then following statements are true:*

• *The optimal relaxation parameter ω_{opt} in SOR(ω) can be computed as*

$$\omega_{opt} = \frac{2}{1 + \sqrt{1 - \tilde{\lambda}^2}}. \tag{12.59}$$

• *The spectral radius $\rho(R_{SOR(\omega_{opt})})$ for ω_{opt} defined by (12.59) can be computed as*

$$\rho(R_{SOR(\omega_{opt})}) = \omega_{opt} - 1 = \frac{\tilde{\lambda}^2}{(1 + \sqrt{1 - \tilde{\lambda}^2})^2}. \tag{12.60}$$

- *The spectral radius $\rho(R_{SOR(\omega)})$ for $0 < \omega < 2$ can be computed as*

$$\rho(R_{SOR(\omega)}) = \begin{cases} 1 - \omega + \frac{1}{2}\omega^2\tilde{\lambda}^2 + \omega\tilde{\lambda}\sqrt{1 - \omega + 0.25\omega^2\tilde{\lambda}^2}, & if\ \omega \in (0, \omega_{opt}), \\ \omega - 1, & if\ \omega \in [\omega_{opt}, 2). \end{cases}$$
$$(12.61)$$

Proof The proof follows from the solution of equation (12.53) for λ. □

Example 12.3 The matrix A in the model problem for the Poisson's equation of Section 8.1.3 is s.p.d. Thus by Theorem 12.4, the values of SOR(ω) for this problem will converge for all $0 < \omega < 2$.

We present the numerical solution of the Dirichlet problem for the Poisson's equation (8.11) in two dimensions using the iterative SOR(ω) method. The setup for our numerical simulations is the same as in Example 8.2 of Chapter 8. The MATLAB® program of Section 1.24 is available for running this test.[4] This program implements two algorithms: the first is Algorithm 12.5, and the second is Algorithm 12.6, or the SOR(ω) method with red–black ordering. We have used the stopping criterion $\|u^{m+1} - u^m\|_2 < tol$, with tolerance $tol = 10^{-9}$.

The results of our numerical simulations are the same as in Fig. 8.1 for the number of inner points $N = 20$ and for the tolerance $tol = 10^{-9}$ in the iterative update (check

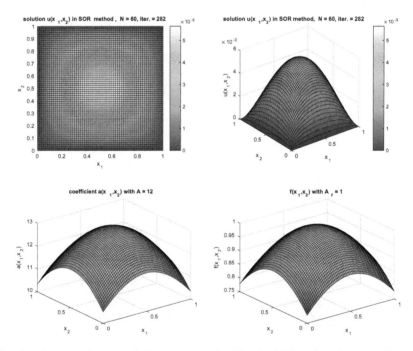

Fig. 12.1 Solution of problem (8.11) in the example of Section 8.1.3 on the unit square with mesh 60×60 points.

---

[4]The MATLAB® programs can be found on the online version of Chapter 1 (doi:10.1007/978-3-319-57304-5_1).

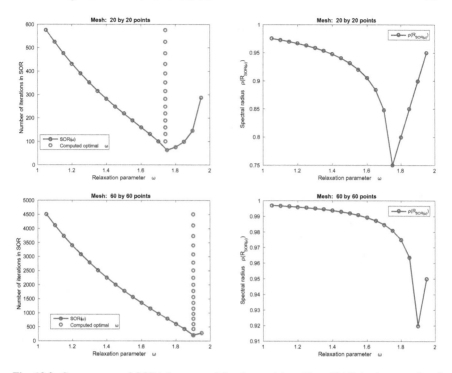

Fig. 12.2 Convergence of SOR(ω) computed for the model problem (8.11) in the example of Section 8.1.3 on the unit square for different discretizations. In the left-hand figures, we plot the number of iterations in SOR depending on ω (*blue line*) and corresponding optimal relaxation parameter ω_{opt} (*red line*). The computed spectral radius $\rho(R_{SOR(\omega)})$ is presented in the right-hand figures.

it by running the MATLAB® programs of Section 1.24).[5] Figure 12.1 presents the numerical solution for the number of inner points $N = 60$, relaxation parameter $\omega = 1.95$, and tolerance $tol = 10^{-9}$.

We perform computations for different relaxation parameters $\omega \in (0, 2)$. We also compute the optimal relaxation parameter ω_{opt}, as well as the spectral radius $\rho(R_{SOR(\omega)})$ using formulas (12.59), (12.61), respectively. Figure 12.2 presents the results of these computations. The left-hand pictures of Fig. 12.2 show the convergence of SOR(ω) for different $\omega \in (0, 2)$ (blue line) and corresponding optimal relaxation parameter ω_{opt} computed by (12.59) (red line). The right-hand pictures of Fig. 12.2 show the spectral radius $\rho(R_{SOR(\omega)})$ computed by (12.61) with ω_{opt} implemented by (12.59) and $\tilde{\lambda}$ given by (12.56). From this figure, we also conclude that the convergence of SOR(ω) is much faster than that of the Gauss–Seidel method and the Jacobi method.

---

[5]The MATLAB® programs can be found on the online version of Chapter 1 (doi:10.1007/978-3-319-57304-5_1).

12.7   Krylov Subspace Methods

Krylov[6] subspace methods are used for the solution of large systems of linear equations $Ax = b$ and for finding eigenvalues of A avoiding matrix–matrix multiplication. Instead, these methods use matrix–vector multiplication.

The *Krylov subspace* generated by the $n \times n$ matrix A and vector b of length n is the linear subspace spanned by powers of A multiplied by b:

$$K_r(A, b) = span\{b, Ab, A^2b, ..., A^{r-1}b\}. \tag{12.62}$$

Let $y_1 = b$, $y_2 = Ab$, $y_3 = A^2b$, $\ldots y_n = A^{n-1}b$ and $A_K = (y_1, y_2, \ldots, y_n)$. Let us write the QR decomposition A_K as $A_K = QR$. We observe that $A_K^{-1}AA_K = R^{-1}Q^TAQR$ from what follows that $Q^TAQ = H$ is an upper Hessenberg matrix (since R^{-1}, R are upper triangular and $A_K^{-1}AA_K$ is upper Hessenberg). We note that for a symmetric matrix A the matrix H in the decomposition $Q^TAQ = H$ will be an upper Hessenberg matrix which also will be a lower Hessenberg and thus a tridiagonal matrix. Writing Q as $Q = (q_1, ..., q_n)$ and using $AQ = QH$, we have

$$Aq_j = \sum_{i=1}^{j+1} h_{i,j}q_i. \tag{12.63}$$

Multiplying both sides of the above expression by orthonormal vectors q_m^T and using the fact that the q_i are orthonormal, we obtain

$$q_m^T Aq_j = \sum_{i=1}^{j+1} h_{i,j}q_m^Tq_i = h_{m,j}, \ 1 \le m \le j. \tag{12.64}$$

We can rewrite (12.63) as

$$h_{j+1,j}q_{j+1} = Aq_j - \sum_{i=1}^{j} h_{i,j}q_i. \tag{12.65}$$

The two formulas (12.64) and (12.65) are used in the Arnoldi[7] algorithm for the reduction of a matrix A to upper Hessenberg form. Let r be the number of columns in the matrices Q and H that we need to compute. We now formulate the Arnoldi algorithm, which performs partial reduction of the matrix A to upper Hessenberg form. The vectors q_j computed in this algorithm are called Arnoldi vectors.

Algorithm 12.9 Arnoldi algorithm.

   Initialization: $q_1 = \frac{b}{\|b\|_2}$

---

[6] Aleksey Krylov (1863–1945) was a Russian naval engineer, applied mathematician, and memoirist.
[7] Walter Edwin Arnoldi (1917–1995) was an American engineer.

for $j = 1$ to r
$z = Aq_j$
for $i = 1$ to j
$h_{i,j} = q_i^T z$
$z = z - h_{i,j} q_i$
end
$h_{j+1,j} = \|z\|_2$
if $h_{j+1,j} = 0$ quit
$q_{i+1} = \frac{z}{h_{j+1,j}}$
end

Let us define $Q = (Q_r, Q_u)$ with $Q_r = (q_1, ..., q_r)$ and $Q_u = (q_{r+1}, ..., q_n)$. We have the following structure of the matrix H after r steps of the Arnoldi algorithm, Algorithm 12.9:

$$H = Q^T A Q = (Q_r, Q_u)^T A (Q_r, Q_u)$$
$$= \begin{pmatrix} Q_r^T A Q_r & Q_r^T A Q_u \\ Q_u^T A Q_r & Q_u^T A Q_u \end{pmatrix} = \begin{pmatrix} H_r & H_{ur} \\ H_{ru} & H_u \end{pmatrix}, \tag{12.66}$$

where H_r is an upper Hessenberg matrix. We know only H_r and H_{ru}, with H_{ur} and H_u still unknown.

For the case of a symmetric matrix A, the Arnoldi algorithm can be simplified, since the matrix H is symmetric and tridiagonal, which means that

$$H = \begin{pmatrix} \alpha_1 & \beta_1 & \cdots & & \cdots \\ \beta_1 & \alpha_2 & \cdots & & \cdots \\ \cdots & \cdots & \cdots & & \cdots \\ \cdots & \cdots & \alpha_{n-1} & \beta_{n-1} \\ \cdots & \cdots & \beta_{n-1} & \alpha_n \end{pmatrix}. \tag{12.67}$$

Rewriting (12.63) for the case of the symmetric and tridiagonal H given by (12.67), we have

$$Aq_j = \beta_{j-1} q_{j-1} + \alpha_j q_j + \beta_j q_{j+1}. \tag{12.68}$$

We note that the columns of Q are orthonormal. Thus, multiplying (12.68) by q_j^T, we get

$$q_j^T A q_j = q_j^T (\beta_{j-1} q_{j-1} + \alpha_j q_j + \beta_j q_{j+1}) = \alpha_j. \tag{12.69}$$

From (12.68), we can obtain an expression for computing q_{j+1},

$$q_{j+1} = (Aq_j - \beta_{j-1} q_{j-1} - \alpha_j q_j)/\beta_j, \tag{12.70}$$

which is used in the Lanczos[8] algorithm. Combining (12.69) and (12.70), we get

---

[8]Cornelius Lanczos (1893–1974) was a Hungarian mathematician and physicist.

the Lanczos algorithm for partial reduction of a symmetric matrix A to symmetric tridiagonal form.

Algorithm 12.10 Lanczos algorithm.

Initialization: $q_1 = \frac{b}{\|b\|_2}$, $\beta_0 = 0$, $q_0 = 0$
for $j = 1$ to r
$z = Aq_j$
$\alpha_j = q_j^T z$
$z = z - \alpha_j q_j - \beta_{j-1} q_{j-1}$ /* no reorthogonalization */
$\beta_j = \|z\|_2$
if $\beta_j = 0$ quit
$q_{i+1} = \frac{z}{\beta_j}$
end

The vectors q_j computed by Algorithm 12.10 are called Lanczos vectors. The vectors q_r computed in the Lanczos or Arnoldi algorithm create an orthonormal basis of the Krylov subspace K_r defined in (12.62). The matrix $H_r = Q_r^T A Q_r$ in both algorithms is called the projection of A onto the Krylov subspace K_r.

Taking into account (12.66), we can write the following structure of the matrix T after r steps of the Lanczos algorithm, Algorithm 12.10:

$$T = Q^T A Q = (Q_r, Q_u)^T A (Q_r, Q_u)$$
$$= \begin{pmatrix} Q_r^T A Q_r & Q_r^T A Q_u \\ Q_u^T A Q_r & Q_u^T A Q_u \end{pmatrix} = \begin{pmatrix} T_r & T_{ur} \\ T_{ru} & T_u \end{pmatrix} = \begin{pmatrix} T_r & T_{ru}^T \\ T_{ru} & T_u \end{pmatrix}. \tag{12.71}$$

We can compute the elements of T_r and $T_{ru} = T_{ur}^T$, since the matrix A is symmetric. However, the elements of T_u are not known.

Our goal now is to use r steps in the Lanczos or Arnoldi algorithm to solve the linear system $Ax = b$. To do so, we seek the best approximation x_r to the exact solution $x = A^{-1}b$ given by

$$x_r = \sum_{j=1}^{r} z_j q_j = Q_r z, \tag{12.72}$$

where $z = (z_1, ..., z_r)^T$. Let us define the residual as $R_r = b - Ax_r$. For the case of an s.p.d. matrix A, we can define the norm $\|R\|_{A^{-1}} := (R^T A^{-1} R)^{1/2}$. We note that $\|R\|_{A^{-1}} = \|x_r - x\|_A$. Thus, the best computed solution x_r will minimize $\|R\|_{A^{-1}}$. An algorithm that can compute such a vector x_r is called the conjugate gradient algorithm (CG).

Theorem 12.7 *Let A be a symmetric matrix, $H_r = Q_r^T A Q_r$, and let the residuals be defined as $R_r = b - Ax_r \forall x_r \in K_r$. When H_r is nonsingular, we can define*

$$x_r = Q_r H_r^{-1} e_1 \|b\|_2, \tag{12.73}$$

where $e_1 = (1, 0, ..., 0)^T$. Then $Q_r^T R_r = 0$.

Let A be also a positive definite matrix. Then H_r must be nonsingular, and x_r defined as in (12.73) minimizes $\| R_r \|_{A^{-1}}$ for all $x_r \in K_r$, where $R_r = \pm \| R_r \|_2 q_{r+1}$.

A proof of this theorem can be found in [23]. Numerical exploration of the convergence of the Lanczos algorithm, Algorithm 12.10, is provided in [23]. We note that rounding errors destroy the orthogonalization property of the Lanczos algorithm 12.10: the vectors q_i can lose orthogonality and become linearly dependent. Below, we present a more expensive algorithm with full orthogonalization for finding eigenvalues and eigenvectors of a symmetric matrix A. We say that the algorithm carries out full orthogonalization because we perform the Gram–Schmidt orthogonalization process twice to be sure that we make z orthogonal to all $q_1, ..., q_{j-1}$; see the discussion in [94].

Algorithm 12.11 Lanczos algorithm with orthogonalization.

Initialization: $q_1 = \frac{b}{\|b\|_2}$, $\beta_0 = 0$, $q_0 = 0$
for $j = 1$ to r
$z = Aq_j$
$\alpha_j = q_j^T z$
$z = z - \sum_{i=1}^{j-1} (z^T q_i) q_i$ /* carry out reorthogonalization twice */
$z = z - \sum_{i=1}^{j-1} (z^T q_i) q_i$
$\beta_j = \|z\|_2$
if $\beta_j = 0$ quit
$q_{i+1} = \frac{z}{\beta_j}$
end

We can see that the r steps of the Lanczos algorithm with full orthogonalization, Algorithm 12.11, takes $O(r^2 n)$ FLOPS compared with $O(rn)$ FLOPS for the Lanczos algorithm, Algorithm 12.10. The selective reorthogonalization process takes advantages of both algorithms and makes Lanczos vectors nearly orthogonal sufficiently cheaply; see details in [23]. To formulate the Lanczos algorithm with selective reorthogonalization, we define the *Rayleigh–Ritz*[9] *procedure*. In this procedure, the eigenvalues of A are approximated by the eigenvalues of $T_r = Q_r^T \Lambda Q_r$ with T_r defined in (12.71), and are called *Ritz values*. Let us define $T_r = V \Lambda V^T$, the eigendecomposition of T_r. The *Ritz vectors* are the columns of $Q_r V$, which are also eigenvector approximations of the eigenvectors of T_r.

The next theorem provides a criterion for selective orthogonalization of Lanczos vectors.

Theorem 12.8 (Paige[10]). *Let the jth step of the Lanczos algorithm 12.10 be written as*

$$\beta_j q_{j+1} + f_j = Aq_j - \alpha_j q_j - \beta_{j-1} q_{j-1}, \tag{12.74}$$

---

[9]Walther Ritz (1878–1909) was a Swiss theoretical physicist.
[10]Constantin Marie Le Paige (1852–1929) was a Belgian mathematician.

where f_j is the rounding error and $\|f\|_2 \leq O(\varepsilon\|A\|)$ with ε represents the machine epsilon. Let $T_r = V\Lambda V^T$ be the eigendecomposition of T_r with orthogonal $V = (v_1, ..., v_r)$, and $\Lambda = \mathrm{diag}(\Theta_1, ..., \Theta_r)$. Let the columns $y_{i,r} = Q_r v_i$ of $Q_r V$ with orthogonal $Q = (q_1, ..., q_r)$ be the Ritz vectors, with the Θ_i the Ritz values. Then

$$y_{i,r}^T q_{r+1} = \frac{O(\varepsilon\|A\|)}{\beta_r |v_i(r)|},$$

where $v_i(r)$ is the rth entry of v_i, $i = 1, ..., r$, and $\beta_r = \|T_{ru}\|_2$ with T_{ru} defined in (12.71).

A proof of this theorem is given in [23].

Using Paige's theorem, Theorem 12.8, we can design the simplest version of the Lanczos algorithm with selective orthogonalization. In this algorithm we check values of $\beta_r |v_i(r)|$ at every step of the algorithm, and then for small values of $\beta_r |v_i(r)|$, we orthogonalize the values of the vector z.

Algorithm 12.12 Lanczos algorithm with selective orthogonalization.

> Initialization: $q_1 = \frac{b}{\|b\|_2}, \beta_0 = 0, q_0 = 0$
> for $j = 1$ to r
> $z = Aq_j$
> $\alpha_j = q_j^T z$
> $z = z - \alpha_j q_j - \beta_{j-1} q_{j-1}$ /* no reorthogonalization */
> for $i \leq r$
> if $\beta_r |v_i(r)| \leq \sqrt{\varepsilon}\|T_r\|$
> $y_{i,r} = Q_r v_i$ /* Ritz vectors */
> $z = z - (y_{i,r}^T z) y_{i,r}$ /* selective reorthogonalization */
> end if
> end for
> $\beta_j = \|z\|_2$
> if $\beta_j = 0$ quit
> $q_{i+1} = \frac{z}{\beta_j}$
> end

Algorithm 12.12 can be improved using a recurrence formula given in [107], since it is not necessary that the condition $\beta_r |v_i(r)| \leq \sqrt{\varepsilon}\|T_r\|$ be checked at every iteration. Thus, many steps in selective reorthogonalization can be eliminated. See also [40], where a shifted block Lanczos algorithm with standard implementation is presented.

However, when the matrix A is asymmetric, then all the Lanczos algorithms considered above are no longer valid. This is because the eigenvalues of A can be complex or badly conditioned. We refer to [5, 6, 20, 100, 102] for the theory and implementation of the Lanczos algorithm for asymmetric matrices.

12.8 Conjugate Gradient Method

In this section we will present the main steps in the derivation of the conjugate gradient (CG) algorithm. We will begin by using the Lanczos algorithm, Algorithm 12.10, and then combine it with formula (12.73). Then using Theorem 12.7, we will conclude that the residuals $R_r = b - Ax_r$ are parallel to the Lanczos vectors q_{r+1}.

Let us introduce *conjugate gradient* vectors p_r. They are called *gradients* because in a single step of the CG algorithm we compute the approximate solution as $x_r = x_{r-1} + \nu p_r$ with some scalars ν (see Algorithm 12.13), and this solution minimizes the residual norm $\|R_r\|_{A^{-1}} = (R_r^T A^{-1} R_r)^{1/2}$. The vectors p_r are called *conjugate*, or more precisely *A-conjugate*, because $p_r^T A p_j = 0$ if $j \neq r$.

Since A is symmetric positive definite, the matrix $H_r = Q_r^T A Q_r$ is also symmetric positive definite. Thus, we can use Cholesky decomposition on H_r to get

$$H_r = \hat{L}_r \hat{L}_r^T = L_r D_r L_r^T, \tag{12.75}$$

where L_r is unit lower bidiagonal and D_r is diagonal. Then using the formula (12.73), we get

$$\begin{aligned}
x_r &= Q_r H_r^{-1} e_1 \|b\|_2 \\
&= Q_r (L_r^{-T} D_r^{-1} L_r^{-1}) e_1 \|b\|_2 \\
&= (Q_r L_r^{-T})(D_r^{-1} L_r^{-1} e_1 \|b\|_2) \\
&\equiv (\tilde{P}_r)(y_r),
\end{aligned}$$

where $\tilde{P}_r \equiv Q_r L_r^{-T}$ and $y_r \equiv D_r^{-1} L_r^{-1} e_1 \|b\|_2$. Let $\tilde{P}_r = (\tilde{p}_1, \ldots, \tilde{p}_r)$. The conjugate gradients p_i will be parallel to the columns $\tilde{p}_i$ of $\tilde{P}_r$.

Lemma 12.3 *The columns p_i of the matrix $\tilde{P}_r$ are A-conjugate. This means that $\tilde{P}_r^T A \tilde{P}_r$ is diagonal.*

Proof We can write

$$\begin{aligned}
\tilde{P}_r^T A \tilde{P}_r &= (Q_r L_r^{-T})^T A (Q_r L_r^{-T}) = L_r^{-1}(Q_r^T A Q_r) L_r^{-T} = L_r^{-1}(H_r) L_r^{-T} \\
&= L_r^{-1}(L_r D_r L_r^T) L_r^{-T} = D_r.
\end{aligned}$$ $\qquad \square$

Let us define the following iterative update for x_r:

$$\begin{aligned}
x_r &= \tilde{P}_r\, y_r = (\tilde{P}_{r-1}, \tilde{p}_r) \begin{pmatrix} y_{r-1} \\ \eta_r \end{pmatrix} \\
&= \tilde{P}_{r-1} y_{r-1} + \tilde{p}_r \eta_r = x_{r-1} + \tilde{p}_r \eta_r.
\end{aligned} \tag{12.76}$$

To use this formula, we need to compute the scalars η_r. Since H_{r-1} is the leading $(r-1) \times (r-1)$ submatrix of H_r, L_{r-1} and D_{r-1} are also the leading $(r-1) \times (r-1)$ submatrices of L_r and D_r, respectively. Thus, we can write

$$H_r = \begin{pmatrix} \alpha_1 & \beta_1 & & \\ \beta_1 & \ddots & \ddots & \\ & \ddots & \ddots & \beta_{r-1} \\ & & \beta_{r-1} & \alpha_r \end{pmatrix} = L_r D_r L_r^T =$$

$$\begin{pmatrix} 1 & & & \\ l_1 & \ddots & & \\ & \ddots & \ddots & \\ & & l_{r-1} & 1 \end{pmatrix} \begin{pmatrix} d_1 & & & \\ & \ddots & & \\ & & d_{r-1} & \\ & & & d_r \end{pmatrix} \begin{pmatrix} 1 & & & \\ l_1 & \ddots & & \\ & \ddots & \ddots & \\ & & l_{r-1} & 1 \end{pmatrix}^T$$

$$= \begin{pmatrix} L_{r-1} & \\ l_{r-1}\hat{e}_{r-1}^T & 1 \end{pmatrix} \mathrm{diag}(D_{r-1}, d_r) \begin{pmatrix} L_{r-1} & \\ l_{r-1}\hat{e}_{r-1}^T & 1 \end{pmatrix}^T,$$

where $\hat{e}_{r-1}^T = (0, \ldots, 0, 1)$ and $\dim \hat{e}_{r-1}^T = r - 1$.

We see also that D_{r-1}^{-1} and L_{r-1}^{-1} are the leading $(r - l) \times (r - 1)$ submatrices of the matrices $D_r^{-1} = \mathrm{diag}(D_{r-1}^{-1}, d_r^{-1})$ and

$$L_r^{-1} = \begin{pmatrix} L_{r-1}^{-1} & \cdots \\ \cdots & 1 \end{pmatrix},$$

respectively. Below we show that the vector defined by $y_{r-1} = D_{r-1}^{-1} L_{r-1}^{-1} \hat{e}_1 ||b||_2$, where $\dim \hat{e}_1 = r - 1$, is the same as the leading $r - 1$ components of the vector y_r. We have

$$y_r = D_r^{-1} L_r^{-1} e_1 ||b||_2 = \begin{pmatrix} D_{r-1}^{-1} & \\ & d_r^{-1} \end{pmatrix} \begin{pmatrix} L_{r-1}^{-1} & \cdots \\ \cdots & 1 \end{pmatrix} e_1 ||b||_2$$

$$= \begin{pmatrix} D_{r-1}^{-1} L_{r-1}^{-1} \hat{e}_1 ||b||_2 \\ \eta_r \end{pmatrix} = \begin{pmatrix} y_{r-1} \\ \eta_r \end{pmatrix}.$$

To use formula (12.76) for computing x_r, we need to derive iterative formulas for the columns of $\tilde{P}_r = (\tilde{p}_1, \ldots, \tilde{p}_r)$ and η_r.

We begin first to derive $\tilde{P}_r = (\tilde{p}_1, \ldots, \tilde{p}_r)$. We observe that L_{r-1}^T is upper triangular, and thus the matrix L_{r-1}^{-T} is also upper triangular, and it forms the leading $(r - l) \times (r - 1)$ submatrix of L_r^{-T}. Therefore, $\tilde{P}_{r-1}$ is identical to the leading $r - 1$ columns of $\tilde{P}_r$:

$$\tilde{P}_r = Q_r L_r^{-T} = (Q_{r-1}, q_r) \begin{pmatrix} L_{r-1}^{-T} & \cdots \\ 0 & 1 \end{pmatrix} = (Q_{r-1} L_{r-1}^{-T}, \tilde{p}_r) = (\tilde{P}_{r-1}, \tilde{p}_r).$$

From $\tilde{P}_r = Q_r L_r^{-T}$, we get

$$\tilde{P}_r L_r^T = Q_r.$$

Now equating the rth columns on both sides of the above equation, we obtain an iterative formula for updating $\tilde{p}_r$:

$$\tilde{p}_r = q_r - l_{r-1}\tilde{p}_{r-1}. \tag{12.77}$$

The iterative formulas (12.77) for updating $\tilde{p}_r$, q_r (from the Lanczos algorithm) and finally (12.76) for computing x_r provide the main steps in the CG algorithm. We will additionally simplify these formulas to obtain the ultimate CG algorithm. By Theorem 12.7, the residuals R_r and vectors q_{r+1} are parallel. Thus, we can replace the Lanczos vectors q_{r+1} with the residuals $R_r = b - Ax_r$. We now multiply both sides of $x_r = x_{r-1} + \eta_r \tilde{p}_r$ by A and subtract from b to obtain

$$R_r = b - Ax_r = b - A(x_{r-1} + \eta_r \tilde{p}_r) = R_{r-1} - \eta_r A\tilde{p}_r. \tag{12.78}$$

The above formula yields the following iterative updates:

$$R_r = R_{r-1} - \eta_r A\tilde{p}_r; \tag{12.79}$$

from (12.76) we get

$$x_r = x_{r-1} + \eta_r \tilde{p}_r; \tag{12.80}$$

from (12.77) we get

$$\tilde{p}_r = q_r - l_{r-1}\tilde{p}_{r-1}. \tag{12.81}$$

The next step is to eliminate q_r. To do this, we substitute $R_{r-1} = q_r\|R_{r-1}\|_2$ and $p_r \equiv \|R_{r-1}\|_2 \tilde{p}_r$ into (12.79)–(12.81) to get

$$R_r = R_{r-1} - \frac{\eta_r}{\|R_{r-1}\|_2} Ap_r \equiv R_{r-1} - v_r Ap_r, \tag{12.82}$$

$$x_r = x_{r-1} + \frac{\eta_r}{\|R_{r-1}\|_2} p_r \equiv x_{r-1} + v_r p_r, \tag{12.83}$$

$$p_r = R_{r-1} - \frac{\|R_{r-1}\|_2 l_{r-1}}{\|R_{r-2}\|_2} p_{r-1} \equiv R_{r-1} + \mu_r p_{r-1}. \tag{12.84}$$

In analyzing (12.82)–(12.84), we observe that we need formulas for the scalars v_r and μ_r. For derivation of v_r, we multiply both sides of (12.84) on the left by $p_r^T A$ and use Lemma 12.3 to get

$$p_r^T Ap_r = p_r^T AR_{r-1} + 0 = R_{r-1}^T Ap_r. \tag{12.85}$$

Multiplying both sides of (12.82) on the left by R_{r-1}^T and using the equality $R_{r-1}^T R_r = 0$ (since the R_i are parallel to the columns of the orthogonal matrix Q) and then (12.85), we obtain

$$v_r = \frac{R_{r-1}^T R_{r-1}}{R_{r-1}^T A p_r} = \frac{R_{r-1}^T R_{r-1}}{p_r^T A p_r}. \tag{12.86}$$

Finally, we derive a formula for μ_r. Multiplying both sides of (12.84) on the left by $p_{r-1}^T A$ and using Lemma 12.3 (by this lemma, p_r and p_{r-1} are A-conjugate), we obtain

$$\mu_r = -\frac{p_{r-1}^T A R_{r-1}}{p_{r-1}^T A p_{r-1}}. \tag{12.87}$$

We can derive alternative formulas for v_r and μ_r. Multiplying both sides of (12.82) on the left by R_r^T, using that $R_{r-1}^T R_r = 0$, and solving for v_r, we get

$$v_r = -\frac{R_r^T R_r}{R_r^T A p_r}. \tag{12.88}$$

Equating (12.86) and (12.88) for v_{r-1} and comparing with Eq. (12.87) yields a different formula for computing μ_r:

$$\mu_r = -\frac{p_{r-1}^T A R_{r-1}}{p_{r-1}^T A p_{r-1}} = \frac{R_{r-1}^T R_{r-1}}{R_{r-2}^T R_{r-2}}. \tag{12.89}$$

Combining (12.82), (12.83), (12.84), (12.86), and (12.89) yields the conjugate gradient algorithm.

Algorithm 12.13 Conjugate gradient algorithm.

    Initialization: $r = 0$; $x_0 = 0$; $R_0 = b$; $p_1 = b$;
    repeat
    $r = r + 1$
    $z = A\, p_r$
    $v_r = (R_{r-1}^T R_{r-1})/(p_r^T z)$
    $x_r = x_{r-1} + v_r p_r$
    $R_r = R_{r-1} - v_r z$
    $\mu_{r+1} = (R_r^T R_r)/(R_{r-1}^T R_{r-1})$
    $p_{r+1} = R_r + \mu_{r+1} p_r$
    until $\|R_r\|_2$ is small enough

A convergence analysis of this algorithm is presented in [23]. From this analysis follows, that

$$\frac{\|R_r\|_{A^{-1}}}{\|R_0\|_{A^{-1}}} \le \frac{1}{T_r(1 + \frac{2}{k-1})}, \tag{12.90}$$

where $k = \frac{\lambda_{max}}{\lambda_{min}}$ is the condition number of A and Tr are Chebyshev polynomials considered in Sect. 12.5. The estimate (12.90) tells us that when the condition number k is close to 1, the term on the right-hand side of (12.90) is small, and we have rapid convergence in the CG algorithm, Algorithm 12.13. If the condition number k is large, then the estimate (12.90) can be rewritten as

$$\frac{\|R_r\|_{A^{-1}}}{\|R_0\|_{A^{-1}}} \le \frac{1}{T_r(1 + \frac{2}{k-1})} \le \frac{1}{(1 + \frac{2}{\sqrt{k-1}})}, \tag{12.91}$$

and convergence is slower.

We note that when the matrix A is simply symmetric, the norm $\|R_r\|_2$ is minimized using the minimum residual algorithm MINRES [93]. When the matrix A is asymmetric, the generalized minimum residual algorithm GMRES for minimization of $\|R_r\|_2$ is used [101].

Example 12.4 We present a numerical solution of the Dirichlet problem for the Poisson's equation (8.11) in two dimensions using the conjugate gradient method. The setup for our numerical experiments is the same as in Example 8.2 of Chapter 8. The MATLAB® program of Section 1.25, which implements Algorithm 12.13, is available for running this test.[11] We used computation of the residual R_r in the stopping criterion $\|R_r\|_2 < tol$, where the chosen tolerance was $tol = 10^{-9}$.

The conjugate gradient method converged in 33 iterations for number of inner points in the computational mesh $N = 20$. Our computed solution is the same as that presented in Fig. 8.1 (check it by running the MATLAB® programs of Section 1.25).

12.9 Preconditioning for Linear Systems

A preconditioning technique is used for the reduction of the condition number of a problem. For the solution of a linear system of equations $Ax = b$, the preconditioner matrix P of a matrix A is a matrix $P^{-1}A$ such that $P^{-1}A$ has a smaller condition number than the original matrix A. This means that instead of the solution of a system $Ax = b$, we will consider the solution of the system

$$P^{-1}Ax = P^{-1}b. \tag{12.92}$$

The matrix P should have the following properties:

- P is an s.p.d. matrix;
- $P^{-1}A$ is well conditioned;
- the system $Px = b$ should be easily solvable.

The preconditioned conjugate gradient method is derived as follows. First we multiply both sides of (12.92) by $P^{1/2}$ to get

---

[11] All MATLAB® programs referenced on this page can be found on the online version of Chapter 1 (doi:10.1007/978-3-319-57304-5_1).

$$(P^{-1/2}AP^{-1/2})(P^{1/2}x) = P^{-1/2}b. \tag{12.93}$$

We note that the system (12.93) is s.p.d., since we have chosen the matrix P such that $P = Q\Lambda Q^T$, which is the eigendecomposition of P. Then the matrix $P^{1/2}$ will be s.p.d. if it is defined as $P^{1/2} = Q\Lambda^{1/2}Q^T$. Defining

$$\begin{aligned}
\tilde{A} &:= P^{-1/2}AP^{-1/2}, \\
\tilde{x} &:= P^{1/2}x, \\
\tilde{b} &:= P^{-1/2}b,
\end{aligned} \tag{12.94}$$

we can rewrite (12.93) as the system $\tilde{A}\tilde{x} = \tilde{b}$. The matrices $\tilde{A}$ and $P^{-1}A$ are similar, since $P^{-1}A = P^{-1/2}\tilde{A}P^{1/2}$. Thus, $\tilde{A}$ and $P^{-1}A$ have the same eigenvalues. Thus, instead of the solution of $P^{-1}Ax = P^{-1}b$, we will present the preconditioned conjugate gradient algorithm (PCG) for the solution of $\tilde{A}\tilde{x} = \tilde{b}$.

Algorithm 12.14 Preconditioned conjugate gradient algorithm.

> Initialization: $r = 0$; $x_0 = 0$; $R_0 = b$; $p_1 = P^{-1}b$; $y_0 = P^{-1}R_0$
> repeat
> $r = r + 1$
> $z = A\, p_r$
> $v_r = (y_{r-1}^T R_{r-1})/(p_r^T z)$
> $x_r = x_{r-1} + v_r p_r$
> $R_r = R_{r-1} - v_r z$
> $y_r = P^{-1}R_r$
> $\mu_{r+1} = (y_r^T R_r)/(y_{r-1}^T R_{r-1})$
> $p_{r+1} = y_r + \mu_{r+1} p_r$
> until $||R_r||_2$ is small enough

Common preconditioner matrices P are as follows:

- Jacobi preconditioner $P = \mathrm{diag}(a_{11}, ..., a_{nn})$. In [108] it was shown that such a choice of preconditioner reduces the condition number of $P^{-1}A$ by about a factor of n of its minimal value.
- Block Jacobi preconditioner

$$P = \begin{pmatrix} P_{1,1} & ... & 0 \\ ... & ... & ... \\ 0 & ... & P_{r,r} \end{pmatrix} \tag{12.95}$$

with $P_{i,i} = A_{i,i}, i = 1, ..., r$, for a block matrix A given by

$$A = \begin{pmatrix} A_{1,1} & ... & A_{1,r} \\ ... & ... & ... \\ A_{r,1} & ... & A_{r,r} \end{pmatrix} \tag{12.96}$$

with square blocks $A_{i,i}$, $i = 1, ..., r$. In [24] it was shown that the choice of preconditioner P given by (12.95) minimizes the condition number of $P^{-1/2}AP^{-1/2}$ within a factor of r.

- The method of SSOR can be used as a block preconditioner as well; see details in [100].
- Incomplete Cholesky factorization [62, 100, 111] with $A = LL^T$ is often used for the PCG algorithm, Algorithm 12.14. In this case, a sparse lower triangular matrix $\tilde{L}$ is chosen to be close to L. Then the preconditioner is defined as $P = \tilde{L}\tilde{L}^T$.
- Incomplete LU preconditioner [100].
- Domain decomposition methods [23].

Some of these preconditioners are implemented in the software package PETSc [98]. An example using PETSc for the solution of the Dirichlet problem for the Poisson's equation (8.11) in two dimensions is presented below.

Example 12.5 In this example we demonstrate how PETSc [98] can be used for the solution of the Dirichlet problem for the Poisson's equation (8.11). The setup for our problem in this and the next example is the same as in Example 8.2 of Chapter 8. The PETSc programs of Section 1.27 are available for running this example.[12] We executed these programs by running the main program `Main.cpp` using PETSc version *petsc* − 3.7.4 on a 64-bit Red Hat Linux Workstation. Here is an example of the Makefile used for compilation of the PETSc programs of Section 1.27:

```
PETSC_ARCH=/sup64/petsc-3.7.4
include ${PETSC_ARCH}/lib/petsc/conf/variables
include ${PETSC_ARCH}/lib/petsc/conf/rules
MPI_INCLUDE = ${PETSC_ARCH}/include/mpiuni
CXX=g++
CXXFLAGS = -Wall -Wextra -g -O0 -c -Iinclude \
 -I${PETSC_ARCH}/include -I${MPI_INCLUDE}
LD=g++
OBJECTS=Main.o CG.o Create.o DiscretePoisson2D.o \
 GaussSeidel.o Jacobi.o PCG.o Solver.o SOR.o
Run=Main
all: $(Run)
$(CXX) $(CXXFLAGS) -o $@ $<
$(Run): $(OBJECTS)
$(LD) $(OBJECTS) $(PETSC_LIB) -o $@
```

The different iterative methods are encoded by numbers 1–7 in the following order: (1) Jacobi's method, (2) Gauss–Seidel method, (3) successive overrelaxation method (SOR), (4) Conjugate gradient method, (5) conjugate gradient method (Algorithm 12.13), (6) preconditioned conjugate gradient method, (7) preconditioned conjugate gradient method (Algorithm 12.14). Methods 1–5 use inbuilt PETSc functions, and

---

[12] All MATLAB® programs referenced on this page can be found on the online version of Chapter 1 (doi:10.1007/978-3-319-57304-5_1).

methods 6 and 7 implement Algorithms 12.13, 12.14, respectively. For example, to run `Main.cpp` with the SOR method, one should run this program, for example, as follows:

```
> hohup Main 3 > result.m
```

The results will be printed in the file `result.m`, and they can be viewed in MATLAB® using the command `surf(result)` (note that before loading the file `result.m` in MATLAB®, one must remove the first two rows in the output file containing information about the method chosen and the number of iterations at which convergence was achieved). An additional file with results called `solution.m` will also be created. By doing so, we simply illustrate different possibilities for the output of results in PETSc. Using the command `surf(solution)` in MATLAB®, the computed solution of the Dirichlet problem for the Poisson's equation (8.11) can be observed.

Example 12.6 In this example we present the numerical solution of the Dirichlet problem for the Poisson's equation (8.11) in two dimensions using the preconditioned conjugate gradient method (Algorithm 12.14) implemented in MATLAB®. The MATLAB® program of Section 1.26 is available for running this test.[13] In this program we can choose among three preconditioners as the preconditioner matrix P in Algorithm 12.14: Jacobi preconditioner, block Jacobi preconditioner, and incomplete Cholesky factorization [62, 100, 111] with $A = LL^T$. We also use the computation of the residual R_r in the stopping criterion $\| R_r \|_2 < tol$, where the chosen tolerance was $tol = 10^{-9}$.

The preconditioned conjugate gradient method converged in 17 iterations for the preconditioner matrix P constructed using incomplete Cholesky factorization for $N = 20$ inner points and for tolerance $tol = 10^{-9}$. We note that by choosing Jacobi or block Jacobi preconditioners, we have the same convergence as in the usual conjugate gradient method. Our final solution is the same as that in Fig. 8.1 (check it by running the MATLAB® programs of Section 1.26).

Questions

12.1 Find values of the real parameter α such that the matrix

$$\begin{pmatrix} 1 & 0 & \alpha \\ 4 & 2 & 0 \\ 6 & 5 & 3 \end{pmatrix} \tag{12.97}$$

(a) has all real values;
(b) has all complex eigenvalues with positive imaginary parts.

---

[13] All MATLAB® programs referenced on this page can be found on the online version of Chapter 1 (doi:10.1007/978-3-319-57304-5_1).

12.2 Let A be an $n \times n$ Hermitian matrix and let λ and μ be such that $\lambda \neq \mu$ are eigenvalues of A with corresponding eigenvectors x and y. Prove that the eigenvectors corresponding to different eigenvalues are orthogonal, i.e., $\bar{y}^T x = 0$.

12.3 Let the $n \times n$ matrix A have block triangular form

$$A = \begin{pmatrix} A_{11} & A_{12} \\ 0 & A_{22} \end{pmatrix}, \tag{12.98}$$

where the block A_{11} is $r \times r$ and the blocks A_{12}, A_{22} are $(n-r) \times (n-r)$. Prove that λ is an eigenvalue of A if λ is an eigenvalue of A_{11} and x is the corresponding eigenvector. Hint: construct a vector y of length $n - r$ such that $(x, y)^T$ is an eigenvector of A for the eigenvalue λ.

12.4 Let the dim $A = n \times n$ matrix A have spectral radius $\rho(A) < 1$.
  (a) Show that the matrix $I - A$ is nonsingular.
  (b) Show that

$$(I - A)^{-1} = \sum_{i=0}^{n} A^i.$$

12.5 Let the $n \times n$ complex Hermitian matrix C be represented as $C = A + iB$, where the matrices A and B are the real and imaginary parts of C, respectively. Let us define the real $2n \times 2n$ matrix $\tilde{C}$ as

$$\tilde{C} = \begin{pmatrix} A & -B \\ B & A \end{pmatrix}. \tag{12.99}$$

  (a) Show that $\tilde{C}$ is a symmetric matrix.
  (b) Let λ be an eigenvalue of C with corresponding eigenvector $x + iy$. Show that λ is an eigenvalue of $\tilde{C}$, where both $(x, y)^T$ and $(-y, x)^T$ are corresponding eigenvectors.

12.6 Find the eigenvalues of the following matrix:

$$\begin{pmatrix} 2i & 1 \\ 1 & 0 \end{pmatrix}. \tag{12.100}$$

How many linearly independent eigenvectors does it have?

12.7 Let λ be an eigenvalue of the orthogonal matrix Q. Show that $|\lambda| = 1$. What are the singular values of an orthogonal matrix?

12.8 Let the matrix A be a real symmetric tridiagonal matrix that has no zero elements on its subdiagonal. Show that the matrix A must have distinct eigenvalues.

12.9 (*Programming*)
  Implement the Arnoldi and Lanczos algorithms, Algorithms 12.9 and 12.10. First, run both algorithms on A with starting vector for q as in Algorithms 12.9, 12.10.

Then run both algorithms on $Q^T A Q$ with the starting vector for q given by $Q^T q$. Confirm that you obtain identical upper Hessenberg matrices H_r in Algorithm 12.9 or tridiagonal matrices T_r in Algorithm 12.10 in both cases.

12.10 (*Programming*)

Implement different versions of the Lanczos algorithm, Algorithms 12.10, 12.11, 12.12. Present the following results graphically depending on the step i of the Lanczos algorithms:

1. some largest and smallest computed eigenvalues $\lambda_i(A)$ of A;
2. the global errors in the computed eigenvalues $\lambda_i(T_r)$ of item 1 given by

$$\frac{|\lambda_i(T_r) - \lambda_i(A)|}{|\lambda_i(A)|};$$

3. the local errors of item 1 given by

$$\min_j \frac{|\lambda_i(T_r) - \lambda_j(A)|}{|\lambda_i(A)|},$$

   where $\lambda_j(A)$ is the eigenvalue of A nearest to $\lambda_i(T_r)$; sometimes these errors are smaller than the global errors;
4. the error bounds of item 1

$$\frac{|\beta_r v_i(r)|}{|\lambda_i(A)|}.$$

12.11 Prove that the conjugate vectors p_r in Algorithm 12.13 are orthogonal with respect to the inner product defined by A.

12.12 Prove that if A, of order dim $A = n \times n$, is symmetric positive definite, then $H_r = Q_r^T A Q_r$, where dim $Q = n \times r$ with full column rank, is also symmetric positive definite. Here the matrix Q has full column rank and is not orthogonal.

12.13 (*Programming*)

Modify the MATLAB® program of Section 1.21, which implements the Jacobi method of Algorithm 12.2, and using this program, solve the model problem for the Poisson's equation of Section 8.1.3 in three dimensions on the unit cube.[14]

12.14 (*Programming*)

Modify the MATLAB® program of Section 1.22, which implements the Gauss–Seidel method of the algorithm 12.4, and using this program solve the model problem for the Poisson's equation of Section 8.1.3 in three dimensions on the unit cube.

---

[14] All MATLAB® programs referenced on this page can be found on the online version of Chapter 1 (doi:10.1007/978-3-319-57304-5_1).

12.15 (*Programming*)

Modify the MATLAB® program of Section 1.24, which implements the SOR(ω) method of Algorithm 12.6 for different ω, and using this program, solve the model problem for the Poisson's equation of Section 8.1.3 in three dimensions on the unit cube.[15] For which ω does the SOR(ω) method converge? Why?

12.16 (*Programming*)

Write a MATLAB® program to implement the SSOR(ω) algorithm, Algorithm 12.8, with Chebyshev acceleration of $x_{i+1} = \hat{R}x_i + c$ for different ω. In [46], it was shown that the choice $\omega = \frac{2}{1+\sqrt{2(1-\rho)}}$ is a good one. In that case, $\rho(\hat{R}) \approx 1 - \frac{\pi}{2N}$, where N is the number of points in the mesh. Compare the performance of SSOR(ω_{opt}) and SOR(ω_{opt}) with optimal ω_{opt} given by (12.59).

12.17 (*Programming*)

Write a program to implement the Arnoldi algorithm , Algorithm 12.9. Test the program on a real symmetric $n \times n$ matrix A with eigenvalues $1, 2, ..., n$. Hint: to generate such a matrix A, first generate an $n \times n$ matrix B with randomly distributed entries in the interval $[0, 1)$ and compute its QR factorization $B = QR$. Then construct the matrix A as $A = QDQ^T$ with diagonal matrix $D = \text{diag}(1, ..., n)$. Run the Arnoldi algorithm, Algorithm 12.9, for n iterations.

12.18 (*Programming*)

Write a program to implement the Lanczos algorithm, Algorithm 12.10. Test the program on a real symmetric matrix A as in Question 12.4. Run Algorithm 12.10 for n iterations.

12.19 (*Programming*)

Modify the MATLAB® program of Section 1.25, which implements the conjugate gradient algorithm, Algorithm 12.13, and using this program, solve the model problem for the Poisson's equation of Section 8.1.3 in three dimensions on the unit cube.

12.20 (*Programming*)

Modify the MATLAB® program of Section 1.25, which implements the preconditioned conjugate gradient algorithm, Algorithm 12.13, and using this program, solve the model problem for the Poisson's equation of Section 8.1.3 in three dimensions on the unit cube.

12.21 (*Programming*)

Modify the PETSc programs of Section 1.27, which solve the Dirichlet problem for the Poisson's equation (8.11) and apply them to solve this problem in three dimensions on the unit cube. See details on running PETSc programs in Example 12.5.

---

[15] All MATLAB® programs referenced on this page can be found on the online version of Chapter 1 (doi:10.1007/978-3-319-57304-5_1).

Erratum to: Numerical Linear Algebra: Theory and Applications

Erratum to:
L. Beilina et al., *Numerical Linear*
Algebra: Theory and Applications,
https://doi.org/10.1007/978-3-319-57304-5

In the original version of the book, belated corrections from author have to be incorporated in Chaps. 8–12. The erratum book has been updated with the changes.

The original version of this book was revised. An Erratum to this book can be found at
DOI 10.1007/978-3-319-57304-5_8

DOI 10.1007/978-3-319-57304-5_9

DOI 10.1007/978-3-319-57304-5_10

DOI 10.1007/978-3-319-57304-5_11

DOI 10.1007/978-3-319-57304-5_12

DOI 10.1007/978-3-319-57304-5

© Springer International Publishing AG 2017
L. Beilina et al., *Numerical Linear Algebra: Theory and Applications,*
https://doi.org/10.1007/978-3-319-57304-5_13

References

1. Allen, G.D.: Lectures on Linear Algebra and Matrices. Texas A&M University. http://www. math.tamu.edu/~dallen/m640_03c/readings.htm
2. Arioli, M., Demmel, J., Duff, I.S.: Solving sparse linear systems with sparse backward error. SIAM J. Matrix Anal. Appl. **10**, 165–190 (1989)
3. Anderson, E., Baiu, Z., Bischof, C., Blackford, S., Demmel, J., Dongarra, J., Du Croz, J., Greenbaum, A., Hammarling, S., McKenney, A., Sorensen, D.: LAPACK Users' Guide Third Edition (2012)
4. Axelsson, O.: Iterative Solution Methods. Cambridge University Press, Cambridge (1996)
5. Bai, Z.: Error analysis of the Lanczoz algorithm for the nonsymmetric eigenvalue problem. Math. Comput. **62**, 209–226 (1994)
6. Bai, Z.: Progress in the numerical solution of the nonsymmetric eigenvalue problem. J. Numer. Linear Algebra Appl. **2**, 210–234 (1995)
7. Bakushinsky, A., Kokurin, M.Y., Smirnova, A.: Iterative Methods for Ill-posed Problems. Inverse and Ill-Posed Problems Series, vol. 54. De Gruyter (2011)
8. Batterson, S.: Convergence of the shifted QR algorithm on 3 by 3 normal matrices. Numer. Math. **58**, 341–352 (1990)
9. Barrett, R., Berry, M., Chan, T.F., Demmel, J., Donato, J.M., Dongarra, J., Eijkhout, V., Pozo, R., Romine, C., Van der Vorst, H.: Templates for the Solution of Linear Systems: Building Blocks for Iterative Methods. SIAM, Philadelphia (1994)
10. Belitskii, G.R., Lyubich, Y.I.: Matrix Norms and their Applications. Birkhäuser Verlag, Basel (1988)
11. Bellman, R.: Introduction to Matrix Analysis. SIAM (1997)
12. Burden, R.L., Douglas Faire, J.: Numerical Analysis, 8th edn. (2004). ISBN-10: 0534392008
13. Byrd, R.H., Schnabel, R.B., Shultz, G.A.: A trust region algorithm for nonlinearly constrained optimization. SIAM J. Numer. Anal. **24**(5), 11521170 (1987). doi:10.1137/0724076
14. Björck, Å.: Numerical Methods for Least Squares Problems. SIAM (1996)
15. Bischof, C.: Incremental condition estimation. SIAM J. Matrix Anal. Appl. **11**, 312–322 (1990)
16. Bunch, J., Kaufman, L.: Some stable methods for calculating inertia and solving symmetric linear systems. Math. Comput. **31**, 163–179 (1977)
17. Calinger, R.: A Contextual History of Mathematics. Prentice Hall (1999). ISBN 978-0-02-318285-3
18. Chan, T.: Rank revealing QR factorizations. Linear Algebra Appl. **88/89**, 67–82 (1987)
19. Ciarlet, P.: Introduction to Numerical Linear Algebra and Optimisation. Cambridge University Press (1989)

20. Cullum, J., Kerner, W., Willoughby, R.: A generalized nonsymmetric Lanczos procedure. Comput. Phys. Commun. **53**, 19–48 (1989)
21. Cuppen, J.J.M.: A divide and conquer method for the symmetric tridiagonal eigenproblem. Numer. Math. **36**, 177–195 (1981)
22. Day, D.: How the QR algorithm fails to converge and how to fix it, Technical report 96–0913J. Sandia National Laboratory, NM (1996)
23. Demmel, J.W.: Applied Numerical Linear Algebra. SIAM (1997)
24. Demmel, J.W.: The condition number of equivalence transformations that block diagonalize matrix pencils. SIAM J. Numer. Anal. **20**, 599–610 (1983). Applied Numerical Linear Algebra. SIAM (1997)
25. Demmel, J., Veselic, K.: Jacobi's method is more accurate than QR. SIAM J. Matrix Anal. Appl. **13**, 1204–1246 (1992)
26. Demmel, J., Dillon, Ren, H.: On the correctness on some bisection-like parallel eigenvalue algorithms in floating point arithmetic. Electronic Trans. Numer. Anal. **3**, 116–140 (1995)
27. Demmel, J., Gragg, W.: On computing accurate singular values and eigenvalues of matrices with acyclic graphs. In: Linear Algebra and Its Applications, pp. 203–217 (1993)
28. Demmel, J., Kahan, W.: Accurate singular values of bidiagonal matrices. SIAM J. Sci. Stat. Comput. **11**, 873–912 (1990)
29. Dhillon, I.S.: A new $O(n2)$ algorithm for the symmetric tridiagonal eigenvalue/eigenvector problem, Ph.D. thesis, Computer Science Division, University of California, Berkeley (1997)
30. Eisenstat, S.: A stable algorithm for the rank-1 modification of the symmetric eigenproblem. Computer Science Department. Report YALEU/DCS/RR-916, Yale University, September 1992
31. Engl, H.W., Hanke, M., Neubauer, A.: Regularization of Inverse Problems. Kluwer Academic Publishers, Boston (2000)
32. Fernando, K., Parlett, B.: Accurate singular values and differential qd algorithms. Numer. Math. **67**, 191–229 (1994)
33. Gantmaher, F.R.: The Theory of Matrices. AMS Chelsea Publishing, vol. 1, vol. 2 (2000)
34. Gantmacher, F.R.: Applications of the Theory of Matrices. Dover Publications (2005)
35. Gelfand, I.M.: Lectures on Linear Algebra, Dover Books on Mathematics (1989)
36. Glazman, I.M., Ljubich, Y.I.: Finite-Dimensional Linear Analysis: A Systematic Presentation in Problem Form. Dover Publications (2006)
37. Godunov, S.K.: Modern Aspects of Linear Algebra. American Mathematical Society (1998)
38. Gu, M., Eisenstat, S.C.: A divide-and-conquer algorithm for the symmetric tridiagonal eigenproblem. SIAM J. Matrix Anal. Appl. **16**, 172–191 (1995)
39. Grcar, J.F.: Mathematicians of Gaussian elimination. Not. Am. Math. Soc. **58**(6), 782–792 (2011)
40. Grimes, R., Lewis, J., Simon, H.: A shifted block Lanczoz algorithm for solving sparce symmetric generalized problems. SIAM J. Matrix Anal. Appl. **15**, 228–272 (1994)
41. Faddeev, D.K.: Lectures in Algebra. Nauka (1984) (in Russian)
42. Fernando, V., Parlett, B., Dhillon, I.: A way to find the most redundant equation in a tridiagonal system. Preprint at California University, Berkeley Center (1995)
43. Fernando, K., Parlett, B.: Accurate singular values and differential qd algorithms. Numer. Math. **67**, 191–229 (1994)
44. Freund, R., Golub, G.H., Nachtigal, N.: Iterative solution of linear systems. Acta Numerica, 97–100 (1992)
45. Hager, W.W.: Condition estimators. SIAM J. Sci. Stat. Comput. **5**, 311–316 (1984)
46. Hageman, L.A., Young, D.M.: Applied Iterative Methods. Academic Press, New York (1981)
47. Halmos, P.: Finite Dimensional Vector Spaces. Van Nostrand, New York (1958)
48. Halmos, P.R.: Linear Algebra Problem Book. The Mathematical Association of America (1995)
49. Hackbusch, W.: Iterative Solution of Large Sparse Systems of Equations. Springer, Berlin (1994)

50. Higham, N.J.: A survey of condition number estimation for triangular matrices. SIAM Rev. **29**, 575–596 (1987)
51. Higham, N.J.: Experience with a matrix norm estimator. SIAM J. Sci. Stat. Comput. **11**, 804–809 (1990)
52. Higham, N.J.: FORTRAN codes for estimating the one-norm of a real or complex matrix. SIAM Rev. **29**, 575–596 (1987)
53. Higham, N.: Accuracy and Stability of Numerical Algorithms, 2nd edn. SIAM (2002)
54. Higham, N.J.: Functions of Matrices. Theory and Computation. Society for Industrial and Applied Mathematics (2008)
55. Horn, R.A., Johnson, C.R.: Topics in Matrix Analysis. Cambridge University Press (1999)
56. Horn, R.A., Johnson, C.R.: Matrix Analysis. Cambridge University Press (2012)
57. Fine, B., Rosenberger, G.: The Fundamental Theorem of Algebra. Springer, Berlin (1997)
58. Forsythe, G.E., Malcolm, M.A., Moler, C.B.: Computer Methods for Mathematical Computations. Prentice Hall, Englewood Cliffs (1977)
59. Ikramov, KhD: Numerical Solution of Matrix Equations. Nauka, Moscow (1984). (in Russian)
60. Ikramov, KhD: Matrix pencils-theory, applications, numerical methods. J. Soviet Math. **64**(2), 783–853 (1993)
61. Kahaner, D., Moler, C., Nash, S.: Numerical Methods and Software. Prentice Hall, Englewood Cliffs (1989)
62. Kelley, C.T.: Frontiers in Applied Mathematics: Iterative Methods for Linear and Nonlinear Equations. SIAM, Philadelphia (1995)
63. Krasnosel'skii, M.A., Vainikko, G.M., Zabreiko, P.P., Rutitskii, Y.B., Stetsenko, V.Y.: Approximate Solution of Operator Equations. Wolters-Noordhoff Publishing, Groningen (1972)
64. Kurosh, A.: Higher Algebra. Mir Publishers (1980)
65. Kuttler, K.: Elementary Linear Algebra. Brigham Young University (2009)
66. Lancaster, P., Tismenetsky, M.: The Theory of Matrices. Academic Press (1985)
67. Lau, D.: Übungsbuch zur Linearen Algebra und analytischen Geometrie. Aufgaben mit Lösungen. Springer, Berlin (2011)
68. Lawson, C.L., Hanson, R.J.: Solving Least Squares Problems. SIAM, Philadelphia (1995)
69. Dongarra, J.J., Bunch, J.R., Moler, C.B., Stewart, G.W.: Linpack Users Guide, 2nd edn. SIAM, Philadelphia (1979)
70. Mal'cev, A.I.: Foundations of Linear Algebra. W.H Freeman and Company (1963)
71. Marcus, M., Min, H.: A Survey of Matrix Theory and Matrix Inequalities. Dover Publications (2010)
72. Meyer, C.D.: Matrix Analysis and Applied Linear Algebra. With Solutions to Problems. SIAM (2000)
73. Mirsky, L.: An Introduction to Linear Algebra. Dover Publications (2011)
74. Moré, : The Levenberg–Marquardt algorithm: implementation and theory. In: Watson, (ed.) Numerical Analysis, Proceedings of the Biennial Conference Held at Dundee, pp. 105–116 (1977)
75. Prasolov, V.V.: Problems and Theorems in Linear Algebra. American Mathematical Society (1994)
76. Shilov, G.E.: Linear Algebra. Dover Publications (1977)
77. Stewart, F.M.: Introduction to Linear Algebra. D. Van Nostrand Company Inc. (1963)
78. Stewart, G.W., Sun, J.: Matrix Perturbation Theory. Academic Press, Inc. (1990)
79. Strang, G.: Linear Algebra and Its Applications. Cengage Learning (2006)
80. Tyrtyshnikov, E.E.: Matrix Analysis and Linear Algebra. FIZMATLIT, Moscow (2007). (in Russian)
81. Varga, R.S.: Matrix Iterative Analysis. Springer, Berlin (2010)
82. Vinberg, E.B.: A Course in Algebra. American Mathematical Society (2003)
83. Voevodin, V.V.: Linear Algebra. Mir Publishers (1983)
84. Voevodin, V.V., Kuznecov, Y.A.: Matrices and Calculations. Nauka, Moscow (1984). (in Russian)

85. Wildon, M.: A short proof of the existence of Jordan normal form. http://www.ma.rhul.ac.uk/~uvah099/Maths/JNFfinal.pdf
86. Zhan, X.: Matrix Inequalities. Springer, Berlin (2002)
87. Zhang, F.: Matrix Theory. Basic Results and Techniques. Springer, New York (1999)
88. Hager, W.: Applied Numerical Linear Algebra. Prentice Hall, Englewood Cliffs (1988)
89. Lau, H.T.: A Numerical Library in C for Scientists and Engineers. CRC Press, Boca Raton (1995)
90. Li, T.-Y., Zhang, H., Sun, X.-H.: Parallel homotopy algorithm for symmetric tridiagonal eigenvalue problem. SIAM J. Sci. Stat. Comput. **12**, 469–487 (1991)
91. Li, T.-Y., Zeng, Z.: Homotopy-determinant algorithm for solving nonsymmetric eigenvalue problems. Math. Comput. **59**, 483–502 (1992)
92. Li, T.-Y., Zeng, Z., Cong, L.: Solving eigenvalue problems of nonsymmetric matrices with real homotopies. SIAM J. Numer. Anal. **29**, 229–248 (1992)
93. Paige, C.C., Saunders, M.A.: Solution of sparce indefinite systems of linear equations. SIAM J. Numer. Anal. **12**, 617–629 (1975)
94. Parlett, B.: The Symmetric Eigenvalue Problem. Prentice Hall, Englewood Cliffs (1980)
95. Parlett, B.: The new qd algorithms. In: Acta Numerica, pp. 459–491. Cambridge University Press, Cambridge (1995)
96. Parlett, B.N., Dhillon, I.S.: Fernando's solution to Wilkinson's problem: an application of double factorization. Linear Algebr. Appl. (1997)
97. Parlett, B.: The construction of orthogonal eigenvectors for tight clusters by use of submatrices, Center for Pure and Applied Mathematics PAM-664, University of California, Berkeley (1996)
98. PETSc, Portable, Extensible Toolkit for Scientific Computation. https://www.mcs.anl.gov/petsc/
99. Rutishauser, H.: Lectures on Numerical Mathematics. Birkhauser, Basel (1990)
100. Saad, Y.: Iterative Methods for Sparse Linear Systems. PWS Publishing Co., Boston (1996)
101. Saad, Y., Schulz, M.H.: GMRES: a generalized minimal residual algorithm for solving nonsymmetric linear systems. SIAM J. Sci. Stat. Comput. **7**, 856–869 (1986)
102. Saad, Y.: Numerical solution of large nonsymmetric eigenvalue problems. Comput. Phys. Comput. **37**, 105–126 (1981)
103. Scilab documentation - number properties - determine floating-point parameters (2013)
104. Skeel, R.D.: Scaling for numerical stability in Gaussian elimination. J. ACM **26**, 494–526 (1979)
105. Skeel, R.D.: Iterative refinement implies numerical stability for Gaussian elimination. Math. Comput. **35**, 817–832 (1980)
106. Skeel, R.D.: Effect of equilibration on residual size for partial pivoting. SIAM J. Numer. Anal. **18**, 449–454 (1981)
107. Simon, H.: The Lanczoz algorithm with partial reorthogonalization. Math. Comput. **42**, 115–142 (1984)
108. Van Der Sluis, : Condition numbers and equilibration of matrices. Numer. Math. **14**, 14–23 (1969)
109. Szegö, G.: Orthogonal Polynomials. AMS, Providence (1967)
110. Tikhonov, A.N., Goncharsky, A.V., Stepanov, V.V., Yagola, A.G.: Numerical Methods for the Solution of Ill-Posed Problems. Kluwer, London (1995)
111. Trefethen, L.N., Bau III, D.: Numerical Linear Algebra. SIAM, Philadelphia (1997)
112. Van Der Sluis, A.: Condition numbers and equilibration of matrices. Numer. Math. **14**, 14–23 (1969)
113. Wilkinson, J.H.: The Algebraic Eigenvalue Problem. Oxford University Press, Oxford (1965)
114. Wright, M.H., Glassman, S.: Fortran subroutines to solve linear least squares problems and compute the complete orthogonal factorization. Technical report, Stanford University, Stanford, CA (1978)
115. Zeng, Z.: Homotopy-Determinant Algorithm for Solving Matrix Eigenvalue Problems and Its Parallelizations, Ph.D. thesis, Michigan State University, East Lansing, MI (1991)

Index

© Springer International Publishing AG 2017
L. Beilina et al., *Numerical Linear Algebra: Theory and Applications*,
DOI 10.1007/978-3-319-57304-5

CPSIA information can be obtained
at www.ICGtesting.com
Printed in the USA
LVHW082131050619
620329LV00005B/358/P

9 783319 861272